Rutger Anthony van Santen and Matthew Neurock

Molecular Heterogeneous Catalysis

Related Titles

Dronskowski, R. V.
Computational Chemistry of Solid State Materials
A Guide for Materials Scientists, Chemists, Physicists and others
2006
ISBN 3-527-31410-5

Cramer, C. J.
Essentials of Computational Chemistry
Theories and Models
2002
ISBN 0-471-48552-7

Rutger Anthony van Santen and Matthew Neurock

Molecular Heterogeneous Catalysis

A Conceptual and Computational Approach

WILEY-VCH Verlag GmbH & Co. KGaA

The Authors

Prof. Dr. R. A. van Santen
Eindhoven University of Technology
P.O. Box 513
5600 MB Eindhoven
The Netherlands

Prof. Dr. M. Neurock
Dept. of Chemical Engineering
School of Engineering and Applied Science
University of Virginia
Charlottesville
VA 22903-4741
USA

Library of Congress Card No.: applied for.

British Library Cataloging-in-Publication Data:
A catalogue record for this book is available from the British Library.

Bibliographic information published by Die Deutsche Bibliothek
Die Deutsche Bibliothek lists this publication in the Deutsche Nationalbibliografie; detailed bibliographic data is available in the Internet at <http://dnb.ddb.de>.

Printed in the Federal Republic of Germany
Printed on acid-free paper

Printing Strauss GmbH, Mörlenbach
Binding J. Schäffer GmbH i.G., Grünstadt
Cover Design SCHULZ Grafik-Design, Fußgönheim

ISBN-13: 978-3-527-29662-0
ISBN-10: 3-527-29662-X

To Edith and Dory

CONTENTS

Molecular Heterogeneous Catalysis. Rutger Anthony van Santen and Matthew Neurock

ISBN: 3-527-29662-X

PREFACE

This book examines the science of heterogeneous catalysis through the eyes of a physical chemist. We follow two main threads of thought throughout our discourse. They include a reductionist approach in which we follow the chemistry in terms of the elementary molecular events that occur on the catalyst surface, and in addition, an integrative approach in which we consider the emergence of kinetic phenomena within the complex catalytic system as interactive networks comprised of their atomic and molecular constituents.

The ability to detect molecules in complex environments and follow their transformations, along with the ability to synthesize atomic scale architectures, has revolutionized research in chemistry and engineering. This molecularization has had a dramatic impact on the science of catalysis. Catalysis has transformed from what was once a qualitative descriptive and empirical area of research, which many termed an art, into more of a predictive science supported by the mechanistic understanding of chemical reactivity.

The scientific discovery of heterogeneous catalysis occurred at the onset of the 20th century during a period of time when chemical thermodynamics was beginning to emerge as a new science, thus enabling the prediction of the process conditions. There was little understanding, at that time, however, of the chemistry of catalysis. Hence, the invention of new catalysts was completely empirical. Mass transfer considerations and reaction kinetics are specific to the structure and composition of the catalyst and thus require the characterization of porosity, surface area and information on the distribution of the catalytically reactive phases which at that point had not been established. This significantly limited any detailed resolution of the chemistry or quantitative understanding of the kinetics. The physical chemistry of catalyst characterization, however, later developed into a major research activity of its own with the arrival of increasingly sophisticated spectroscopic techniques. The resolution of catalyst structure and reactivity has dramatically changed over the past few decades from micron to nanometer to the molecular scale.

The tremendous advances that have occurred in spectroscopy over the past decade now make it possible to resolve molecular intermediates on surfaces through the course of reaction. Similarly, the advances that have occurred in the development of theoretical methods and also computational power have made it possible to begin to calculate elementary step kinetics. These advances in theory and spectroscopy taken together allow us to begin to identify molecular intermediates and establish fundamental mechanistic reaction routes. This detailed level of mechanistic information can subsequently be used to simulate molecular transformations that occur over the catalyst surface. Theory and simulation can, thus, be used in a predictive and hierarchical manner. Chemical events that occur in complex systems at longer time and length scales can therefore be deduced from integrating knowledge from well-defined interacting subsystems that act over shorter time and length scales.

Catalytic kinetics is intrinsically complex since the active sites on the catalyst must be regenerated after each reaction cycle. In this book, we attempt to follow the elementary chemical bond making and bond breaking processes along with intrinsic diffusion events. These processes occur over very short time scales. This has to be integrated with surface reconstruction and self organization phenomena that allow the system to regenerate itself, but occur on much longer time scales. The material presented herein is based on our current understanding of catalysis as it follows mainly from theoretical studies. The key experimental information on which many of the concepts are based are presented in this context.

Molecular Heterogeneous Catalysis. Rutger Anthony van Santen and Matthew Neurock

ISBN: 3-527-29662-X

In order to bridge the knowledge between elementary reaction steps and their kinetics and dynamics under more realistic operational conditions requires one a range of different methods which enable one seamlessly to span time and length scales. This hierarchical coupling is critical to describing complex multiscale catalytic systems but is still in its infancy in modeling catalysis. Herein we describe some of the earliest efforts which, for the most part, have decoupled the different time and length scales and pass on appropriate knowledge between the them.

Ab initio quantum mechanical methods can be used to establish the electronic structure for model surfaces and clusters, and to predict chemical bonding and reactivity of different molecular reactants, intermediates and products on and within different model systems. Elementary reaction steps that are activated require time scales that are typically on the order of 10^{-4} sec or longer. Non-activated processes such as diffusion are typically faster. Transition-state reaction rate theory can, therefore, be used to establish the kinetics for these systems. For systems where the diffusion is on the order of the reaction time scale or slower, molecular dynamics methods have proven to be very useful provided one knows or can develop the interatomic interaction potentials between the intermediates and the surface. The full simulation of dynamics, however, can become intractable if the time scales of interest are significantly greater than the time scales for diffusion. Systems that contain many degrees of freedom and consist of many interacting molecules, therefore, typically require more coarse-grained methods to advance to longer time and length scales. Monte Carlo statistical methods are typically used to determine the lowest free energy states and the chemical potential of complex systems. Dynamic Monte Carlo techniques, on the other hand, are used to simulate complex kinetic behavior. We review the application of each of these techniques to many catalytic systems in this book. An introduction to each of these methods is provided in the Appendix.

Classical heterogeneous catalysts are the subject of four major chapters in the book. We elaborate in detail on our current understanding of the molecular events that underline their catalytic phenomena and attempt to deduce from these results important catalytic reactivity concepts. This detailed understanding provides a basis for the comparison of the mechanistic principles between heterogeneous, enzyme, and homogeneous organometallic cluster catalysis.

We review the governing mechanisms for many of these systems thus having in mind the leading scientific question: "What are the fundamental similarities and differences between these systems?" We explore the use of these insights towards the design of new catalytic systems. Of particular interest is the comparison of traditional heterogeneous systems with biochemical systems for specific reactions. This leads to an understanding of the fundamental differences between enzymes and chemo-systems, which tend to relate to the differences in the adaptability of the enzyme to different stereochemical requirements as the reaction proceeds to that of the typically non-adaptable chemical-based systems. In addition, the biochemical systems are of great interest due to their internal complexity, which appears to be responsible for formation, replication and metabolism of living cellular systems.

Insights on chemo-evolutionary theories of proto-cellular systems are given in the final chapters of the book, with the purpose of defining criteria or discover conditions by which a catalytically active protocell can be designed. The biochemical catalyst design exploits immunoresponse or evolutionary recombinatorial cell growth techniques. It has been discovered that the processes fundamental to the formation of the microporous zeolites with their well defined cavities and channel systems have a number of similarities to such biochemical processes.

The comparison of these interesting features between biochemical and chemical catalysts, along with the theories on the origin of the metabolic protocellular systems, we hope will be an inspiration for endeavors to design new catalytic systems that are able to self assemble themselves for use in a particular catalytic application.

The book is targeted at the readership at the graduate student level. We hope that the book will help to convince the research community of the importance of molecular level research and the fruitfulness of theoretical approaches.

The stimulating environment of the Schuit Institute of Catalysis and the University of Virginia and the collaboration with many colleagues and coworkers over the past 15 years have been invaluable to us. We are grateful for the input this has provided to our work and have used the opportunity when writing this book to use as examples many of the results produced in our laboratories. We have placed these results in the context of the most relevant advances in catalysis research by the international research community. Of necessity a selection had to be made for which, we take full responsibility.

This book could have never been realized without the invaluable assistance of Joop van Grondelle, who did most of the editing of the book. MN would like to acknowledge the valuable input from past and present students and academic as well as industrial colleagues. In particular, he offers special thanks to Professor Robert Davis, Michael Janik, Dr. Randall Meyer, Chris Taylor and Dr. Sally Wasileski for their valuable input to different sections in the book. RAvS appreciates the elucidating comments from past and present TUE colleagues, especially Dr. A.P.J. Jansen, Professor M. Koper, Professor J.W. Niemantsverdriet, Dr. X. Rozanska, Dr. N. Sommerdijk and Dr. E. Hensen. MN would also like to thank his wife Dory and daughters Nicole and Sabrina for their unending love and support and their understanding of why Dad was not able to play. RAvS thanks his wife Edith especially for her patience and companionship on the many working visits on behalf of the book.

Most importantly, without the love and encouragement of our spouses Edith and Dory, we could not have embarked on or finished this endeavor.

August 2005

Rutger A. van Santen, Eindhoven
Matthew Neurock, Charlottesville

CHAPTER 1
Introduction

1.1 Importance of Catalysis

Catalysis is ubiquitous to life as well as to society. Catalysts are used in the production of the foods that we eat, the clothes that we wear, the energy necessary to heat and cool our homes, the enzymatic transformations that occur throughout our body to provide function to nearly every organ, the purification of the air that we breathe, the fuels used in our cars, and the fabrication of the materials used in and around our homes and offices. Catalysts are at the heart of nearly all biological as well as many chemical transformations of molecules and mixtures into useful products. Enzymes in our body, for example, carry out nearly all of the biological conversions necessary for us to live. They are critical in fighting off infection, building DNA, digesting foods, moving muscles, stimulating nerves and aiding breathing. In terms of chemical conversions, catalysts are responsible for the production of over 60% of all chemicals that are made and are used in over 90% of all chemical processes worldwide[1−2]. This accounts for 20% of the Gross Domestic Products in the USA. Catalyst manufacturing alone accounts for over $10 billion in sales worldwide and is spread out across four major sectors: refining, chemicals, polymerization, and exhaust emission catalysts[1−2]. Refining is the largest sector with the production of catalysts for alkylation, cracking, hydrodesulfurization, fluid catalytic cracking, hydrocracking, isomerization, and reforming chemistry. The value derived from catalyst sales, however, is really only a very small fraction of the total value derived from catalysis overall, which includes the value of the products that are produced, i.e. chemical intermediates, polymers, pesticides, pharmaceuticals, and fuels. The overall impact of catalysis is estimated to be $10 trillion per year[1]. The intermediates made by catalysis are used in the production of materials, chemicals, and control devices that cross many different manufacturing industries including petroleum, chemicals, pharmaceuticals, automotives, electronic materials, food and energy[1−3].

As we look to the future, catalysis holds the promise of eliminating, or at least substantially reducing, pollution from chemical and petroleum processes, electronics manufacturing, pharmaceutical synthesis, and stationary and vehicular emission sources. Heterogeneous catalysis is at the heart of many of the proposed green chemical processes targeted to reduce emissions dramatically. A catalyst, by definition, is a material that is used to convert reactants to products without itself being consumed. The goal then is to tailor atomically the structure of an active catalyst so as to convert reactants directly to products without the production of by-products along the way which typically go on to become waste. Catalysts then by nature would help eliminate the production of side products, thus eliminating most waste.

Fossil fuels currently make up the backbone of the US energy economy. The processing of these fuels leads to considerable levels of CO_2 production. An estimated 1.5 billion tons of carbon in the form of CO_2 is emitted each year. About 40% is produced in the conversion of fuel into electricity. Inefficient chemical processes can also be added to the list of major energy consumers. For example, petroleum reforming and ammonia synthesis both consume considerable amounts of resources in order to provide the heat necessary to drive their respective reactions. In addition, they operate at high temperatures, which tends to lead to the greater production of combustion products and thus lower overall selectivities. The design of catalysts which are more active would lower the temperature of

Molecular Heterogeneous Catalysis. Rutger Anthony van Santen and Matthew Neurock

ISBN: 3-527-29662-X

operation thus dramatically reducing the energy demands. There is currently a worldwide research effort aimed at the development of new catalytic materials to reduce energy consumption.

A third environmental issue concerns the generation of toxic waste solvents used to carry out various liquid acid catalytic conversions. Many of today's petroleum refining processes carry out acid-catalyzed isomerization and alkylation reactions using corrosive or toxic liquid acids such sulfuric acid and hydrofluoric acid. These solvents pose significant environmental concerns. Various solid acid materials have been targeted to replace these corrosive liquids. While there are at least three new patents on processes that can use solid acid catalysts, nearly all alkylation processes are still carried out using liquid acids.

Perhaps one of the greatest challenges facing society over the next quarter of a century will the production of energy resources necessary to sustain the 10^{10} people on the planet. This situation has been clearly outlined by Professor Richard Smalley of Rice University[4]. This will require a minimum of 10 terawatts of power from cheap, clean and potentially renewable energy sources. Smalley indicates that this problem will likely transcend many other societal issues such as water and food shortages since a solution to the energy problem could be used in solution strategies to these others. While there is no current solution to this major challenge to energy, catalysis is likely play a pivotal role. In particular, novel catalytic materials will be required for the advancement of three major areas. The first involves the development of catalysts for the photocatalytic reduction of CO_2. Any solution strategy that uses combustion or oxidation of hydrocarbons to provide hydrogen still face the great challenge of dealing with CO_2 emissions. New catalytic methods that can reduce CO_2 will clearly be necessary. The second area involves the development of novel photocatalysts that could activate water leading to the direct production of hydrogen and oxygen. The third area requires the development of inexpensive, highly reactive electrocatalysts that are resistant to poisoning in order to advance significantly the deployment of fuel cells. The biggest issues are associate with fuel cells are the high overpotentials that exist at the cathode as well as the anode. This is directly tied to their sluggish catalytic activity and the inherent dependence on Pt-based catalysts which are rather expensive[5–6]. Meeting the energy demands of the 21st century is clearly an unsolved dilemma where catalysis will play an important role.

As we look further into the future, it may become clear that the current practice of the production of chemicals may be antiquated. We currently use severe temperatures in order refine petroleum feedstocks into a range of hydrocarbon chemical intermediates. We then subsequently attempt to add selectively oxygen or nitrogen functionality back into the molecule by selective oxidation or amination processes, respectively, in order to produce valuable functionalized intermediates. Nature, however, already starts with many of these functional intermediates trapped inside the structures of carbohydrates and other natural occurring feedstocks. In the production of specialty chemicals, it might make more sense to try to carve out selectively the chemical and stereochemical functionality needed directly from nature. Indeed, there is currently a strong and growing effort in what some call the Bioindustrial Revolution[7]. The current goal is to design chemical and biological catalysts that can convert bio-based feedstocks into chemicals and fuels. These feedstocks are renewable and, therefore, very attractive from an environmental standpoint. In addition, they contain a wide range of intermediates that could lead to new products provided that we can design inexpensive catalytic materials and processes to carry out these conversions. Some would argue though that traditional petrochemical processes will always be cheaper and that the production of crude oil will last for many years into the future. Despite these arguments there are a number of processes and plants built around the concepts of bio-renewable chemicals and energy.

It is clear that catalysis plays an important role in society today and will be a critical technology for advancing our future.

The inception of industrial heterogeneous catalysis started early in the 20th century with the invention of a continuous process to produce ammonia from nitrogen and hydrogen[8]. This provided a very low cost route to produce ammonia, a major ingredient of dynamite and agricultural fertilizers. Ammonia is currently one of the largest commodity chemicals produced worldwide. The Born–Haber process to produce ammonia was a technological breakthrough since nitrogen together with hydrogen gas could simply be passed through a tube that contained an inorganic solid, thus promoting the two to react to generate ammonia, which was collected at the end of the tube in a continuous fashion. Such continuous heterogeneous catalytic processes were revolutionary since they could be readily scaled up in order to produce desired product yields. This was in sharp contrast to the conventional batch processes, which up to that point in time were the main modes of industrial production.

The key to the Born–Haber process was the inorganic packing material inside the tube, for reaction would not occur in the absence of this material. The preferred catalytic material for this process was a fused iron doped with potassium. The nitrogen and hydrogen reactant gases were converted to ammonia without a change in the macroscopic performance of the catalyst material over the course of the reaction. This agrees with the classic Berzelius definition of a catalyst: a material which will increase or decrease the rate of a particular reaction without itself being consumed in the process.

1.1.1 Additional Suggested Textbooks on Heterogeneous Catalysis

The specific aim of this book is to provide a molecular basis and in-depth understanding of the mechanisms involved in heterogeneous catalysis. The presentation is at an advanced level. There are many important books on catalysis that, in general, provide either an introductory and explanatory view of catalysis or that are focused on specific aspects of catalysis such as kinetics or synthesis or related to industrial catalysis. Here we provide a short list of selected relevant books for the interested reader.

These texts cover different catalytic principles and disciplines along with their application to industrial practice. Collectively they span a wide range of material including basic concepts in heterogeneous catalyst synthesis, characterization, kinetics, reaction engineering and their application to industrial catalytic systems.

1. J.M. Thomas, W.J. Thomas, *Priciples and Practice of Heterogeneous Catalysis,* Wiley-VCH, First Edition (1967), Second Edition (1997).
2. B.C. Gates, J.R. Katzer G.C.A. Schuit, *Chemistry of Catalytic Processes,* McGraw-Hill (1979), integrating chemical understanding of catalytic processes.
3. C. N. Satterfield, *Heterogeneous Catalysis in Industrial Practice,* (1991).
4. B.C. Gates, *Catalytic Chemistry,* Wiley (1992).
5. R.J. Farrauto, C.H. Bartholomew, *Fundamentals of Industrial Catalytic Processes.*, Blackie-Chapman and Hall (1997).
6. M. Bowker, *The Basis and Application of Heterogeneous Catalysis,* Oxford Science Publishers (1998).

7. R.A. van Santen, P.W.N.M. van Leeuwen, J.A. Moulijn, B.A. Averill, *Catalysis, an Intergrated Approach,* Elsevier (1999).
8. J. Hagen *Industrial Catalysis,* Wiley-VCH (1999).
9. I. Chorkendorff, J.W. Niemantsverdriet, *Concepts of Modern Catalysis and Kinetics,* Wiley-VCH (2003).

There are various books which cover in detail the fundamental principles established from surface science and their application to heterogeneous catalysis:

10. R.I. Masel, *Principles of Adsorption and Reaction on Solid Surfaces,* Wiley (1996).
11. G.A. Somorjai, *Surface Chemistry and Catalysis,* Wiley (1994).
12. K. W. Kolasinski *Surface Science: Foundations of Catalysis and Nanoscience,* Wiley (2001).

The following two texts describe the fundamental kinetics and modeling of heterogeneous catalytic systems:

13. M. Boudart, *Kinetics of Chemical Processes,* Prentice-Hill (1968).
14. J.A. Dumesic, D.F. Rudd, L. M. Aparicio, J.E. Rekoske, A.A. Trevino, *Micro Kinetics of Heterogenous Catalysis,* American Chemical Society (1992).

There are two modern compilations on heterogeneous catalysis that are comprised of a series of volumes which cover many aspects of catalysis. Individual sections are written by leading experts. Both series are highly recommended as general references:

15. G. Ertl, H. Knözinger, J. Weitkamp, *Handbook of Heterogeneous Catalysis* Wiley-VCH (1997).
16. I.T. Horvath, *Encyclopedia of Catalysis,* Wiley International (2003).

1.2 Molecular Description of Heterogeneous Catalysis

The ability to predict catalyst performance as a function of chemical composition, molecular structure and morphology is the foundation for the science and technology of catalysis. We aim to describe the use of currently available theoretical and computational methods for both qualitative and quantitative predictions on the molecular events on which the catalytic reaction is based. This relates to the prediction of catalyst structure and morphology as well as the simulation of dynamic changes that occur on the catalyst surface as the result of reaction.

We will provide the reader with an introduction to fundamental concepts in catalytic reactivity and catalyst synthesis derived from the results of computational analysis along with physical and chemical experimental studies. The tremendous advances in nanoscale materials characterization, *in-itu* spectroscopy to provide atomic and molecular level resolution of surfaces and adsorbed intermediates under reaction conditions, predictive ab initio quantum mechanical methods and molecular simulations that have occurred over the past two decades have helped to make catalysis much more of a predictive science. This has significantly enhanced the technology of catalysis well beyond the historical ammonia synthesis and petrochemical processes.

Herein we attempt to highlight advances in the molecular science of heterogeneous catalysis. We will focus on the mechanistic phenomena that make catalysis possible. This enables one to begin to answer the chemist's questions: What are the fundamental processes that occur at the catalyst surface and how do they act to control its remarkable behavior? What are the molecular system parameters that control rate and selectivity for a specific catalytic process?. The ultimate goal would be the prediction of the catalytic behavior of an arbitrary material along with an arbitrary catalytic reaction system. We introduce the concepts along with a methodology that is fundamental to catalyst design, based on mechanistic analyses. This cross-cuts a range of different sub-fields of chemistry including physical, synthetic, organometallic, inorganic, coordination, theoretical, biochemical and solid state and biochemistry as well as chemical engineering.

While it is important to devise strategies that may help to predict material features that could improve catalyst performance, let us not forget that the ability to synthesize materials that contain these features presents yet an even greater challenge. Synthesis is still somewhat of an art that requires marrying the knowledge and skills of the inorganic chemist with those of the solid state chemist. The increased molecular understanding along with its application will require increased precision in the molecular design of catalysts and their specific features. This moves catalyst synthesis from solid state colloidal chemistry into molecular inorganic chemistry as ideas and techniques from coordination chemistry and organometallic chemistry now play much more important roles. The link between well-defined molecular complexes used as homogeneous catalysts in the liquid phase and heterogeneous catalysts applicable to gas phase catalytic reactions can be made when synthetic approaches are developed to immobilize the organometallic complexes. This is an important field of current research.

Significant progress has been made in terms of the design of organometallic and inorganic complexes that provide molecular models of the reactive centers as identified by spectroscopy. These models can therefore be manipulated at the atomic scale to help establish the necessary structural and chemical features required for the molecular design of these systems. One of the early pioneering inventions in heterogeneous catalysis was the discovery that increases in the catalyst surface area lead to significantly improved catalytic efficiency. Solid heterogeneous catalysts are active because much of their surface is exposed to the reacting molecules. Hence, the rate of the reaction increases with increases in the surface area of an inorganic material. The high temperatures often applied in the catalytic process make the application of powders difficult, since they sinter and, hence, will rapidly lose surface area. Well-dispersed nanoscopic inorganic particles, however, can be stabilized on high surface area porous inorganic supports which can yield substantially improved catalytic performance. The activity of these materials may be significantly higher, due to the high surface areas afforded by these supported nanoparticles.

In the middle of the last century, synthesis techniques were developed that enabled the fabrication of well-defined and highly regular microporous silica materials, such as zeolites, with pore sizes comparable to the size of the molecules one would like to catalyze[9]. The appropriate matching of size and shape of these micropores with shape and size of reactant, intermediate or product molecules has been demonstrated to be an important factor in the control of catalytic performance. This is analogous to the lock and key reactivity principle which was developed in the early part of the last century as a way to describe the activity of enzymes, the biochemical proteins in living systems. This process involves matching the shape and size of the catalytic cavity with that of the reactant molecules.

The mechanisms which control zeolite catalytic systems show features similar to those known in biochemistry and lead to the formulation of a more general question: What are the basic differences between chemocatalysis for reactions carried out in man-made catalytic systems and biocatalysis for reactions as they occur in biochemical systems?

Over the past century we have witnessed an impressive increase in our understanding of the molecular aspects that control heterogeneous catalytic systems due to major advances in *in-situ* spectroscopy, theoretical methods and computational power. Molecular and atomic scale descriptions of the fundamental physicochemical steps involved in the overall catalytic cycle are thus becoming possible. This same level of description has also advanced the description of biochemical systems. As we will see later in the book, detailed comparisons between chemical and biochemical systems can be very useful for aiding our understanding of the common features governing both systems and also the primary differences. Nature is not only elegant in the materials it creates but in its process of discovery. The comparisons between biochemical and heterogeneous catalytic systems may therefore provide a wealth of information not only in ideas for new biomimetic materials but also in the development of novel catalyst discovery processes.

A detailed analysis of the chemistry alone reveals that catalysis is not comprised of a single elementary event, but is a complex phenomenon in which the elementary reaction as well as the catalyst can both take part in feedback cycles. We examine this unique catalyst feedback cycle by probing the possible mechanisms involved in the conversion of a catalytic material from its initial state towards its catalytically active state.

Earlier we gave the classical Berzelius definition for a catalyst as is found in most textbooks, i.e. a material which enhances a particular reaction but is not used or consumed in the process. Decades of fundamental research have shown that the working catalyst is a dynamic entity that can continuously change. The reactive surface structure is metastable and, in some cases, even mobile. Active sites are formed and consumed dynamically either through surface reactions or through structural changes in catalyst surface topology. Some have described the catalyst as a living system whereby active sites and ensembles must be continuously reborn.

During the course of a catalytic reaction many sites, regions on the catalyst surface or actual catalytic particles can interact. Under some conditions, these sites or regions can communicate with one another and thus lead to self-organizing phenomena that occur in both space and time. This provides information on the complexity in catalytic reaction system, that once again can be related back to phenomena that are well known in biocatalytic systems.

We will discuss the molecular basis of chemocatalysis in comparison with related aspects of biocatalytic systems. This will enable us to elucidate the molecular foundation and mechanisms for catalytic reactions and also provide a broader perspective to these principles. In many instances, the refinement of the biological system is often lacking in the chemo-systems. Current challenges in chemocatalysis are related to a more complete understanding of these issues.

The complexity of the catalytic reaction is a common thread through most of the chapters that follow. We describe the issues associated with the different time and length scales that underpin the chemical events that constitute a catalytic system. For example, a typical time scale for the overall catalytic reaction is a second with characteristic length scales that are on the order of 0.1 micron. The time scales for the fundamental adsorption, desorption, diffusion and surface reaction steps that comprise the overall catalytic cycle, however, are often 10^{-3} sec or shorter. The time scales associated with the movement of atoms, such as that which must occur for surface reconstruction events, may be on the order of a nanosecond. The vibrational frequencies for adsorbed surface intermediates occur at time scales on the order of a few picoseconds. The different processes that occur at these time scales obey different physical laws and, hence, require different methods in order to calculate their influence on reactivity. In this book we will show how the description of these processes should be integrated in order to provide predictions on the performance of catalytic systems.

A proper kinetic description of a catalytic reaction must not only follow the formation and conversion of individual intermediates, but should also include the fundamental steps that control the regeneration of the catalyst after each catalytic turnover. Both the catalyst sites and the surface intermediates are part of the catalytic cycle which must turn over in order for the reaction to remain catalytic. The competition between the kinetics for surface reaction and desorption steps leads to the Sabatier principle which indicates that the overall catalytic reaction rate is maximized for an optimal interaction between the substrate molecule and the catalyst surface. At an atomic level, this implies that bonds within the substrate molecule are broken whereas bonds between the substrate and the catalyst are made during the course of reaction. Similarly, as the bonds between the substrate and the surface are broken, bonds within the substrate are formed. The catalyst system regenerates itself through the desorption of products, and the self repair and reorganization of the active site and its environment after each catalytic cycle.

This reorganization may be governed or, at least, aided, by self organization phenomena. In heterogeneous catalytic systems, molecular events that occur at different positions on the catalyst surface can interact through diffusion or surface strain. These interactions can lead to complex self organization phenomena when the catalytic cycle proceeds through the elementary reaction steps that constitute this cycle. The time and length scales for these self organization phenomena are typically on the order of seconds and microns, respectively. Self organizing spatial patterns can be formed due to the coupling of autocatalytic reaction steps and diffusion concentration gradients. Wave fronts in the form of spirals or pulses can also result. Ertl and co-workers were the first to demonstrate this experimentally for reactions catalyzed by single crystal transition-metal surfaces[10]. Self organizing systems are well known in biology and are related to systems that self replicate. Computational approaches suitable to the simulation of these processes are based on the concept of cellular automata, a technique which is just beginning to be used to simulate the kinetics for chemical catalytic systems.

As discussed in Chapter 9, perhaps the most elegant biocatalytic system which demonstrates the next higher level of organization is that for the immunoresponse system. This is a biological system that operates through a combinatorial chemical process. A particular catalytic antibody is selected and amplified in response to a specific reacting molecule, the antigen. If the shape of the antigen molecule is similar to the shape of the transition state for of the reaction intermediates involved in a chosen reaction, it can induce the production of catalytically active antibodies with high selectivity for a chosen particular reaction.

Interestingly, the inorganic chemistry and kinetics for template-directed zeolite synthesis is rather analogous to that of the biocatalytic system. Aluminosilicate oligomers are consumed during zeolite synthesis through the formation of a template aluminosilicate complex. The complex is unique for the zeolite system and subsequently crystallizes. The template can be chosen to resemble the transition state for a specifically desired reaction, thus serving as a catalyst designed to enhance the rate of a pre-selected reaction. This is directly analogous to an antigen-induced antibody synthesis. The difference between the biological and chemical systems is that that the chemical system does not replicate.

1.3 Outline of the Book

While the primary focus of the book is on the molecular basis of heterogeneous catalysis, a final chapter (Chapter 9) is partially devoted to the description of self organizing catalytic systems that lead to the formation of (proto) cellular systems. The design of self organizing and replicating catalytic systems can be considered one of the ultimate goals of molecular heterogeneous catalysis. This chapter is concluded with a description of the physical chemistry of biomineralization and analogous processes to produce microporous systems with widely varying organization of their micropores.

Between the initial chapter and the final chapter, we present a more conventional treatment of molecular heterogeneous catalysis, with a focus on surface catalytic elementary reaction steps and their connection to overall catalytic behavior for a series of different substrates including metals, zeolites, metal oxides and metal sulfides.

As an introduction to the principles in molecular heterogeneous catalysis, we focus in Chapter 2 on basic elementary concepts in heterogeneous catalysis including catalytic reaction cycles, mechanisms for the regeneration of the catalytic active site and the Sabatier principle, which relates catalytic activity with free energy associated with reactant-catalyst interactions. These general concepts provide the constructs necessary to begin to establish molecular level theories and models. The lock-and-key model along with specific modifications to this model are described and used to compare with theories for steric control in homogeneous and heterogeneous catalysis. Steric control in homogeneous catalysis is established by ligand choice, an art that has became increasingly refined over the past few years due to the drive to design homogeneous catalysts with high enantiomeric selectivities. Steric control in heterogeneous surface catalysis is assisted by the coadsorption of shaped adsorbates. In an organizational sense, these shaped adsorbates can be considered as a primitive version of ligands attached to single-site organometallic catalysts.

A very useful procedure to analyze transition states in heterogeneous catalysis relates changes in the activation energy for a particular elementary reaction step with changes in the overall reaction enthalpy for that step over a family of similar catalytic materials. This is the Brønsted–Evans–Polanyi relationship. It is also analogous to the Hammett relationship pioneered in physical organic chemistry that linearly relates the activation barriers for a family of substituted aromatics with a substituent parameter. The Brønsted–Evans–Polanyi relationship as well as other linear free energy relationships hold when the molecules or catalytic materials fall within the same family. The entropic changes across a reaction family are either considered negligible in comparison with the enthalpic contributions or are linearly related to the changes in enthalpy. The entropic changes can be estimated by an understanding of the type of transition for the reaction in question and the difference between early and late transition states.

The final section in Chapter 2 deals with the molecular aspects of transition-metal catalysis. It serves as an introduction to Chapter 3. A characteristic feature of the transition-metal surfaces under catalytic conditions is their potential to restructure. Adsorbate overlayer adsorption can induce the surface to reconstruct with rapid diffusion of the metal as well as the overlayer atoms. The state of the surface may start to resemble that of a solid state compound. The state of the surface is not only strongly influenced by the composition of the reactant gas, but can also be strongly affected by the addition of promoters or other modifiers, that can result in alloy formation or new complex surface phases.

In order to describe the active sites and the associated kinetics, two predominant theories ascribed to Langmuir and Taylor have prevailed in heterogeneous catalysis. In the

Langmuirian view, the active catalytic surface is comprised of a uniform distribution of static sites that do not interact with one another. This is sharply contrasted by the Taylor view, which proposes vacancies and topologically unique surface atom configurations as the centers of reactivity. The Langmuirian idea of a catalytically reactive surface leads to the ensemble effect that ascribes the changes in the selectivity for an alloy surface to the dilution of multi-atom surface ensembles in the alloy induced by mixing inert components into the active surface. In this view, the selectivity of a particular reaction depends predominantly on the number of reactive surface atoms that participate in elementary reaction events.

Recent surface science discoveries, however, demonstrate that step edges and defect sites display markedly lower activation barriers than terrace sites, and thus promote the Taylorian view of catalysis. The selectivity can be strongly influenced by the specific poisoning of these step edge sites. For a number of hydrocarbon conversion processes, these steps will be the most active and lead to potential C–H and C–C bond breaking steps which can ultimately result in deactivation via the formation of surface graphene overlayers.

In Chapter 3, we extend the general concepts developed in Chapter 2 on chemisorption and surface reactivity to establish a fundamental set of theoretical descriptions that describe bonding and reactivity on idealized metal substrates in Chapter 3. There is an extensive treatment of the adsorbate transition-metal surface bond, its electronic structure, bond strength and its influence on its chemical activity. Attention is given to periodic trends in the interaction energy as a function of transition metal and also on the dependence in transition-metal structure.

To illustrate the use of this type of information, an extensive analysis of C–H and C–C bond activation and formation reactions is given. The chapter is concluded with a section that focuses on experiments and theories that explicitly consider lateral effects between adatoms and molecules.

We transfer some of the general concepts developed for the chemical bonding on metals in Chapter 3 to describe the bonding and reactivity that occur in zeolites in Chapter 4. Zeolites are mesoporous systems that have well-defined atomic structures, in contrast with the ill-defined structures of supported metals. This well-defined structure allows them to benefit greatly from the close ties between theory and spectroscopy. This combination of theory and experiment helps to provide for a more detailed understanding of the intrinsic and extrinsic factors that control catalytic reactivity. In the analysis of reactions carried out on zeolites, it becomes clear that the micropores of the zeolite play an important role in dictating their catalytic performance. In order to understand the mechanistic factors that control the sorption and reactivity in zeolites, we focus on two general features in our analysis: the nature of solid acid acidity and the influence of the micropore size and shape on catalysis. Ab initio quantum mechanical calculations now allow for a detailed analysis of reaction intermediates and transition states for reactions of practical interest along with more realistic models of the active sites that capture the full pore cavity. Some of the key concepts developed in this chapter include the importance of pre-transition state stabilization, the screening of the charge separation when charged protons activate bonds, and physical effects that relate to adsorption and diffusion.

Complexity in zeolite catalysis can take on various forms. We focus on two of the challenging issues herein. The first involves the complexity of treating multicomponent systems. There is a strong non-additive behavior for the adsorption isotherms for multicomponent mixtures found in zeolites. This can dramatically affect the selectivity of the zeolite for specific reactions. The complexity also obscures a fundamental analysis based on isotherms of individual molecular fragments. The second challenge relates to the analysis of catalysis by cationic complexes in zeolites which show a strong dependence on the chemical state of the cationic complex, as well as self organizing features. Theory and simulation have helped to gain insight into both of these challenges by its ability intrinsically to simulate multicomponent systems and the chemistry of cationic complexes in zeolites, respectively.

The results from this chapter on zeolite catalysis provide a good reference point for the discussion presented later Chapter 8 where we compare heterogeneous catalysis and biocatalysis. The similarity between the Michaelis–Menten kinetic expression for enzyme catalysis and the Langmuir–Hinshelwood kinetic models for heterogeneous catalysis are noted. This ultimately derives from the conservation in the number of active reaction centers for both systems. However, the more refined synergy of the activation of molecular bonds by the enzyme will become apparent as a major difference between the two.

This general understanding of the similarities, as well as the differences, between biological and heterogeneous catalysts has been the basis for numerous attempts in the literature to synthesize novel chemical systems that mimic enzymes. In Chapter 7, we will review some of the major advances that have taken place, which will also help to highlight further routes for new research. Another important characteristic for enzyme systems is that they are often part of an electrochemical bio-systems and can therefore be considered as bio-electrodes. This implies that electron-transfer catalysis governs their performance. Hydrogenation, for example, tends to occur in biosystems via the combination of electron transfer and reaction with protons. We present a short discussion on the bio-catalytic reduction of nitrogen, and subsequently compare it with the traditional heterogeneous nitrogen reduction by the iron heterogeneous catalyst.

We extend our understanding of the concepts of chemical bonding and reactivity learned in Chapter 3 on metals and Chapter 4 on zeolites to catalysis over metal oxides and metal sulfides in Chapter 5. The features that lead to the generation of surface acidity and basicity are described via simple electrostatic bonding theory concepts that were initially introduced by Pauling. The acidity of the material and its application to heterogeneous catalysis are sensitive to the presence of water or other protic solvents. We explicitly examine the effects of the reaction medium in which the reaction is carried out. In addition, we compare and contrast the differences between liquid and solid acids. We subsequently describe the influence of covalent contributions to the bonding in oxides and transition to a discussion on the factors that control selective oxidation.

Selective oxidation requires an understanding of the active metal cations in complex solid state matrices and their changes in complex reaction environments. The active systems are controlled by the oxidation state and the coordination number of the metal cations, the interaction between the oxide and the support, the domain size of the active cluster, the electronic properties of the active domain and the influence of the oxide support, the presence of defect sites, the surface morphology and surface termination. The metal cations are analogous to the single metal cation centers used in homogeneous catalysts but are now influenced by the presence of the oxide media in which they exist.

Many selective oxidation reactions demonstrate a strong synergy of combining both acid/base and redox functionality. Nature does this in a transparent way by strategically placing these functions in unique positions so as to enable both functions to act synergistically. The key to many oxidation reactions will likely also require a delicate balance between the strength as well as the specific spatial arrangements of acid, base and oxidation sites. The reactivities of reducible oxides or sulfides have many similarities with one another and also with coordination complexes. The close collaboration between theory, fundamental surface science studies and industrial experiments has helped to reveal the nature of the active sits for hydrodesulfurization (HDS). The current model of the active sulfide surface under practical sulfiding conditions is that the Mo edge and the sulfur edge of supported MoS_2 particles are coordinatively saturated with sulfur. However, metal atoms which are just inside of this edge are shown to be partially reduced with vacant sulfur sites on the surface. These sites form a brim at the surface whereby sulfur-containing molecules can adsorb and undergo desulfurization. The addition of Co to the supported MoS_2 is well know to promote HDS activity. The promotional effects have been speculated to be the result of the weaker Co–S bond over that of the Mo–S bond.

Electrocatalysis by chemical systems is extensively discussed in Chapter 6. The chapter provides a direct comparison between reactions occurring at the gas-solid interface with those occurring at the liquid-solid interface. Systems are specifically chosen that have been studied at both the gas-solid interface as well as in the liquid phase only. The latter case involves catalytic reactions carried out with organometallic coordination complexes. This provides an opportunity to compare catalysis at the surface-liquid interface with homogeneous catalysis with coordination complexes. More specifically, we describe the oxidative acetoxylation of ethylene to form vinyl acetate. Results from both homogeneous catalytic reactions carried out in solution as well as over ideal metal surfaces exist to provide guidance and help to interpret computational results. We subsequently transition into electrocatalysis over a metal substrates. In order to illustrate the comparison between reactions that are run in the gas phase with those run electrocatalytically, we specifically examine ammonia oxidation since there is an extensive data base for both gas phase as well as electrocatalytic reactions. In the presence of protic or aqueous medium, the solvent can enhance catalytic reactions by stabilizing polar transition states as reported in most physical organic chemistry text books. The solvent molecules can also specifically participate in the reactions themselves. An important example is the proton-transfer reaction. Consequences of the latter effect are extensively discussed.

The concluding chapters in the book attempt to draw analogies between the concepts regarding the self-assembly and the catalytic propagation of the elementary structures in the origin of life and the analogous active self-sustaining features required for heterogeneous catalysis. As mentioned earlier, an important reason to include this topic in the book is that it offers insights into novel strategies for the development of new catalytic systems. More specifically, we describe biomineralization strategies that have been used to synthesize silica materials with pore sizes architecture over different length scales.

The book is concluded with a final chapter that summarizes the key concepts presented in the book in a concise way. To some extent, Chapter 10 can be used as a glossary highlighting the important concepts presented throughout the book that govern catalysis.

1.4 Theoretical and Simulation Methods

One of the ultimate goals in modeling heterogeneous catalytic reaction systems would be the development of a multiscale approach that could simulate the myriad of atomic scale transformations that occur on the catalyst surface as they unfold as a function of time, processing conditions and catalyst structure and composition. The simulation would establish all of the elementary physicochemical paths available at a specific instant in time, determine the most likely reaction paths by which to proceed and then accurately calculate the elementary kinetics for each process along with the influence of the local reaction environment internal on the simulation. In addition, the simulation would predict how changes in the particle size, shape, morphology, chemical composition and atomic configurations would influence the catalytic performance, including activity, selectivity and lifetime. Modeling the spatial surface and gas phase composition along with the temperature would enable us to follow self organization phenomena also.

This is obviously well beyond what we can currently simulate. Even subsystems of this would be quite difficult to carry out with any meaningful accuracy. This does not mean, however, that theory is of little use and should be abandoned. On the contrary, one of the primary goals of this book is to highlight the impact that theory has made in establishing governing catalytic principles important for the science of catalysis. Many of these ideas could not have been conceptually or quantitatively obtained without the help of state of art computational methods.

The detailed prediction on the state of adsorbed species can be validated by comparison with experimental studies on well-defined model surfaces and model catalytic systems under controlled reaction conditions for which adequate theoretical modeling techniques are available. This includes the prediction of adsorbate surface structure, their properties and their reactivity. This can be determined by comparing the surface structure of adsorbed intermediates under idealized conditions measured through scanning tunneling microscopy (STM), and low energy electron diffraction (LEED), vibrational frequencies from high resolution electron energy loss spectroscopy (HREELS) or reflection adsorption infrared spectroscopy (RAIRS), and their adsorption and reactivity measured from temperature programmed desorption (TPD) and temperature programmed reaction (TPR) spectroscopy, or microcalorimetry. This provides quantitative information on the elementary adsorption and reaction steps that occur on these model surfaces.

The understanding of catalysis, however, will require modeling the differences in surface structure between the ideal single crystal surfaces studied under ultrahigh vacuum (UHV) conditions and those likely present for the supported particles used industrially. This is typically called the materials gap in surface science. In addition, the understanding of catalysis will also require a move from the ideal conditions of the UHV and those modeled quantum mechanically to industrially relevant conditions. The difference in pressure for experiments carried out under UHV and those under catalytic conditions can be as high as ten orders of magnitude. This can significantly alter the surface coverages and composition and thus lead to significant changes in the rate. This is known as the "ressure gap".

The disparate time and length scales that control heterogeneous catalytic processes make it essentially impossible to arrive at a single method to treat the complex structural behavior, reactivity and dynamics. Instead, a hierarchy of methods have been developed which can can be used to model different time and length scales. Molecular modeling of catalysis covers a broad spectrum of different methods but can be roughly categorized into either quantum-mechanical methods which track the electronic structure or molecular simulations which track the atomic structure (see the Appendix).

The ability to calculate the intrinsic catalytic reactivity of bond-breaking and bond-making events requires a full quantum-mechanical description of these events. The simulation of catalyst structure and morphology or reaction kinetics, on the other hand, would be more easily simulated via atomistic scale simulations, provided the appropriate interatomic potentials or intrinsic kinetic data exist. Over the past decade, it has become possible to derive such data from ab initio calculations, thus allowing for a hierarchical approach to modeling.

There are a number of excellent reviews and discussions about the advances that have taken place in quantum-mechanical method development and their ability to calculate a host of different material properties[11]. We therefore, do not go into this in detail in this book. Instead, we present a short overview of covering salient features of the different theoretical and computational methods and their application to catalysis. This is presented in the Appendix along with references to more detailed reviews on the different methods.

The path taken over the past decade for most reseachers modeling catalytic systems has been to move to first-principle quantum mechanical methods to describe bond-breaking and -making steps since they provide a reliable degree of accuracy. While there is still work on the development of semiempirical methods to describe transition metals more accurately, many of those modeling catalysis have abandon semi-empirical approaches, at least for the near future. High-level coupled cluster ab initio methods which attempt to simulate the wavefunction accurately exist and, in principle, provide the highest level of intrinsic accuracy as they can predict the heats of formation that are on the order of 4 kJ/mol or less in terms of accuracy. They can only treat, however, systems with less than about 10 heavy atoms. This is the level of accuracy that is required in order to predict rates of reaction with a significant degree of confidence. The computational requirements necessary even to come close to this level of accuracy for reliable catalytic models are currently well outside the reach of even today's fastest multiprocessor computers.

Two other approaches have been taken in order to model the active site and its environment. The first has been to use somewhat less accurate quantum-chemical methods to obtain a more qualitative understanding of the key surface states, reaction pathways and mechanism. The key parameters can then be refined by the use of high level theory and/or experiments on model systems which are much smaller. The main benefit of theory then has been the design of a physically justifiable microscopic description of the catalytic system, with a qualitatively correct conceptual understanding.

The second approach has been to develop a model of the interactions that occur between the reactant intermediates and the catalyst surface using a force field that has been empirically or theoretically obtained using a well-defined model system. Molecular mechanics and molecular dynamics studies can then be used to simulate properties of the system which can be compared with experiment. This is the more conventional approach in enzyme catalysis[12].

Most of the calculations on heterogeneous catalytic systems today use ab initio density functional theoretical methods. DFT (density functional theory) is fairly robust and allows a first-principle-based treatment of complex metal and metal oxide systems whereby electron correlation is included at significantly reduced CPU cost. DFT can be used to calculate structural properties and typically reports accuracies to within 0.05 Å and 1-2°, overall adsorption and reaction energies that are typically within 20-35 kJ/mol and spectroscopic shifts that are within a few percent of experimental data.

A comparison between experimental adsorption energies for different adsorbates on different metal surfaces estimated from UHV temperature-programmed desorption studies and those calculated using density functional theory is shown in Fig. 1.1a. Although this is a very useful first step, the differences are certainly not within the 5 kJ/mol engineering accuracy that one would like. Figure 1.1b shows a comparison between HREELS and DFT calculated vibrational frequencies for maleic anhydride adsorbed on Pd(111).

The success in modeling catalytic systems depends not only on the accuracy of the

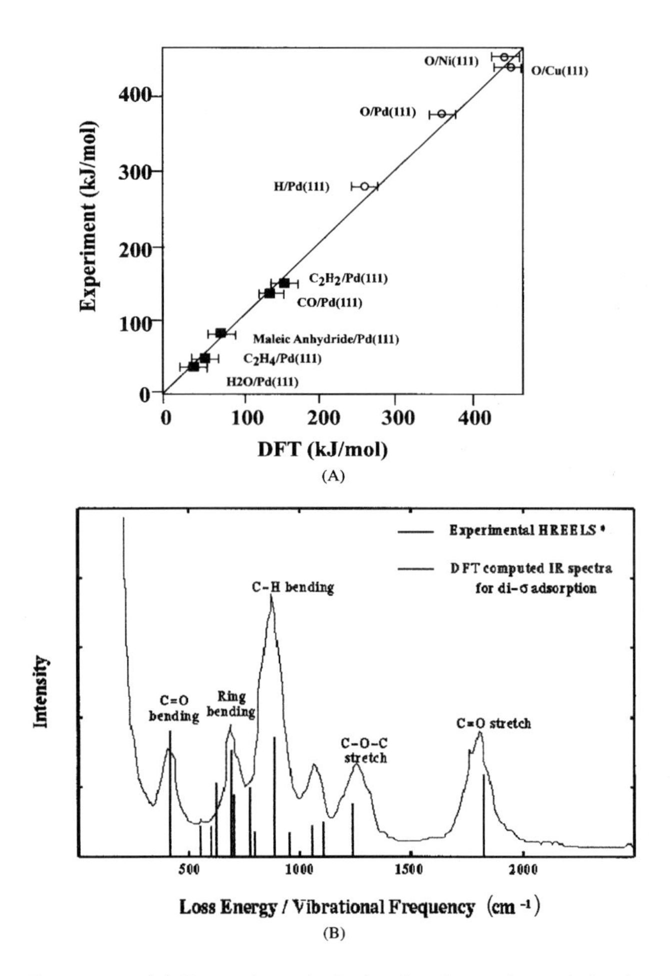

Figure 1.1. (a) Comparison of calculated and experimental chemisorption energies for different adsorbates on different metal surfaces and (b) vibrational frequencies for surface adsorbates such as maleic anhydride bound to Pd(111)[13].

methods employed, but also on the reality of the model chosen to mimic the actual reaction system studied. A single metal atom, for example, would be a poor choice for modeling a transition-metal surface regardless of the accuracy of the method used. The model can not capture the metal band structure. This leads to errors that are at least as large as those from the accuracy of the method.

There are three different techniques that are currently used to model the structure at the active site, known as cluster, embedded cluster, and periodic methods. Each method has its own set of advantages and disadvantages. Characteristic models for each of these systems are presented in Fig. 1.2.

In the cluster approach, a discrete number of atoms is used to represent only the very local region about the active site. The basic premise is that chemisorption and reactivity are local phenomena, primarily affected only by the nearby surface structure.

In the embedded cluster approach, a rigorous QM method is used to model the local region about the active site. This primary cluster is then embedded into a much larger model which simulates the external structural and electronic environment. The outer model employs a much simpler quantum-mechanical treatment or an empirical force field to simulate the external environment but still tends to treat the atomic structure explicitly. This minimizes cluster-size artifacts. The outer model can subsequently be embedded in yet a third model, which is made of point charges in order to treat longer range electrostatic interactions and the Madelung potential.

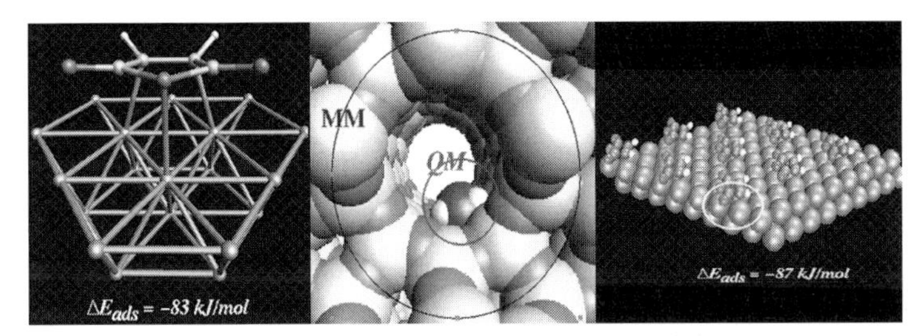

Figure 1.2. Three approaches and examples for modeling chemisorption and reactivity on surfaces. (Left) cluster approach, maleic anhydride on Pd; (center) embedding scheme: ammonia adsorption in a zeolite cage; (right) periodic slab model: maleic anhydride adsorption on Pd(111).

The last approach is the periodic slab method. In this approach one defines a unit cell which comprises a large enough surface ensemble. Periodic boundary conditions are then used to expand the cell in the x, y, and/or z directions, thus providing the electronic structure for linear, slab (surface), and bulk materials, respectively.

In later chapters we will meet applications of all of these approaches in the solution of a number of the example systems described.

On transition metals it has been suggested that a well-chosen cluster of 20-30 atoms enables one to simulate the interaction of an isolated molecule with a transition-metal surface provided that there are no atoms in the surface which form bonds to the adsorbate that are left coordinatively unsaturated.[14]

In ionic systems it is essential to chose clusters that are electrostatically neutral, otherwise electrostatic boundary effects tend to dominate computed results. In the application of embedded methods care has to be taken to correct properly for boundary effects between cluster and medium. Madelung electrostatic field simulations have to be done with proper choice of charges and dielectric constants. For these reasons periodic calculations, when feasible, are typically preferred for ionic systems.

Ab initio quantum-chemical methods can be used to calculate a range of relevant properties for homogeneous and heterogeneous catalysts. The size of the system that can be examined, however, is still quite small in comparison with the features that make up the actual system.

Structural Monte Carlo simulations can explore significantly larger system sizes due to the fact that they only treat interatomic interactions with no focus on the electronic structure[15]. The interatomic potentials that are necessary for structural simulations can be derived either from experiment or from rigorous QM methods. Potentials of choice for ionic systems are typically additive potentials such as the rigid ion potentials, that depend on charges, bond distances, and bond angle or shell model potentials that also contain terms that describe polarization. Potential parameters can be deduced by comparison with theory[16] or experiment.

Both the embedded atom method (EAM) and effective medium theory (EMT) have been used with some success to model the structure and composition of metal systems. In EAM, the potential is not linear in the number of neighbors and depends on the electron density. The parameters for both theories can be deduced from experiments[17] or theory[18]. These systems, for the most part, have been limited to looking only at the properties of only the metal atoms without adsorbates since there were few metal-adsorbate potentials. Recent simulations by van Beurden and Kramer[18] indicate that with some effort modified embedded atom (MEAM) potentials can be established for adsorbates in order to simulate the dynamics of the species on metal substrates.

Statistical mechanical Monte Carlo[19] as well as classical molecular dynamic methods can be used to simulate structure, sorption, and, in some cases, even diffusion in heterogeneous systems. Kinetic Monte Carlo simulation is characteristically different in that the simulations follow elementary kinetic surface processes which include adsorption, desorption, surface diffusion, and reactivity[20]. The elementary rate constants for each of the elementary steps can be calculated from ab initio methods. Simulations then proceed event by event. The surface structure as well as the time are updated after each event. As such, the simulations map out the temporal changes in the atomic structure that occur over time or with respect to processing conditions.

A much more in-depth description of the full range of different ab initio quantum mechanical methods, free energy and kinetic Monte Carlo methods, molecular dynamics, ab initio molecular dynamics and linear scaling methods is given in the Appendix.

References

1. *New Chemical Science and Engineering Technology. Vision 2020 Catalysis Report*, Council for Chemical Research (1997) see; http://www.ccrhq.org/vision/index /roadmaps/catrep.html
2. National Research Council, Commission on Physical Sciences, Mathematics, and Applications, *Catalysis Looks to the Future*, National Academies Press, Washington, DC (1992)
3. Opportunities for Catalysis in the 21st Century: A Report from the Basic Energy Sciences Advisory Committee (May 2002)
4. R.E. Smalley, Rice University Department of Chemistry, presentation at Columbia University (September 2003), *Our Energy Challenge* (see http:smalley.rice.edu)
5. *Basic Research Needs to Assure a Secure Energy Future: A Report from Basic Energy Sciences Advisory Committee* (February 2003)
6. *Basic Research Needs for the Hydrogen Economy, Basic Energy Sciences Workshop on Hydrogen Production, Storage and Use* (May 2003)
7. D. Miller, *NSF Workshop Report: Catalysis for Biorenewables Conversion* (April 2004)
8. W.G. Frankenburg, in *Hydrogenation and Dehydrogenation Catalysis*, Vol. 3, P.H. Emmett (ed.), Reinhold, New York (1995)
9. S.M. Auerbach, R.A, Currado, P.K. Dutta, *Handbook of Zeolite Science and Technology*, Marcel Dekker, New York (2003);
   H. van Bekkum, E.M. Flanigan, J.C. Jansen, *Introduction to Zeolite Science and Practice, Stud. Surf. Sci. and Catal.* Vol. 58, Elsevier, Amsterdam (1991)
10. G. Ertl, *Adv. Catal.* 37, 213 (1990);
    R. Imbihl, G. Ertl, *Chem. Rev.* 95, 697 (1995)

CHAPTER 2
Principles of Molecular Heterogeneous Catalysis

2.1 General Introduction

The three predominant criteria for the development and the design of new catalytic materials are that they:

1. are highly active
2. produce only the desired products and
3. maintain activity for prolonged periods of time.

All three of these factors are controlled by the kinetics of the elementary steps in the overall catalytic cycle. The activity is defined as the rate at which the overall catalytic cycle turns over and is dictated by the rate-limiting step or steps in the overall process. These steps can include adsorption, desorption, specific surface reaction steps and even surface diffusion. The selectivity of a reaction is defined as the percentage of desired product molecules that are part of the product spectrum. A 100% selectivity is always desired but never fully realized. The lifetime of catalysts is related to the selectivity of a desired coproduct poisoning the catalyst. As discussed in the previous chapter, theory along with fundamental experiments have advanced considerably over the past decade and can be used to determine the energetics for elementary reaction steps as well as the overall thermodynamics. This information can be used to help determine the kinetics by the application of theories or kinetic principles in physical chemistry including the Sabatier principle, transition state theory, and the Brønsted–Evans–Polanyi relationship , thus allowing one to relate overall thermodynamic parameters to the rates of catalytic reactions. In this chapter we describe the development and application of these principles to modeling heterogeneous catalytic systems.

Catalytic reactivity is controlled by the combination of intrinsic chemical reactivity and the extrinsic heat and mass transfer effects related to the catalyst morphology and reactor configuration. We will mainly focus on the intrinsic activity except in the cases where it is difficult to uncouple these phenomena such as in the reaction and diffusion in porous materials, which is covered in Chapter 4.

The kinetics of a catalytic system are dictated by the nature of the chemical complex formed as the result of the interactions between atoms in the catalyst and the adsorbate. The reactivity of the complex is controlled by local as well as long range chemical effects. A classical question in heterogeneous catalysis is whether the catalytic sites can be described with either a Langmuir[1] or Taylor[2] model. These models refer to local descriptions of the active site. According to the Langmuir model, all sites are considered to be similar. In the Taylor model, however, there is a distribution of different sites, and catalysis occurs at a small fraction of unique sites.

A second important question involves the range of the chemical interactions. It is now well known that the nature of the reaction environment about the active catalytic site can be just as important in describing and potentially controlling the catalytic performance as the intrinsic chemical interactions in the catalytic complex. The reaction environment includes the influence of the solvent media; solid state matrix, i.e. the effects of the cavity, the support, alloy composition and structure, and defects at the catalyst surface; long-range electrostatic forces between the catalyst and the reactive complex; relaxation and reconstruction of the surface; promoters and lateral interactions between surface adsorbates that change with reaction conditions.

Molecular Heterogeneous Catalysis. Rutger Anthony van Santen and Matthew Neurock

ISBN: 3-527-29662-X

The third predominant issue refers to whether the active site and its environment can be treated statically or whether one must follow their dynamics. Interesting physical effects and phenomena can arise from the coupling of the surface mobility along with intrinsic chemical reactivity which include kinetic oscillations and self organized pattern formation under particular catalytic conditions. As an organizing principle for the chapter, we use the concepts outlined above, and for this reason, different catalytic systems will be discussed under the same headings.

This chapter proceeds with a general discussion of the overall catalytic cycle and Sabatier's principle in order to illustrate the comparison of relative kinetic and thermodynamic steps in the overall cycle. This is followed by a fundamental discussion of the intrinsic surface chemistry and the application of transition state theory to the description of the surface reactivity. We discuss the important problem of the pressure and material gap in relating intrinsic rates with overall catalytic behavior and then describe the influence of the "tatic" reaction environment including promoters, cluster size, support, defects, ensemble, coadsorption and stereochemistry. Lastly, we discuss the transient changes to the surface structure as well as intermediates and their influence on catalytic performance.

2.1.1 The Catalytic Cycle

2.1.1.1 The Sabatier Principle

When the concept of catalysis was first formulated, the idea that the catalytic reaction is actually a catalytic cycle was not at all obvious. In 1836 Berzelius defined "catalytic force" as the process responsible for catalysis in which the decomposition of bodies was caused by the action of another simple or compound body. Faraday later showed that a catalytically reactive surface was chemically altered by contact with reacting gases. It was not, however, until after chemical thermodynamics had been developed that a more scientific understanding of catalysis was formulated. In 1896 Van't Hoff demonstrated that the rate of a catalytic reaction depended upon the amount of catalyst. Soon after Ostwald defined a catalyst to be a substance that changes the velocity of a reaction without itself being altered in the process. A catalyst, however, must operate within the thermodynamic limits of the reacting system[3].

The reaction conditions for a specific system are defined by the overall thermodynamics of the reaction. The catalyst facilitates the adsorption of the reactants and their subsequent conversion into products. An important feature, however, is that the products must be rapidly removed from the surface in order to regenerate active surface sites. These ideas led to the concept that the catalytic reaction is comprised of a "cycle" which is made up of elementary physicochemical processes. At the most basic level, catalysis is comprised of at least five elementary steps:

- Chemisorption
- Dissociation/activation
- Diffusion
- Recombination
- Desorption

The incorporation of these steps into a cycle and the overall concept of the catalytic cycle are illustrated for the catalytic decomposition of N_2O over a catalytic substrate in Fig. 2.1. N_2O is an environmentally detrimental molecule. It is produced as an undesirable product

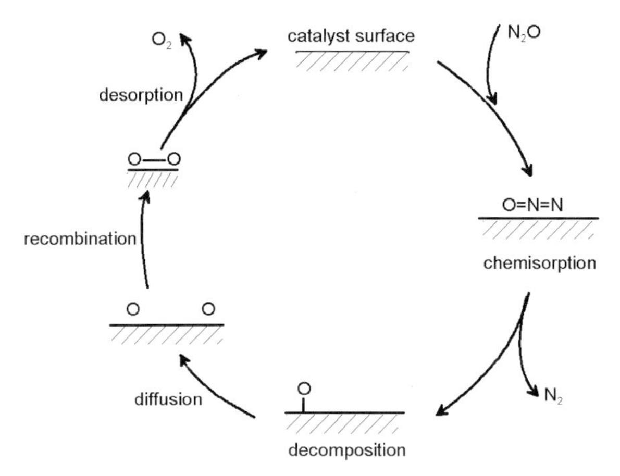

Figure 2.1. Schematic illustration of the catalytic cycle of reaction events for the decomposition reaction of N_2O.

in automotive exhaust catalysis as well as in nitric acid synthesis. It is a precursor to nitric acid production in the stratosphere. Conventional catalysts for N_2O decomposition are the transition metals and the transition-metal oxides.

Chemisorption is defined as the adsorption of reactants or intermediates on the catalyst substrate with an interaction energy with the surface which is strong enough to form chemical bonds between the adsorbate and the surface and to weaken internal bonds within the adsorbate. This helps to aid dissociation processes. In order for the N_2O molecule to decompose it must bind strongly to the catalyst surface, thus weakening the internal N–O bonds. The interaction energy must therefore be strong enough to overcome the N–O bond energy in N_2O. When N_2O decomposition is the rate limiting step, the rate of the catalytic reaction should increase with increase of the N_2O-catalyst interaction energy. On the other hand, the fragment molecules generated on the surface by decomposition of N_2O have to desorb in order to regenerate the free catalytic sites necessary to continue the cycle. Hence, when the interaction between product molecules and the catalyst becomes too strong the desorption of product molecules becomes rate limiting. The rate will then decrease with increasing interaction energy of products with the catalyst. This ultimately leads to a balance of the catalyst-adsorbate bond strength so as to activate the adsorbate but avoid poisoning of the surface. This balance results in a volcano-type plot of rate against reactant-interaction strength whereby the rate increases up to a particular interaction strength known as the Sabatier maximum and then decreases. This shape of the plot is a consequence of the Sabatier principle (see Fig. 2.2). The rate of a catalytic reaction is maximized at an optimum interaction strength of the reactants with catalyst[4]. This provides a rational strategy for the optimization of catalytic rates. For instance, by comparing different catalysts for the same reaction and measuring the reactant-adsorption energies and catalytic rate, one can determine whether the reaction rate increases or decreases with adsorption energy of the reactants. If one then uses trends in measured or computed adsorption energies as a function of material, one can select surfaces that might improve catalytic performance.

The order for a monomolecular reaction changes from positive in the reactant concentration to the left of the Sabatier maximum to zero or negative order in the reactant

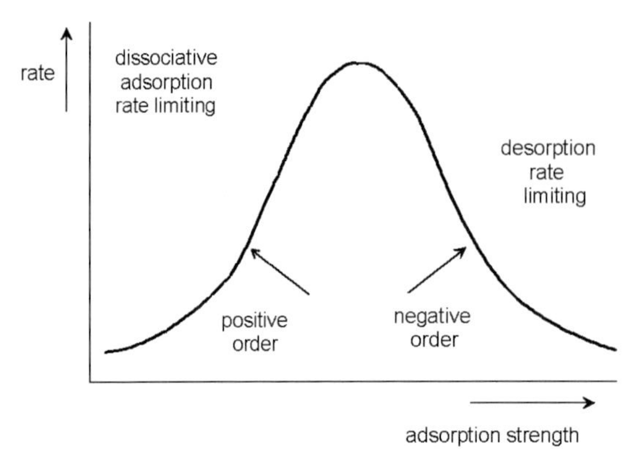

Figure 2.2. Sabatier principle. Catalytic rate is a maximum at optimum adsorption strength.

concentration to the right of the Sabatier maximum. This corresponds to a low surface concentration of adsorbed intermediates on the left side of the optimum and an increasing surface coverage to the right side.

Sabatier-type volcano plots have been constructed for a number of different commercially relevant systems[3]. A simple kinetic expression that simulates the Sabatier result is found when one realizes that the decomposition of molecules requires a vacant site for molecular fragments to adsorb on. For instance, in the N_2O decomposition reaction, the dominant surface species (most abundant reaction intermediate)[5] is atomic oxygen (O), which is in equilibrium with the gas phase. When the slow step in the reaction is dissociative adsorption of N_2O, the mean-field kinetic rate expression for N_2O decomposition, normalized per unit surface area of catalyst, becomes:

$$r_{N_2O} = k_{dec}[N_2O](1 - \theta_O) \tag{2.1}$$

where k_{dec} is the rate constant for N_2O decomposition, θ_O is the coverage of surface oxygen and nitrogen atoms and $[N_2O]$ is the gas phase concentration of N_2O.

While kinetic rate expressions such as Eq. (2.1) are widely used, they are considered as very approximate. Generally the rate constants, as well as the equilibrium constants, cannot be considered concentration independent and therefore are only effective parameters. Therefore, when they are measured in the laboratory, they will inherently be a function of the conditions under which the reaction was performed. This will be described more extensively in Chapter 3 on metal catalysis.

On transition metals or transition metal oxides, the rate of O_2 desorption (k_{des}) competes with the rate of N_2O decomposition (k_{diss}). Each adsorbed oxygen atom blocks a site for N_2O decomposition. Nitrogen will adsorb only weakly, but oxygen is effectively a poison to the reaction. Expression (2.1) can be rewritten as a function of equilibrium and rate constants to give

$$r = \frac{k_{diss}[N_2O]}{1 + \frac{k_{diss}[N_2O]}{k_{des}} + K_{eq}(O_2)[O_2]^{1/2}} \qquad (k_{des} \ll k_{diss}[N_2O]) \tag{2.2}$$

When the interaction energy of the adsorbate with the transition metal increases k_{diss} will increase. As we will see later in Section 2.3, the activation energy of a reaction decreases when the adsorption energy increases. On the other hand, the rate for the associative desorption of O_2 will also decrease. Hence if Eq. (2.2) is studied as a function of the adsorbate-surface interaction energy, it will follow the Sabatier volcano curve. The rate will increase with increase in the surface interaction energy up to a specific interaction energy whereby there is a maximum rate. The maximum occurs when k_{des} and k_{diss} N_2O balance one another. Increasing the adsorbate-surface interaction energy beyond this point decreases the rate of reaction as the surface becomes covered with an oxygen overlayer.

An important topic in this book is the prediction of reaction mechanisms. Elucidating the mechanism is enhanced by the construction of reaction energy diagrams which follow the energy changes of the different reaction intermediates as the reaction proceeds through its reaction cycle.

The kinetic expressions which govern different reaction mechanisms are usually very different. We will illustrate this by comparing expression (2.2) with the kinetic expression of N_2O decomposition found for zeolites. The N_2O decomposition reaction has a very different reaction sequence when catalyzed by isolated Fe cations in a zeolite, as compared with the reaction sequence found for catalysis by transition metals.

As we will see in Chapter 4, in the zeolite system we find that reaction occurs in two steps:

$$N_2O + Fe^{3+}[*] \xrightarrow{k_a} FeO^{3+} + N_2 \uparrow \tag{a}$$

$$N_2O + FeO^{3+} \xrightarrow{k_b} Fe^{3+}[*] + O_2 \uparrow \tag{b}$$

where $[*]$ denotes a vacancy site. Both steps are thought to be irreversible. In the presence of water the reaction sequence dies due to the formation of the non-active $Fe(OH)_2^+$ intermediate. The equilibrium constant for H_2O adsorption is K'. In the presence of water the kinetic rate expression for N_2O decomposition becomes[6]

$$r_{\mathrm{zeolite}} = \frac{k_a k_b}{\left[k_a\left(1 + K'[H_2O]\right) + k_b\right]}[N_2O] \tag{2.3}$$

The rate of N_2O decomposition here remains linear in N_2O partial pressure, also at high pressures. This is quite different to what is found in Eq. 2.2 for N_2O decomposition over the metal. Once again Sabatier-type behavior is expected as a function of the metal-oxygen interaction energy. The rate of reaction a (k_a) increases with an increase in the metal-oxygen bond energy, whereas the rate of reaction b (k_b) decreases.

More generally, for the reaction of R to product P which proceeds through the formation of adsorbed intermediates I_1 and I_2:

$$R \underset{k_{\mathrm{ads}}}{\overset{k_{\mathrm{des}}}{\leftrightarrows}} I_1 \xrightarrow{k_r} I_2 \xrightarrow{k_{\mathrm{des}}^{(2)}} P$$

one can deduce the following steady-state expressions:

$$\frac{\mathrm{d}P}{\mathrm{d}t} = k_{(\mathrm{des})}^{(2)} \theta_{I_2} \tag{2.4a}$$

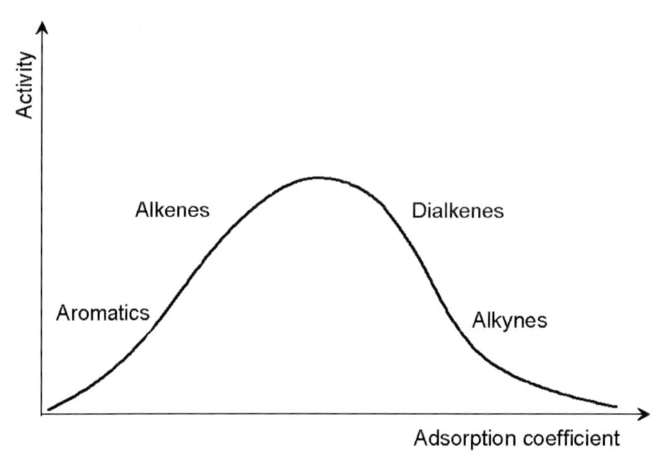

Figure 2.3. Schematic graph of the activity of hydrocarbon hydrogenation as a function of the adsorption coefficients of the respective substrate molecules[7a].

$$= k_r \frac{K_{\text{eq}} P}{1 + K_{\text{eq}} \cdot P} \tag{2.4b}$$

As the interaction between reactant and surface increases, the rate according to Eq. 3b will increase linearly at lower pressures and subsequently saturate to a constant value at higher pressures. The production of P shows a maximum which is defined here as Sabatier's maximum when

$$k_r \approx k_{\text{des}}^{(2)}$$

To the left of the Sabatier maximum the surface is predominantly covered by the reactant, to the right of the maximum the surface is predominantly covered by products. To the right of the maximum the rate decreases because of the decrease in the desorption rate.

An interesting illustration of this concept is given by a comparative study of the hydrogenation of unsaturated hydrocarbons presented in Fig. 2.3. The rate of hydrogenation of different hydrocarbons is given as function of their adsorption equilibrium constants. One notes the volcano-type dependence found experimentally. This behavior implies that the rate of hydrogenation is a direct function of the adsorption constant of the reactant molecule. Mittendorfer et al.[7b] have shown theoretically that the binding energies for C_4 to C_6 intermediates on Pt(111) increase in following order:

Benzene < Butene < Butadiene < Butylene

This ordering is consistent with the changes that occur along the x-axis in Fig. 2.3.

Sabatier's principle provides a kinetic understanding of the catalytic cycle and its corresponding elementary reaction steps which include adsorption, surface reaction, desorption and catalyst self repair. The nature of the catalytic cycle implies that bonds at the surface of the catalyst that are disrupted during the reaction must be restored. A good catalyst has the unique property that it reacts with the reagent, but readily becomes liberated when the product is formed. This will be further discussed in Section 2.2, where we describe the kinetics of elementary surface reactions and their free energy relationships.

2.1.1.2 Reaction Cycles; Intermediate Reagents

We will note in Chapter 7 that biochemical systems often consist of reaction cycles where the key molecular intermediate is regenerated after the catalytic reaction. The intermediate itself can be a catalytic reagent. In the biochemical reaction cycle, enzymes catalyze reactions between different molecular components in order to convert and create the catalytic reagent. For instance, in the citric acid cycle, see Fig. 2.4, which produces energy, an activated acetyl unit reacts with oxaloacetate and CO_2 is produced by the oxidation of molecular fragments[8]. The overall reaction involves the conversion of acetyl and the water to CO_2 and H_2 and metabolic energy.

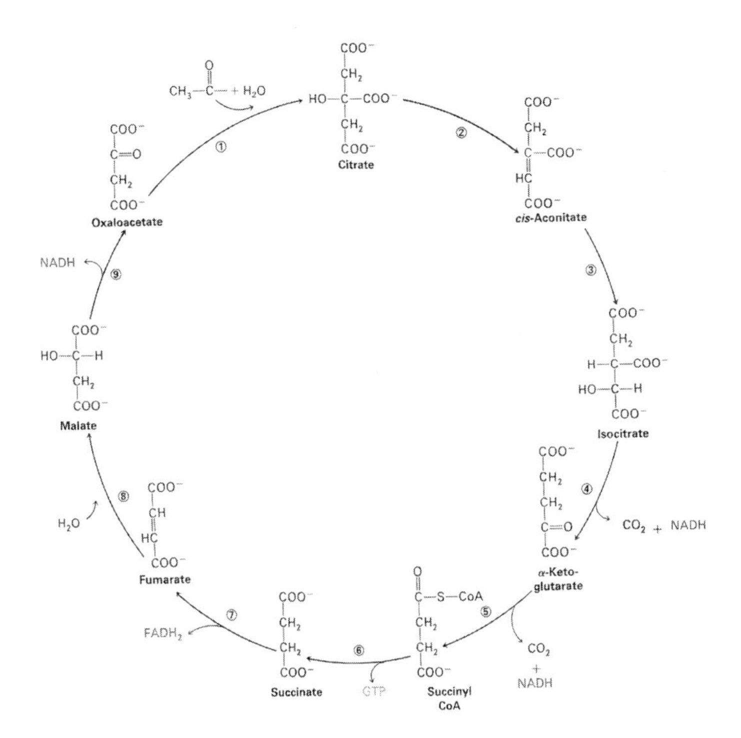

Figure 2.4. Citric acid cycle. This series of reactions is catalyzed by the following enzymes as numbered in the diagram: (1) Citrate synthase, (2) Aconitase, (3) Aconitase, (4) Isocitrate dehydrogenase, (5) α-Ketoglutarate: dehydrogenase complex, 6) Succinyl CoA syntetase, (7) Succinate dehydrogenase, (8) Fumarase and (9) Malate dehydrogenase. Adapted from L. Stryer[8].

Each step in the citric acid cycle is catalyzed by an enzyme. Two CO_2 molecules are split off in reaction steps 4 and 5. The addition of H_2O helps to regenerate the catalytic oxaloacetate intermediate. While each step in the citric acid cycle in itself is catalytic, the regeneration of the catalytic reagent oxaloacetate within the cycle is critical for the overall cycle to proceed.

In chemocatalysis, an analogous cycle would involve the formation of a surface intermediate (catalytic molecule) within the reaction cycle by the catalytic reaction with reactant molecules. Free radical reaction schemes, as discussed in Chapter 4, can be considered analogues of the biochemical cycle, where a reactive chemical intermediate is regenerated typically by catalytic steps. They also share the feature that reaction steps are often autocatalytic in this intermediate.

An interesting organic-chemical analogue to that of the biochemical system in which a particular molecule acts as a selective oxidation catalyst is tetramethylpiperidin-N-oxyl or TEMPO[9]. This molecule acts as catalytic reagent and is regenerated by two coupled catalytic cycles as illustrated in Fig. 2.6. In one cycle, the oxygen is donated by the nitro-oxide catalyst to the reacting substrate. The TEMPO molecule is regenerated by oxidation over a metal catalyst.

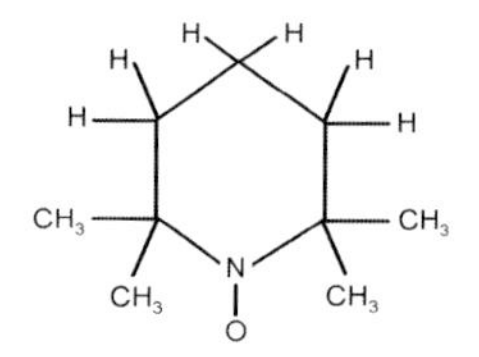

Figure 2.5. 2,2,6,6-tetramethylpiperidine-N-oxyl, the TEMPO molecule[9].

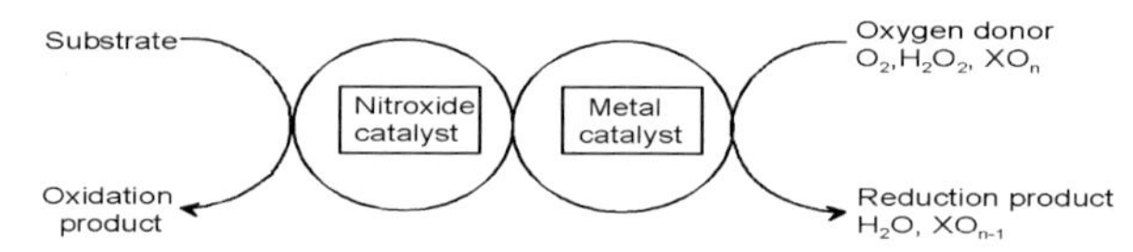

Figure 2.6. In-situ catalytic regeneration of the TEMPO molecule.

In a second example, we focus on an industrially relevant Wacker reaction system. In the homogeneous Wacker system, Pd^{2+} is the active intermediate that generates atomic oxygen from H_2O. Cu, however, is necessary to act as a redox couple in order to reoxidize the Pd^0 that forms with air:

$$\begin{aligned} Pd^{2+} + H_2O + H_2C{=}CH_2 &\longrightarrow H_3C\text{–}CHO + Pd^0 + 2H^+ \\ O_2 + 2Cu^+ &\longrightarrow 2Cu^{2+}O \\ 2Cu^{2+}O + Pd^0 &\longrightarrow 2Cu^+ + Pd^{2+} \end{aligned}$$

The Cu redox cycle shown above is the catalytic system necessary to regenerate the catalytic active Pd^{2+} reagent.

There are many examples where electrochemical oxidation or reduction is used to regenerate the key reagent when it cannot be regenerated directly. An interesting example is the electro-catalytic system that oxidizes higher alkenes to epoxides[10].

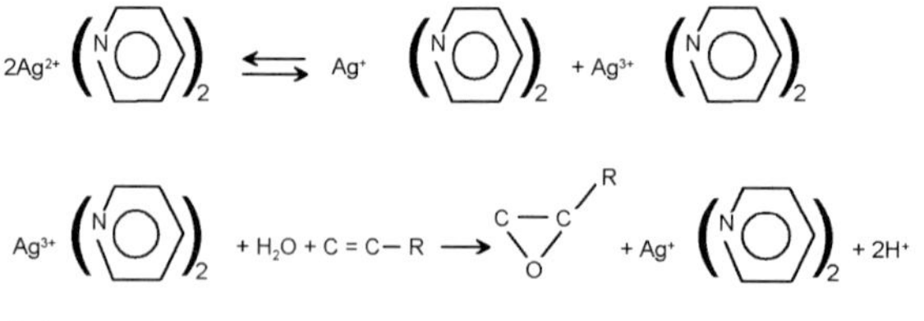

2Ag²⁺ (pyridine)₂ ⇆ Ag⁺ (pyridine)₂ + Ag³⁺ (pyridine)₂

Ag³⁺ (pyridine)₂ + H₂O + C = C– R ⟶ (epoxide C–C(R)–O) + Ag⁺ (pyridine)₂ + 2H⁺

Scheme 2.1

The Ag^{2+} is regenerated by anodic oxidation.

The most important issue that has to be addressed in the design of these cycles is that of selectivity. The catalyst that regenerates the catalytic reagent should not induce undesired reactions with reactant molecules or products.

2.2 Physical Chemistry of Intrinsic Reaction Rates

2.2.1 Introduction

According to the Sabatier principle introduced in Section 2.1.1.1, the rate of a catalytic reaction is a maximum when the interaction between reactant and catalyst is at an optimum value. The key to the Sabatier principle then involves understanding the free energies of adsorption as a function of catalyst structure and composition. In the 1960 's, Tanaka and Tamaru[11] noted that trends in the interaction energies of different molecules with varying catalyst surfaces are often very similar. Theoretical surface studies provide a justification as to why. As we will discuss in detail in Chapter 3, the strength of the surface-chemical bond directly depends on the electronic structure of the metal surface. In the 1980 's it had been proposed that electronic properties such as the local density of states at the Fermi level or d-electron occupation are parameters that critically control catalysis. More recent studies by Hammer and Nørksov[12] show that it is possible to tie the reactivity to the interaction with the relative positioning of the d-valence band center with respect to the Fermi level at the surface. The Hammer–Nørskov model has rapidly become the standard probe of surface reactivity and is being used to design surfaces with specific reactivity. As we will illustrate in the next chapter, these changes in surface electronic structure can be related to the degree of coordinative unsaturation of the surface atoms. In Section 2.2.2 we discussed the point that an optimum catalyst has maximized the balance between the rates of intermediate forming reactions and the rates of reactions by which the product molecules leave the surface.

While the relationship between the electronic properties and the reaction enthalpy is important in understanding energetics, the more important thermodynamic feature to focus on is the free energy. Indeed, in Chapter 4 the maximum for the rate of a zeolite-catalyzed reaction is not found for the zeolite with the smallest pore size (maximum adsorption enthalpy) but for medium-sized micropores where adsorbates have a higher entropy, and as a consequence, their concentration is a maximum. The gain in entropy often balances the loss in adsorption enthalpy.

In the subsections that follow we will focus on the factors that maximize the rate constant for elementary surface reaction steps. Again we will stress the need explicitly to include entropic contributions. According to transition-state reaction rate theory[13a], the rate of the elementary conversion step is defined as

$$r_{TST} = \Gamma \frac{kT}{h} e^{-\frac{\Delta G^{\ddagger}}{kT}} \tag{2.5}$$

where $\Delta G^{\ddagger}$ refers to the free energy difference between the transition state and the initial reactant state and Γ is the transmission factor. Expression (2.5) is a maximum when $\Delta G^{\ddagger}$ is a minimum, which implies a maximum stabilization of the transition state. A condition for the transition-state reaction rate expression to be valid is that the vibrational and rotational modes in the transition state are in equilibrium with the corresponding modes of the reactant ground state. This implies that energy transfer between these modes is fast compared with the reaction rate. This condition is usually satisfied for intramolecular reactions. Reactions that occur on surfaces between adsorbed reaction intermediates can also be considered to belong to this category of reactions. When molecules collide in the gas phase or collide with a surface, their contact time can be short compared with the energy transfer time. In such a case, energy transfer can become rate limiting. The transition-state theory rate expression then predicts an upper bound to the reaction rate. For a proper treatment, a molecular dynamics approach should be used. This situation is unusual but can occur for dissociation reactions of gas-phase molecules impinging upon a surface. The residence time will be short when the interaction potential is small. This can be the case for small coordinatively saturated molecules such as methane [13b,15].

2.2.2 The Transition-State Theory Definition of the Reaction Rate Constant; Loose and Tight Transition States

In order to appreciate the use of transition-state rate expressions, it is important to be reminded of the different time scales of the processes that underpin the chemistry we wish to describe. The electronic processes that define the potential-energy surface on which atoms move have characteristic times that are of the order of femtoseconds, 10^{-15} sec, whereas the vibrational motion of the atoms is on the order of picoseconds, 10^{-12} sec. The overall time scale for bond activation and formation processes that control catalysis vary between 10^{-4} and 10^{2} sec. This implies that on the time scale of the elementary reaction in a catalytic process, many vibrational motions occur. If energy transfer is efficient, then the assumption that all vibrational modes except the reaction coordinate of the chemical reaction are equilibrated is satisfied. Kramers[14] defined this condition as $E_b > 5kT$. Under this condition the transition state reaction-rate expression applies:

$$r_{\mathrm{TST}} = \Gamma \cdot \frac{kT}{h} \frac{Q^{\ddagger}}{Q_0} \mathrm{e}^{\frac{E_b - E_0}{kT}} \tag{2.6}$$

where $E_b - E_0$ refers to the barrier height of the reaction as computed from electronic structure calculations. Except when quantum-mechanical tunneling or reaction dynamics become important, the transmission factor Γ can be assumed to be one. This is a reasonable assumption for most surface reactions which have activation barriers that satisfy the Kramers condition (see Ref. 15 for more details). $Q^{\ddagger}$ and Q_0 are the partition functions of transition state and initial state, respectively.

The pre-exponential factor, ν_{eff}, and the activation barrier, E_{act}, are the kinetic parameters that are necessary to describe a reaction, deduced from the Arrhenius reaction-rate expression in the following way:

$$r_{\mathrm{Arr}} = \nu_{\mathrm{eff}}\, \mathrm{e}^{\frac{E_{\mathrm{act}}}{kT}} \tag{2.7}$$

$$E_{\mathrm{act}} = -k \frac{\partial}{\partial\left(\frac{1}{T}\right)} \ln r \tag{2.8a}$$

$$= kT^2 \frac{\partial}{\partial T} \ln r \tag{2.8b}$$

The pre-exponential factor ν_{eff} can be identified with the initial terms in Eq. (2.6)

$$\nu_{\text{eff}} = \Gamma \frac{ekT}{h} e^{\frac{\Delta S_0^{\ddagger}}{k}} = \Gamma \frac{kT}{h} e^{\frac{\Delta S_0^{\ddagger}}{k}} \tag{2.9}$$

where

$$\Delta S_0^{\ddagger} = k \ln \frac{Q^{\ddagger}}{Q_0} \tag{2.10}$$

The activation energy is defined as

$$E_{\text{act}} = (E_b - E_0) + 1/2h\left(\sum_i \nu_i^{\ddagger} - \sum_{i'} \nu_{i'}^{0}\right) + kT \tag{2.11}$$

The second term in the equation for E_{act} is due to the zero point vibrational correction.

The partition function is made up of the translational, rotational and vibrational contributions defined by the following:

$$Q = Q_{\text{trans}} \cdot Q_{\text{rot}} \cdot Q_{\text{vibr}} \tag{2.12}$$

with

$$Q_{\text{trans}} = \left(\frac{2\pi m_{c.m} kT}{h^2}\right)^n \tag{2.13}$$

where n is the number of the degrees of translational motion and $m_{c.m}$ the corresponding center of mass.

For a hetero-diatomic molecule, the partition function of rotational motion equals

$$Q_{\text{rot}} = \frac{8\pi^2 IkT}{h^2} \tag{2.14}$$

where I is the moment of inertia, μR_{eq}^2, μ is the reduced mass and R_{eq} is the atomic distance.

For harmonic frequencies, the expression to use for Q_{vibr} is

$$Q_{\text{vibr}} = \prod_i \left(\frac{1}{1 - e^{-\frac{h\nu_i}{kT}}}\right) \tag{2.15}$$

In the calculation of the vibrational entropy of the transition state, the reaction coordinate itself is not included. The transition state is the saddle point on the potential energy surface that occurs along the reaction coordinate. This is sketched in Fig 2.7. The use of Eq. (2.15) for evaluating the partition function is only valid when the zero point frequency corrections to the barrier energy have been included in the calculation of the transition-state energy.

To help illustrate the process of calculating reaction rates, we will describe the definition of the transition state for the dissociation of a CO molecule on the terrace of a transition-metal surface. CO is initially bound perpendicular to the surface. To activate the CO bond, it must first stretch and bend with respect to the surface normal in order to accept

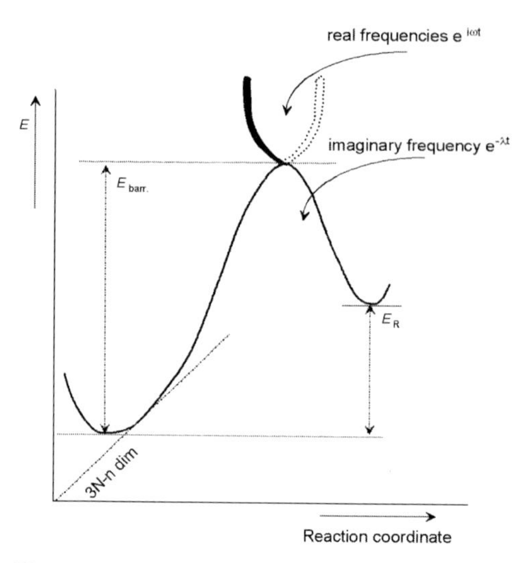

Figure 2.7. Schematic illustration of reaction coordinate and saddle point definition of the transition state[15].

electrons into its antibonding π^* orbital and initiate dissociation. After dissociation, the atomic C and O intermediates are each strongly adsorbed to the metal surface. The difference in the projection of the O coordinates from the C coordinates onto the surface can be taken as the measure of the actual reaction coordinate. The reaction coordinate is zero before reaction and after reaction it equals the C and O distance between the adsorbed C and O atoms.

When the C–O bond length increases from its initial length (in its adsorbed state), the potential energy of the system increases until a maximum is reached. Further increases in the reaction coordinate lower the potential energy until it reaches the equilibrium value of the fully dissociated system. The barrier for the transition state is defined as the saddle point on the potential energy surface. The derivatives of the energy with respect to all degrees of freedom at this point are zero, and at a minimum for all modes, except that which corresponds to the reaction coordinate for which the potential energy is at a maximum.

Γ is defined as the probability that once the system reaches the top of the transition-state barrier, it actually passes over it. In transition-state reaction rate theory Γ is usually taken to be equal to one. This is often a reasonable approximation. However, when the reaction system is viscous, as in a solution, or has diffusional restrictions, Γ can be substantially less. Molecular dynamics simulations for the motion of the system near the transition state can be carried out in order to determine Γ.

By way of example, we show how the transition-state rate expression can be used to determine the rates for both surface desorption and the dissociation of CO at low surface coverage on the terrace sites of the transition-metal substrate. This also allows for an illustration of the concepts of tight and loose transition states and their respective definitions[15,16].

Transition states which contain a high degree of entropy are typically called loose, whereas those with a low degree of entropy are considered tight. The entropy of a CO molecule adsorbed to a metal surface is not negligible, but is quite small since the main contribution to the entropy is the liberating motion of the oxygen atom versus the surface normal. The entropy of the surface atoms is essentially zero. The strong surface bonds give rise to fairly high vibrational frequencies and, hence, the vibrational partition functions are nearly equal to one.

A diatomic molecule such as CO will only dissociate with a kinetically acceptable barrier if the C–O bond is significantly weakened in the transition state. This implies a strong electronic interaction between metal-surface electrons and the lowest unoccupied states of CO. In the transition state, the molecular axis of CO is nearly parallel to the metal surface such that the overlap between molecular π and π^* orbitals of CO and transition-metal surface orbitals is maximized. This weakens the molecular bond due to back-donation of surface electrons into the anti-bonding π^* states on CO. In the transition state, the C and O atoms have nearly the same distance as those in the separared product state, and thus their interaction with the surface is quite large. The entropy of this transition state is very similar to the final state and hence essentially equal to zero. This defines the transition state of the dissociating molecule on the transition-metal surface as a tight transition state.

The transition state for a molecule desorbing from the surface is quite different to that of the reacting molecule. The desorbing molecule in its transition state is nearly free from the surface. The molecule can therefore be considered to freely rotate and to move freely parallel to the surface. Hence the activation entropy for desorption is quite large. The ratio of the rate of CO bond cleavage, r_{diss}, over the rate of desorption is

$$\begin{aligned}\frac{r_{\mathrm{diss}}}{r_{\mathrm{des}}} &= \frac{Q^{\ddagger}_{\mathrm{diss}}}{Q^{\ddagger}_{\mathrm{des}}}\,\mathrm{e}^{-\frac{E'_b(\mathrm{diss})-E'_b(\mathrm{des})}{kT}} \\ &= 10^{-4}\,\mathrm{e}^{-\frac{E'_b(\mathrm{diss})-E'_b(\mathrm{des})}{kT}}\end{aligned}$$

The values for the activation energies, E'_b, include the zero point vibrational-frequency corrections. The transition state for desorption can be characterized as a loose transition state. The entropy of this state is quite high, which is the result of the weak interaction of the desorbing molecule with the catalyst.

On most of the group VIII metals the heats of adsorption are comparable with their heats of dissociation. This implies that the adsorbed molecule will desorb with a rate which is 10^4 times faster than that for dissociation owing to the difference in the activation entropy. The reaction path that proceeds through a transition state of maximum entropy is the preferred path. For catalysis, the implication is that dissociation of adsorbed molecules will occur at those pressures and temperatures where equilibrium between the gas phase and surface maintains a significant surface concentration of the dissociated molecules. The temperature has to be chosen such that the rate constant for dissociation is in the proper time regime. The observation that for a transition-metal surface reaction, the transition entropy is low, but that for a reaction step between the surface and gas phase is large, is quite general.

2.2.3 The Brønsted–Evans–Polanyi Reaction-Rate Expression Relations

In order to establish equations which describe the reaction rate from first principles, theories based on non-equilibrium statistical mechanics have to be used[15]. However, useful empirical relations, which have some theoretically justification, exist. They linearly relate the activation energies for a reaction with some property of the reaction. Hammett, for example, showed a direct correlation between the rate of reaction for a family of different substituted benzoic acid species with a measure of the electronic properties of the substituent (the Hammett substituent parameter in physical organic chemistry)[17a]. Hammett, therefore, established a linear relationship between the logarithm of the rate and the substituent parameter. The correlations were extended over a full range of different types of reactions whereby changes in the reaction were described by changes in the slope of the line. A similar relationship for catalysis was establish by Brønsted, who showed that the rate constant of an acid-catalyzed reaction was correlated with the catalyst's equilibrium acid dissociation constant, K_A[17b]. Evans–Polanyi[17c] showed that the molecular reaction rates could be directly correlated with the overall reaction enthalpy. The Evans–Polanyi and Brønsted relationships are quite similar as the Brønsted relationship is simply a subset of the more general Evans–Polanyi relationship. Some have combined the two in what is named the BEP relationship, attribut to the three authors Brønsted, Evans, and Polyani. The BEP relationship is perhaps the most widely used linear relationship in all of catalysis. The BEP relationship is valid when one compares related elementary reaction steps that proceed through nearly the same intermediate structures and have similar reaction coordinates.

For the simplified parabolic energy curves of reactant and product states shown in Fig. 2.8, if one shifts the reaction energies but holds the coordinates of the final and reactant states fixed one can derive the following expression[15]:

$$\delta E^{\ddagger} = \alpha \delta E_R \qquad (0 < \alpha < 1) \tag{2.16a}$$

with the BEP coefficient

$$\alpha = \frac{1}{2}\left(1 + \frac{E_R}{E_0^{\ddagger}}\right) \tag{2.16b}$$

$E^{\ddagger}$, the activation energy, and ΔE_R the reaction energy are defined in Fig. 2.8. $\Delta E_0^{\ddagger}$ is the activation energy of the system when the overall reaction energy $\Delta E_R = 0$. Expressions (2.16) are valid as long as $|\frac{E_R}{E_0^{\ddagger}}| \ll 1$.

In order to discuss the physical meaning of α, we need to introduce the concept of early and late transition states. In the previous section we discussed in detail the transition state for CO dissociation over transition-metal surfaces and described the reaction as an example of a late transition state. The transition state is late along the reaction coordinate since the transition-state structure is close to the final dissociated state. Transition states which are early along the reaction coordinate are called early transition states and thus resemble the initial reaction states (see Chapters 4 and 7 for the definition of the pre-transition state). The activation energies for the protonic zeolite reactions correlate with deprotonation energies (see Fig. 2.9) and are examples of intermediate transition states that also vary with the energies of the initial states[18,19]. When:
$\alpha \ll 0.5$ ($\alpha \approx 0$), the transition state is early; $\Delta S \approx 0$
$\alpha \gg 0.5$ ($\alpha \approx 1$), the transition state is late;
and
$\Delta S < 0$ for surface reactions (tight transition states);
$\Delta S > 0$ for desorption (loose transition state),

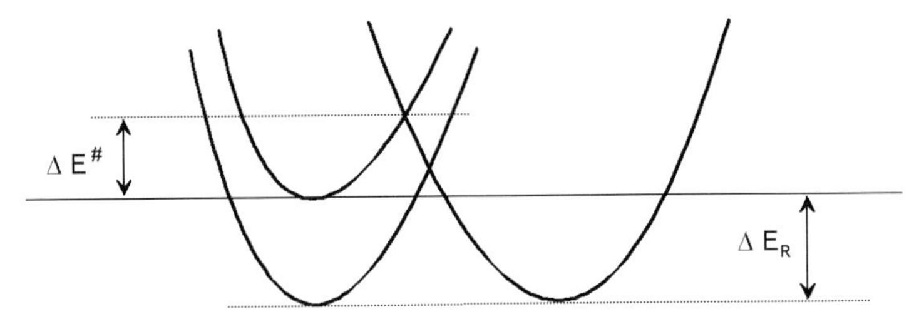

Figure 2.8. Parabolic energy curves of reactant and product states that differ in reaction energies. The transition states are defined as the cross-section of the parabolas.

the general expression for the activation energy becomes:

$$\Delta E^{\ddagger} = \Delta E_0^{\ddagger} + \alpha\delta\Delta E_R \tag{2.17}$$

Because of microscopic reversibility, the microscopic balance of the elementary reaction steps gives a relation between the forward (denoted with +) and backward reaction (denoted with -):

$$\Delta G_+^{\ddagger} - \Delta G_-^{\ddagger} = \Delta G_R \tag{2.18}$$

so that

$$\begin{aligned} \alpha_+ + \alpha_- &= 1 \\ \alpha_+ &= 1 - \alpha_- \end{aligned} \tag{2.19}$$

As a result, we have the following three important relations for the activation free energies of an elementary reaction step:

$$\Delta S_+^{\ddagger} - \Delta S_-^{\ddagger} = \Delta S_R^{\ddagger} \tag{2.20a}$$

$$\Delta E_+^{\ddagger} = \Delta E_0^{\ddagger} + \alpha\delta\Delta E_R \tag{2.20b}$$

$$\Delta E_-^{\ddagger} = \Delta E_0^{\ddagger} - (1-\alpha)\delta\Delta E_R \tag{2.20c}$$

An interesting consequence of these relationships is that they help us to understand the trends in reactivity for selected reactions.

Let us discuss this again for the case of the CO dissociation reaction. In the next chapter on the reactivity of transition-metal surfaces, we will help to justify why the bond energy for surface atoms decreases from left to right across a given row of the periodic table. In addition, we demonstrate that this change in the adsorption energy is much less for adsorbed molecules than it is for adsorbed atoms. The adsorption energy of a molecule is much smaller than that of an adsorbed atom. The reaction energy ΔE_R for the dissociation of an adsorbed CO molecule can change from exothermic to endothermic. It decreases when the transition-metal to which adsorption occurs changes within a row from the left to the right in the periodic table. The corresponding activation energy for dissociation increases ($\alpha \approx 1$) from left to right across a row of transition metal elements in the periodic table. The rate of recombination, which is the microscopic reverse reaction, however, will have a nearly constant activation energy ($\alpha \approx 0$) and be rather independent of transition metal.

The other important elementary step for CO is its desorption from the surface. The activation energy for desorption is equal to the adsorption energy and, hence, the trend for desorption will be opposite of that for dissociation. The desorption of a molecule becomes easier as one moves across a row of the periodic table from left to the right where the binding energy of the molecule to the surface is weakest. This also has an interesting consequence for trends in selectivity for different catalytic systems. For instance, the reaction products for the hydrogenation of CO are methanol, methane and/or hydrocarbons (i.e. Fischer–Tropsch type of selectivity). The selectivity of these reactions is governed by the reactivity of CO and its fragments, since hydrogen will readily dissociate on most metals without an appreciable activation energy. In order to produce methanol, the intramolecular CO bond should not be broken and, hence, the favorable transition metals are those that are overall endothermic ($\Delta E_{R_{\mathrm{diss}}} > 0$) for CO activation such as Cu and Pd. Once CO dissociates the selectivity for methanation versus hydrocarbon growth is determined by the rate of CH_4 formation and desorption versus the rate of carbon-carbon coupling. Here there is competition in the requirements for the metal substrate.

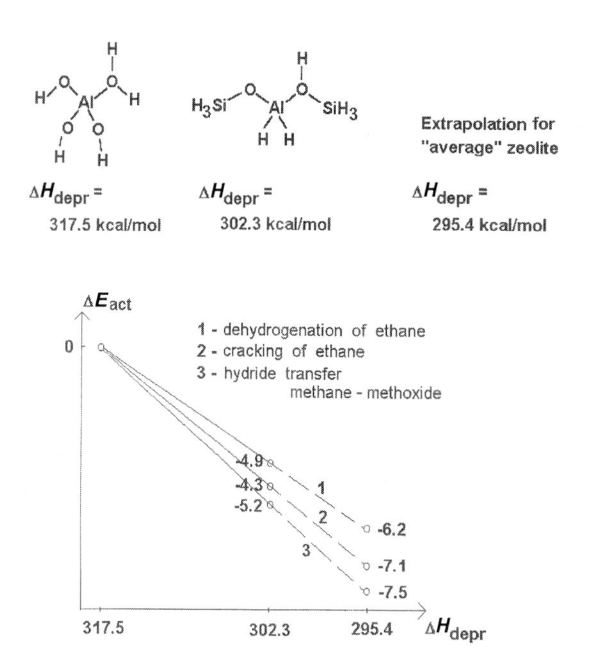

Figure 2.9. The activation energies of ethane activation as a function of deprotonation energies as computed for different clusters modeling the zeolite proton (see also Chapter 4). ΔE_{act} and ΔH_{depr} are in kcal/mol.

The interaction between the metal and the reactant has to be strong enough to dissociate CO, but weak enough so that CH_4 is readily formed and desorbs from the surface. Nickel is the preferred metal for methanation. Fe, Co and Ru, on the other hand, are the preferred Fischer–Tropsch catalysts mainly because they efficiently dissociate CO and suppress methane formation because of the stronger metal-carbon bond. The formation of C–C bonds is expected to be almost independent of the metal ($\alpha \approx 0$). CO dissociation has the lowest barrier on the metal with highest reactivity ($\alpha \approx 1$). Metals such as Fe, Co, and Ru are preferred for the Fischer–Tropsch reaction. For ethylene hydrogenation the site requirement varies as a function of coverage. At low coverage ethylene adsorbs to two surface-metal atoms and at high coverage it reacts from a position where it is only bound to a single metal atom.

To apply the BEP relation, one has to be careful to select properly the elementary reaction step to which one wishes the relation to apply. In zeolite-catalyzed reactions an important elementary reaction step is the protonation of a hydrocarbon, to give as the reaction intermediate, a protonated carbenium ion. The proton activates the molecule to rearrange and after reaction the proton is back-donated to the solid. Often intermediate cationic molecular fragments can stabilize themselves by forming covalent bonds with the zeolite framework oxygen atoms. This will be explained in much more detail in Chapter 4. Figure 2.9 shows that for several such reactions there is a linear relationship between the deprotonation energy of a particular cluster representing the zeolite proton and computed activation energies. The proportionality constant between the activation energy and the deprotonation energy was calculated to be 1/3. This implies that the intermediate carbocation is stabilized by its interaction with the negatively charged zeolite wall. It will be clear in Chapter 4 that the intermediate is a transition state and that the product state has a strong bond with the zeolite oxygen atom. This interaction varies also with the O–H proton energy and explains the low proportionality constant.

2.3 The Reactive Surface-Adsorbate Complex and the Influence of the Reaction Environment

2.3.1 Introduction

Catalysis is controlled by the rate at which the active surface complex turns over in the catalytic reaction cycle. This involves making and breaking of bonds between an adsorbate and the surface site to which it is bound as well as within the adsorbate. For a bimolecular reaction, this involves the bonds between two adsorbates and between the adsorbates and their respective surface sites. The active surface site for the reaction of small molecules typically contains anywhere between one atom such as a metal atom redox site in a homogeneous organometallic complex or in a metal-loaded zeolite such as the Fe^{2+} cation in ZSM-5 (Chapter 4, page 193) on up to 7-10 atoms for a special metal surface site such as the special C_5 site for Fe involved in the activation of N_2 in the synthesis of ammonia from nitrogen and hydrogen[20] (Chapter 7, page 333). The active sites and their environments for these examples are shown in Fig. 2.10. The conversion of larger molecules in enzymes, on the other hand, may involve multiple activating contacts between the adsorbate and the cavity surface. The type, number and the strength of bonds in the active adsorbate surface complex are all important in controlling the overall rate at which the catalytic cycle turns over (Chapter 7, page 319). The principles of the adsorbate surface chemical bonds and their transformations are followed in detail in Chapters 3, 4, 5, and 6 for metals, zeolites, metal oxides and sulfides, and enzymes, respectively.

As was presented in the Introduction to this chapter, two different views on the relationship between the active site and the catalytic cycle currently exist. The first follows the view taken by Langmuir suggesting that all of the sites on the surface are same and

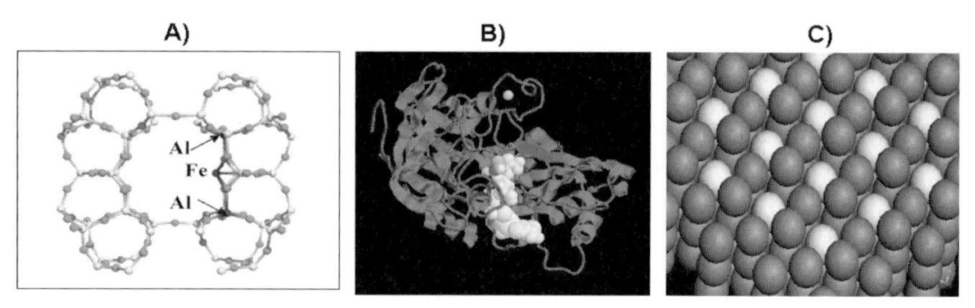

Figure 2.10. The active site and its local environment for different catalytic systems. (A) The single monoatomic zeolitic cation site (Fe in ferrierite); (B) the single site enzyme and its local cavity which provide multiple points of additional contact (methanol oxidase) the ensemble of metal atoms, (C) site for Fe-catalyzed ammonia synthesis.

contribute equally to the catalytic cycle. The second follows the view put forth by Taylor indicating that the sites are not equivalent, and that only a selective few actually control the rate. In either case, what controls the activity and selectivity are the nature and the strength of the chemical bond that forms in the adsorbate surface complex. The detailed atomic structure or topology of the surface can also be important for controlling reactivity. Some reactions are quite sensitive to the topology whereas others are not. These are better known as structure-sensitive and structure-insensitive reactions.

The local chemical interactions predominantly control the activity of the reaction. Environmental effects around the active site are essential for stereochemical selectivity. Examples include the influence of the cavity on controlling reaction selectivity in zeolites and enzymes. The effects of ligands in homogeneous catalysis belong to this same category of sterochemical control. In addition, the external environment can also considerably influence the activity. The environment in which the active site sits can be rather complex, offering a variety of features that could alter the intrinsic kinetics. Catalytic reactions carried out over supported metal particles in the presence of an aqueous medium, for example, can demonstrate changes in the rate and selectivity based on the changes in the transition metal used, the metal particle size and morphology, the exposed surface facets, defect sites on the metal, the support that is used, the interaction between the metal and the support, the composition and relative position of a second metal, the solution phase and its properties near the active metal substrate. Many of these features are captured in Figure 2.11.

In order to elucidate which of the structural features control catalytic performance, it will be critical to establish not only the local bond-making and bond-breaking events that drive the catalytic reaction but also the influence of the active site environment. We examine some of the general ideas on the influence of the environment in the next few sections of this chapter.

We start by distinguishing between two major effects. The first involves the influence of the medium n which the reaction is carried out, while the second refers to the influence of the solid-state matrix in which the active site is embedded. We first summarize cavity, ligand and solvent effects which comprise the catalytic medium. This is then followed by a brief discussion on the different matrix effects.

Zeolites form their own category of heterogeneous systems in that they have well-defined ordered micropores with dimensions that are similar to those of substrate molecules.

For reactions in zeolites and enzymes, the cavity created by the inorganic framework of

Figure 2.11. The complexity of the catalytic center in its reaction environment is illustrated here with a schematic of the hydgrogenation of glucose over carbon-supported Pd particles that are alloyed with Ru in an aqueous medium. The interaction between the metal and the support, the surface composition of Pd and Ru and possible nucleation of Ru clusters, the aqueous medium and its wetting of the metal and the support, the presence of other metal and support surface intermediates such as hydroxyl groups can all act to influence the catalytic reaction.

the zeolite and the peptide backbone of the enzyme can offer specific stereochemical control of the reaction. For reacting molecules adsorbed in these cavities, the intermolecular bonding can be broken down into interactions between the framework and the adsorbate that lead to activation of bonds in the substrate molecules and interactions of the substrate with the framework through physical forces such van der Waals and electrostatic interactions and hydrogen bonds that affect the positioning of the substrate molecules in the cavity. For larger molecules, the latter forces control the matching of size and shape of the cavity with the size and shape of the reacting molecules. Enzymes, as was discussed above, have many point contacts between the protein backbone that comprises the reaction cavity and the substrate molecule. More detailed discussions on the influence of the cavity on both activity and selectivity in zeolites and enzyme catalysis will be presented in Chapters 4 and 7 respectively.

On two-dimensional surfaces the stereochemical control can be moderated by the addition of enantiomeric molecules that play a similar role to the enzyme cavity in providing steric control induced by the weak physical interactions in three dimensions between the enantiomeric coadsosorbed molecule and the reacting substrate molecules. These interactions are analogous to those that occur between the ligands and the substrate molecules that control the selectivity in organometallic homogeneous catalysis. Moderation of heterogeneous catalytic surfaces and ligand effects in homogeneous catalysis are discussed in Sections 2.3.6 and 2.4.

Solvent effects play a role similar that of the cavity in influencing the catalytic activity by providing a medium with a different dielectric constant that can stabilize or destabilize polar transition states. The difference here, however, is the lack of stereochemical control. Protonic solvents open up new reaction channels involving proton transfer-mediated reactions. In polar solvents, the electrochemical activation of molecules can also become possible thus, allowing the surface chemical potential as an additional variable in tuning the activity. The influence of solvent and electrochemical modulation on the catalytic activity is described in more detail in Chapter 7.

In order to understand the influence of the solid-state matrix on activity, some of the relevant structural aspects of the different heterogeneous catalytic systems will first be reviewed. The catalytically active material for many heterogeneous systems is usually dispersed over an inert, high surface area support in order maximize the surface area and stabilize the particle size of the active material on the support. The chemical interactions between the support and the active particle often cannot be ignored as they can influence catalytic activity. This interaction with the support occurs for a number of heterogeneous systems including metallic, bimetallic, metal oxide and metal sulfide particles.

The size and shape of the active particles as well as their activity can be greatly affected by the interaction between the active particle and the support. In addition, the structure of the support including its shape, size and porosity can have a significant influence on the overall catalytic behavior, predominantly due to extrinsic factors that result from mass and heat transfer phenomena. These extrinsic support effects are not explicitly covered in this book. The interested reader is referred to the books by Thomas and Thomas[3] and Farrauto and Bartholomew[21].

For heterogeneous systems, the solid-state matrix can influence the activity by altering both electronic and structural features about the active site. We can distinguish two types of solid-state matrix effects. The first involves the embedding of the active site within the catalytically active particle and the indirect changes that arise from the interaction of the active particle and the support. Examples of the direct effects include the overall size, shape and morphology of the metal particle and the composition and the specific atomic arrangements in alloy particles.

The indirect effects related to the particle-support interactions include the wetting of the particle on the support, which influence both the shape and size of the particles that form, and the stress or strain that the support imparts on the electronic structure of the particle.

A second effect involves the chemical nature of the interface between the particle and the support, which can influence the reactivity of the active particle. For instance, in metal particles, the metal atoms at the interface are often not completely reduced. This can lead to unique activity at the interface and can lead to a perturbation in the chemical reactivity of active centers removed from the interface. As a third effect, the support can impart unique properties to the particle due to charge transfer between the active particle and the support and electronic perturbations due to structural defects in the support which would influence its reactivity.

We conclude this introductory with a short outline of the sections that follow. In Section 2.3.2, we offer a brief introduction to pressure and material gap problems that arise when model catalytic systems and conditions are used to emulate working catalytic systems. In Section 2.3.3, we describe the local aspects of the catalytically reactive sites in the section titled "Ensemble effects and defect sites". This is followed by four sections on environmental influences on the reactivity entitled "Cluster size and metal supports", "Cooperativity", "Surface moderation by coadsorption of organic molecules" and "Stereochemistry of homogeneous catalysts". The chapter is concluded with a section on surface kinetics, dealing with surface reconstruction and transient reaction intermediates.

2.3.2 The Material- and Pressure-Gap Problem in Heterogeneous Catalysis

It is often observed that a catalytic reaction proceeds quite differently under ultra-high-vacuum (UHV) conditions then under the high-pressure conditions used in practical catalysis. The materials gap refers to the differences in reactivity that arise between single crystal surface and an operating industrial catalyst. When the same reaction conditions are used, the discrepancy between the performance of the practical system and the model system is typically related to the differences in the structure of the exposed catalytic surfaces. The pressure-gap problem refers to the often very different experimental conditions used in the model UHV experiments compared with those under operating conditions used in practical catalysis. Often very different kinetics are observed that are ascribed to the formation of different surface phases. We will illustrate these effects using different examples. For the example of the material-gap problem, we describe results for the methanation reaction. For the pressure-gap problem, we explore the methanation reaction along with examples from oxidation catalysis and olefin hydrogenation catalysis.

The aim of many surface science experiments is to provide the fundamental detail of a reaction over a well-characterized single crystal surface in order to establish structure-reactivity relationships. Supported catalytic particles, on the other hand, may have various exposed surface facets along with defect sites. A choice then has to be made as to which single crystal surface will provide the most accurate representation of the active surface facets of the support particles. In order to address the similarities or differences in the rate over ideal surfaces and those over supported particles, Kelly and Goodman[22] compared the rate of methane formation from CO and H_2 catalyzed by an Ni(100) single crystal with that over a supported catalyst taken under the same conditions (see Fig. 2.12).

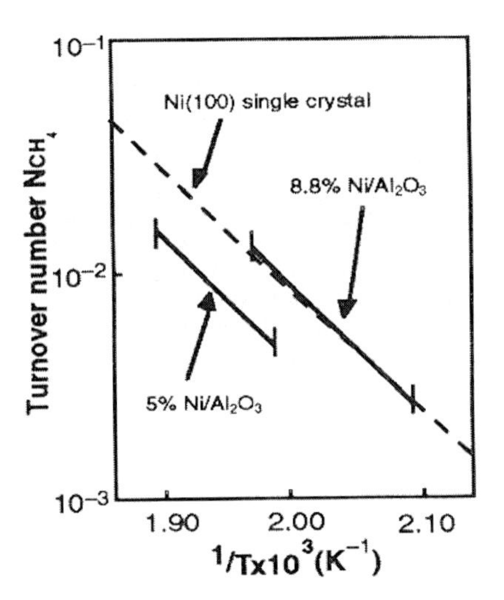

Figure 2.12. A comparison of rates of CO hydrogenation to methane over a single crystal and over supported catalysts, showing similar activation energies. Adapted from R.D. Kelly and D.W. Goodman[22].

For this particular reaction, the comparison of activation energies shows very similar results for the two systems. The agreement between the activation energies suggests that the practical catalysts and the single crystal catalytic surfaces have similar sites that determine their catalytic kinetics. Various general explanations may be possible that help to explain this comparison of the rates between the two different surfaces. One explanation might be that the dominant exposed Ni surface on the supported catalyst may just so happen to be the same as that for the Ni(100) surface. Alternatively, during reaction there might be a surface reconstruction that acts to expose the same surfaces under reaction conditions on both the model single crystal and the supported particles. Alternatively, the reaction may by catalyzed by surface defect sites that are similar on both the model and actual catalytic surfaces. As we have learned in Section 2.2, CO dissociation preferentially at a step defect site. This is considered the rate-limiting step for the methanation reaction and may play an important role. We will elaborate more on this issue in the next section.

In turning to the pressure-gap problem, we note that this same reaction carried out under UHV would show a marked decrease in the rate of reaction as measured by the catalytic turnover number. The turnover number is defined as the rate of a reaction normalized per reactive surface site.

The difference in the rates is due to the difference in the pressure when the reaction under atmospheric conditions is compared with vacuum conditions (10^{-6} bar). In general, the desorption energy of adsorbed CO from a transition metal surface tends to be close to the activation energy for its dissociation (see Section 3.3 for details). Since the activation entropy for a surface reaction is low (tight transition state), but the activation entropy for desorption is high ($\Delta S^{\ddagger}\text{surf}/\Delta S^{\ddagger}_{\text{des}} \approx 10^{-1}$), the rate of CO desorption will be orders of magnitude faster than that of dissociation under UHV conditions. Only at significantly higher pressures, such as performed under atmospheric conditions, will the steady-state surface concentration of CO remain high enough at the temperatures necessary to dissociate CO. Under atmospheric conditions, CO dissociates with an overall rate fast enough to compete with desorption.

High reaction pressures are needed for many other systems as well in order to convert the surface into a uniquely reactive state such as has been found for ethylene epoxidation. The epoxidation reaction of ethylene catalyzed by silver shows a distinct pressure gap. Higher oxygen pressures are needed in order to convert the silver surface into a silver-oxide overlayer where weakly adsorbed oxygen atoms are formed, that selectively epoxidize ethylene[23].

Higher surface coverages will not only influence the rate but can also change the selectivity for a reaction. An interesting example is the low-temperature oxidation of ammonia over Pt (see Sections 3.3 and 6.1). Single crystal studies of ammonia oxidation demonstrate the presence of only two products, N_2 and NO, when a Pt single crystal is exposed to molecular beams of ammonia and oxygen. Ammonia oxidation studies carried out under atmospheric conditions, however, reveal the formation of N_2O. N_2O can be formed from NO at step edges, which may not be present on the ideal single crystal surfaces studied, or requires weakly adsorbed intermediates not formed under UHV conditions but only at high coverage. Unique intermediates such as NO_3^- may also form when the surface is slightly oxidized owing to its exposure to higher oxygen pressure.

An example of such high-pressure effects are the studies of ethylene hydrogenation. Hydrogenation of ethylene[24] by a Pt(111) surface under atmospheric conditions occurs by a weakly adsorbed ethylene species (π-bonded), that predominantly forms only under reaction conditions, when the surface is nearly completely covered with strongly adsorbed (σ-bonded) ethylene. Similar results were also found computationally over Pd(111) with the exception that both π and weakly held di-σ intermediates were found to be active.

2.3.3 Ensemble Effects and Defect Sites

In zeolites as well as in enzymes, the cavity shape and form determine the rate of conversion and selectivity of the catalytic reaction (Chapters 4 and 7). The selectivity of a reaction is defined as the percentage of a particular product molecule formed with respect to the total amount of product molecules.

The planar surfaces of transition-metal catalysts, however, have few steric possibilities to influence selectivity, since the steric constraints are absent. As a consequence, such heterogeneous catalysts very often only have a high selectivity when limited product options exist. On such surfaces, the differences in rates of competitive reaction pathways are controlled by differences in activation entropies and energies of the elementary rate constants r_i. These are controlled by the intrinsic chemistry of the reactions. One geometric parameter that can critically affect the reaction rate is the size of the atomic surface ensemble necessary to activate bonds in molecules.

In Chapter 7 we discuss the unique seven-atom surface-ensemble cluster on the Fe(111) surface (shown in Fig. 2.10C) that is optimum for N_2 activation. Early suggestions that surface ensembles with a particular number of atoms are necessary for a particular reaction to occur are deduced from alloying studies of reactive transition-metal surfaces, with catalytically inert metals such as Au, Ag, Cu or Sn[25]. For example, the infrared spectrum of CO adsorbed on Pd shows the characteristic signature of CO adsorbed one-fold, two-fold or three-fold to surface Pd atoms[26,27]. Alloying Pd with Ag, to which CO only weakly coordinates, dilutes the surface ensembles. One observes a decrease of the three-fold and the two-fold coordinated CO and the one-fold coordinated CO becomes the dominant species. The effect of alloying a reactive metal with a more inert metal is especially dramatic when one compares hydrocarbon hydrogenation reactions with hydrocarbon hydrogenolysis reactions[28].

As an illustration we present in Fig. 2.13 the classical results of Sinfelt et al.[28] on the effect of alloying of Cu with Ni.

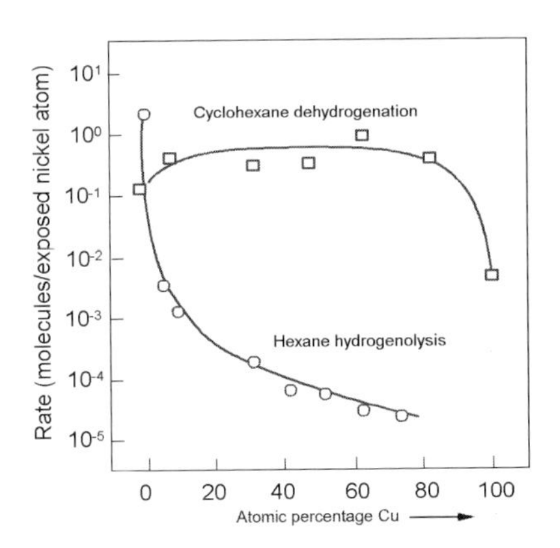

Figure 2.13. Cyclohexane dehydrogenation and hexane cracking conversion as a function of the Cu/Ni ratio of the catalyst[28].

One notes the small dependence of the dehydrogenation rate of cyclohexane on Cu concentration. Actually the dehydrogenation reaction rate shows a slight initial increase. In contrast, there is a strong decrease in the rate of the hydrogenolysis of hexane with increasing Cu concentration. The dehydrogenation of cyclohexane requires only a small ensemble of reactive atoms, whereas the hydrogenolysis requires a large ensemble of atoms. To explain the latter, the intermediates sketched in Fig. 2.14 have been proposed. The bond cleavage between atoms 1 and 2 along with the subsequent hydrogenation of the intermediates that form result in the formation of hydrocarbon fragments adsorbed on the transition-metal surface atoms.

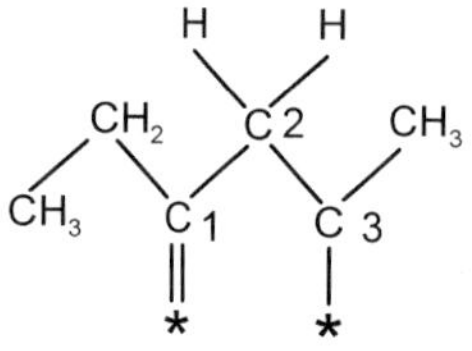

Figure 2.14. Multi-point adsorption of heptane on a metal surface. Surface ensemble requirement for the hydrogenolysis reaction.

The ensemble effect is typically inferred when one realizes that a molecule that dissociates requires at least two different surface sites to accommodate the molecular fragments. For instance, the rate constant for CO dissociation can in an elementary fashion be written as:

$$r_{\text{diss}} = k_{\text{diss}}\, \theta_{\text{CO}}\, (1 - \theta_{\text{CO}}) \tag{2.21}$$

where k_{diss} is the elementary rate constant for the dissociation of CO. This expression predicts a strong dependence of the rate on CO coverage. This is the result of the ensemble-size requirement for the dissociating the CO molecule. Experimental confirmation of this coverage dependence can be deduced from measurements of the rate of methane formation from pre-adsorbed CO, as for instance illustrated in Fig. 2.15. As was discussed earlier, the rate of methanation is typically controlled by the dissociation of CO. We can, therefore, use the rate of methane formation to help probe CO activation.

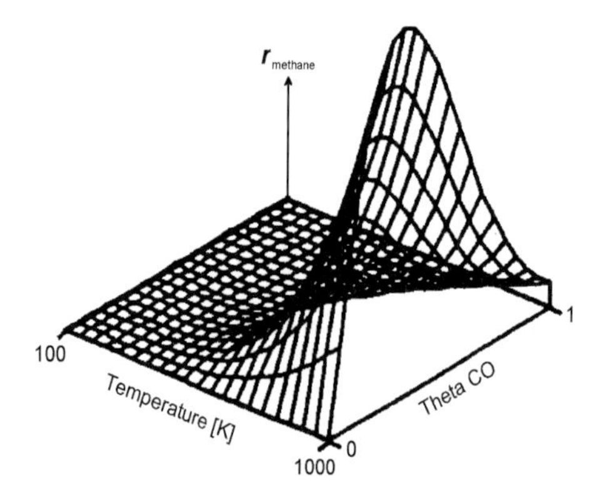

Figure 2.15. The rate of methane formation reflects the rate of CO dissociation as a function of CO coverage on the surface of a rhodium catalyst[29].

Using the Langmuir adsorption expression for CO, Eq. (2.21) can be rewritten as:

$$r_{\text{diss}} = k_{\text{diss}} \frac{K_{\text{eq}}^{\text{ads}}\, p(\text{CO})}{\left[1 + K_{\text{eq}}^{\text{ads}}\, p(\text{CO})\right]^2} \tag{2.22}$$

One recognizes here Sabatier behavior in the volcano-type dependence of the reaction rate as a function of the adsorption equilibrium constant for CO. One can also note the positive order of the rate constant in CO pressure at the left of the Sabatier maximum and the negative order in CO pressure to the right of the Sabatier maximum.

The structure insensitivity of hydrogenation or dehydrogenation reactions may be the result of competing influences. One has to distinguish changes in adsorbate metal-surface atom bond energies by electronic changes on the metal atoms, due to their altered environment (Scheme 2.2a) from changes in the adatom bond energies due to changes in the specific coordination of the adatom (Scheme 2.2b).

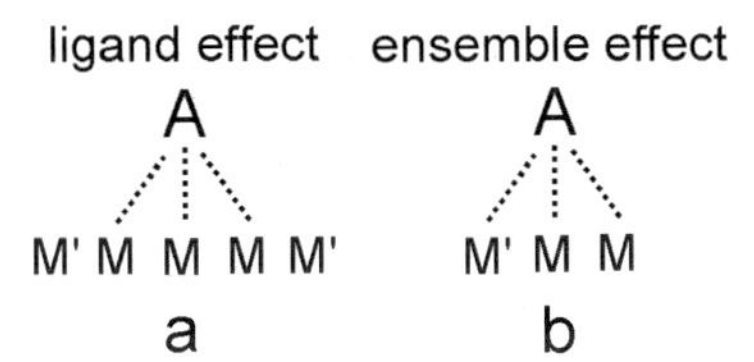

Scheme 2.2 (a) The interaction of A with M changes with a change in the electronic structure of M due to neighboring atoms M′. (b) The interaction A with the surface changes because one of the coordinating atoms M is replaced by M′.

Experimental[30] and theoretical[12,31] results are available that indicate a general ligand-type surface chemical effect occurs upon alloying. The ligand effect refers to change in the environment about the adsorption site without changes to the geometric structure of the local adsorption site. While the structure and composition of the adsorption site remain the same (Scheme 2.2a), the local environment of the M atoms within the adsorption complex have different electronic properties owing to their external environment that arises from its interaction with M′. This subsequently alters the A–M interaction.

While the ligand effect can change the nature of the adsorbate-metal surface bonding and reactivity, it is an indirect electronic effect and is therefore less predominant than more direct changes in the A–M adsorption complex. For instance, in Scheme 2.2b, when M′ is substituted by atom M″ and the M–M″ bond energy is larger than the M–M′ bond energy, the adatom surface-metal cluster interaction energy will decrease much more strongly than if M′ were only at a neighboring site. This is in agreement with qualitative deduction based on the principle of Bond Order Conservation (BOC)[32]. According to BOC principles, more extensively discussed in Chapter 3, the interaction energy of an adsorbate with a surface atom depends on the number of bonds and their bond strength with neighboring atoms of the surface atom. The interaction energy increases with a decrease in the bond order of the interacting surface-metal atoms with their embedding environment. The ensemble effect (Scheme 2.2b) ascribes changes in adsorption energy to direct differences in atoms within the adsorption complex.

Neurock and co-workers[33] used ab initio calculations to determine the influence of both Ag and Au on Pd for the hydrogenation reaction of ethylene and acetylene.

The electronic ligand effect due to alloying Pd with Au or Ag is significantly smaller than the geometric (ensemble) effect but does play a role. The activation barriers are reduced by about 10 kJ/mol over the alloy owing to weakening of the metal-hydrogen and metal-carbon bonds as the result of the electronic effect[33b]. The binding energies for both ethylene and hydrogen , however, are reduced more substantially if Au (or Ag) is actually part of the adsorption complex. This is known as the geometric or ensemble effect (Scheme 2.2b). The weaker metal-hydrogen and metal-hydrocarbon bonds enhance the intrinsic hydrogenation activity. This promotional effect, however, is offset by the fact the hydrogen-Au interactions are so weak that they limit the amount of hydrogen on the surface. The rate of hydrogenation is typically first order or lower in hydrogen. Therefore, decreases in the hydrogen surface coverage will inherently decrease the intrinsic rate. Upon alloying, the order of the reaction rate in hydrogen tends to increase.

The ab initio results were used to develop an ab initio kinetic Monte Carlo scheme to follow the rates of ethylene hydrogenation and the influence of Au[33a,c,d]. The details of the kinetic Monte Carlo method and its application to ethylene are presented in more detail at the end of Chapter 3. The simulations explicitly followed the adsorption, surface reaction, desorption and diffusion steps, surface site specificity, and lateral interactions between surface adsorbates throughout the simulation. The simulations examined the steady-state catalytic hydrogenation of ethylene at temperatures and pressures of interest. The surface was then alloyed with different compositions and different atomic configurations of Au to examine its effects on the molecular transformations and on the overall rate and selectivity to specific pathways. A snapshot from one of the simulation runs which captures the atomic level detail of the simulation is shown on the cover of this book. The hydrogen atoms in the simulation predominantly sit at the three-fold Pd sites. This significantly lowers the surface coverage of hydrogen, which should act to lower the rate since the reaction is first order in hydrogen. The intrinsic activation barrier for hydrogenation over the alloy, however, is also lowered owing to the presence of Au. This increases the likelihood for each reaction. Two of these effects (lower hydrogen surface coverage and weaker adsorbate bonding) should off-set one another, maintaining more of a constant rate. Indeed, the simulated turnover for ethylene hydrogenation showed little change upon changing the relative amount of Au or its specific location. These simulation results are consistent with experiments carried out by Davis and Boudart[34] that showed that alloying Pd with Au had little effect on the measured turnover frequency.

Whereas there is little change in catalytic activity, the selectivity over the alloy is significantly improved. The hydrogenation of ethylene can produce significant amounts of ethylidyne or other carbonaceous intermediates on pure Pd surfaces especially at higher temperatures where competing dehydrogenation and carbon-carbon bond breaking steps become more prevalent. These paths are typically much more structure-sensitive since they require larger surface ensembles. Ab initio based simulations showed that the addition of group IB metal such as Au or Ag can act to minimize the number of these larger ensembles and thus reduce the unselective decomposition paths.[33a,33c,33d]

There are a range of different catalytic reactions where explicit changes in the ensembles upon alloying can change the activity and the selectivity. An understanding of how the surface-ensemble size, morphology and specific atomic arrangement influence activity and selectivity could ultimately be used in the "design" of specific arrangements to optimize catalytic performance. Vinyl acetate synthesis, which involves the acetoxylation of ethylene in the presence of oxygen carried out over Pd and PdAu alloys, is significantly influenced by alloy composition. Kinetic experiments in which ethylene, acetic acid and

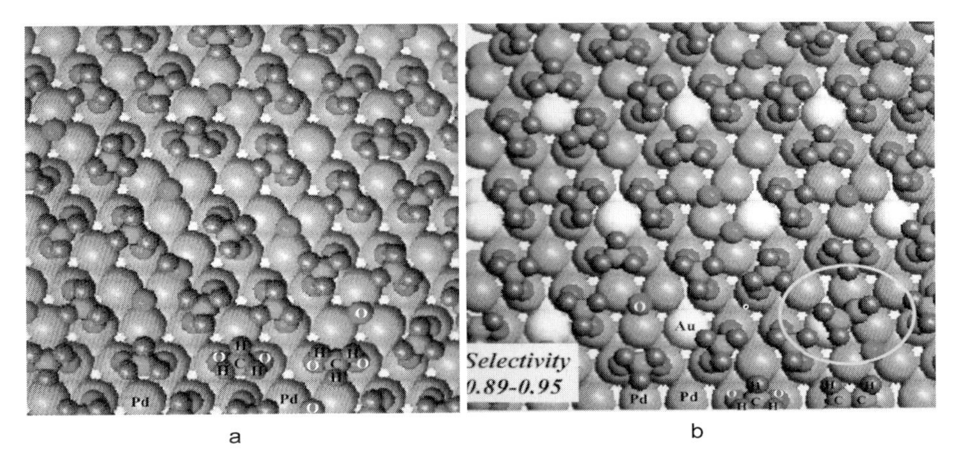

Figure 2.16. (a) Snapshot from the steady-state simulation of VAM synthesis over pure Pd(111). (b) Simulation of surface coverage for vinyl acetate synthesis in the presence of PdAu alloys[33].

oxygen were pulsed over supported Pd and PdAu alloys to elucidate the temporal synthesis of vinyl acetate indicate that both the activity and the selectivity of this reaction were improved by alloying Pd with Au[35]. Ab initio calculations were coupled with kinetic Monte Carlo simulations to examine how changes in the surface composition, ensemble size, shape and specific structural arrangements of Pd and Au for model substrates influence the simulated activity and selectivity[33]. The simulation of vinyl acetate synthesis over Pd(111) results in a very low production of vinyl acetate. The surface is essentially poisoned by the acetate and oxygen intermediates that form. At the steady-state, ethylene has a difficult time finding free Pd sites available on which to adsorb. Higher pressures of ethylene are required in order to adsorb ethylene to any appreciable degree. This is consistent with experimental results.

Alloying Au into the surface opens up surface sites for ethylene to adsorb, since ethylene is only 20 kJ/mol more weakly bound to Au than to Pd. Acetate and oxygen, however, are much more strongly bound to the surface and will therefore tend to bind selectively only to the Pd sites. Au decreases the surface coverage of acetate, and even more importantly, provides sites where ethylene can adsorb exclusively. Well-dispersed Au will therefore create ensembles where ethylene is surrounded by acetate intermediates, thus creating more active local environments. Figure 2.16, for example, shows the steady-state surface population (coverage) over a well-dispersed PdAu (111) alloy surface.

Acetate and oxygen adsorb exclusively on the bridge and three-fold Pd sites, respectively. Ethylene, however, adsorbs on Au sites where it reacts with neighboring Pd to form vinyl intermediates. These vinyl groups react with neighboring acetate to form vinyl acetate. The simulated activity and selectivity in this system was improved by a factor of 2 and 3%, respectively. The size and shape of the Pd and Au ensembles were subsequently tailored to enhance the self assembly of ethylene and acetate under reaction conditions and optimize the simulated activity and selectivity. The Au island sizes in Fig. 2.16, for example, were found computationally to improve the activity by an order of magnitude and the selectivity by 7%. Whether these surfaces can be made and whether or not they are stable is an important issue. Note that qualitatively very similar results were found for simulations which follow the mechanism where ethylene first couples with acetate and then subsequently undergoes a β-hydride elimination from the acetoxyethyl intermediate to form vinyl acetate. This mechanism has shown experimentally to be more favorable than the route through vinyl.

In addition to the topological surface-ensemble effect, there is increasing awareness of the importance of the activation of molecules by surface defect sites such as kinks or steps compared with their reactivity over terrace sites. For example, studies on single crystal surfaces of Ni demonstrated that for the steam reforming reaction $CH_4 + H_2O \rightarrow CO + 3H_2$[36], CH_4 activation is rate limiting. Model single crystal surface experiments have shown that the D_2/CH_4 exchange reaction tends to proceed at step sites where the activation barrier which is 82 kJ/mol, is significantly lower than the barrier of 101 kJ/mol which is found at the terrace sites. In addition, the C atoms that form prefer to bind at the surface edges where their recombination leads to deactivating graphene-carbon formation. Alloying Ni with Au tends selectively to poison the coordinatively unsaturated sites which occur at both kink and step sites. Indeed, the nickel catalyst is promoted by alloying it with Au, which results in a significant increase of the activation energy for methane activation (CH_4 activation now only occurs at the terraces of Ni). In addition, there is an increase in the stability of the catalyst since coke formation is now suppressed.

The difference in surface energies of the metal components that form the alloy helps to determine the level of enrichment with the low surface energy component at coordinatively unsaturated surface atom sites. The enrichment depends exponentially on the degree of coordinative unsaturation of the surface atoms. For the statistical mechanical ideal solution model of an alloy, one can derive the surface composition, x_s, to be

$$\frac{x_s}{1 - x_s} = \frac{x_b}{1 - x_b}\, \mathrm{e}^{\frac{1}{4}m(\epsilon_{11} - \epsilon_{22})/kT} \tag{2.23}$$

where x_s and x_b are respectively surface and bulk concentration, m is the coordinative unsaturation of a surface atom and ϵ_{11} and ϵ_{22} are the bond energies between two atoms in the pure components[37]. The energy expression used to deduce Eq. (2.23) is derived from the elementary broken-bond model, that assumes the energy to be linear in the number of neighbor atoms.

Relation (2.23) implies that in alloys such as Ni with Cu, there is substantial enrichment at the surface where the low surface energy component, or a complex or a phase on a metal oxide, segregates to the surface. This enrichment preferably occurs at the more coordinatively unsaturated step edge and kink sites.

The degree of surface segregation can change significantly in the presence of a reactive environment. Adsorbates that bond strongly to the surface can readily lead to surface reconstruction as well as changes in the segregation behavior. This is the result of differences between the bond strengths between the adsorbate and metal 1 and the adsorbate and metal 2. A more accurate analysis of the surface composition would require following the dynamics of surface segregation, and surface reactivity all at the same time.

The difference in the dependence of the two reactions shown in Fig. 2.13 (of dehydrogenation and hydrogenolysis) on the degree of alloying may also be due to the differences in the reactivity of steps and kinks versus terrace sites. The C–C bond cleavage reaction will occur preferentially at step edges (see Chapter 3, page 139). The very rapid decrease in hexane hydrogenolysis rate with Cu alloying is then explained by preferential poisoning of the step edges, which are only present in small amounts. This point of view is supported by experiments by Blakely and Somorjai[38] that show on Pt single crystal surfaces with varying step density structural independence for the hydrogenation reaction but structure dependence for the hydrogenolysis reaction. As we will describe in the next section, as the particles become less than some critical size less than 1 nm, the chemistry may be controlled by unique electronic and structural properties of the particle that are dictated by metal-support interactions.

The highly reactive step edges also provide adsorption sites which have strong binding energies. Therefore, these sites are most likely to be readily poisoned during reaction. Experiments using radiochemical labels[39] have provided significant evidence for the formation of carbonaceous overlayers on transition-metal catalysts. It was found, for example, that the small Pt particles used in bifunctional zeolite catalysts to establish hexane-hexene equilibrium are only selective in this reaction after being deactivated with a carbonaceous residue. This carbonaceous residue is thought to poison step edge or kink type sites that are selective towards the hydrogenolysis reaction. The terrace atom sites, however, appear to be responsible for carrying out dehydrogenation/hydrogenation reactions.

The two points of view that we have outlined here so far are the Langmuirian uniform surface view[1] versus the Taylorian[2] specific reactive site view. Clearly, the demonstration of the unique properties of step and kink sites versus terrace sites supports the Taylorian view that catalysis occurs by uniquely active sites, that are sometimes only present in very small numbers.

2.3.4 Cluster Size Effects and Metal-Support Interaction

2.3.4.1 Metal-Support Effects and Promotion; Relation to Catalyst Synthesis

We continue with our investigation of the features of the extrinsic environment which influence the intrinsic kinetics of the active site. It is well established that the particle size and shape as well as the support on which they sit can significantly influence their performance for structure sensitive reactions. In order to provide a perspective of the physical chemistry of supported clusters, we first provide a brief overview here of the basic catalyst preparation methods. These methods ultimately dictate the surface composition of the support, and thus control the size, shape, morphology and adhesion of the metal particles that form on the support. Understanding the chemistry that occurs at the metal-support interface is therefore important for improving the preparation of heterogeneous catalysts. The aim of catalyst synthesis is usually to produce a catalyst with a high dispersion of catalytically active surface components that therefore have a small particle size. In catalyst synthesis, the catalytically active precursor complexes are first dissolved in an aqueous solution and then contacted with the catalyst support. After impregnation, the catalyst is dried and subsequently activated. In the aqueous phase, the catalyst support is typically covered with terminal hydroxyl groups of varying basicity and acidity. Their relative concentration is determined by the colloid chemistry of the system. The pH of the solution with respect to the isoelectric point of the oxide plays a key role (see, for example, Farauto and Bartholomew[40] and van Santen et al.[41]). The choice of the catalyst precursor complex relates to whether basic hydroxyl or acidic groups cover the surface of the support. For example, as will be explained in Chapter 5, the commonly used alumina supports are dominated by basic hydroxyl species whereas silica supports contain only weakly acidic hydroxyl surface species. An ion exchange with surface hydroxyls creates surface complexes with single metal cations or small metal clusters, and hence provides a good strategy to synthesize catalysts that are highly dispersed. The catalyst precursor used depends on the nature of the surface hydroxyls that dominate on the support surface. For example, to prepare a highly dispersed Pt on alumina catalyst a negatively charged Pt complex is typically used such as $PtCl_4{}^{2-}$. This complex can ion exchange with the basic hydroxyls of the alumina support. The initial complex on the catalyst support is considered to be a surface complex such as (AlOH)($PtCl_4$).

A very different system is used, however, in preparing a catalyst supported by silica, which contains only weakly acidic silanol groups. In preparing a Pt on silica catalyst, $Pt(NH_3)_4{}^{2+}$ is typically used to carry out the ion-exchange reaction with silanol protons.

In subsequent catalyst activation steps, a complex set of reactions can take place, and depend on catalyst loading as well as the chemistry. For surface complexes that are not easily reduced, such as for Co^{2+} or Co^{3+} reacted with alumina or silica, the support-metal particle interface may exist as a Co-aluminate or -silicate layer. Under reducing conditions, this interface is covered with reduced Co. In addition, subsequent catalyst preparation steps carried out on a reactive support such as TiO_2 can increase the interface with a reduced metal particle by partially covering the reduced metal particle with the oxide or by an increased wetting of the particle surface which will increase the interfacial area[42a,b].

The chemical reactivity of the catalyst support may make important contributions to the catalytic chemistry of the material. We noted earlier that the catalyst support contains acidic and basic hydroxyls. The chemical nature of these hydroxyls will be described in detail in Chapter 5. Whereas the number of basic hydroxyls dominates in alumina, the few highly acidic hydroxyl groups also present on the alumina surface can also dramatically affect catalytic reactions. An example is the selective oxidation of ethylene catalyzed by silver supported by alumina. The epoxide, which is produced by the catalytic reaction of oxygen and ethylene over Ag, can be isomerized to acetaldehyde via the acidic protons present on the surface of the alumina support. The acetaldehyde can then be rapidly oxidized over Ag to CO_2 and H_2O. This total combustion reaction system is an example of bifunctional catalysis. This example provides an opportunity to describe the role of promoting compounds added in small amounts to a catalyst to enhance its selectivity or activity by altering the properties of the catalyst support. To suppress the total combustion reaction of ethylene, alkali metal ions such as Cs^+ or K^+ are typically added to the catalyst support. The alkali metal ions can exchange with the acidic support protons, thus suppressing the isomerization reaction of epoxide to acetaldehyde. This decreases the total combustion and improves the overall catalytic selectivity.

The chemistry at the interface of a transition metal and a reactive metal oxide, such as TiO_2, can be quite different to that carried out over large metal particles alone. Reducible oxide supports such as TiO_2 or V_2O_5 can help to promote the chemistry on the metal and behave quite differently to than oxides such as alumina. The dissociation of CO, for example, is usually considerably enhanced at the interface of the metal and a reactive oxide, where it can dissociate leaving a carbon atom attached to the metal and the oxygen at a cation site such as Ti or V on the metal oxide support.

Consecutive reactions with hydrogen or CO lead to the removal of this oxygen as H_2O or CO_2, respectively. In reactions such as Fischer–Tropsch , where CO dissociation is rate-limiting, the addition of such promoters helps to enhance the activity and even the selectivity for chain growth reactions. The increase in selectivity is the result of increasing the concentration of reactive carbon atoms on the transition metal.[43a,b].

Another example of the influence of the metal/metal oxide interface on the chemistry of the metal refers to the promotional effects that take place in the presence of alkali metal oxides on transition-metal catalyzed reactions. Alkali and alkaline earth metal oxides are known to promote the catalytic activation of different molecules such as CO (in Fischer–Tropsch) and N_2 (in ammonia synthesis) over supported metal particles. Coadsorbed alkali metal generates an electrostatic field that favors electron donation from the transition metal to the adsorbate. These effects can be understood by generally lowering of the work function in the presence of the alkali and alkaline metal oxide promoters[43c,d,e]. This lowers the activation energy of the bond dissociation reaction. For more details on the chemistry of the promoter effects we refer to Thomas and Thomas[3], Somorjai[44], and Diehl and McGrath [45].

To maximize the rate of a reaction, one needs the maximum exposure of metal- or catalytically active atoms to the reactants. Hence there is a great desire to stabilize small particles on catalyst supports. In the next two subsections on transition metals we will provide a detailed description of changes in the chemical reactivity of transition metals when the particle size decreases. This provides a short background to aid in understanding the effects of particle size on catalysis. In the next subsection we discuss cluster size dependence effects and in the subsections that follow we will summarize the specific effort on supported Au clusters.

2.3.4.2 Cluster Size Dependence

One can distinguish at least three different characteristic regions for transition metal particles[46] and their catalytic activity:

(a) Specific molecular structures with sizes that are less than 40 atoms. These are comprised predominantly of surface atoms. There are very few, if any, bulk atoms.

(b) The nanoparticle region falls in between that for the molecular structures and that for bulk particles. The relative energies of the particles are dominated by differences in the surface energies. Structures different from the bulk may become stabilized. Surface energies are of the same order of magnitude as the differences in the bulk energies.

(c) Crystallites of bulk lattice structures, with faceted morphologies.

We will discuss in detail cases (a) and (b). The shapes of crystallites in catagory (c) are controlled by bulk-metal energies and therefore do not require separate treatment.

(a) Molecular clusters

To introduce the subject, we examine a range of different Rh_x clusters and their corresponding energies. Figure 2.17 depicts 21 highly symmetric cluster shapes, that small metal atoms can assume. DFT calculated formation energies for each structure, normalized per Rh atom, are also given in Fig. 2.17.

While the bulk formation energy of metallic Rh is −555 kJ/at, the most stable Rh_{13} cluster results in a corresponding value of only −299 kJ/at. The differences may reflect the much lower average coordination number of the cluster atoms compared with the bulk. One also notes that a few selected clusters such as Rh_3, D_{3h}; Rh_4, D_{h4}; Rh_4, D_{3h}; Rh_9, O_h; Rh_{13}, I_h have similar formation energies, whereas the atoms in these clusters have very different coordination numbers. The planar configurations appear to be the preferred clusters at least for the smaller sized clusters.

Quantum chemical bonding details determine the relative stability and structure of these clusters. Because of their lower stability, the small metal clusters can be expected to be generally highly active. The reactivities usually show a maximum at a particle size between three and seven metal atoms[48,49]. Three parameters appear to be important in controlling the reactivity of these clusters: the coordinative unsaturation of the surface atoms, the availability of enough cluster atoms to bind with an adsorbing atom or

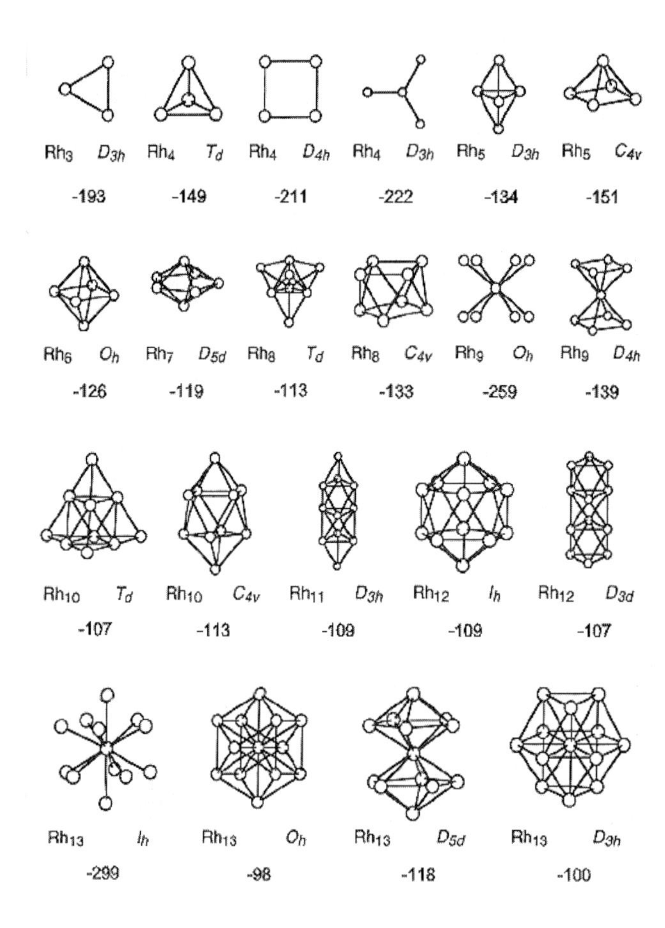

Figure 2.17. The topologies of 21 small spin-optimized Rh_x rhodium clusters and their corresponding formation energies per atom in kJ/at[47].

molecule, and the ionization potential or electron affinity of the cluster. A low reaction barrier for dissociative adsorption of a gas-phase molecule requires electron donation from the metal cluster into an antibonding orbital of the dissociating bond.

For the alkali metals[50] and noble metals (i.e. Cu, Ag and Au), the differences in the relative stability and electronic structure of these clusters as a function of cluster size N can be understood by using a spherical free electron model. The clusters contain well-defined orbital structures analogous to those found for atoms and molecules rather than the band structure found for bulk metal systems. Electron-shell closure and maximum stability are predicted for N = 8, 20, 34, 40, 58, and 92 atom clusters.

An example of an experimentally measured and simulated sequence of measurements for Au clusters[51] is shown in Fig. 2.18.

It appears that on practical catalysts which contain reactive hydroxyl groups or co-adsorbed water, small metal particles are highly reactive towards oxygen and, hence, are difficult to reduce. Temperatures for reduction of small metal oxides may differ from

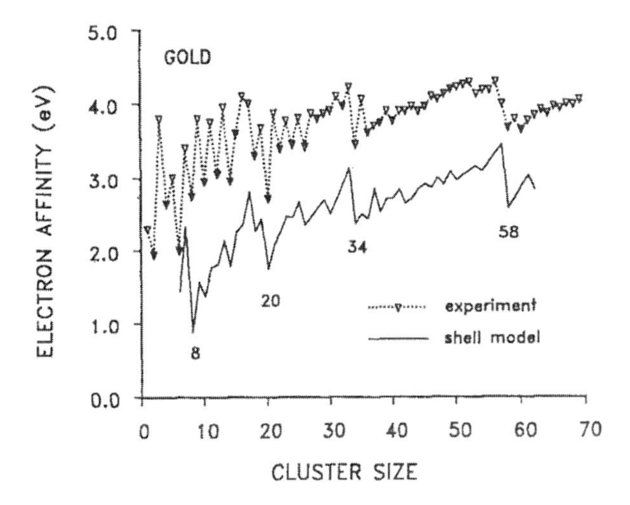

Figure 2.18. Electron affinities of Au_{1-79}. Predicted shell closings at 8, 20, 34, and 58 are observed[51]. The gradual increase in electron affinity, and also the decrease in ionization potential can be understood electrostatically. When a conductive sphere increases in size, the charge can be better accommodated because it can distribute over a larger surface. The low electron affinities are found at the shell closure values, because the empty orbitals have a high energy with respect to the occupied orbitals. The differences in energy between the odd- and even-number clusters reflect differences in energy between the corresponding HOMO (Highest Occupied Molecular Orbital) and LUMO (Lowest Unoccupied Molecular Orbital).

temperatures for reduction of the corresponding metals by several hundred of Kelvin. Also during reaction, the highly reactive small particles may form cluster compounds with unique reactivity determined by cluster-complex chemistry.

(b) The Intermediate Nanoparticle Region

This regime is of most interest to practical catalysis, since most of the supported metal particles used industrially fall in this size range. The nanoparticles that fall within this size range tend to form three basic types of structure: cuboctahedron (Fig. 2.19a); decahedron (Fig. 2.19b) and icosahedron FCC metals tend to take on the cuboctahedron cluster as it usually works out to be a minimum in energy following closely the structure of atoms in the infinite bulk structure. This structure is similar to the D_{3h} Rh_{13} cluster in Fig. 2.19. The icosahedron structure can be recognized as I_h Rh_{13} and the D_{5d} decahedron as Rh_{13}, respectively. Crossover between the different structures occurs as a function of particle size. Such crossovers are due to an increase in the relative importance of the surface energies when the size of a particle decreases. The simple expression for the energy of a droplet-like cluster with one type of surface can be used to help understand the relative stability of different sizes. With symmetrical potentials between the atoms it can be written in the form (Fig. 2.19c).

$$\frac{U}{N} = A + BN^{-1/3} + CN^{-2/3}$$

where N refers to the number of atoms in the cluster, and U is the potential energy of

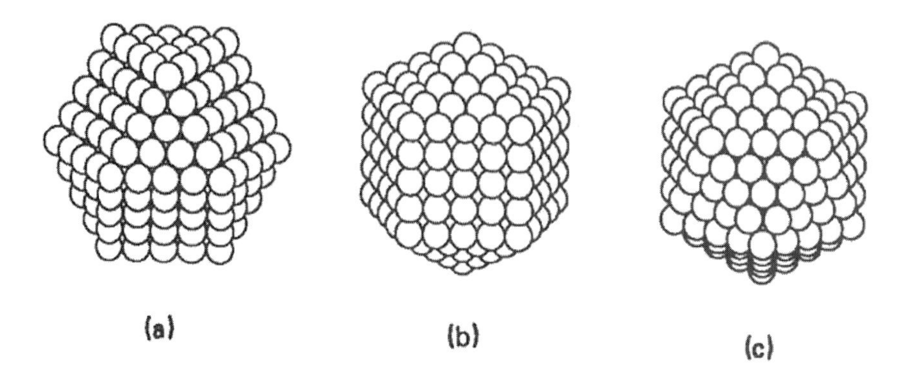

Figure 2.19. Structures of 309 atom clusters: (a) fcc cuboctahedron; (b) decahedron; and (c) icosahedron.

the cluster. In this polynomial function A describes the volume (bulk) contribution, B the surface contribution and C the edge contribution to the energy.

In the structures shown in Fig.2.19, the bulk energies, A, are different. Hence, when crossover from one particle structure to another occurs, the bulk energy A also changes. Table 2.1 summarizes the calculated cluster sizes at which crossover takes place for various different metals [52]. The results were calculated for different metals using molecular mechanics simulations.

Table 2.1. Crossover numbers between the different morphologies for the structures in Fig. 2.19[52]

Crossover	Ag	Pd	Ni	Au
fcc↔icos	5286	1948	11175	550
icos↔deca	3739	1388	7382	393
deca↔fcc	83905	19022	121371	704

The equilibrium shape of a particle is determined by the Wulff rule[53], according to which the ratio of the surface areas are inverse proportional to their surface energy.

For a metal such as Co, there is a phase transition from the low-temperature bulk hcp structure to the high-temperature fcc structure. In moving to small Co particles, the fcc structure now becomes stabilized because of the low surface energies of the fcc faces as compared with those of the hcp particles. The crossover between hcp to fcc occurs when the particles become smaller than 175 nm.

An extensive number of studies have examined the equilibrium shapes of Au clusters in the range of 50–5000 atoms.[46]. A broad range of different structural forms have been analyzed. This includes truncated octahedra in addition to the structures given in Fig. 2.20. This figure summarizes the computed predictions.

The equilibrium shape of an fcc particle is the truncated octahedron with "magic" numbers of $N = 38, 201, 585, \cdots$. In Fig. 2.20 these sizes are recognizable as the points on a drawn line. Over a wide range of sizes the decahedral clusters with varying ratios of their surface edges are found to be most stable.

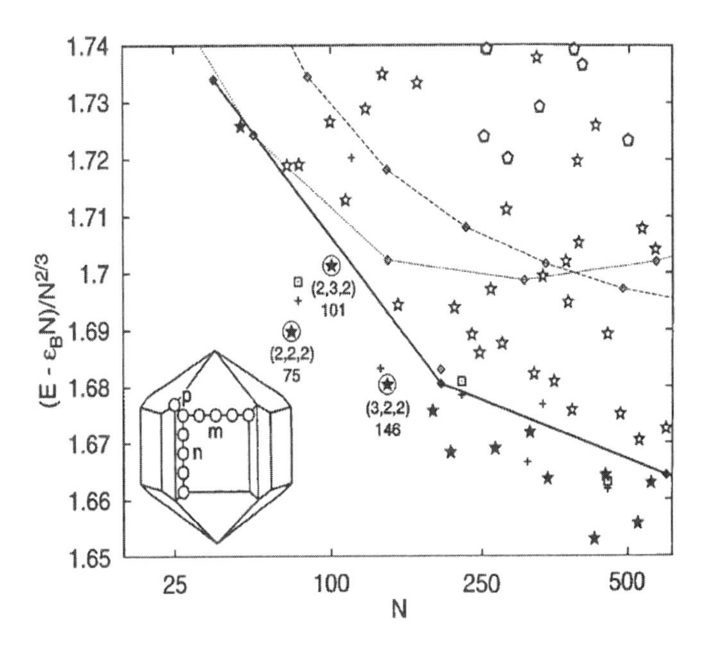

Figure 2.20. Energies of structurally optimized Au_n ($N \leq 520$) clusters plotted as $(E - \varepsilon_B N)/N^{2/3}$ vs N (on an $N^{1/3}$ scale), where $\varepsilon_B = 3.93$ eV is the cohesive energy of an atom in bulk Au. Various structural motifs are denoted as Oh(- -◇ - -), Ih (...◇...), TO (–∗–), t–TO (◇), To$^+$ (+), t–TO$^+$ (□), i-Dh (•), and m-Dh (⊗), with the ⊙ denoting m-Dh clusters in the enhanced stability region. The 75, 101, and 146 atom m-Dh clusters corresponding to particular stable structural sequences are denoted by ⊙. The (m, n, p) indices of m-Dh are shown in the inset for a (5,5,2) cluster. Adapted from C.L. Cleveland et al.[46].

2.3.4.3 Gold Catalysis; an Example of Coordination, Particle Size and Support Effects

The importance of metal particle size and metal support effects on catalytic reactivity is probably best illustrated by examining the current flurry of work in the literature on the catalysis of supported Au nanoparticles. Pioneering work by Haruta et al.[54] led to the first discovery of the unique catalytic performance of nanometer-sized Au particles on various different supports including α-Fe_2O_3, Co_3O_4 and NiO. It is now well established that finely dispersed Au nanoparticles are highly active for a range of different catalytic reactions including CO oxidation, H_2 oxidation, water gas shift, hydrogenation of unsaturated hydrocarbons and alkene epoxidation[55–58]. The nature of the support in these systems plays a very important role since bulk Au is quite noble and hence non-reactive. Since the first discovery by Haruta et al.[54] in 1989, the research efforts on Au catalysis have increased exponentially.

Despite the substantial experimental and theoretical efforts, the nature of the active site and the features which control its reactivity are still intensely debated in the literature. There are three predominant explanations for the low-temperature activity of supported Au. The unique activity is attributed to: (1) Changes in the specific particle's shape, atomic structure or size which ultimately controls the relative ratios of edge, corner and terrace sites[55,59–62]. The coordinatively unsaturated edge and corner sites are defect sites which result in stronger bonding to the adsorbates. This lowers the barriers to break adsorbate bonds and can therefore help activate various reactions. In other cases, the bonds with the adsorbate are too strong and can thus lead to deactivation. (2) Quantum size effects that occur when the particle size is below a few nanometers[55,60]. In moving from large metal particles to nanometer size particles, the electronic structure changes from one which is comprised of valence and conduction bands to one which is made up of individual molecular states, thus leading to a metal-insulator transition. (3) Electronic or structural influence as the result of the metal-support interaction. In particular, the support can induce strain[64] on the metal–metal bonding at the surface. In addition, there can be electron transfer to or from the active Au particle, thus influencing its behavior. The role of neutral Au metal atoms[59,62,65–69], Au cations[56,70–72] and Au anions [73–75] have all been suggested as active surface sites that carrier out the chemistry. The interaction between the metal and the support can also lead to the formation of unique active sites that can form at the metal-support interface.

2.3.4.4 Structural Effects

As long as the number of terrace atoms is large compared with the number of edge atoms, the shape of large metal particles can be predicted from the Wulff–Kaichev construction[53,76]. Depending on the surface energy of the crystal facets and the metal oxide adhesion energy, the Wulff equilibrium polyhedra truncate. Molina and Hammer[68] computed an adhesion energy of 0.52 J/m^2 for the Au(100)–MgO(100) interface. This can be compared with the Au(100) surface energy, which is 0.84 J/m^2. The equilibrium shape of the Au particle deforms so as to favor an increased surface area of Au exposed at the Au–MgO interface. This is illustrated in Figs. 2.21. The Wulff construction ignores the formation energy of the edge and the corner atoms. Therefore, it is really only valid for large particles.

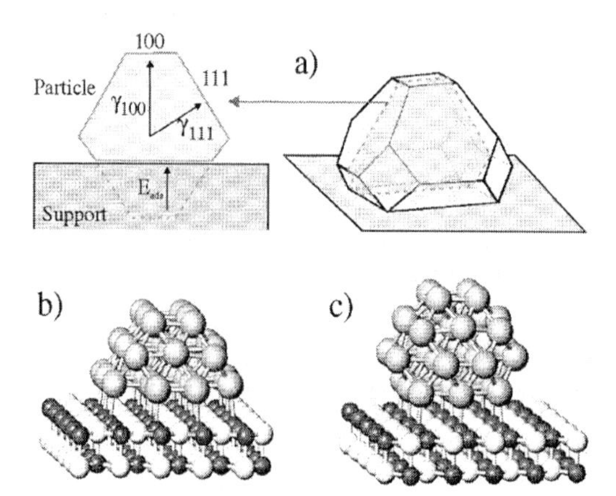

Figure 2.21. (a) Schematic illustration of a truncated Wulff octahedron on a support. (b)and (c) are DFT (PW91 self consistent) derived, relaxed structures of Au supported on MgO(100) with a decreasing degree of partial wetting behavior[68].

Molina and Hammer[68] studied the size of small Au particles dispersed on the MgO(100) surface by DFT. The structure shown in Fig. 2.21b is found to be more stable by 0.1 eV than the structure given in Fig. 2.21c. The latter corresponds to the equilibrium shape sketched in Fig. 2.21a.

A decrease in particle size increases the relative ratio of surface atoms and atoms which have lower coordination numbers. Therefore, the reactivity of these particles tends to increase. Interestingly, often the activity per surface atom also changes. For supported catalysts, the turnover number (TON) typically goes through a maximum[77]. Smaller particles lead to increased coordinatively unsaturated sites. The adsorption energies at these sites are typically the highest. Dissociative addition reactions, therefore preferentially occur on such sites also for reactions that are positioned to the left of the Sabatier maximum. This is consistent with an increase in TON. Second, the active sites are sometimes positioned at the interface of metal particle and catalyst support. With a decrease in particle size, this interface between the metal and support increases, which should also increase the overall activity.

Decreasing the particle size to very small atomic ensembles , however, can sometimes lead to reduced activity. This is related to two factors. Metal atoms in the cluster which have very low coordination numbers form very strong bonds with the adsorbate to compensate for the smaller number of metal–metal bonds. This increased binding energy between the metal and the adsorbate leads to higher reduction temperatures. This shifts the reactivity patterns. There is therefore a shift in the position along the Sabatier curve. Larger Au particles tend to lie closer to the left of the Sabatier maximum whereas these smaller particles tend to lie to the right of the maximum whereby the products or other intermediates begin to inhibit the surface reactivity. In addition to the enhanced metal-adsorbate bonding, the bonding between the metal and the oxide support becomes stronger, and as such, may begin to deactivate the metal clusters. For example, metal atoms bound to an oxide support tend to transfer electrons to the support and become oxidized. Bogicevic and Jennison [78] have shown that the nature of the metal-oxygen bond for single metal atom and small metal clusters at very low coverages on the oxide support is primarily ionic regardless of the metal chosen. For larger particles there is a trade-off between metal–metal and metal-oxygen bonding.

Results of DFT calculations predict that Pd, as well as Pt, on the ideal MgO(100) surface will tend to form clusters rather then isolated ions[79]. Metals that lie closer to the right in the periodic table, such as Cu, form much weaker metal–metal bonds. Copper, silver and gold tend to prefer isolated ions which tend wet the surface. For a detailed review on metal-support interactions we refer to the review by Campbell[80].

Figure 2.22 illustrates the particle size effects for Au particles of nanometer size supported by TiO_2. The catalytic properties of Au are altered in a unique fashion. Au particles between 2 and 4 nm supported on titania show unique activity for the low temperature oxidation of CO, whereas large particles are non-reactive.

As the particles become smaller, the fraction of metal atoms in the cluster that reside at the surface increases. This increases the ratio of corner and edge atoms over terrace atoms. For a few different surfaces the adsorption energies of O and CO are computed and plotted as a function of coordination number of the metal surface-atoms. The lower the degree of coordination, the higher is the degree of coordinative unsaturation representative of edge atoms. The computed adsorption energies, given as a function of the Au surface-atom coordination number, are shown in Fig. 2.23[59]. The increased degree of coordinative unsaturation will act to increase the rate up to the point where the adsorbate binding is too strong (Sabatier maximum). The rate will then decrease owing to the inability to remove CO from the surface. This is analogous to conventional CO oxidation catalysts

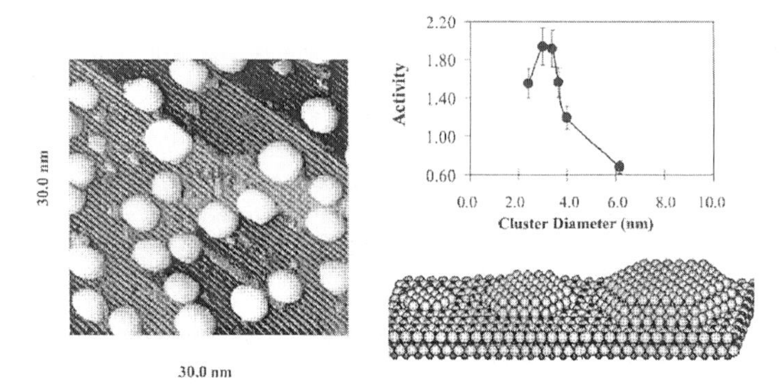

Figure 2.22. Effects of particle size on the activity of titania-supported Au for the oxidation of CO[63].

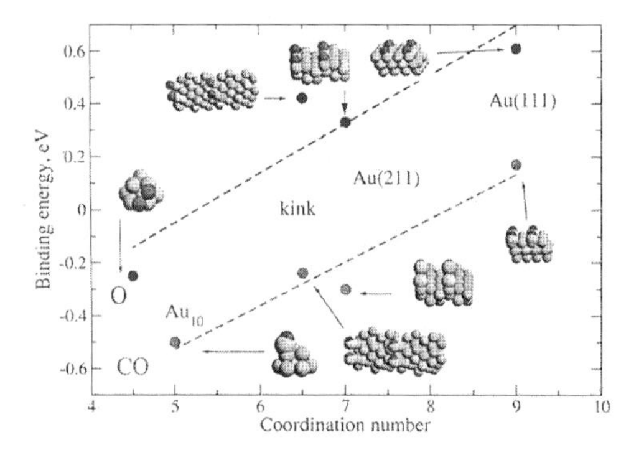

Figure 2.23. The correlation between the binding energies, for CO molecules and O atoms, with respect to the coordination number of Au atoms in a series of environments. Binding energies, reported in eV, here are referred to gas-phase CO and O_2[59].

that demonstrate suppression of the rate by CO inhibition at low temperatures.

The adsorbate bond energy increases with increase in the degree of coordinatively unsaturated metal atoms. This is due to the decrease in the localization energy of electrons on the Au surface atoms for structures with fewer neighboring atoms.

As we will discuss in more detail in Chapter 3, the delocalization of electrons is proportional to the square root of the number of coordinating atoms[37]. One would therefore expect adsorbate binding energies to increase with decreasing particle size, owing to the increased number of coordinatively unsaturated surface atoms. The reactivity of these particles with respect to cluster size will then depend the position of the adsorbate bond energy with respect to the Sabatier curve maximum.

2.3.4.5 Quantum Size Effects

As was briefly mentioned above, the increased reactivity of small Au nanoparticles may be the result of the unique electronic characteristics of small Au clusters. There is a clear shift from the bulk metal properties which readily allow electron transfer between the valence and conduction bands of the metal due to small energy differences between these states. In reducing the size of a metal particle to the nanometer size scale, we lose the band structure of the metal and form more of a molecular-like structure with discrete energy differences between molecular orbitals. This results in the formation of a more appreciable energy gap between the highest occupied and lowest unoccupied states and thus the transition from metallic to insulator electronic properties. This is know as a quantum size effect. The influence of quantum size effects on the unique low-temperature activity of Au has been speculated by Valden et al.[63], who demonstrated unique reactivity for 3.5 nm Au particles supported on model TiO_2 substrates. The size was consistent with the transition from metal to insulator. Theoretical studies by Mills et al.[60] also suggest quantum size effects. A more extensive description of these effects can be found in Section 2.3.4.2.

2.3.4.6 Support Effects

The support clearly affects the rate of some Au-catalyzed reactions. The support can play various roles. First, the support can change the nature of the metal particle adhesion to the surface, and thus change the metal particle size that forms, as was discussed above. Second, the support can act to strain the metal–metal bonds, which would significantly change the electronic properties of the metal atoms near the interface and thus their catalytic properties. Third, there can be electron transfer between the metal and the support, which would change the electronic properties of the metal. Neutral and positively and negatively charged Au clusters have all been proposed to be catalytically active for specific reactions in the literature. Lastly, the interface between the metal and the support can act to create unique bifunctional sites which demonstrate enhanced reactivity. We discuss the last three effects below. The effect of particle size on the catalytic performance was discussed in detail in the previous section.

Mavrikakis et al.[64] have nicely shown that the strain induced on the metal–metal bonds by the misalignment of the metal lattice to the registry of the oxide support leads to a shift in the center of the d-band. This change in the electronic structure alters the adsorbate bond strength at these sites, which ultimately dictates the reactivity of the metal. While these effects may die off for large particles on the support, they can clearly play a role for smaller nanoparticles that are in direct contact with the support.

2.3.4.7 Elucidating Mechanisms and the Nature of Active Sites

Much of the current work on the mechanisms responsible for the unique Au activity have focused on understanding CO oxidation. CO oxidation is generally agreed to proceed by the adsorption of both CO and oxygen, the activation of oxygen, and the subsequent formation and desorption of CO_2.[55,62,66,77,81] It is still debated as to whether atomic or molecular oxygen is the reactive oxygen species. A fair amount of the current evidence appears to point to a bifunctional-type site where CO is adsorbed to the Au at the Au/oxide interface whereas molecular oxygen is activated on a nearby site on the oxide.[55,62,66,77,81] This, however, is still debated. The mechanism, however, is strongly dependent upon the charge state of the metal. We will, therefore, discuss the current thoughts on the mechanisms in the subsections that follow and more specifically analyze the different charged states of the metal.

2.3.4.8 Electron Transfer Effects

The unique properties of small Au particles responsible for the low-temperature catalytic activity have not been given a definitive explanation[55]. Neutral and negatively and positively charged gold particles have been identified on different metal oxide supports and speculated in catalyzing different reactions. The formation of neutral and positively and negatively charged gold clusters is clearly influenced by the nature of the support and its ability to transfer or accept charge. The reaction conditions can also act to influence charge transfer and modify the oxidation state of the metal. The sensitivity of the reaction conditions, the support used and the reaction examined suggests that active sites for different reactions may actually be different. That is, the active site for CO oxidation may be different to that for water gas shifts and alkene hydrogenation. We discuss experimental evidence along with theoretical results for each of the three different Au charged states as potential reaction sites.

2.3.4.9 Neutral Au Clusters

While much of the work in the literature has speculated on the presence of neutral Au clusters, there has been very little in-situ experimental evidence to support these ideas. Calla and Davis[65] claerly showed the presence and the activity of neutral Au clusters under reaction conditions. They followed the CO oxidation reaction over Au supported on Al_2O_3 using in-situ X-ray absorption spectroscopy as well as with transient isotopic labeling studies. Their results indicate that Au^{3+} is reduced during calcination and that the active species throughout the reaction appears to be metallic Au[65].

The results from many of the theoretical studies on idealized models of TiO_2 and MgO surfaces indicate that Au metal clusters remain nearly neutral. Some of the earliest theoretical studies suggested that the increased activity of Au was the result of the increased presence of coordinatively unsaturated sites such as step edges and corner sites for small Au clusters. CO, as well as oxygen, showed significantly enhanced adsorption energies at these sites, as depicted in Fig. 2.23. The increased binding energies at these sites were thought to enhance the rate of elementary steps involved in CO oxidation[59].

In subsequent studies, Liu et al.[66] calculated the activation barriers for CO oxidation over Au(211) slabs supported on TiO_2. The barriers for the oxidation of CO by atomic (O^*) and molecular (O_2^*) oxygen were calculated to be 25 and 60 kJ/mol, respectively[66]. The barrier required to dissociate O_2 to form atomic oxygen, however, was found to be over 100 kJ/mol. The reaction path involving molecular O_2, therefore, appears to be the favored route. On small metal particles, the increased adsorption energies at step edges increases the surface coverage of CO. O_2, on the other hand, adsorbs at the positively charged Ti sites and results in a charge transfer from the supported Au into the antibonding 2π orbital of O_2, thus activating O_2 for reaction[66,67].

Theoretical studies of Au-supported clusters on MgO by Molina and Hammer[68] support this same view that the adsorption of molecular oxygen is stabilized at the interface between Au and the MgO support. Molecular oxygen is found to be the reactive oxygen intermediate. The active interface is shown in Fig. 2.24 which identifies different proposed CO binding configurations and reaction intermediates for CO oxidation on various shaped Au[68] clusters supported on the model MgO(100) surface.

More recent results by Remediakis et al.[62] identify two possible mechanisms for CO oxidation. The first suggests that the reaction takes place between CO on the metal O_2 which is bound to the titanium support at the Au interface. This is similar to the studies by both Liu et al.[66,67] and Molina et al.[68,82] This mechanism strongly depends on the nature of the support and its ability to stabilize O_2 at the interface. The second mechanism, however, is one which is nearly independent of the support. This mechanism proceeds solely over Au. Low coordination sites are found to be important for activating CO and O_2. The presence of these two routes agrees with the current experimental evidence which shows strong support dependence for some systems with little to no dependence on the

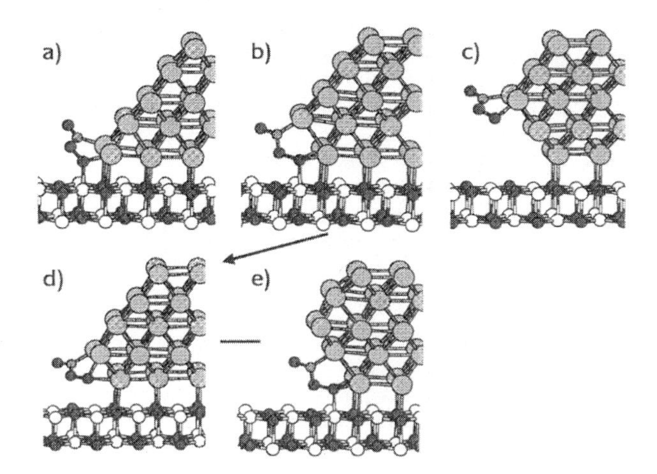

Figure 2.24. Proposed CO–O_2 binding configurations on various shaped Au[68] clusters on Mg(100). CO adsorption at the interface between the support and cluster is disadvantageous owing to steric effects as the CO is repulsed by the support. Oxygen is bound to both the cluster and the support, ultimately forming a CO–O–O complex at the cluster's edge with the most favorable arrangement shown in (b). When the complex rearranges as depicted in (d), the barrier to reaction is lowered significantly and CO_2 is readily formed. The second oxygen remaining on the gold particle is even more easily reacted with CO. (e) shows an alternative arrangement of the reaction interface. However, in this case the adsorbates are not bound to low-coordinated gold as in (b) or (d) so the reaction is less favorable[68].

support for other reactions.

Gold catalysts dispersed on TiO_2 also demonstrate unique catalytic reactivity for the epoxidation of propylene with hydrogen and oxygen. Hydrogen peroxide is proposed to be the active intermediate. It is also thought to be formed at the interface of Au and TiO_2 in a mechanism very similar to that proposed for the reactive O_2/CO complex involved in CO oxidation.[69].

2.3.4.10 Negatively Charged Au Clusters

Studies on the soft landing of Au clusters on an MgO support taken together with ab initio calculations suggest that anionic Au is the active surface species necessary for oxdizing CO. [73–75] The authors suggest the presence of F-center defects, formed as the result of oxygen vacancies enable charge transfer from the support to the metal. Hakkinen et al.[73] et al., for example, concluded that the unique reactivity of an Au_8 particle attached to the MgO(100) surface is the result of its adsorption to an MgO defect site. Due to the presence of an oxygen vacancy (F center) the cluster is charged via (partial) electron transfer from the oxidic support. Whereas low-temperature reaction channels show chemistry consistent with the Molina and Hammer results, Hakkinen et al. find that at high temperatures defect-rich MgO(100) shows strongly enhanced reactivity. CO_2 is formed initially with CO adsorbed on the top facet of the Au_8 cluster and the peroxo O_2 molecule bonded to the periphery of the interfacial layer of the cluster. As we have learned in the section on cluster size effects, Au_8 is a non-reactive species due to electron shell closure. On the other hand, Au_8^- is a highly reactive intermediate because it will have a very low ionization potential. This is consistent with the need for electron donation to Au_8 in order to produce its high activity.

More recent ab initio calculations carried out by Pacchioni et al.[83], however, indicate that the formation of the anionic Au particles may be due to electron transfer via tunneling from the underlying Mo substrate (used to grow the films) to the Au particles through the very thin MgO films used experimentally. This would suggest that defect sites may not be necessary, and that the application of these elegant surface results may not translate to the unique catalytic particles of supported particles. Recent DFT calculations by Molina and Hammer[84] show that the presence of F centers does not appear to lead to appreciable charge transfer to the metal particle. Their results do, however, show that artificially charging the tetrahedral Au_{20} cluster on the MgO support results in a significant improvement in the CO oxidation kinetics.

2.3.4.11 Positively Charged Au Clusters

Cationic gold species have also been speculated as well as detected in-situ for various different reactions. The unique reactivity of cationic Au clusters supported on ceria has been suggested to be important in catalyzing the water gas shift (WGS) reaction. Fu et al.[70] deposited small Au clusters on an La-doped reducible CeO_2 substrate and demonstrated unique, highly reactive Au particles for WGS. The surface was subsequently washed with a basic NaCN solution in order to leach out metallic Au. Despite the removal of metallic Au, the WGS activity over these leached systems remained the same, suggesting that the active catalytic sites were comprised of cationic gold.

The reactivity of cationic gold particles has also been speculated for Au supported on $Mg(OH)_2$ and Fe_2O_3 as well as for Au atoms in zeolites. Gusman and Gates[71] used X-ray absorption spectroscopy to identify single Au^{3+} atoms as the active form of Au present in the zeolite and responsible for ethylene hydrogenation. As we will see in Chapter 3, metal cations with partially empty d-shells are active catalysts for insertion reactions; such a reaction is, for instance, the formation of adsorbed ethyl from ethylene and adsorbed hydrogen. The reaction between surface CO and surface hydroxyl intermediates has also been suggested to occur over cationic Au thus leading to the formation of formate species[56].

In a more recent study, Guzman et al.[81] used in-situ time-resolved XANES and Raman spectroscopy to probe the nature of the active oxidation state of the metal along with the active oxygen form for CO oxidation over Au nanoclusters supported on CeO_{2-x}. Their results indicate that there is a Ce(4+)/Ce(3+) redox couple. In-situ Raman spectroscopy shows evidence for molecular oxygen bound as both an η_1 superoxide as well as a peroxide surface species form at defect sites in the oxide. The η_1 superoxide appears to be the reactive form. Au helps to promote O_2 on the oxide support. XANES data suggest that cationic Au is responsible for carrying out the chemistry.

Others have speculated that cationic Au along with metallic Au must be present[72]. Hydroxylated Au^+ species can form at the periphery of the Au/oxide interface and may be the active species.

We have seen that the chemistry of the catalytic system strongly depends on the Au particle size, the support used and the reaction studied. The relation between catalytic reactivity and the charge state of the reactive Au center may depend strongly on the size of the Au particle. A second important effect is the interaction between the metal particle and the support. Here again, the metal particle size as well as the reducibility of the oxide can critically impact the catalytic behavior. Theory has offered valuable insights into each of these proposed effects by providing elementary reaction energetics along with detailed features of the reaction mechanism.

2.3.5 Cooperativity

In heterogeneous catalysis, there are many examples where addition of a second component can change the overall catalytic reactivity in the system by changing its solid-state chemistry. An example of this includes the addition of Co^{2+} to the MoS_2 and NiS_2 systems discussed in Chapter 5. The mixed metal sulfides offer significantly increased activity due to changes in the chemical reactivity of the sulfide surface. We introduce here the solid-state chemistry of oxide catalysts (see also reference 3). A more detailed discussion on mixed metal oxides is presented in Chapter 5.

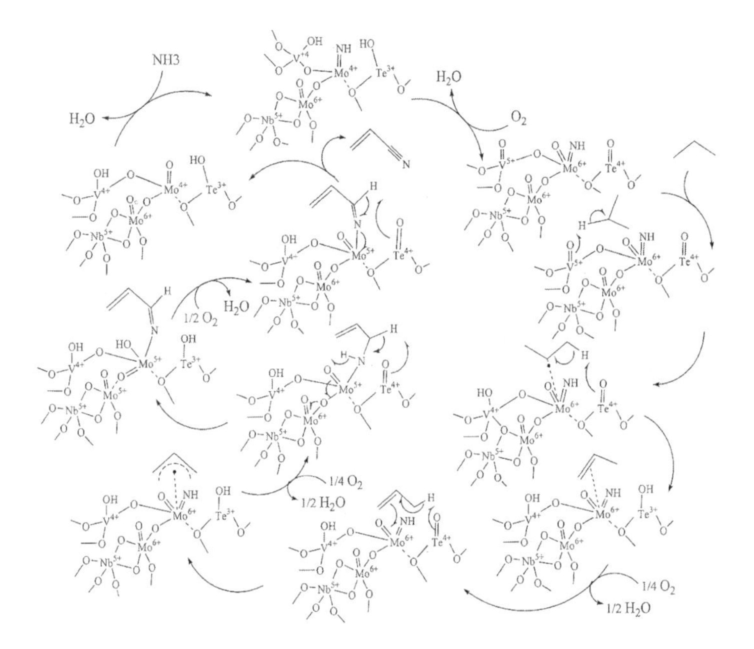

Figure 2.25a. Proposed propane ammoxidation mechanism[85] over Mo–V–Nb–Te–O_x catalysts.

The inorganic chemistry of a multi-component heterogeneous catalyst is often very complex, as it is quite difficult to obtain structural information at the molecular level to help establish the fundamental processes. As an example, we discuss the chemistry of the complex mixed metal oxide catalyst $Mo_{7.5}V_{1.5}NbTeO_{29}$ shown in Fig. 2.25, which is known to catalyze the ammoxidation of propane to acrylonitrile. The active centers in this system are multifunctional metal oxide assemblies that are spatially isolated from one another owing to their unique crystal structures.

The Nb^{5+} centers are thought to stabilize the primary active Mo and V centers by structurally isolating them. This site isolation appears to be a prerequisite for selective oxidation. The presence of excess adsorbed oxygen would otherwise tend to lead to total combustion. The activation of propane is thought to proceed via the abstraction of the methylene hydrogen by a V^{5+} center, which subsequently changes its oxidation state to V^{4+}.

$$^{5+}V{=}O \longleftrightarrow {}^{4+}V{-}O\bullet$$

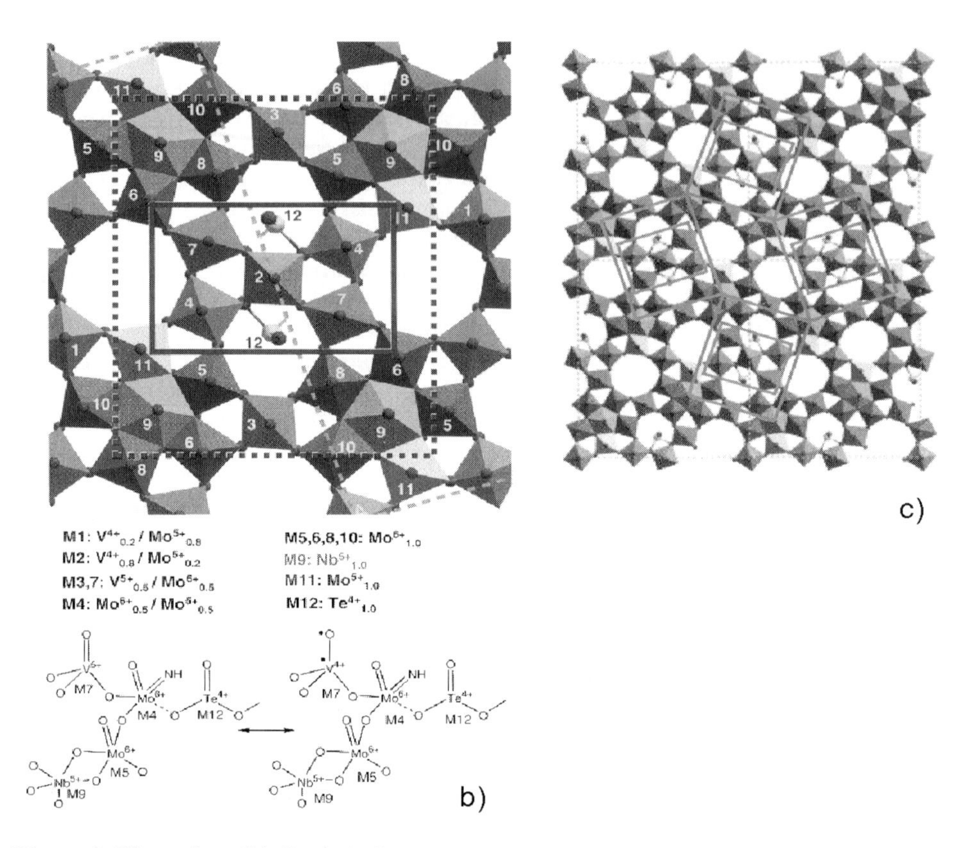

Figure 2.25b and c. (b) Catalytically active center of $Mo_{7.5}V_{1.5}NbTe_{29}$ in [001] projection and schematic depiction of the active site. (c)2 x 2 unit cell structure model of $Mo_{7.5}V_{1.5}NbTeO_{29}$ in [001] projection showing four isolated catalytically active sites[85].

Paraffin activation is thought to be free radical in character. H-atom abstraction by O radical centers is a facile process. The adsorbed propyl radical can then lose a methyl-H to oxygen at an adjacent $Te^{4+(6+)}$ center, thus forming propylene. The catalyst is considered to be bifunctional in character whereby the ammoxidation of propylene subsequently occurs at the Mo^{6+} centers. The reduced Mo and V centers are regenerated by lattice oxygen originating at a reoxidation site that is physically separated from the active site. Dioxygen dissociates at the redox site and can then be incorporated as lattice oxygen. Hence the ammoxidation sites of the reaction are regenerated by oxygen atoms, formed at other sites on the surface via the dissociative adsorption of O_2. The oxide medium helps to facilitates the transport of this oxygen to the catalytically reactive and selective centers. This is known as the Mars–van Krevelen mechanism[86]. The well-defined crystallographic structure of these catalysts, Fig 2.25(c), and in particular in the organic chemistry of their synthesis, allows them to be viewed as self-assembled multifunctional catalysts in which covalently bonded intermediates are exposed to topologically optimum atomic configurations for catalysis. The catalyst can, in some sense, be considered as a solid-phase analogue of the Wacker catalyst in its ability to simulate the catalytic cycle and regenerate the active redox site by the remote activation of oxygen.

In multifunctional catalysts, reactions occur in consecutive steps where each step can be catalyzed by a different active site. Typically many reactions are at equilibrium and one or two of the steps are rate-limiting. The selectivity then is controlled by the relative composition of the different reaction sites which catalyze specific steps. For example, oxidation catalysts are typically optimum when the V=O sites that catalyze the alkane to alkene reaction are present in a high enough concentration that an alkane/alkene equilibrium is reached in the particular oxidation system of interest. The subsequent catalytic oxidation steps which are involved in the functionalization of the alkene then become rate limiting.

2.3.6 Surface Moderation by Coadsorption of Organic Molecules

The coadsorption of surface moderating organic molecules can be used to induce significant steric control of the reaction product selectivity if the size and shape of coadsorbed molecules will form three-dimensional structures a the site of reaction on the 2D surface (such as shrubs or trees emerging from the bottom of a forest). These groups induce preference for reactant molecules to adsorb or form intermediates due to their optimal fit. Coadsorption of enantiomeric molecules have been used to "design" heterogeneous catalytic surfaces that show enantiomeric selectivity[87]. The best known is the effect of the addition of the cinchonidine complex to Pt. The cinchona-modified Pt/Al2O$_3$ catalysts have been designed with enantioselectivity higher than 90% for the hydrogenation of α-keto-esters. The higher selectivity observed here is thought to be executed by analogues of the large organometallic clusters immobilized on a transition metal surface. On model catalysts comprised of enantiomeric tartaric acid adsorbed on the Cu(110) surfaces, the surface overlayer is found to be enantiomerically active and creates chiral channels that expose bare metal atoms[88]. Vayner et al.[89] proposed a mechanism for the reduction of pyruvate that proceeds through an intermediate covalently bonded to the chinonidine (Scheme 2.3).

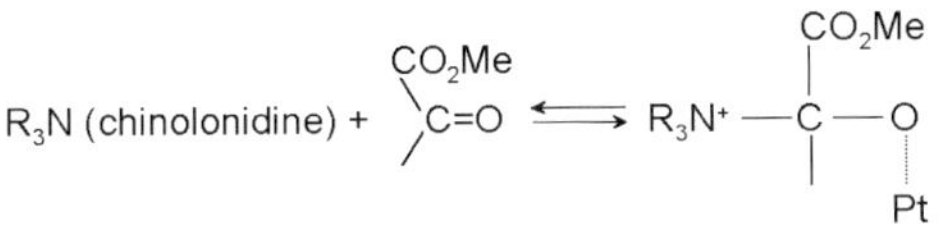

$$R_3N \text{ (chinolonidine)} + (CO_2Me)C{=}O \rightleftharpoons R_3N^+{-}C(CO_2Me){-}O\cdots Pt$$

Scheme 2.3

This bond is subsequently broken through a hydrogen addition step. The stereochemical arrangement and interaction between the catalyst ligands and the reactant molecule can result in important selectivity preferences. The impact on selectivity is especially true for homogeneous catalysts whereby catatalyst design for homogeneous catalysts often involves screening a variety of different ligands. These same selectivity influences are also seen in enzymes. Both homogeneous, and enzyme catalysts typically involve single metal atom centers which impart stereochemistry by the topological positioning of the ligands and the reactant molecules about the active center.

Heterogeneous supported transition-metal catalysts, on the other hand, lack individual molecular centers and instead are comprised of surfaces with different metal ensembles. Surfaces that contain step edges can begin to exploit the stereochemistry of the step to induce topologically controlled stereoselectivity. For instance, enantioselective catalysis has been realized electrochemically by the use of oriented transition-metal catalysts with optically active steps[90]. Horvath and Gellman showed similar behavior over specific single crystal surfaces under UHV conditions[91]. Similarly, optically active polymorphs of zeolite crystals (and even quartz) have been used as catalyst supports for enantioselective catalysis, albeit with limited success. Another heterogeneous stereoselective catalyst is the $TiCl_3$-based alkene polymerization catalyst. In this catalyst, Ti^{3+} is the reactive center present at the edge of the $TiCl_3$ surface. The catalytic site here can be considered a coordination complex. Reactant molecules adsorb to the Ti^{3+} centers similarly to the ligands in organometallic complexes. The Ti^{3+} center at the surface edge is four coordinated (see Fig 2.26).

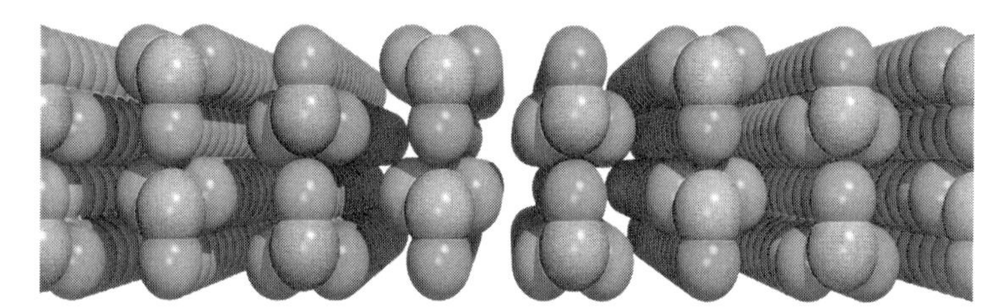

Figure 2.26. Side view of the $TiCl_3$ surface.

An elegant polymerization experiment that illustrates the importance of these edges is shown in Fig. 2.27.

Figure 2.27. Electron micrograph of a hexagonal crystal of α-$TiCl_3$. The dots are believed to represent polypropylene chains growing on sites located along a crystal growth spiral[93]. Stereoregular polymerization occurs by selective self control of the growing polymer chain that leads to preferred alkene insertion[92]. This is very similar behavior to that found in the corresponding homogeneous single-site cyclopentadienyl-Ti or -Zr catalysts. The $TiCl_3$ edge sites can be considered the inorganic surface analogues of homogeneous organometallic complexes and display very similar chemistry. Improved heterogeneous catalysts have been designed using selected basic coadsorbates to optimize stereoregularity.

The layered MoS_2-based catalyst (discussed in Chapter 5) that are used in the hydrodesulfurization of crude oil can also be considered a solid-state chemical surface analogue of organometallic coordination complexes. Indeed, a wealth of knowledge on HDS has come out of well-defined studies using metal sulfide clusters and complexes [94–100]. Similarly, the surfaces of heterogeneous solid-state metal oxide catalysts can also be analogues of corresponding coordination complexes. These catalysts operate as sulfide or oxide phases since these are the stable phases under reaction conditions.

We have now seen that specific steric factors can control catalytic selectivity by guiding specific reactants to specific products. Most heterogeneous catalysts, with the exception of zeolites, can be expected to be intrinsically less selective, since they typically do not contain a three-dimensional architecture that can help to guide the formation of specific products. Therefore, they are optimum for those applications where thermodynamics prescribes the formation of one particular product over the others. The careful choice of transition metals and promoters, however, can appreciably alter chemical reactivity and bias specific reactions, thus altering the relative rates of competing elementary reaction steps.

2.3.7 Stereochemistry of Homogeneous Catalysts. Anti-Lock and Key Concept

The selectivity of organometallic complexes used in homogeneous catalysis can be significantly improved by changing the metal center, its oxidation state, or by manipulating the structure as well as the electronic properties of the ligands about the active metal center. The bulkiness of these ligands can even be tuned to help develop more highly selective enantiomeric catalysts. The Noyori hydrogenation catalyst provides a good example. The transition-metal cation in this catalyst is coordinated to the enantiomeric BINAP catalyst (Fig. 2.28).

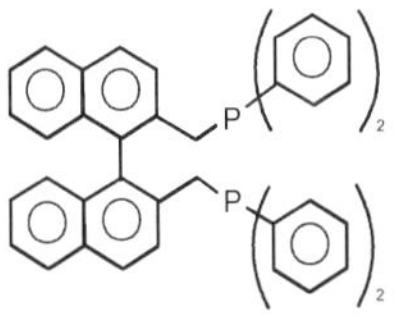

Figure 2.28. The BINAP ligand[101].

Since the BINAP ligand is optically active, the catalytic reaction is enantiomerically selective. The rigidity of this phosphine ligand derives from the connectedness of the phosphine groups.

Stereochemical control is typically due to the interaction between reactants and the bulky ligands. Stereochemical control by enzymes can be comparable to the highly selective organometallic catalysts. Stereochemical control of a catalytic reaction is another example of the use of molecular recognition to control the relative adsorption strength of reaction intermediates. One of the greatest challenges in chemocatalysis is to design catalytic systems that combine different catalytic functions in a controlled fashion, so as to integrate different reactions into a single catalytic system.

A successful homogeneous system involves the catalytic hydroaminomethylation of internal alkenes to produce the linear products over an organometallic Rh complex which contains the rigid bulky diphosphorus ligand[102]. It helps to catalyze the combined sequence of bond isomerization, CO insertion and amination at the Rh center with stereochemical control. The unique feature of this catalyst is that all three of these different

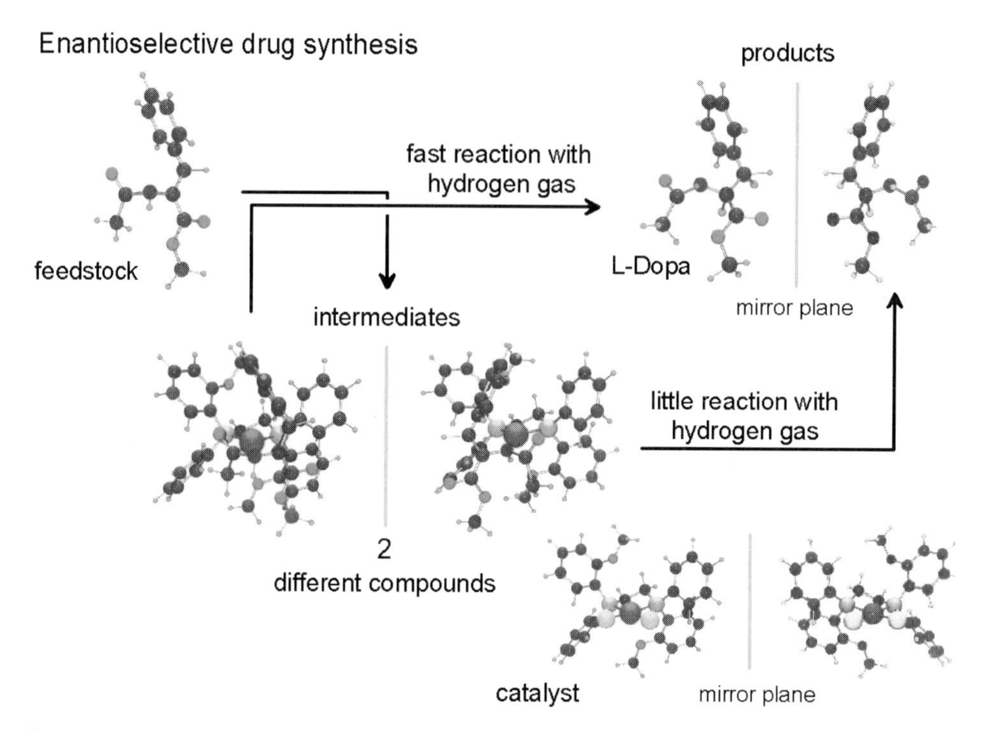

Figure 2.29. The enantioselective hydrogenation by homogeneous Rh–BINAP catalyst.

reactions are catalyzed by the same reaction center. Chiral phosphine ligands attached to cationic Rh catalysts are efficient enantiomeric hydrogenation catalysts. The non-bonding interactions between the reactant alkenes adsorbed to the metal center and bulky ligands force the reactant molecules to adsorb in a strongly preferred orientation. Figure 2.29 illustrates this for enantioselective hydrogenation to produce the L-Dopa product molecule. Mechanistic studies by Landis and Halpern[103] helped to elucidate the physical organic chemistry of this reaction. They performed kinetic studies on the hydrogenation of methyl(Z)-acetamidocinnamate by the Rh(DiPAMP) catalyst (Fig. 2.30) and proposed two specific pathways, namely the hydride and the alkene pathways. According to the hydride pathway H_2 initially dissociates over the Rh center to form a alkene Rh hydride. This is subsequently followed by the reaction between an alkene and dihydride. The alternative pathway is termed the alkene pathway. Hydrogenation now occurs after alkene binding. This actually appears to be the preferred pathway (see Fig. 2.31).

Interestingly, the major product comes from the very rapid hydrogenation of the less stable diastereomer with H_2. This feature that the least-stable intermediate is actually the most reactive has been called anti-lock and key behavior. The enantioselectivity of the reaction is affected by the competition between the rate of hydrogenation and the interconversion of the two diastereomers.

An increase in the hydrogen pressure suppresses the enantiomeric yield. Figure 2.32 [105], shows the computed free energy diagrams for this hydrogenation reaction. The calculations were performed using hybrid methods, involving DFT calculations embedded into a Molecular Mechanics force field to describe properly the interaction between reactants

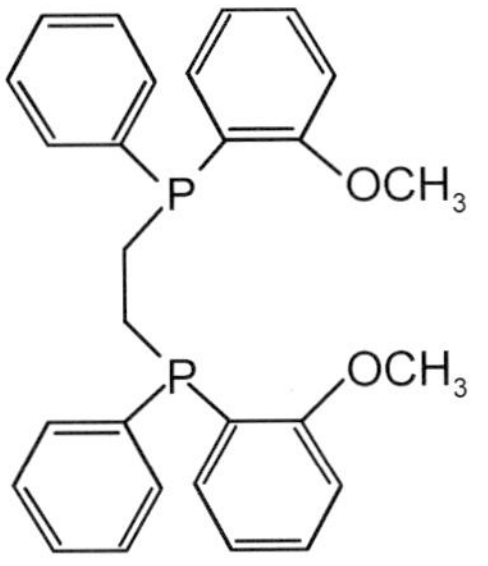

Figure 2.30. DiPAMP ligand.

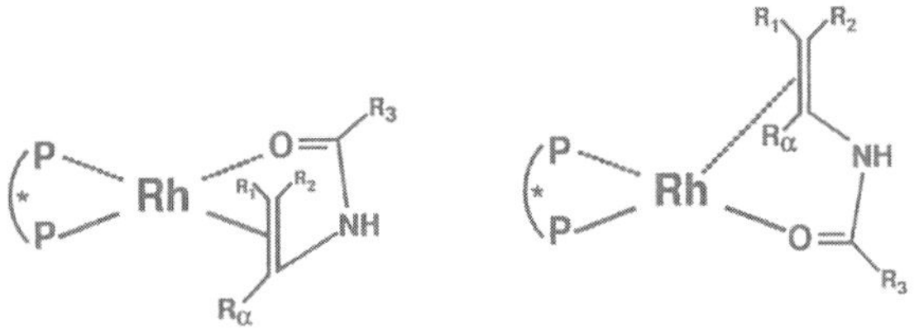

Figure 2.31. DiPAMP diphosphine coordinated to cationic rhodium[105].

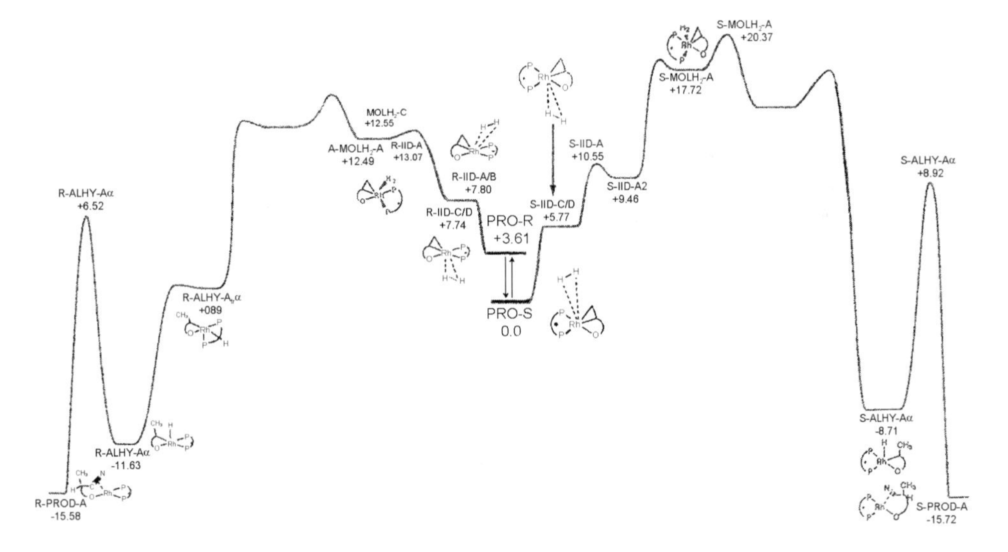

Figure 2.32. Free energy surface (in kcal/mol) for reaction of catalyst enamide diastereomers with hydrogen (simplified). Adapted from S. Feldgus and C.R. Landis[105].

and ligands. The results presented in Fig. 2.32 indeed demonstrate that the intermediate (PRO-R) with the highest free energy reacts with the lowest activation barrier.

The results can be best explained as being representative of a loose transition state where the barrier height is dominated by the need to weaken the interaction between catalyst and substrate. The unfavorable adsorption state appears to be the most reactive. It is also the adsorption mode in which the entropy is a maximum.

2.4 Surface Kinematics

2.4.1 Surface Reconstruction

The adsorption of adatoms, and also other strongly bound surface intermediates, weakens the metal–metal bonds in the surface layer and between surface and subsurface. The bond strength between the adatom (A) and the surface-metal atom (M) is sensitive to the coordinative unsaturation of the surface metal atoms. The combination of these two effects can result in a rearrangement of the surface metal atoms to increase the surface energy when the surface is covered by an overlayer of adsorbed atoms.

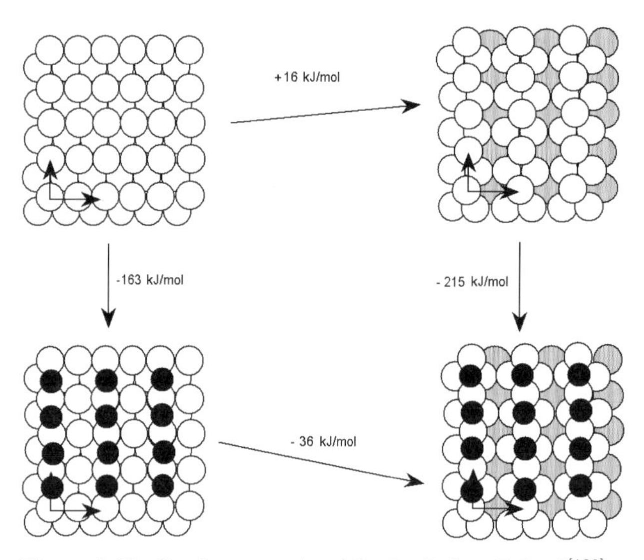

Figure 2.33. Overlayer energies of O adsorbed on Cu(110)[129].

Figure 2.33, for example, illustrates how the removal of a row of Cu atoms on the Cu(110) surface releases the stress in the strained overlayer of adsorbed atomic oxygen, thus enhancing surface reconstruction.

The details of the surface reconstruction depend on the surface as well as the adsorbate overlayer composition. Figure 2.34 illustrates this for the adsorption of CO, H_2 and O_2 on the Pt(110) surface. Each different adsorbate overlayer generates a different surface metal atom topology. Surface reconstruction has several important consequences. When such a reconstruction occurs, the density of the surface atoms changes. Hence, there are local regions on the surface where there is accumulation of surface atoms and other regions where there is a depletion. This creates defects such as edge and kink sites. As we mentioned in the previous section, the edge or kink sites are often the sites where dissociative adsorption occurs. Reconstruction of surfaces can significantly affect the reactivity of catalytically active surfaces because of this generation of highly reactive defect sites.

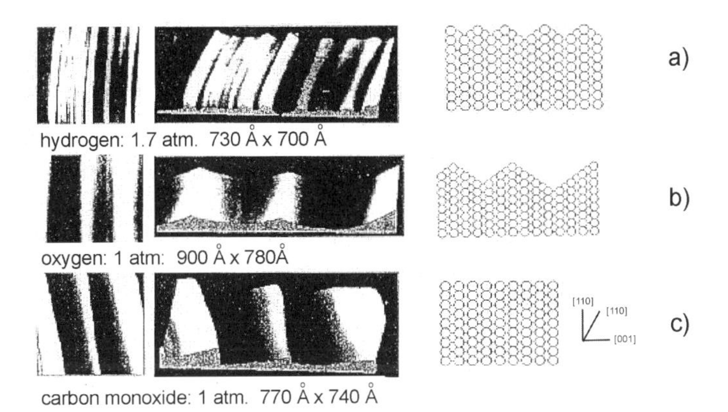

Figure 2.34. The reconstruction of the Pt(110) surface upon exposure to H_2, O_2 and CO. (a) nested missing-row reconstructions; (b) (1:1) micro incets; (c) unreconstructed (1 x 1) terraces separated by multiple height steps. Adapted from B.J. McIntyre et al.[106].

A predictive theory of heterogeneous catalysis, therefore, should include the prediction of the restructuring phenomena that occur when a catalyst is brought into its reactive phase. This is especially important when surface roughening occurs with the creation of reactive steps and kinks. Surface reconstruction is driven by the surface's desire to minimize its surface energy. Sometimes a surface which is free of adsorbates will reconstruct its bulk terminated surface, as for instance the hexagonal reconstruction of the Pt(100) surface. Surface reconstruction of clean surfaces occurs predominantly on transition metals with spatially extended d-valence atomic orbitals and with high electron occupations such as Pt and Au.

According to van Beurden and Kramer[107], the clean surfaces of these metals reconstruct because of the low values for vacancy formation as compared with the respective cohesion energies. The surface atom density changes upon reconstruction, therefore the heat of reconstruction, ΔH_r, is given by[108]

$$\Delta H_r = (E_n + \Delta N_{\mathrm{coh}} - E_r)/N_0 \tag{2.24}$$

where ΔN is the difference in the number of atoms in the reconstructed layer and N is the total number of atoms in the reconstructed layer, E_n is the total energy before reconstruction and E_r is the total energy of the reconstructed system. The heat of reconstruction is seen to be explicitly dependent upon the cohesive energy. The large spatial extent of the d-valence atomic orbitals, characteristic of the group 5d transition metals, and the nearly complete electron occupation generate a strong repulsive contribution to the metal–metal bonds to be overcome by the attractive contribution of the free–electron type s- and p-valence electrons. Vacancy formation becomes easier because it reduces these repulsive interactions. A similar reduction of the repulsive interactions between the highly occupied d-valence atomic orbitals assists the stabilization of the longer bond distances found in the reconstructed surfaces.

Covering the surface with an adsorption layer changes the surface energy. For Langmuir adsorption, the change in the surface energy is given by the expression [109,110a,b]

$$\gamma = \gamma_0 + \frac{RT}{a}\ln(1-\theta) \tag{2.25a}$$

$$= \gamma_0 - \frac{RT}{a} \ln(1 + Kp) \qquad (2.25b)$$

where γ is the surface energy of the surface layer covered with adsorbate, γ_0 is the surface energy of the free surface, a is the surface area of the adsorbate, θ is the coverage, p is the partial pressure of adsorbate and K is the adsorption equilibrium constant. Surface reconstruction occurs when the increased surface energy of the surface metal atoms with low coordination is compensated for by the increased adsorption energies. This is illustrated schematically in Fig. 2.35, where the changes of the two surface energies are plotted as a function of partial pressure of the adsorbate. Surface (2) is the less stable surface, with the higher adsorption energies. If the two curves cross, as indicated in the figure, there is a critical pressure p_c beyond which there is a driving force for reconstruction.

The existence of a critical pressure p_c, beyond which surface reconstruction occurs, defines a critical coverage θ_c for each surface that is related by p_c through the adsorption isotherm.

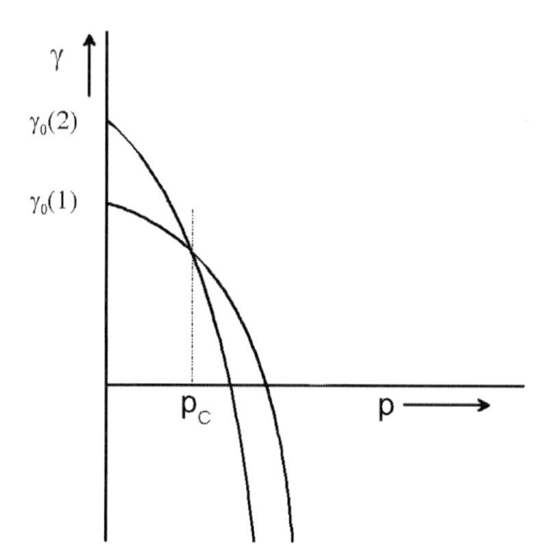

Figure 2.35. Schematic illustration of the dependence of the surface energy on partial pressure. Surface (2) has surface atoms of lower coordination than surface (1), and hence is most reactive. p_c is the partial pressure beyond which there is a driving force for reconstruction of surface (1) to surface (2).

Theory can be used to compute surface energies as well as the adsorption energies and entropies. Hence a theoretical prediction of Fig. 2.35 can be made for any two surfaces. Surfaces covered with strongly bonded atoms such as C are often found to have a critical coverage beyond which there is a driving force for reconstruction.

A theoretical analysis of the surface reconstruction dynamics is possible using first principle calculations. For example, Molecular Dynamics was recently used to study the reconstruction of the Pt(111)hex surface phase by CO as a function of time[107,111]. These simulations employed embedded atom potentials that were developed based on quantum-mechanical calculations of the surface.

The local surface concentration of molecules and adatoms can change over the course of a catalytic reaction. This is intrinsic to the catalytic reaction cycle. Reconstruction can deactivate the catalyst surface due to presence of strongly bound adatoms or the formation of a non-reactive surface. In addition, reconstruction can drive particular elementary steps that lead to autocatalytic behavior. For example, surface reactions in which a product molecule is formed and subsequently desorbs, often require a vacancy to activate a reactant molecule and usually create more vacancies than were initially present (see Chapter 8). Such reactions are autocatalytic in the number of vacancy sites as is shown in the following equation

$$\mathrm{R}* + * \longrightarrow \mathrm{P} + 2*$$

where R and P refer to the gas-phase reactant and product, respectively, and $*$ and $R*$ refer to free and reactant-covered sites.

Under particular conditions, the combination of deactivation that results from surface reconstruction and the activation that occurs via autocatalytic steps in the surface overlayer ultimately lead to transient collective self organizing pattern formation such as moving spirals, pulsars or other patterns[112], known as Turing structures[113] (see Section 8.3). These structures are named after the mathematician Turing, who discovered such self organized structures when diffusion of the participating components is very different. In the Turing system, one chemical component is autocatalytic. The product itself enhances the rate of its own production. It also catalyzes the production of another chemical component. The second chemical, however, inhibits production of the first. Self organizing phenomena in catalysis will be extensively discussed in Chapter 8.

In the heterogeneous catalytic system, the reacting molecules, and the metal-surface atoms, can be quite mobile. This leads to locally ordered structures when synchronization or self organization phenomena are present however, disorder in the surface layer prevails when these phenomena are absent. For example, it has been proposed that the catalytic oxidation over mixed metal oxides, discussed in Section 2.3.5, actually occurs in the disordered overlayer that forms at the surface under reaction conditions.

The epoxidation of ethylene which is catalyzed by Ag and promoted by chlorine compounds, for example, is thought to occur in a surface overlayer that has features similar to a melt of Ag ions. The silver-oxychloride reactive surface layer requires Ag^{3+} ions (as in the electrochemical system, see Scheme 2.1) to enhance the overall selectivity. Reduced Ag clusters, however, are required to activate molecular oxygen. Dynamic events between these two states are necessary to close the catalytic cycle. Chlorine in combination with Cs is added to promote the Ag catalyst. Eutectic melting points of this phase are close to the reaction temperature[114].

An interesting phenomenon that nicely illustrates the consequence of the dynamic surface events is shown in Fig. 2.36, which shows the structure of an Ir-surface, after exposure at 1000 °C to an oxidizing mixture of methane giving CO, CO_2 and H_2O. Within half an hour the initial surface, which was flat, is transformed into a mountainous landscape with altitude differences on the order of 1 μm. Kramer[115] has estimated that in processes that result in these transformations each surface atom can jump at a rate of 10^6/sec, whereas the elementary reaction rates that occur under these conditions occur at a rate of 10^4–10^5/sec. The most adequate picture of this reactive surface was suggested to be that of a dynamically changing surface that reacts with a quasi-static ensemble of adsorbed molecules or molecular fragments. The surface etching process is the result of the balance between momentous stable surface reconstruction and destabilization of

Figure 2.36. An interference–microscopic photograph of a iridium surface after one hour of catalyzing the CPO process at 1000 °C. The different color densities indicate differences in height of about 1 μm[115].

the surface due to reaction events. In the case of the Turing instability, this leads to self organization effects that are stationary during reaction. In the above example, no balance between the two is reached. As a consequence, the reconstruction processes persist in an irreversible manner until finally a stationary situation is reached.

A second well-known example where strong restructuring of the surface occurs is the high-temperature (~1200 K) Ostwald process in which ammonia is oxidized to NO over a Pt/Rh gauze. Under industrial conditions, so-called cauliflower structures develop. These changes occur during the first hours of operation and are assumed to play an important role in improving catalyst performance[40]. Restructuring occurs in two stages with initial formation of parallel facets followed by growth of microcrystals[116].

In the previously discussed Fischer–Tropsch reaction catalyzed by Co, adsorbed reaction intermediates were able to lead to reconstruction, thus resulting in the formation of a roughened surface which contains edge, kink and hollow sites[117,118].

As we will discuss in Chapter 3, two factors control the unique reactivity of the step versus the terrace. The first deals with the fact that the step provides the ability to separate the reaction fragments so as they do not share bonds to the same metal atoms in the transition state. This will favor both bond-making and bond-breaking reactions. CO dissociation and carbon–carbon coupling are two examples of bond-breaking and bond-making reaction steps in FT catalysis that are significantly enhanced at the step edges as a result of lowering metal atom sharing. These elementary surface reaction steps are sensitive to the surface topology in both directions.

The second feature of the step is the under-coordinative saturation of the metal atoms at the top edge of the step. For reactions that only proceed over the edge atom, these sites tend to enhance bond-breaking steps but may slow bond-making steps owing to the enhanced binding of the fragments. For reactions that proceed through a transition state whereby the reaction fragments share a metal atom, the rate of dissociation is enhanced at the step due to the decrease in metal atom sharing and the enhanced binding of the product fragments. The rate of bond-making will depend on the degree of metal-atom sharing in the transition state since the enhancement due to the decrease in metal atom sharing now competes with the impedance due to stronger metal-adsorbate bonds of the reactant at the step edge. When the transition state is late, only the forward bond cleavage reaction step will be sensitive to the surface topology. The backward reaction will not depend on the surface topology. Methanation reactions, for example, are typically suppressed at the step edge whereas carbon–carbon coupling is enhanced.

The above arguments illustrate the importance of edge and kink sites in catalysis. As a consequence, reconstruction phenomena that change also the edge and kink site distribution can have a large effect not only on the overall rate of a catalytic reaction but also on its selectivity. The latter occurs when competing elementary reaction steps have different sensitivities.

Surface reconstruction is inherent to surface oxidation and sulfidation chemistry. In involves essentially surface corrosion and surface compound formation phenomena. The state of a surface can change from a metallic state to that of a solid oxide, sulfide, carbide or nitride depending upon the reaction environment. The surface of the epoxidation catalyst, discussed earlier, in the absence of Cl or Cs, for example, has a composition similar to AgO in the oxidizing reaction environment of the epoxidation system. The oxidation of CO over Ru can readily lead to the formation of surface RuO_2 (see Chapter 5). In desulfurization reactions the transition-metal surface is converted to a sulfide form. The reactivity of the surface in these systems begins to look chemically more similar to that of coordination complexes. This we will illustrate in Chapter 5 for the CoS/MoS_2 system.

2.4.2 Transient Reaction Intermediates in Oxidation Catalysis

Earlier in the section on the pressure gap, we saw that adspecies with unique reactivity may appear when the surface state changes as a function of coverage. Weak and highly reactive species that are not stable at low coverage may develop. The π-bonded ethylene intermediate which forms at higher surface coverages present under actual reaction conditions on different metal surfaces is one such intermediate which has been found to be important in the hydrogenation of ethylene. Notwithstanding the short residence time of the weakly bonded reactive intermediate, the rate of reaction is finite owing to the increased concentration at high pressure. The residence time is equal to ${k_{des}}^{-1}$. The rate depends on the surface coverage of the reactive intermediate, θ_r, which is a strong function of the surface state.

Short-lived, highly reactive intermediates can form on a surface upon molecule dissociation. Carley et al.[119], for example, demonstrated that a uniquely reactive atomic oxygen (O^-) is formed in the oxidation of ammonia by oxygen over Cu, Zn, and Mg. This unique O^- develops the instant that O_2 dissociates. The oxygen atom that forms does not fully equilibrate with the surface to form O^{2-}. Upon dissociation, molecular oxygen moves through a high energy state to overcome the reaction barrier. Immediately after reaction, the atoms that result are not energetically equilibrated, and move over a metal atom position before they adsorb and equilibrate. These "hot" O^--type atoms have been shown to have unique reactivity. Their short lifetime relates to the time required to equilibrate[13b,120] the high energy-atoms to the surface. This will be typically on the order of a few picoseconds. At low pressure this time is independent of the pressure of the reaction gas. It will become shorter at higher pressures where gas-phase molecules collide with the hot atoms. In oxidation catalysis the reactivity of such hot atoms has to be distinguished from the reactivity of short-lived molecular oxygen species such as ${O_2}^-$, that are highly reactive but have a short residence time not because they desorb, but because they readily dissociate. The surface concentration of such reactant molecules has to be high enough and the activation barrier low enough in order for these species to compete with dissociation. Owing to their short residence times, it is usually difficult to isolate such intermediates spectroscopically or identify their role in the mechanism. The decomposition of NH_3 by oxygen is an interesting example that has been studied in detail over different surfaces by Au and Roberts[121].

$$NH_{3ads} + (O_2^-)_a \longrightarrow H_2O + H_{ads}^- + NO$$

Theoretical calculations have been performed to demonstrate their importance of the O_2^- intermediate for this reaction over Cu[122].

On oxide catalysts O_2^- species have often been proposed to be the intermediates for total combustion. However, it is not always easy to distinguish between the role of short-lived O_2^- intermediates or the presence of uniquely reactive oxygen atoms. Unraveling the selective epoxidation mechanism by silver is a good example[123] of the difficulty in establishing the reactive oxygen form.

In the earlier mentioned ethylene epoxidation reaction catalyzed by Ag, initial spectroscopic isotope exchange and chemisorption data indicated that adsorbed O_2 species were responsible for epoxide formation, and that the oxygen adatoms were responsible for activating C–H bonds, hence leading to total combustion[124]. These conclusions have been disputed on the basis of kinetic experiments by Force and Bell[125] that lead to the interpretation that the reactivity of adsorbed oxygen atoms to Ag strongly depends on the state of the oxygen overlayer. Low concentrations of adsorbed oxygen create electronegative oxygen adspecies that help to activate the C–H bond; oxygen adsorbed at high oxygen coverage is much more electrophilic and therefore prefers to insert into the ethylene π bond. The main evidence[123] now supports the Force and Bell point of view. The intermediate leading to oxidation is proposed to be the oxymetallocycle[126] as sketched in Scheme 2.4.

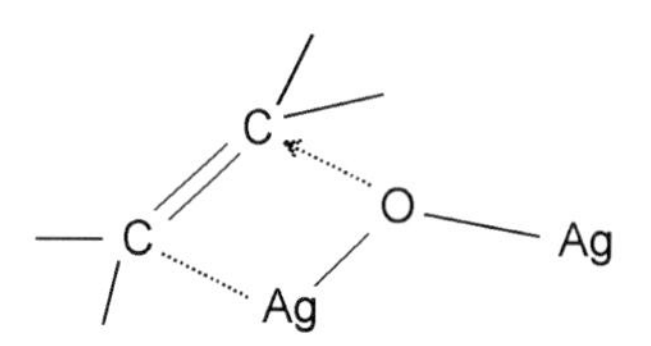

Scheme 2.4 The oxymetallocycle was proposed by Barteau[128] to be the active intermediate for ethylene epoxidation. Surface experiments by Bocquet et al.[127] indicate that electron-deficient Ag atoms coordinate with adsorbed ethylene with moderate energy. This is the precursor state to the oxymetallocycle complex. Oxygen activation can also occur directly on the oxide overlayer.

An interesting example where the O_2^- intermediate forms and appears to play an important role in the catalytic reactivity is the high-temperature oxidation of methane over La_2O_3. The reactive O_2^- intermediate here is not generated directly from adsorption of O_2 on the surface, but indirectly by a unique dissociation of molecular oxygen over the lattice oxygen atoms of La_2O_3. Lanthanum oxide is non-reducible oxide which has a high affinity for oxygen which helps to make this path possible. Calculations by Palmer et al.[128] have demonstrated that the activation of oxygen given in Scheme 2.5 occurs with low endothermicity of 50 kJ/mol and an activation barrier of 132 kJ/mol.

Owing to the endothermicity of the reaction, the equilibrium concentration of the resulting O_2^- will be low and their residence time short. In Scheme 2.5, the reaction of O_2 over the La_2O_3 surface converts the surface O^{2-} anion which is attached to La to a reactive surface superoxide O_2^- intermediate. These anions are of radical type character and can subsequently activate CH_4 to produce a CH_3 radical and OOH. This reaction

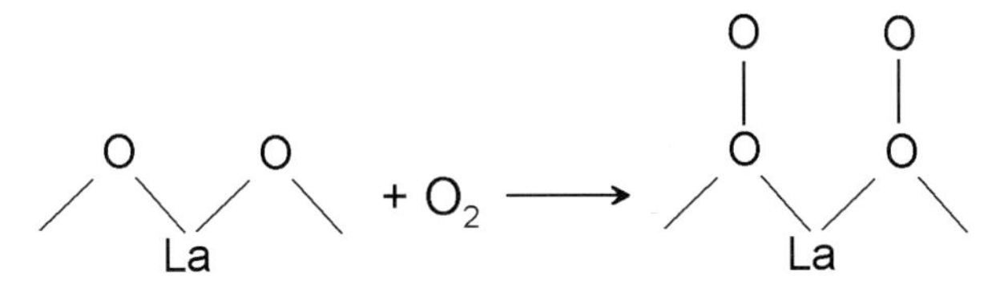

Scheme 2.5 The reaction of O_2 over La_2O_3 to create reactive O_2^- surface peroxide intermediates.

was found to 116 kJ/mol endothermic. Subsequent water formation from the recombination of surface OH radicals is an exothermic process. The reactivity of the oxygen atoms of the stoichiometric La_2O_3 surface was found to be very low, as can be deduced from the endothermicity of the reaction with CH_4 (370 kJ/mol). In contrast, the reactivity of surface-defect centers is high, but such centers require high energy to be generated.

2.5 Summary: Concepts in Catalysis

In this chapter we introduced the basic physical chemistry that governs catalytic reactivity. The catalytic reaction is a cycle comprised of elementary steps including adsorption, surface reaction, desorption, and diffusion. For optimum catalytic performance, the activation of the reactant and the evolution of the product must be in direct balance. This is the heart of the Sabatier principle. Practical biological, as well as chemical, catalytic systems are often much more complex since one of the key intermediates can actually be a catalytic reagent which is generated within the reaction system. The overall catalytic system can then be thought of as nested catalytic reaction cycles. Bifunctional or multifunctional catalysts realize this by combining several catalytic reaction centers into one catalyst. Optimal catalytic performance then requires that the rates of reaction at different reaction centers be carefully tuned.

To predict catalyst performance, one needs to predict the rates of the elementary reaction steps at the catalyst surface. This must ultimately be integrated into a kinetic simulation which treats the interactions between the many different adsorbates present on the catalyst surface. In this chapter, we presented rate expressions derived from transition state reaction rate theory as a bridge to connect ab initio quantum mechanical information to reaction rate predictions. In Chapter 3, we present a more extensive treatment of kinetic simulations including many-body interactions and their influence on the catalytic performance.

We use the constructs of transition state theory in order to define the Brønsted–Evans–Polanyi (BEP) relationship, which relates the equilibrium thermodynamics (reaction enthalpy or free energy) with non-equilibrium thermodynamic features, namely the activation energy and activation entropy. A small value of the proportionality parameter in the BEP relationship, α, is identified with an early transition state, whereas values of α that are close to 1 relate to a late transition state. Microscopic reversibility ensures that if the forward reaction is an early transition state then the backward reaction must be a late transition state and vice versa.

The rates of reaction that proceed through early transition states are rather insensitive to changes in the reaction enthalpy, and hence variations in the catalyst, provided that there is no change in the reaction path. On the other hand, reactions with late transition states depend strongly on the reaction energy and hence are quite sensitive to variations in the catalyst. Using the BEP theory, the reactivity of a surface with respect to a particular elementary reaction step can be determined by the difference in the energy of the adsorbed reagents before reaction and the energies of their product fragments after reaction.

The transition-state entropies for a surface reaction tend to be small because of the need for tight contact with the catalytic surface atoms. On the other hand, changes in the activation entropies are large for elementary reaction steps in which the reactants desorb from the surface.

The transition-state entropy may play an important role in chemo-selective and enantio-selective reaction steps in which subtle steric interactions are often mediated through weak van der Waals electrostatic interactions or hydrogen bonding. The lock and key model, which suggests an optimal fit of the transition-state conformation into a potential cavity, is then a very useful concept. The reaction enthalpy is typically lowest for the transition-state configurations that maximize weak intermolecular attractive interactions with the cavity. If the fit within the cavity, however, is too tight there will be a large loss in entropy at the expense of the enthalpy. In many cases the favored reaction path is then one in which the activation energy is slightly larger so as to minimize the loss in entropy.

An important aim of theoretical catalysis is to develop the rules that relate catalyst performance to catalyst structure and composition. In this chapter, we introduce various general rules that concern this relationship. We have already referred to steric control which can be due to the interaction between the ligands in a homogeneous catalyst, organic overlayer on a heterogeneous surface, or the cavities within zeolites. The last will be extensively discussed in Chapter 4.

The energy of adsorption on a surface atom increases with increasing coordinative unsaturation of the surface metal atom(s). This agrees with ideas proposed by the Bond Order Conservation Principle, which would indicate that the strength of the chemical bond increases when the number of atoms which share bonds to different adorbates decreases. As we will learn in Chapter 3, this affects the adsoption strength of the surface atoms more than that of the molecules. Hence more open surfaces are often much more reactive than the dense closely packed surfaces which are comprised of atoms that are close to being coordinatively saturated.

Similarly, steps or kink sites are often sites that are uniquely reactive. As will also be explained in Chapter 3, it is important to analyze in detail the geometry of the transition states. One has to distinguish reactions with transition states in which the reaction fragments share bonding with other surface metal atoms from transition states in which there is no such sharing of surface metal atoms. Within the latter, transition-state structures, both association and dissociation reactions will proceed with low energies. The activation of CO and N_2, for example, demonstrate these features.

Other reactions such as C–H activation usually proceed through transition states of the former type, in which the reaction fragments share metal atoms. The step edges will be more favorable sites for C–H activation but the reverse reaction will now have an increased activation energy.

Particle size effects are important since they can influence the ratio of different surface facets along with the ratio of step, kink and terrace sites. In addition, as the particle sizes becomes smaller than a critical size, they can take on unique behavior owing to quantum size effects. When a molecule adsorbs there is an attractive interaction between the molecule and the atoms at the catalyst surface. Bonds within the molecule, as well as bonds within the metal cluster, tend to weaken. The overall interaction energy is then the sum of these three terms. Cluster size effects specifically alter the response of the chemical bonding within the cluster to the adsorbed molecule. We elaborate much more on this in Chapter 3. Support effects tend to become much more important for small clusters, because the interaction between the clusters atoms and the support tends to be at its largest. Strong covalent interactions with the support will decrease the cluster reactivity. On the other hand, clusters within particular charge states may become stabilized, which can have the opposite effect. Insights into the detailed chemistry of such systems is important.

When the surface becomes covered with an overlayer, the lateral interactions between adsorbed molecules become important. These interactions are reviewed in Section 3.3. The resulting many-body effects in the surface overlayer may lead to changes in the molecular arrangement at the surface including the formation of ordered overlayers, disordered structures or phase-separated regions.

Often surface reconstruction occurs at higher adsorbate surface concentrations within the overlayer. Reconstruction can lead to more reactive surface phases. As we will see in Chapter 3, the kinetic implication is that mean-field theory does not always apply since the reactions now predominantly occur at the boundaries of the different overlayer phases present on the catalyst. Similarly Chapter 5 treats, in detail, examples from oxide and sulfide catalysis which show the importance of surface phase changes in relation to catalytic activity.

Lateral interactions will alter the binding energies of the reactants as well as the products to different degrees. As such, they also influence the activation barriers for surface reaction steps. At high coverages, unique adsorbate bonding states may become possible only present in significant concentrations at high pressures. This may result in significant differences between the reactivity at low coverages under typical UHV conditions and the activity at higher coverages found for reactions carried out at much higher pressures. This is known as the pressure gap problem. Therefore, simulations which include these interactions ultimately allow us to understand the differences that may result between surface science experiments performed under UHV conditions and those performed at more realistic operating pressures. As an example, we discussed the unique reactivity of the π-bonded ethylene intermediate. In other systems, such precursor states may be quite general. They are difficult to access experimentally owing to their short lifetimes. The nature of these short-lived states and their influence on the overall catalytic performance can be modeled using dynamic Monte Carlo simulation methods that explicitly treat inter-adsorbate interactions on the surface and thus changes due to changes in operating pressures.

References

1. I. Langmuir, *Trans. Faraday Soc.* 17, 62 (1921)
2. H.S. Taylor *Proc. Roy. Soc. A* 105, 9 (1925)
3. J.M. Thomas, W.J. Thomas, *Heterogeneous Catalysis*, Wiley, New York (1997)
4. P. Sabatier, *La Catalyse en Chimie Organique*, Libraire Polytechnique, Paris (1913)
5. M. Boudart, G. Djega-Mariadassou, *Kinetics of Heterogeneous Catalytic Reactions*, Princetown Univ. Press, Princetown, NJ (1984)
6. A. Heyden, A. Bell, F.J. Keil, *J. Catal.* 233, 26 (2005)
7. (a) S. Hub, L. Milaire, R. Touronde, *Appl. Catal.* 63, 307 (1988);
   (b) F. Mittendorfer, C. Thomazeau, P. Raybaud, H. Toulhoat, *J. Phys. Chem. B* 107, 12287 (2003)
8. L. Stryer, *Biochemistry*, Freeman, New York, p. 513 (1995)

9. P.L. Bragd, H. van Bekkum, A.C. Besemer, *Top. in Catal.* 27, 49 (2004)
10. J.M. van der Eyk, Th. J. Peters, N. de Wit, H.A. Colijn, *Catal. Today*, 3, 259 (1988)
11. K.-I. Tanaka, K. Tamaru, *J. Catal.* 2, 366 (1963)
12. B. Hammer, J.K. Nørskov, *Adv. Catal.* 45, 71 (2000)
13. (a) S. Glasstone, K.J. Laidler, H. Eyring, *The Theory of Rate Processes*, McGraw-Hill, New York (1941);
    (b) G. Ertl, *Adv. Catal.* 45, 1, (2000)
14. H.A. Kramers, *Physica* 7, 284 (1940)
15. R.A. van Santen, J.W. Niemantsverdriet, *Chemical Kinetics and Catalysis*, Plenum, New York, p. 199 (1995)
16. (a) O.K. Rice, *Statistical Mechanisms, Thermodynamics and Kinetics*, Freeman, New York (1967);
    (b) K.J. Laidler, *Theories of Chemical Reaction Rates*, McGraw-Hill, New York (1969)
17. (a)L.P. Hammett, *J. Am. Chem. Soc.* 59, 96 (1937);
    (b) J.N. Brønsted, *Chem. Rev.* 5, 231 (1928);
    (c) M.G. Evans, N. Polanyi, *Trans. Faraday Soc.* 32, 1333 (1936)
18. G.J. Kramer, R.A. van Santen, C.A. Emeis, A.K. Novak, *Nature*, 363, 529 (1993)
19. R.A. van Santen, G.J. Kramer, *Chem. Rev.* 95, 637 (1995)
20. R. van Hardeveld, F. Hartog, *Surf. Sci.* 15, 189 (1969)
21. R.J. Farauto, C.H. Bartholomew, *Fundamentals of Industrial Catalytic Processes,* Blackie Academic, Chapman and Hall, London (1997)
22. R.D. Kelly, D.W. Goodman, in *The Chemical Physics of Solid Surfaces and Heterogeneous Catalysis*, Vol. 4, p. 36 (1982)
23. R.A. van Santen, C.P.M. de Groot, *J. Catal.* 98, 530 (1986)
24. P.S. Cremer, X. Su, Y.R. Shen, G.A. Somorjai, *J. Am. Chem. Soc.* 118, 2942 (1996); *Catal. Lett.* 40, 143 (1996)
25. W.M.H. Sachtler, R.A. van Santen, *Adv. Catal.* 26, 69 (1977)
26. Y. Soma-Noto, W.M.H. Sachtler, *J. Catal.* 32, 315 (1974)
27. Y. Soma-Noto, W.M.H. Sachtler, *J. Catal.* 34, 162 (1974)
28. J.H. Sinfelt, J.L. Carter, D.J.C. Yates, *J. Catal.* 24, 283 (1972)
29. T. Koerts, W.J.J. Welters, R.A. van Santen, *J. Catal.* 134, 1 (1992)
30. J.A. Rodriguez, D.W. Goodman, *Science* 257, 897 (1992)
31. M. T.M. Koper, R.A. van Santen, M. Neurock, in *Catalysis and Electrocatalysis at Nanoparticle Surface*, A. Wieckowski, E.R. Savinova, C.G. Vayenas (eds.), Marcel Dekker, New York (2003), Chapter 1, p. 133 (2003)
32. E. Shustorovitch, *Adv. Catal.* 37, 101 (1990)
33. (a) M. Neurock, *J. Catal.* 216, 73 (2003);
    (b) P. Sheth, M. Neurock, C.M. Smith, *J. Phys. Chem. B* 109, 12449 (2005);
    (c) M. Neurock, D. Mei, *Top. in Catal.* 20, 1, 5, (2002);
    (d) D. Mei, E. W. Hansen, M. Neurock, *J. Phys. Chem. B* 107, 798 (2003)
34. R.J. Davis, M. Boudart, *Catal. Sci. Technol.* 1, 129 (1991)
35. W.D. Provine, P.L.Mills, J.J. Lerou, *Stud. Surf. Sci. Catal.*, 101, 191 (1996)
36. H.S. Bengaard, J.K. Nørskov, J. Sehested, B.S. Clausen, L.P. Nielsen, A.M. Molenbroek, J.R. Rustrop-Nielsen, *J. Catal.* 209, 365 (2002)
37. R.A. van Santen, *Theoretical Heterogeneous Catalysis*, World Scientific, Singapore p. 342 (1991)
38. D.W. Blakeley, G.A. Somorjai, *J. Catal.* 42, 181 (1970)

39. (a) R.A. van Santen, B.G. Anderson, R.H. Cunningham, A.V.G. Mangnus, L.J. van IJzendoorn, M.J.A. de Voigt, *Angew. Chem. Int. Ed. Engl.* 35, 2785 (1996); (b) S.J. Thomas, G. Webb, *J. Chem. Soc. Chem. Commun.* 526 (1976); (c) G.A. Somorjai, F. Zaera, *J. Phys. Chem.* 86, 3070 (1982)
40. R.J. Farrauto, C.H. Bartholomew, *Fundamentals of Industrial Catalytic Processes,* Chapman and Hall, p. 481 (1997)
41. R.A. van Santen, P.W.N.M. van Leeuwen, J.A. Moulijn, B.A. Averill, *Catalysis, an Integrated Approach,* 2nd edn. Elsevier, Amsterdam, Ch. 9 and 10 (1999)
42. (a) T. Huizinga, R. Prins, *J. Phys. Chem.* 85, 2156 (1981); R.T.K. Baker, E.B. Preshidge, R.L. Garten, *J. Catal.* 56, 390 (1979); (b) D.W. Goodman, *Catal. Lett.* 99, 1 (2005)
43. (a) S.J. Tauster, *Acc.Chem. Res.* 20, 389 (1987); (b) D.E. Resonco,G.L. Haller, *Adv. Catal.* 36, 173 (1989); (c) R.A. van Santen, in *Proc. 8th Int. Congress Catalysis,* Springer-Verlag, Berlin p. 97 (1984); (d) E. Wimmer, C.L. Fu, A.J. Freeman, *Phys. Rev. Lett.* 55, 2618 (1985); (e) S. Holloway, J.K. Nøskov, N.D. Lang, *J. Chem. Soc. Faraday Trans.* I83, 135 (1987)
44. G.A. Somorjai, *Introduction to Surface Chemistry and Catalysis,* Wiley, New York (1994)
45. D.Diehl, R. McGrath, *Surf. Sci. Rep.* 23, 2, 43 (1996)
46. C.L. Cleveland, U. Landman, Th. G. Schaaff, M.N. Shafigullin, P.W. Stephens, R.L. Whetten, *Phys. Rev. Lett.* 79,1873 (1997)
47. W. Biemolt, PhD. Thesis, Technical University of Eindhoven, (1995).
48. M.E. Geusic, M.D. Morse, R.E. Smalley, *J. Chem. Phys.* 82, 590 (1985)
49. R.L. Whetten, D.M. Cox, D.J. Trevor, A. Kalder, *Phys. Rev. Lett.* 54, 1494 (1985)
50. W.D. Knight, K. Clemenger, W.A. de Heer, A. Saunders, *Phys. Rev. Lett.* 52, 2141 (1984)
51. K.J. Taylor, C.L. Pettiette-Hall, O. Cheshnovsky, R.E. Smalley, *J. Chem. Phys.* 96, 3319 (1992)
52. J. Uppenbrink, D.J. Wales, *J. Chem. Phys.* 96, 8520 (1992)
53. G. Wulff, *Z. Kristallogr.* 34, 449 (1901)
54. M. Haruta, N. Yamada, T. Kobayashi, S. Ijma, *J. Catal.* 115, 301 (1989)
55. R. Meyer, C. Lemire, Sh. Shaikhatdina, H.-J. Freund, *Gold Bull.* 37, 72 (2003)
56. G.C. Bond, D.T. Thompson, *Gold. Bull.* 33, 61 (2000)
57. M. Haruta, N. Yamada, T. Kobayashi, S. Ijma, *J. Catal.* 115, 301 (1989)
58. G.C. Bond, D. Thompson, *Catal. Rev. Sci. Eng.* 41, 319 (1999)
59. N. Lopez, T.V.W. Janssens, B.S. Clausen, Y. Xu, M. Mavrikakis, T. Bligaard, J.K. Nørskov, *J. Catal.* 223, 232 (2004)
60. G. Mills, M.S. Gordon, H. Metiu, *Chem. Phys.* 359, 493 (2002)
61. N. Lopez, J.K. Nørskov *J. Am. Chem. Soc.* 124, 11262 (2002)
62. I. Remediakis, N. Lopez, and J.K. Nørskov, *Angew. Chem. Int. Ed.* 44, 1824 (2005)
63. M. Valden, X. Lai, D.W. Goodman, *Science* 281, 1647 (1998)
64. M. Mavrikakis, P. Stoltze, J.K. Nørskov *Catal. Lett.* 64, 101 (2000)
65. J. Calla, R.J. Davis, *J. Phys. Chem. B*, 109, 2307 (2005)
66. Z. P. Liu, P. Hu, A. Alavi, *J. Am Chem. Soc.* 124, 14770 (2002)
67. Z. P. Liu, X.Q. Gong, J. Kohanoff, C. Sanchez, P. Hu, *Phys. Rev. Lett.* 91, 266102 (2003)

68. L.M. Molina, B. Hammer, *Phys. Rev. Lett.* 90, 206102 (2003)
69. T.A. Nijhuis, B.J. Huizinga, M. Makkee, J.A. Moulijn, *Ind. Eng. Chem. Res.* 38, 884 (1999)
70. Q. Fu, H. Saltsburg, M. Flytrani-Stephanopoulos, *Science* 301, 935 (2003)
71. J. Guzman, B.C. Gates, *Angew. Chem. Int. Ed.* 42, 69203
72. H. Kung, M.C. Kung, C. K. Costello, *J. Catal.* 216, 425 (2003)
73. H. Hakkinen, S. Abbet, A. Sanchez, U. Heiz, U. Landman, *Angew. Chem. Int. Ed.* 42, 1297 (2003)
74. I.D. Socaciu, J. Hagen, T.M. Bernhardt,L. Woste, U. Heiz, H. Hakkinen, U. Landman, *J. Am. Chem. Soc.* 125, 10437 (2003)
75. A. Sanchez, S. Abbel, U. Heiz, W.D. Schneider, H. Häkinnen, R.N. Barnett, U. Landman, *J. Phys. Chem. A*, 103, 9573 (1999)
76. W.L. Winterbottom, *Acta Metall.* 15, 303 (1967)
77. C.R. Henry, in *Catalysis and Electro Catalysis at Nano Particle Surfaces*, A. Wieckowski, E.R. Savinova, C.G. Vayenas (eds.), Marcel Dekker, New York (2003)
78. A. Bogicevic, D.R. Jennison, *Surf. Sci.* 515, L481 (2002);
D.R. Jennison, A. Bogocevic, *Surf. Sci.* 414(, 108 (2000);
A. Bogocevic, D.R. Jennison, *Phys. Rev. Lett.* 82, 4050 (1999)
79. K.M. Neyman, V.A. Innham. R. Kosarev, N. Rosch, *Appl. Phys. A*, 78(6), 823 (2004);
A.V. Matveev, K.M. Noynun, I.V. Yudanov, N. Rosch, *Surf. Sci.* 426, 123 (1999);
A.V. Matveev, K.M. Neyman, G. Pacchioni, N. Rosch, *Chem. Phys. Lett.* 299, 603 (1999)
80. C.T. Campbell, *Surf. Sci. Rep.* 27, 1 (1997)
81. J. Guzman, S. Carrettin, J.C. Fierro-Gonzalez, Y. Hao, B.C. Gates, A. Corma, *Angew. Chem. Int. Ed.* 44, 4778 (2005)
82. L. Molina, M.D. Rasmussen, B. Hammer, *J. Chem. Phys.* 120, 7673 (2004)
83. G. Pacchioni, L.Giordano, M. Baistrocchi, *Phys. Rev. Lett.* 94, 226104 (2005)
84. L.M.Molina, B. Hammer, *J. Catal.* 233, 399 (2005)
85. R.K. Grasselli, J.D. Burrington, D.J. Buttrey, P. DeSanto Jr, C.G. Lugmair, A.F. Volpe Jr, T. Weingand, *Top. Catal.* 23, 5 (2003)
86. P. Mars, D.W. van Krevelen, *Chem. Eng. Sci. Suppl.* 3, 41 (1954)
87. Y. Orito, S. Imai, S. Niwa, *J. Chem. Soc. Jpn.* 4, 670 (1980);
H.U. Blaser, H.P. Jalett, J. Wiehl, *J. Mol. Catal.* 68, 215 (1991)
88. M.Ortega Lorenzo, C.J. Baddeley, C. Muryn, R. Ravel, *Nature*, 404, 376 (2000)
89. G. Vayner, K.N. Houk, Y.K. Sun, *J. Am. Chem. Soc.* 126, 199 (2004)
90. G.A. Attard, *J. Phys. Chem. B*, 105, 3158 (2001)
91. J.D. Horvath, A.J. Gellman, *J. Am. Chem. Soc.* 123, 7953 (2001)
92. B.C. Gates, J.R. Katzer, G.C.A. Schuit, *Chemistry of Catalytic Processes*, McGraw-Hill, New York 1979
93. L.A.M. Rodriguez, J.A. Gabant, *J. Polym. Sci. C*, 4, 125 (1963)
94. R.J. Angelici, *Organometallics* 20, 1259 (2001)
95. P.A. Vecchi, A. Ellern, R.J. Angelici, *Organometallics*, 24, 2168 (2005)
96. R.J. Angelici, *Acc. Chem. Res.* 21, 387 (1988)
97. R.J. Angelici, *Coord. Chem. Rev.* 105, 61 (1990)
98. C. Bianchini, A. Meli, *J. Chem. Soc. Dalt. Trans.* 801 (1996)

99. C. Blonski, A.W. Meyers, M.S. Palmer, S. Harris, W.D. Jones, *Organometallics*, 16, 3819 (1997)
100. D.A. Vicic, A.W. Meyers, W.D. Jones *Organometallics* 16, 2751 (1997)
101. A. Miyashita, A. Yasuda, H. Takaya, R. Toriumi, T. Ito, T. Souchi, R. Noyori, *J. Am. Chem. Soc.* 102, 7932 (1980)
102. A. Seayad, M. Ahmed, H. Klein, R. Jackstell, T. Gross, M. Beller, *Science*, 297, 1676 (2002)
103. C.Landis, J. Halpern, *J. Am. Chem. Soc.* 109, 1746 (1987)
104. W.S. Knowles, *J. Chem. Educ.* 63, 222 (1986)
105. S. Feldgus, C.R. Landis, *J. Am. Chem. Soc.* 122, 12714 (2000)
106. B.J. McIntyre, M.B. Salmeron, G.A. Somorjai, *Catal. Lett.* 14, 263 (1992)
107. P. van Beurden, G.J. Kramer, *Phys. Rev.* B63, 165106 (2001)
108. V. Fiorentine, M. Methfessel, M. Scheffler, *Phys. Rev. Lett.* 71, 1051 (1993)
109. E.A. Guggenheim, *Thermodynamics,* North-Holland, Amsterdan (1959)
110. (a) R. Defray, I. Prigorine, A. Bellemans, D.H. Everett, *Surface Tension and Adsorption,* Longmans, Harlow (1966);
(b) R.A. van Santen, *Theoretical Heterogeneous Catalysis,* World Scientific, Singapore (1991)
111. P. van Beurden, G.J. Kramer, *J. Chem. Phys.* 121, 2317 (2004)
112. M. Eiswirth, P. Möller, K.Wetzl, R. Imbihl, G. Ertl, *J. Chem. Phys.* 90, 510 (1989)
113. A.M. Turing, *Phil. Trans. Roy. Soc.* 237, 37 (1952)
114. R.A. van Santen, The Active Site of Promoted Ethylene Epoxidation Catalysts, *Proc. 9th Internat. Congres on Catalysis,* M.J. Phillips, M. Ternan, (eds.), Vol. 3, 1152 (1988)
115. G.J. Kramer, personal communications
116. R.W. McCabe, T. Pignet, L.D. Schmidt, *J. Catal.* 32, 114 (1974);
M.R. Lyunbovsky, V.V. Barelko, *J. Catal.* 149, 23 (1994)
117. J. Wilson, C. de Groot, *J. Phys. Chem.* 99, 7860 (1995)
118. H. Schulz, *Top. Catal.* 26, 73 (2003)
119. A.F. Carley, P.R. Davies, R.V. James, K.R. Harikumar, G.U. Kulkarni, M.W. Roberts, *Top. Catal.* 11/12, 299 (2000)
120. R.A. van Santen, J.W. Niemantsverdriet, *Chemical Kinetics and Catalysis,* Plenum, New York p. 184 (1995)
121. C.T. Au, M.W. Roberts, *Nature*, 319, 206 (1986)
122. M. Neurock, R.A. van Santen, *J. Am. Chem. Soc.* 116, 8860 (1994)
123. R.A. van Santen, H.P.L.E. Kuipers, *Adv. Catal.* 35, 265 (1987)
124. P.A. Kilty, W.M.H. Sachtler, *Catal. Rev. Sci. Eng.* 10, 1 (1974)
125. E.L. Force, A.T. Bell, *J. Catal.* 38, 440 (1975)
126. S. Linic, M.A. Barteau, *J. Catal.* 214, 200 (2003)
127. M.L. Bocquet, P.Sautet, J. Cerda, C.I. Carlisle, M.J. Webb, D.A. King, *J. Am. Chem. Soc.* 125, 3119 (2003)
128. M.S. Palmer, M. Neurock, M. Olken, *J. Am. Chem. Soc.* 124, 8452 (2000)
129. F. Fréchard, R.A. van Santen, *Surf. Sci.* 407, 200 (1998)

CHAPTER 3
The Reactivity of Transition-Metal Surfaces

3.1 General Introduction

The chemical bond that forms between an adsorbate and a solid surface, and their strength, are critical to the chemical and physical behavior of the adsorbate. They control whether or not the adsorbate will desorb from the surface or diffuse along the surface, and, in addition, determine whether or not the adsorbate will decompose into product fragments or associate with other surface adsorbates to form new products. The strength of these bonds controls the relative kinetics for adsorption, desorption, diffusion and surface reaction and thus controls the reaction rate and selectivity. As was discussed in the previous chapter, the maximum catalytic rate is determined by an optimum in the interaction energy between the adsorbate and catalyst surface. Understanding the chemical bonding parameters that determine the trends in the surface adsorbate interaction energy with varying catalyst surfaces is therefore a prerequisite to any predictive theory of catalysis.

In this chapter, we focus first on the basic concepts of chemical bonding for simple gas-phase species. We demonstrate the power of using molecular orbital diagrams and orbital population analyses in the interpretation of chemical bonding. These concepts are subsequently extended to the analysis of adorbate-surface interactions in Section 3.3. We probe well-established chemical bonding concepts such as hybridization, electron donation, electron back-donation, and Pauli repulsion in order to understand the bonding of different adsorbates and how they change as we change the metal substrate to which they bond. In particular, we try to establish how these changes correspond to changes across the periodic table. Many of these concepts have been described in terms of formal chemisorption theory or tight-binding quantum mechanical methods to provide an understanding as they elegantly capture the salient features that control the chemistry. We demonstrate these concepts using simple probe molecules that typify donation, back-donation, and rehybridization such as NH_3, CO, and ethylene. Since the nature of these interactions is related to the binding of the atomic species, we describe in Section 3.4 the features that control the binding of adatoms and how they change across the periodic table. Many surface–adsorbate interactions also directly relate back to the bonding in organometallic and coordination complexes for which there are well-prescribed rules. We therefore also compare the bonding principles of adsorbates on solid surfaces with ligand-metal interactions in organometallic and coordination complexes in section 3.3.1.

3.2 Quantum Chemistry of the Chemical Bond in Molecules

As an introduction to the more complicated surface chemical bonding, we first present the chemical bonding principles in simple molecular systems. These same concepts are subsequently used to begin to analyze to the adsorbate-surface bonds.

Elementary bonding theory[1] teaches us that the chemical bonds in a molecule are comprised of the direct attractive and repulsive interactions between the atomic orbitals. When the atomic orbitals are located on different atoms, bonding and antibonding molecular orbitals are formed. The total bond energy depends on the way that electrons are distributed over the bonding and antibonding molecular orbitals. Electron occupation of bonding orbitals leads to attractive interactions that strengthen the chemical bond, whereas the occupation of antibonding orbitals weakens this bond. A molecular orbital is

Molecular Heterogeneous Catalysis. Rutger Anthony van Santen and Matthew Neurock

ISBN: 3-527-29662-X

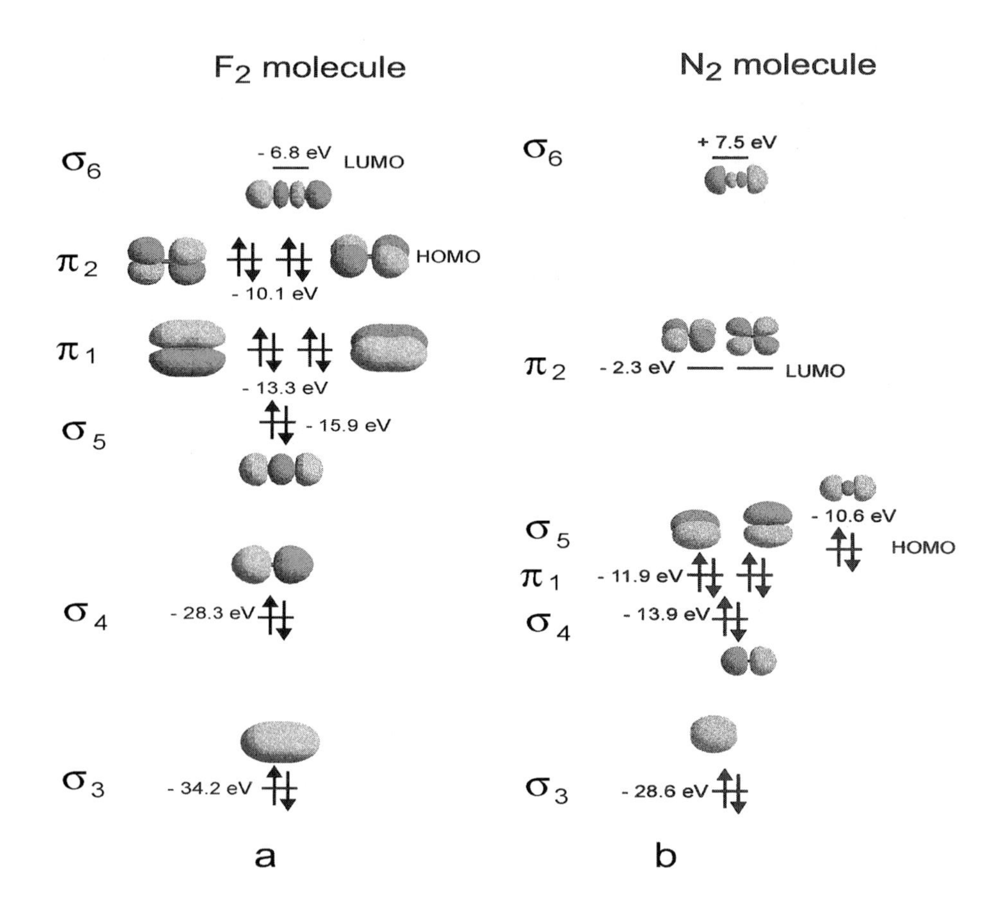

Figure 3.1. (a) Molecular orbital scheme of F_2 as computed by Density Functional Theory. $R_{eq} = 1.43$Å; $E_{bonding} = -2.6$ eV. (b) Molecular orbitals and orbital energies of N_2 as computed by Density Functional Theory. $R_{eq} = 1.11$Å; $E_{bonding} = -10.0$ eV.

considered bonding or antibonding when, with respect to a symmetry axis or the plane of a bond, the orbital is symmetric or antisymmetric, respectively. As a direct consequence, the wave character of a chemical bond is higher in energy as more nodes appear in the wavefunction. This is illustrated for the F_2 molecule in Fig. 3.1a.

The F_2 molecule has a plane of symmetry which is perpendicular to the F–F bond and an axis of rotation along the F–F bond, both of which contain molecular orbitals which are symmetric and asymmetric with respect to these axes. The symmetric orbitals are termed σ-type, while the asymmetric orbitals are denoted π-type. We will consider only electronic structure changes in the valency region. The inner core (1s) electrons are therefore not included in the following discussion. For F_2, the three lowest occupied orbitals in the valence-region are σ-type, whereas the four highest occupied orbitals are of π-type. The lowest unoccupied orbital is again a σ-type. Considering the σ-type orbitals only, one observes an alternation of bonding and antibonding orbitals with increasing orbital energy along with an increase in the number of nodes. For π-type orbitals, the orbitals of lowest energy are bonding in character, the π-orbitals that are higher in energy are antibonding.

Each pair of atomic orbitals that combine results in the formation of both bonding and an antibonding molecular orbitals. For instance, in F_2 the two lowest orbitals are σ_3 and σ_4, which are bonding and antibonding molecular orbitals, respectively, that are formed by the combination of the fluorine 2s atomic orbitals. The σ_5 and σ_6 orbitals can be considered the bonding and antibonding components of the $2p_z$ atomic orbitals. Similarly, the degenerate π_1 molecular orbitals are comprised of bonding combinations of $2p_x$ and $2p_y$ atomic orbitals. The π_2 orbitals serve as the antibonding counterparts to the π_1 orbitals.

When the bonding and the antibonding counterparts are occupied, the result is a repulsive interaction. This repulsive energy is the so-called Pauli repulsion energy. The corresponding mathematical expressions are given here using the tight-binding or Extended Hückel theory formulations. The molecular orbitals ψ_i can be written as linear combinations of atomic orbitals (φ_k):

$$\psi_i = \sum_{k=1}^{n} c_k^i \, \varphi_k \tag{3.1}$$

where c_k^i are the coefficients which relate the atomic orbitals to the molecular orbitals and are found by solving the secular equations

$$\sum_{l=1}^{n} (h_{kl} - \epsilon_i \, S_{kl}) = 0 \qquad (k = 1...N;\, S_{kl} = 1,\, k = l; |S_{kl}| < 1,\, k \neq l) \tag{3.2}$$

Within the orbital interpretation view presented for F_2, $n = 2$ for each atomic orbital pair of 2s, $2p_z$ and $2p_y$, $2p_x$, respectively. The diagonal matrix elements h_{kl} $(k = l)$ represent the energy of an electron in an atomic orbital α the non-diagonal matrix elements h_{kl} $(k \neq l)$ are the overlap energy integrals β

$$h_{kk} = \alpha \tag{3.3a}$$

$$h_{kl} = \beta \qquad (k \neq l) \tag{3.3b}$$

S_{kl} is the overlap integral between the atomic orbitals k and l:

$$S_{kl} = \int \mathrm{d}^3 \vec{r} \, \varphi_k(\vec{r}) \varphi_l(\vec{r}) \tag{3.4}$$

The binding energy E_b within the tight binding approach equals

$$E_b = \sum_{i=1}^{N} \nu_i \, \epsilon_i - \sum_{k=1}^{n} \nu_k^0 \alpha_k \tag{3.5}$$

where ν_i and ν_k^0 refer to the occupation numbers of the molecular and atomic orbitals, respectively. The molecular orbital energies are the eigenvalue solutions of the secular equations (3.2) and have the general form

$$\in_i = \sum_{k=1}^{N} |c_k^i|^2 \, \alpha_k \; + \; 2 \sum_{k<l} c_k^i c_l^i \beta_{kl} \tag{3.6}$$

The first term here is the energy contribution due to the residence of an electron in a particular atomic orbital and the second term represents the interference of the atomic wavefunctions. Waves can annihilate or have positive or negative interference. For the 2 x 2 secular matrix equation of F_2 the bonding and antibonding energies are given by the expression

$$\in_i^{\pm} = \frac{\alpha_i \pm \beta_i}{1 \pm S_i} \tag{3.7}$$

where β, which is the overlap energy, is attractive and has a negative value. The positive sign (bonding) lowers the energy and the negative sign (antibonding) increases the energy contribution of the denominator. The corresponding molecular orbital expressions are

$$\psi_i^{\pm} \; = \; \frac{1}{\sqrt{2 \pm 2S_i}} \left[\varphi_1(i) \pm \varphi_2(i) \right] \tag{3.8}$$

The sign in front of the second atomic orbital is negative for the antibonding orbitals. Returning to the orbitals in F_2, we recognize σ_3 and σ_4 as bonding-antibonding pairs, σ_5 and σ_6 as bonding and antibonding pairs and the same for π_3 and π_4. When only bonding orbitals are occupied, the bond energy becomes (as is for instance the case in H_2)

$$\Delta E_b(\nu^+ = 2) \; = \; 2 \in^+ - 2\alpha \tag{3.9a}$$

$$= \; 2\Delta(1 - S) \tag{3.9b}$$

with

$$\Delta \; = \; \frac{\beta - \alpha S}{1 - S^2} \qquad \Delta < 0 \tag{3.9c}$$

where 2Δ is the energy difference $\in^+ - \in^-$ between the bonding and corresponding antibonding orbitals. The parameter that mainly controls Δ is β.

We observe that the larger the value of Δ, the larger is the bond energy. This is an important result since Δ, the difference in bonding and antibonding orbital energies, in principle can be measured spectroscopically and, hence, spectroscopic measurements can provide indirect information on bond energies.

The expression for the interaction energy is very different when two atomic orbitals are occupied, as is the case for the imaginary He_2 molecule. Then:

$$\Delta E_b \; = \; (\nu^+ = 2, \nu^- = 2) \; = \; 2 \in^+ + 2 \in^- - 4\alpha \; = \; -4S \cdot \Delta \tag{3.10}$$

Now, within the tight-binding model, the interaction energy is repulsive. It is again approximately proportional to the square of the overlap energy as well as the atomic orbital overlap. As a general result, the expression for the total binding energy of a homopolar chemical bond is

$$E_b(\text{total}) \; = \; \sum_{i_p} - (n_{i_p}^+ + n_{i_p}^-) S_i \cdot \Delta_i + \sum_j n_j^+ \cdot \Delta_i \cdot (1 - S_j) \tag{3.11}$$

where i_p sums the contribution to the energy of the pairs of occupied and the corresponding antibonding molecular orbitals and j sums the contribution to the bond energy of the occupied bonding orbitals. The contribution due to occupation of bonding as well as corresponding antibonding orbital pairs is repulsive. The occupation of the bonding orbitals is only attractive.

As an illustration of this, the difference between the computed bond energies of F_2, which is –260 kJ/mol, and N_2, which is –1000 kJ/mol, is analyzed. In the F_2 molecule both the bonding and antibonding $\sigma(2s)$ and $\pi(2p_x, 2p_y)$ orbital are occupied, thereby resulting in repulsive interactions. The only pair of molecular orbitals where electrons exclusively occupy only the bonding orbital is the σ_5 orbital constructed from the $2p_z$ atomic orbitals. This results in an overall attractive contribution to the chemical bond. The attractive orbital overlap contribution which is equal to $2\Delta(2p_z)$=-9.1 eV is counteracted by the Pauli-repulsive interactions due the $\sigma(2s)$ and $\pi(2p_x,2p_y)$ orbital pairs.

The N_2 molecule, on the other hand, has 4 electrons less. The N_2 molecular orbitals, illustrated in Fig. 3.1b, are very similar to those of F_2, but the relative energies are shifted. The σ_5 energy is now higher than that of the π_1 orbitals. The primary difference between F_2 and N_2 is that for N_2, the antibonding π_2 orbitals are not occupied and, hence, the π systems changes from being repulsive in F_2 to being attractive in N_2. This explains the large differences in bond energies between N_2 and F_2.

As a prelude to our discussions on chemisorption, we will now discuss orbital changes that occur when an additional bond to the dimer is formed. We will use as an example the hydrogen bond in HCN. The molecular orbitals for CN^- and HCN along with their energies are shown in Figs. 3.2a and b, respectively. The CN^- ion is iso-electronic with N_2 and hence the two antibonding π_2 type orbitals are unoccupied. The strong CN bond corresponds to three bonding orbitals being occupied. The bond order, therefore, is three. Figure 3.3b shows the electronic structure of HCN which contains the same number of electrons, but a bond is now formed between the C atom and the proton. The result is an overall downshift in energy of all the molecular orbitals and the generation of altered σ-type orbitals. The hydrogen atomic orbital is symmetric and interacts with the C 2s and C $2p_z$ atomic orbitals.

Before we continue further with the discussion of HCN, it is important to note the upward shift of the σ_5 orbital in CN^- (Fig. 3.2a) and N_2 (Fig. 3.1b) with respect to the π orbital system, compared with its relatively low position in F_2 (Fig. 3.1a). The background to this is the much smaller difference in atomic orbital energies for the 2s and 2p states in N and C than in F. In fluorine, the $[3\sigma, 5\sigma]$ subset of orbitals are in essence constructed of only 2s or $2p_z$ atomic orbitals. This, however, is not true for N_2 or CN^-, where the 3σ and 5σ orbitals have the same symmetry and therefore have the same general structure:

$$\psi_\sigma^+(1) = \lambda\Big[\varphi_{2s}(1) + \varphi_{2s}(2)\Big] + \mu\Big[\varphi_{2p_z}(1) + \varphi_{2p_z}(2)\Big] \tag{3.12}$$

$$\psi_\sigma^+(2) = -\mu'\Big[\varphi_{2s}(1) + \varphi_{2s}(2)\Big] + \lambda'\Big[\varphi_{2p_z}(1) + \varphi_{2p_z}(2)\Big] \tag{3.13}$$

Analogous expressions are valid for the antibonding 2s and $2p_z$ combinations $\psi_\sigma^-(1)$ and $\psi_\sigma^-(2)$. Orbital σ_5 is to be identified with $\psi_\sigma^+(2)$, which explains its upwards shift in N_2 compared to F_2. In F_2, the mixing of orbitals is virtually absent. The decrease in density between the N atoms indicates an increasing localization of electrons in positions to the left or right of N_2. The mixing of the 2s and $2p_z$ orbitals is seen to lead to hybridization

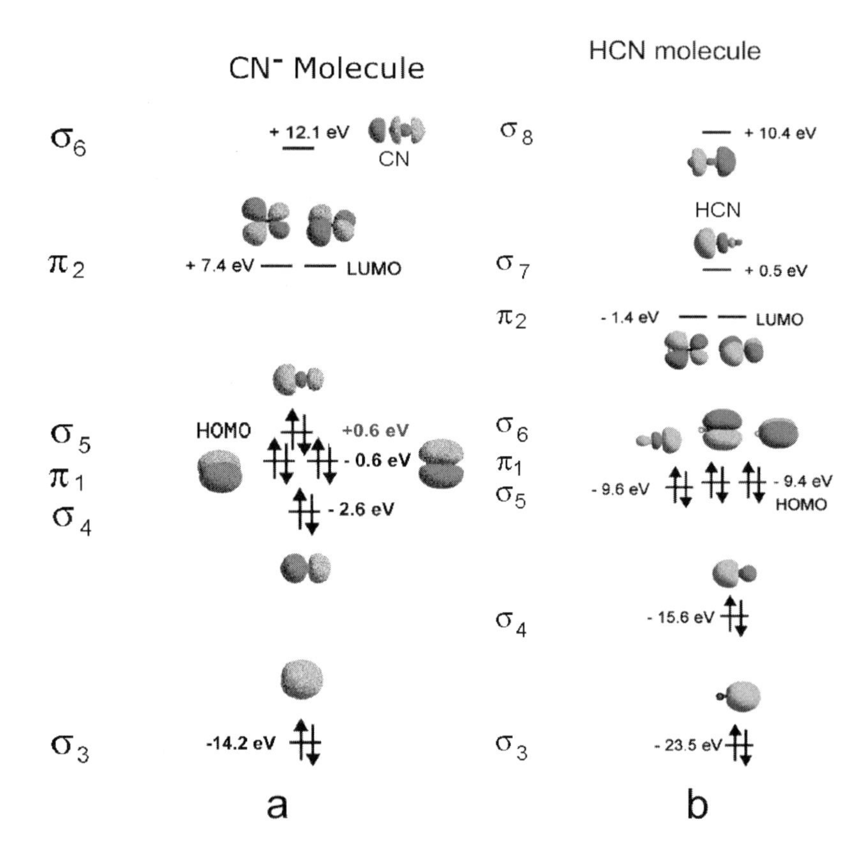

Figure 3.2a. (a) Molecular orbital scheme and respective energies of the CN^- ion. (b) Molecular orbital scheme and respective molecular orbital energies of the HCN molecule.

(for details on hybridization, see Addendum 3.11). The σ_5 orbital can be considered as on the way to form lone pair orbitals on N_2 directed away from the atoms. The antibonding orbitals σ_4 and σ_6 are identified with the $\psi_\sigma^-(1)$ and $\psi_\sigma^-(2)$ orbitals. Again one notes the decrease in density in the σ_6 orbital between the two N atoms. The σ_6 orbital in N_2 can be considered an antibonding combination of N_2 lone pair orbitals. The σ_4 orbital is predominantly an antibonding N≡N orbital. In CN^- the higher occupied molecular orbital now has a slightly increased density in the C lone-pair orbital. The corresponding σ_3 molecular orbital has increased density on nitrogen.

We now return to HCN. By inspecting the σ-type orbitals we note significant changes. There are now 5 instead of the 4 σ-type orbitals in CN^-. There is a small contribution of the hydrogen 1s orbital to σ_3, the stronger inner σ bonding orbital in CN. The bonding interaction between H and C is clear in the σ_4 orbital, which becomes antibonding between C and N. The antibonding C–H σ orbital can be recognized as the σ_7 orbital, that is unoccupied. The σ_5 orbital in HCN becomes the empty lone pair orbital on nitrogen, whereas in CN^- it is primarily comprised of the lone pair orbital on carbon. This extensive analysis of the orbital nature changes in HCN on the addition of H^+ to CN^-, which illustrates the significant rehybridization that occurs in a molecule when strong new bonds are formed. This is important to realize, since similar rehybridization is observed when molecules interact with surfaces, as will be seen for CO in subsequent sections.

3.3 Chemical Bonding to Transition-Metal Surfaces

In this section we introduce principles of the surface chemical bond. First principle ab initio computational results are analyzed using basic quantum-chemical concepts. In this section, we analyze the adsorption of molecules. In the following section, we analyze the adsorption of atoms. The adsorption of ammonia and CO is discussed first since they are known to interact predomenantly through donation and back-donation interactions, respectively. This will subsequently lead into the analysis of the stronger bonds that form between adatoms and a surface. We note the similarities in chemical bonding of these adsorbates to surfaces, clusters and organometallic complexes, and in addition describe some of the differences.

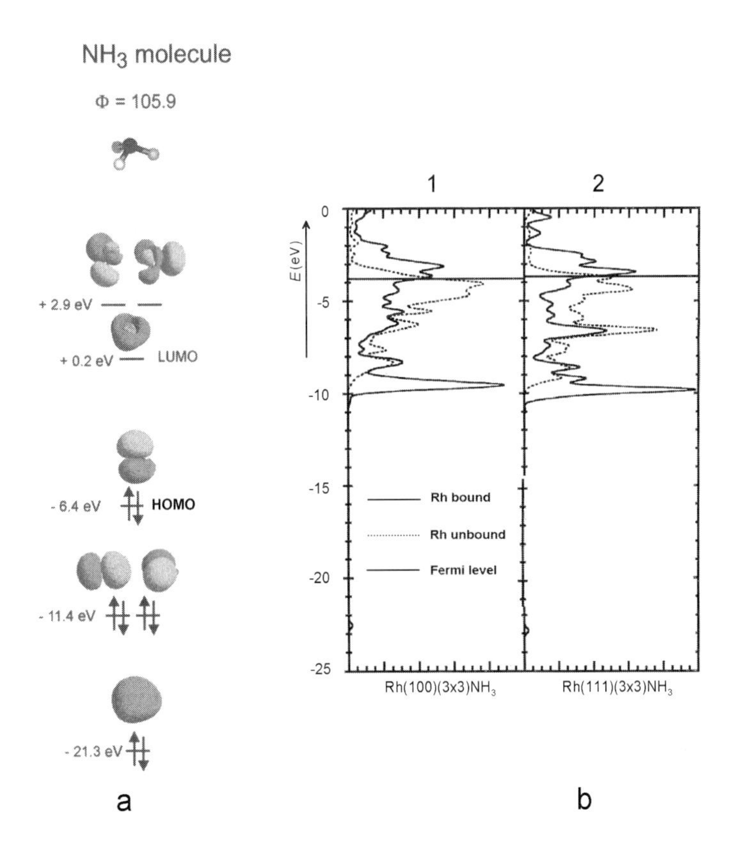

Figure 3.3. (a) Molecular orbital and energy scheme of NH_3. $R_{NH} = 1.03Å$; $\Phi = 105.9°$. (b) The electronic local density of states $\rho(E)$ of an adsorbed free Rh atom and a Rh atom bound to NH_3 on Rh(100) (1) and Rh(111) (2) respectively[2].

The electronic structure of NH_3 is shown in Fig. 3.3a. The angle between the NH bonds is 105.9°. The N–H chemical bonds are very close in character to the 2p N atomic orbitals and the H atomic orbitals of which they are comprised. The N $2p_z$, 2s and symmetric combination of hydrogen atomic orbitals are σ symmetric with respect to the NH_3 z-axis. The lower σ-type orbital is predominantly comprised of the 2s N orbital stabilized by a bonding interaction with the symmetric combination of three H s-atomic orbitals. The highest occupied molecular orbital (HOMO) for NH_3 is predominantly non-bonding $2p_z$ in character located on the N atom. The lowest unoccupied molecular orbital (LUMO) of NH_3 is the antibonding analogue of the lower σ N–H bonding orbital.

The primary interaction between NH_3 and a metal surface is predominantly a donative one which occurs via the transfer of electrons from the doubly occupied nonbonding $2p_z$ lone-pair type orbital on N. The corresponding states on the metal depend upon the metal.

In a transition metal, the valence electrons available for bonding are of nd and $(n + 1)$s and $(n + 1)$p character. The metal-electron energies are continuously distributed in valence-electron bands, between upper and lower energy bounds as is shown schematically in Fig. 3.5b. The d-electrons form a narrow band of states with a bandwidth of a few electronvolts. The electronic structure of these states can be rather well described within the tight-binding formalism introduced in the previous section. The s and p electrons, on the other hand, behave more as free electrons. The interaction between adsorbate electrons and the sp electrons of the metal is usually bonding and does not vary much between different metals. The major variation in binding stems from the interaction between electrons of the adsorbate with the valence d-electrons of the metal. This is quite sensitive to d-valance electron occupation. The interaction between adsorbate orbitals and transition-metal states leads to the formation of bonding and antibonding surface-adsorbate orbitals. The bond energy depends on the distribution of electrons over these orbitals, and the changes that occur in adsorbate and transition metal electronic structure.

Figure 3.3b illustrates the electron density distribution for NH_3 adsorbed atop Rh(100) and Rh(111) surfaces. For symmetry reasons when NH_3 adsorbs atop a surface atom, its NH_3 $2p_z$-type HOMO orbital interacts only with the Rh $4d_{z^2}$-atomic orbital of the metal d-atomic orbitals.

The electron distribution (Partial Density of States, PDOS) within the $4d_{z^2}$ state on one of the surface atoms of the pristine Rh(100) surface is shown in Fig. 3.3b(1). The electron distribution within the $4d_{z^2}$ state of the Rh(111) surface is shown in Fig. 3.3b(2). The coordination number of the surface atoms in the Rh(100) surface is 8 whereas that of the Rh(111) surface is 9. The width of the d-valence electron band is smaller for the Rh atom on the (100) surface, which has the smaller number of metal neighbors. In addition, the average energy of the valence band has been shifted slightly upwards. For an atom with an s-valence-electron distribution, it can be shown that the valence bandwidth is approximately proportional to $\sqrt{N_n}$, N_n being the number of nearest-neighbor atoms. The delocalization of the electrons increases with increase in the number of nearest-neighbor atoms[3].

The bandwidth is also a measure of the average difference in energy between bonding and antibonding orbitals. Hence the bonding contribution to the stability of surface atoms also increases with $\sqrt{N_n}$. This suggests that surface atoms with fewer neighbors are more reactive. This was briefly discussed this in Chapter 2. We show in this chapter that this is generally the case.

Valence electron band narrowing increases the average energy of the electrons, because it increases the repulsive electron-electron energy. There appears to be a nearly linear relation between this increase in average d-valence electron energy and number of nearest-neighbor electrons. Therefore, a nearly linear relation in average local d-valence electron energy and the adsorption energy is often found[4].

In the adsorption of ammonia on the surface, the NH_3 $2p_z$ lone-pair molecular orbital interacts with a transition–metal surface atom to form bonding and antibonding orbital fragments. The resulting $4d_{z^2}$ electron distributions are also shown in Fig. 3.3b.

The sharp peak at the bottom of Fig. 3.3b(1) at 9 eV represents the bonding orbital fragment between the Rh $4d_{z^2}$ state and the NH_3 σ-type orbital. There appears an upward shift of the average d-valence electron distribution in the antibonding regime. As a consequence of the upward shift above the Fermi level (the highest occupied metal orbital), the total electron occupation of the d–valence orbitals decreases. The antibonding orbital fragments between the Rh d_{z^2} and NH_3 σ-type orbital become less occupied and overall bonding becomes less repulsive or slightly bonding. A stronger interaction between Rh d_{z^2} and its metal neighbor atoms results in a weaker interaction between the Rh d_{z^2} state with the ammonia σ orbital. The average position of the d_{z^2} orbital energy on Rh(111) is lower than that on the Rh(100) surface. The average upwards shift is less, so that less empty d_{z^2} orbital density now appears above the Fermi level. Since the Fermi level refers to the highest occupied molecular orbital energy in the bulk, it is the same for the (111) or (100) surface. Therefore, the repulsive interactions of NH_3 with the Rh d_{z^2} orbital on the Rh(111) surface are larger.

Ammonia adsorption is stronger to the (100) surface (–91 kJ/mol) than (111) surface (–82 kJ/mol). Ammonia prefers atop adsorption over two- or three-fold adsorption on both surfaces. Adsorption at these higher-fold coordination sites is less favored on the (111) surface [E_{ads} (two-fold) (111) surface = –13 kJ/mol; (100) surface = –36 kJ/mol]. This is due to the much larger repulsive interaction of the doubly occupied ammonia lone pair orbital when ammonia is adsorbed to more surface atoms on the Rh(111) surface.

Calculations for NH_3 chemisorbed to Cu clusters which simulate the Cu (100) surface illustrate the change in bonding character of the adsorbate bond with the metal surface nicely. Figure 3.4 shows the orbital interaction of the NH_3, lone pair orbital with Cu(4s), Cu(4p) and Cu($3d_z^2$) orbitals. Both the OPDOS (the Overlap Population Densities of States) and Local Density of States (LDOS) are shown.

The OPDOS $\pi_{ij}(E_k)$ is defined as

$$\pi_{ij}(E_k) = c_i^k \, c_j^k \, S_{ij} \tag{3.14}$$

where $\pi_{ij}(E_k)$ is proportional to the interference term in the interaction part of the orbital energy. Hence, when $\pi_{ij}(E_k)$ is positive the orbital interaction is attractive when $\pi_{ij}(E_k)$ is negative this interaction is repulsive. The LDOS are the electron densities in the interacting fragment orbitals which are proportional to $(c_i^k)^2$.

The OPDOS curves, Figs. 3.4a and b, show that the interaction of the NH_3 σ_{2p_z} orbital with the metal d-orbitals changes from attractive (positive) to repulsive (negative) in comparing the values for π_{ij} at increasing orbital energies. The Bond Order Overlap Population (BOOP) between fragment orbitals i and j is defined as

$$P_{ij} = 2\sum_k^{occ} \pi_{ij}\,(E_k) = 2\sum_k^{occ} c_i^k \, c_j^k \, S_{ij} \tag{3.15}$$

The BOOP provides a measure for the strenght of bonding. The BOOP P_{ij} for an adsorbate with the Cu d-atomic orbitals at the atop adsorption site is –0.005, while that for

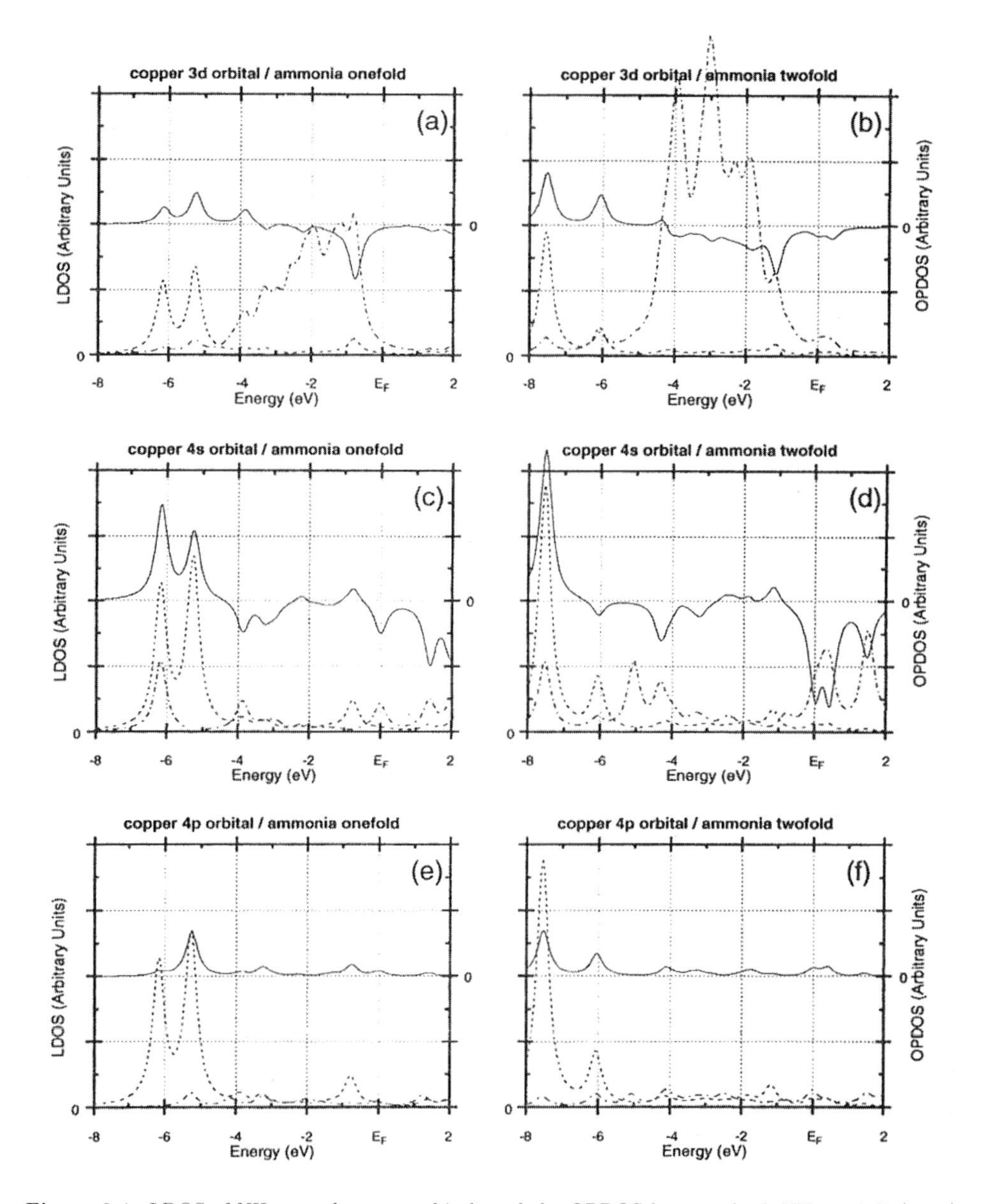

Figure 3.4. LDOS of NH_3 σ and copper orbitals and the OPDOS between both NH_3 and Cu(9,4,5), and Cu(8,6,2), respectively, for one-fold and two-fold NH_3 adsorption. Central copper orbitals: (a, b) 3d; (c, d) 4s; (e, f) 4p; NH_3 position; (a, c, e) one-fold, (b, d, f) two-fold. Each graph shows - - - - - - LDOS of NH_3 σ orbital, -.-.-.-. LDOS of central copper orbitals, and —— OPDOS[5].

the adsorbate at the two–fold site is –0.157. This larger repulsive interaction at two-fold position is the result of the larger repulsive interaction when two Cu atoms with doubly occupied d-atomic surface orbitals interact with the doubly occupied NH_3 σ-orbital. The computed NH_3 interaction energies at the two-fold sites on the (111) and (100) surfaces are E_{ads} (two-fold)= –13 and –36 kJ/mol, respectively.

For s-type orbitals the Pauli repulsion energy increases linear with the number of neighbors:

$$E_{\text{Pauli}} \approx ZS^2 \tag{3.16}$$

where Z is the coordination number of the adsorbate orbital with surface atoms. The attractive contribution to the bond energy, as mentioned earlier, is proportional to $\sqrt{Z}$.

The attractive interaction between NH_3 and the Cu surface arises from the interaction with the free electron-type Cu(4s) and Cu(4p) orbitals. These metal orbitals contain one electron per metal atom.

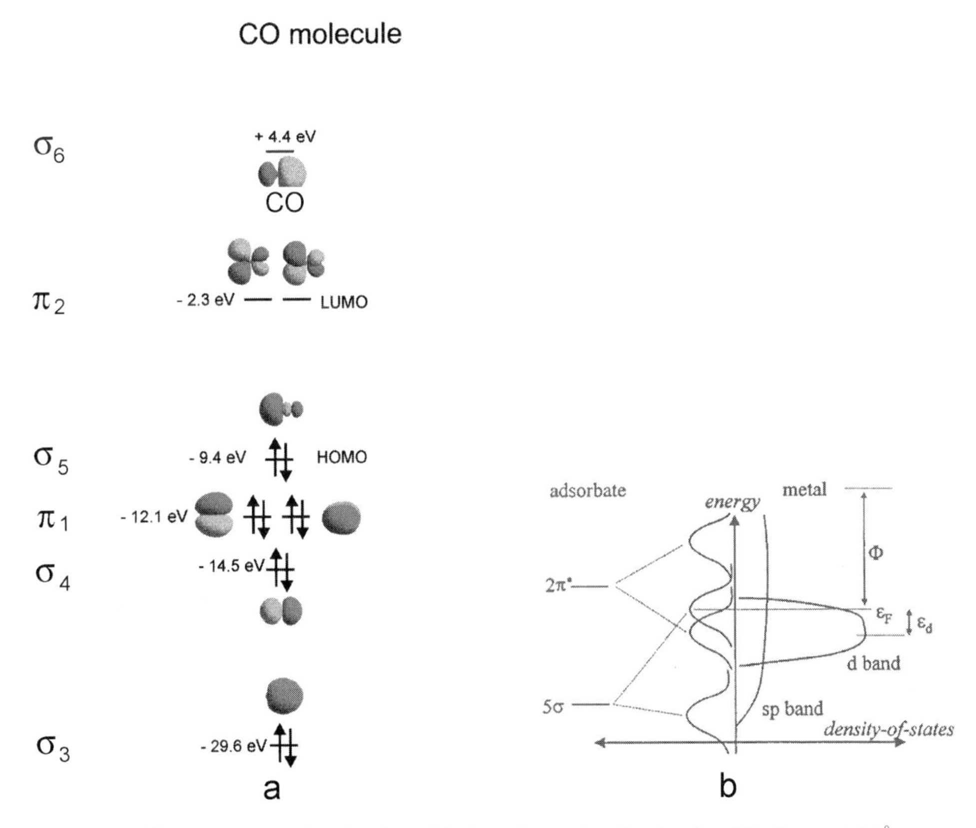

Figure 3.5. (a) DFT-computed molecular orbitals and energies of molecular CO. $R_{\text{CO}} = 1.14$Å; $E_{\text{bonding}} = -11.9$ eV. (b) Left-handside: orbital energy-level diagram for metal chemisorption system of adsorbed CO; σ_5 to be identified with 5σ and σ_4 to be identified with 4σ; π_2 to be identified with $2\pi^*$ and π_1 to be identified with 1π. Right-hand side: schematic representation of the electronic structure of the transition metal.

The interaction of CO and a metal is very different to that for ammonia, since the CO has lower antibonding energy π^* states that are unoccupied and can accept electrons from the metal, i.e. back-donation. The electronic structure of CO is shown in Fig. 3.5a. The hybridization of 2s and $2p_z$ atomic orbitals in CO is clearly noted in its highest occupied 5σ lone-pair orbital, that is mainly localized on carbon. The 5σ-orbital is antibonding with respect to the C–O bond, and predominantly comprised of C 2s, $2p_z$ and O $2p_z$s character (see also Fig. 4.19). In between the occupied 4σ- and 5σ-orbitals there are the two occupied 1π-orbitals. These four occupied orbitals will strongly interact with the transition metal surface and, similarly as in the case of the lone-pair orbital of NH_3, are expected to have a repulsive interaction with the occupied transition metal d–orbitals. The lowest unoccupied $2\pi^*$-orbitals have a much lower energy than the antibonding unoccupied orbitals in NH_3. Therefore, in CO the interaction between the CO $2\pi^*$-orbitals and surface states becomes important. The frontier orbital scheme for the interaction of CO with the transition–metal surface is illustrated in Fig. 3.5b. In this elementary scheme only the interactions of the HOMO and LUMO of CO with transition-metal electrons are considered. The interaction of the frontier 5σ and $2\pi^*$ orbitals creates bonding as well as antibonding orbital fragments. The antibonding orbital fragments of the 5σ orbital are partially occupied, which leads to a repulsive interaction. Only bonding orbital fragments of the $2\pi^*$ and surface d-orbitals are occupied, thus resulting in an attractive contribution.

Compared this elementary bonding scheme, OPDOS calculations of CO adsorbed atop or three-fold to an Ru_{19} cluster simulating the Ru(0001) surface modify this bonding picture in an essential way (see Fig. 3.6). The binding energies for CO adsorbed atop and three-fold are –192 and –116 kJ/mol, respectively. The orbital pictures as well as the overlap population density of states (OPDOS) between the C-atomic orbital CO and the nearest-neighbor metal d-states are shown for CO at atop and three-fold hollow sites of the Ru in Fig. 3.6. The corresponding LDOS are shown in Fig. 3.7. The comparison of Figs. 3.6 and 3.7 allows for an assignment of the LDOS peaks.

We will first consider the interaction of the metal states with the CO σ-orbitals. We recognize in Fig. 3.6a the 4σ orbital and the 5σ orbitals of CO at –14 and –11 eV, respectively. The features above –10.2 eV correspond to the antibonding 4σ–d interactions. In addition to the elementary Blyholder picture[6] that considers only the 5σ CO lone pair interactions, we now note that the interaction with the 4σ orbital is also important and that this interaction is also repulsive. A comparison between the orbitals for adsorbed CO and those for the free CO indicate that there is a rehybridization between 4σ and 5σ orbitals, with a shift of electron density towards the oxygen atom. Hybridization effects have been confirmed recently by XPS measurements[7]. Figure 3.8a illustrates that mixing of the two σ-CO orbitals with a metal-surface atom leads to bonding, nonbonding and antibonding orbital interactions. The nonbonding interaction is distinguished by strong localization of density on the oxygen atom.

If one compares the σ-type interaction at the atop and three-fold adsorption sites in Fig. 3.6, one notes a much greater interaction for CO at the three-fold site, especially for that of the 4σ CO orbital with the transition-metal surface, thus leading to much stronger repulsive interactions than in the atop adsorption mode.

Let us now analyze the π-type interactions. In Fig. 3.6a we note at –10.6 eV a strong bonding interaction between CO-1π and surface d_{xz} and π orbitals. A nonbonding π-type orbital is present at –9.2 eV which is localized on oxygen. The bonding interactions with the $2\pi^*$ orbital of CO are apparent around –6 eV (–1.5 eV below the Fermi level) and

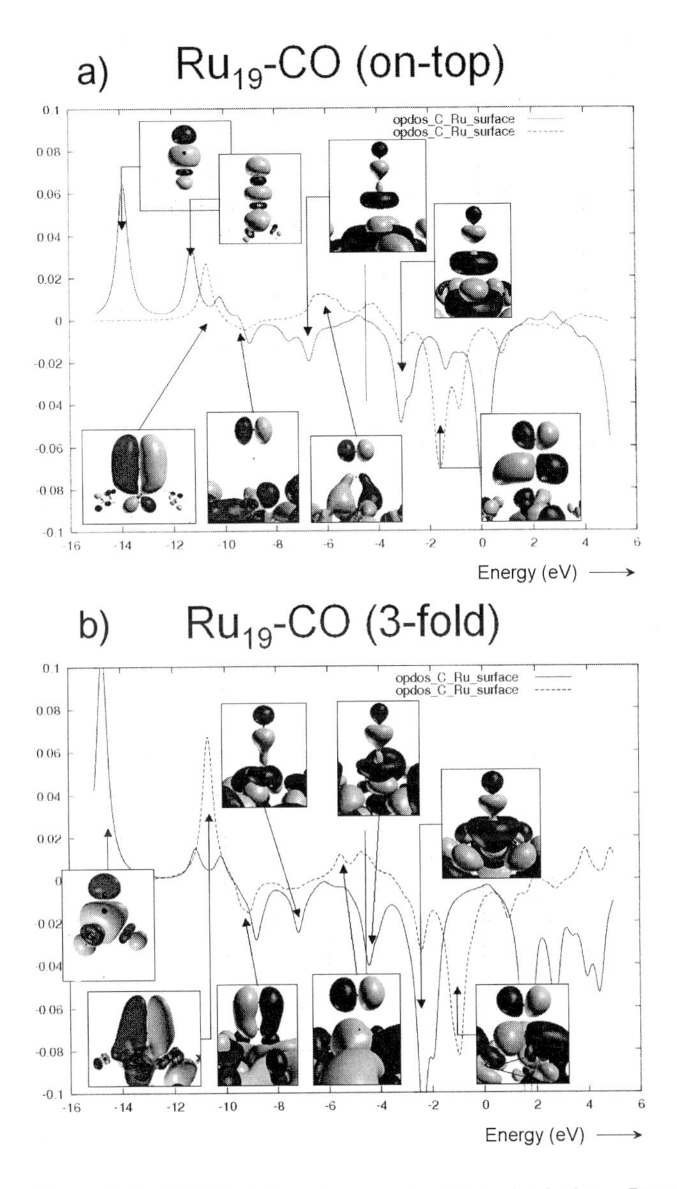

Figure 3.6. OPDOS of the carbon atomic orbital adsorbed on a Ru surface with Ru d-valence electrons. Also the orbitals with maximum density are shown. (a) CO adsorbed atop; (b) CO adsorbed three-fold. —, σ-Symmetric orbital interactions; ..., π-symmetric orbital interactions. The Fermi level is at –4.8 eV.

antibonding interactions around –2 eV (2.5 eV above the Fermi level). The shift of the electron density towards the oxygen atom occurs in the rehybridization regime of 1π and $2\pi^*$ orbitals (–10 to –6 eV). This agrees very well with Scheme 3.8b. The interaction of the molecules 1π, metal d, molecules $2\pi^*$ states leads to a bonding, nonbonding and antibonding fragment orbitals. The nonbonding orbital mainly has electron density located on the oxygen atom. For CO bound at the three-fold sites, the corresponding positions

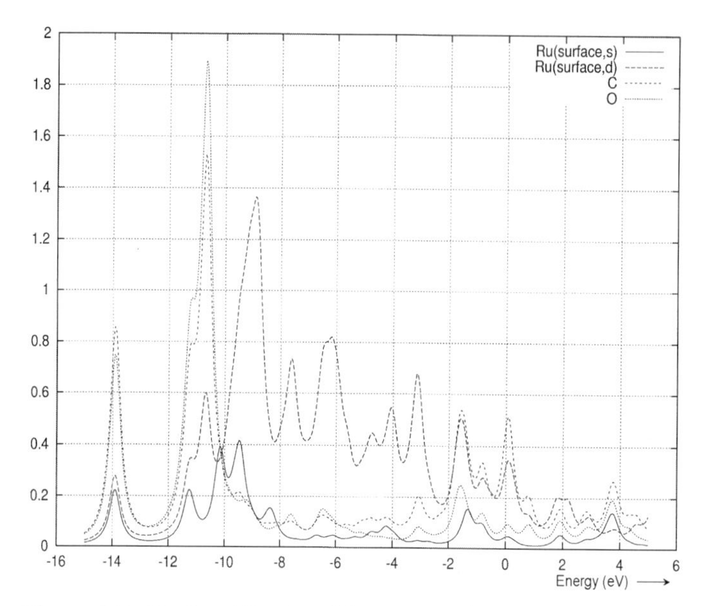

Figure 3.7. The Local Density of States on surface and adsorbate atoms corresponding to Fig. 3.6a.

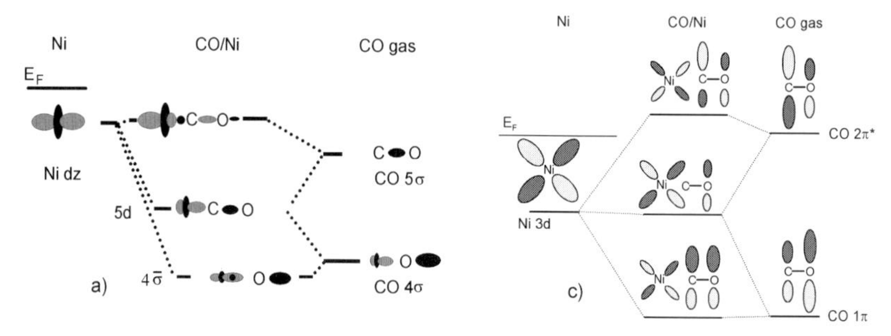

Figure 3.8a. (a) Schematic diagram of the σ-interaction. The model is based on adsorption of CO on a single Ni atom[7]. (b) Schematic orbital diagram of the π-orbital structure. An allylic configuration is formed in the model of a single CO molecule intracting with a Ni atom[7].

for the π and $2\pi^*$ orbitals are -11 eV and -9 eV and -6 eV. For CO adsorbed at the three-fold site, the strong rehybridization of the $2\pi^*$ and 1π states create a weak bonding π-type contribution near the Fermi level. The repulsive interactions appear to dominate. A comparison of the total BOOP's of atop and three-fold coordinated CO with the metal d orbitals gives as a result

One-fold: BOOP(σ) $= -0.07$; BOOP(π) $= 0.02$
Three-fold: BOOP(σ) $= -0.11$; BOOP(π) $= -0.03$

The CO molecule also interacts with the transition-metal s, p valence electron band, which is partially occupied with one electron per atom. The main contribution to the attractive interaction with CO is due to the interaction with the s, p valence electrons. Figure 3.7 shows the LDOS of the Ru valence s-electrons, with peaks that correspond to the interaction with CO orbitals. In the three-fold coordination site the Pauli-repulsive interactions dominate so strongly that atop adsorption is actually preferred.

The above analysis illustrates the importance of including the interaction of the natal states with the 4σ and 1π in addition to the 5σ and $2\pi^*$ molecular orbitals of CO to describe the surface chemical bond.

Electrochemical experiments[8] show that changes in local electric field can shift the coordination of CO. When the potential is biased such that the difference in energy between empty $2\pi^*$-CO orbitals and Fermi level decreases, electron back-donation is increased and CO can be shifted from atop to three-fold coordination. The larger back-donation into the CO-$2\pi^*$ antibonding orbital in three-fold coordination lowers the bond strength of CO. This illustrates the general result that back-donation into molecular orbitals that are asymmetric with respect to the surface normal favors bonding in high coordination sites. As will be discussed in more detail later, this is due to the increased interaction with antisymmetric group orbitals of s- or d-atomic orbitals located on different atoms in higher-fold coordination sites. This interaction, however, is very small or absent at atop sites.

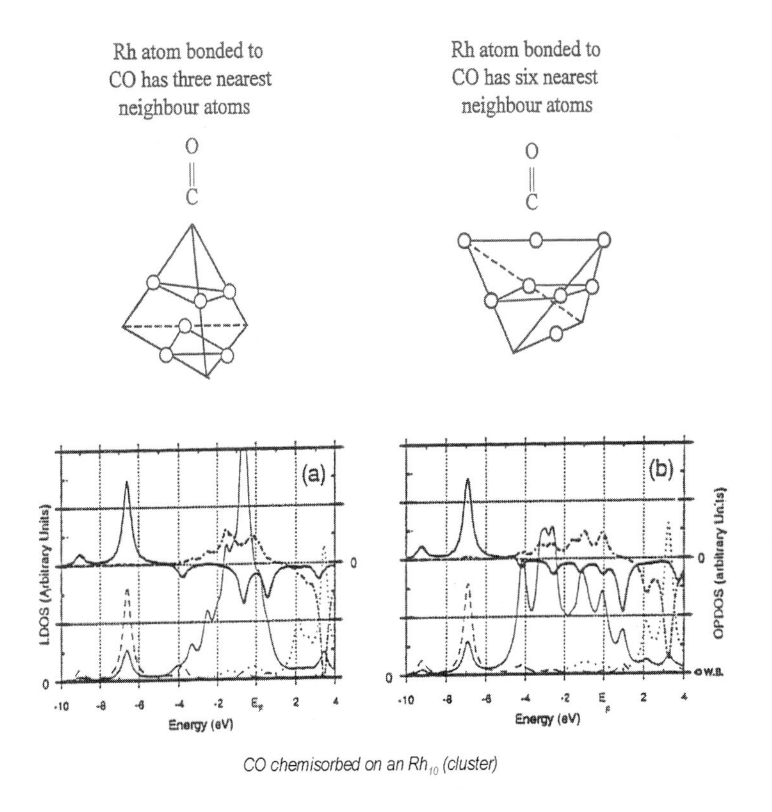

Figure 3.9. The local densities of states (LDOS) of $Rh(4d)_z$ and $CO(2\pi^*)$ and the Overlap Density of States (OPDOS) (π_{ij}) between the rhodium and CO orbitals for atop CO adsorption on Rh_{10}. Adsorption on Rh (with three nearest neighbors) is shown (a) and adsorption on Rh(with six nearest neighbors) is shown (b)[9]. Each graph shows: —— LDOS of Rh, – – – LDOS of CO(5σ), - - - - - - LDOS of CO($2\pi^*$), —— BOOPD between Rh–CO(5σ), — - — - BOOPD between Rh–CO($2\pi^*$).

The orbital interactions of CO adsorbed atop two different Rh atoms in an Rh_{10} cluster are shown in Fig. 3.9 a and b. In these calculations, only the C and O positions of CO have been optimized. The apex atom of Rh in this cluster has only three neighbors, whereas atoms at the edge of this cluster have six neighbors. CO is expected to bind more weakly to an atom in the latter. Indeed the bond energy is computed to be –86 kJ/mol at the apex atom and only –65 kJ/mol at the edge atom. Interestingly the bond energy for CO adsorbed atop the Rh(111) surface atom is nearly twice that of these computed cluster values. The lower adsorption energy calculated on the cluster relates to the relatively high ionization potential of the Rh cluster (6 eV). In Fig. 3.9b the LDOS and OPDOS interactions are shown for the initially doubly occupied 5σ and empty $2\pi^*$ orbitals of CO and the Rh 4d orbitals. We again note the bonding nature of the 5σ lone-pair orbital interaction at low energy (the double peak at low energy in OPDOS is the result of mixing between the d_{z^2} transition-metal orbital and the CO(4σ) and CO(5σ) states). There is a strong antibonding interaction which appears in the energy regime of the metal d-electrons. On the other hand, the interaction with the CO-$2\pi^*$ orbital and atomic d_{xz} and d_{yz} orbitals is attractive.

We find, as a general result, that the attractive interaction between CO and metal d-electrons is due to the bonding interaction with the $2\pi^*$ orbitals. Repulsive interactions are due to the interactions between doubly occupied metal orbitals and those in the molecule. The molecular orbitals involved are the 4σ, 5σ and 1π orbitals.

Again we note that the average energy for the local density of states on the d-electrons on the atom with lowest metal-atom coordination number N has increased, compared with the average position of the d-electron LDOS on the atom with the larger number of neighbors. This is expressed in the following relation:

$$\overline{\epsilon}_d = \overline{\epsilon}_d^{\,0}(N_m = \max) + \gamma\left(N_m^{1/2} - N^{1/2}\right)(\gamma > 0, N_m - N > 0) \qquad (3.17)$$

where N_m and N are the number of nearest neighbours of a metal atom, respectively.

The electronic interactions between the molecular orbitals of adsorbate and d states of the metal that result upon adsorption are shown schematically in Figs. 3.10 and 3.11 within the context of tight-binding theory. We first analyze the NH_3 $n\sigma$ lone-pair interaction and the CO 5σ-type interaction with the d-valence electrons. It is important to focus on the interaction with the d-valence electrons since their interaction with the adsorbate is mainly responsible for differences in reactivity between different metals.

If the d–valence electron band and the 5σ–orbital are both completely filled with electrons, the interaction energy will be repulsive since Pauli repulsion is proportional to the overlap of S and $|\beta|$. When the d-electron valence bond is partially empty, this repulsive interaction is decreased because the antibonding orbital fragments become less occupied. The decrease in repulsive energy with a decrease in number of metal-atom neighbors of the surface atom involved in the adsorbate bond relates to an increase in the number of empty antibonding orbitals, determined by the electron density between $\epsilon'_{\max}$ and E_F (see Fig. 3.10).

In the chemisorbed state the adsorbate surface-orbital fragments are made up of a mix of adsorbate σ and metal valence-electron molecular orbitals. The σ-electron occupation of the adsorbate, which was originally 2 electrons, decreases upon adsorption. This type of interaction is therefore referred to as an electron-donative interaction.

DFT cluster results on the interaction between metal d states and the unoccupied CO-$2\pi^*$ orbitals indicate significant $2\pi^*$ electron density in the d–valence electron regime

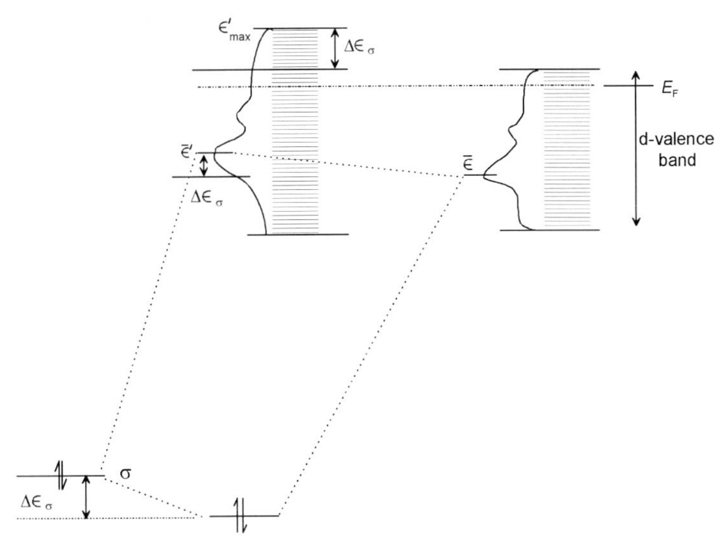

Figure 3.10. Orbital interaction of doubly occupied $n\sigma$ lone pair orbital with d-valence energies $\Delta \in_\sigma = \left| -1/2\sqrt{(\in_{n\sigma} - \overline{\in})^2 + 4Z\beta^2} + 1/2(\in_{n\sigma} - \overline{\in}) \right|$ (Z is the surface-atom coordination, β adsorbate–surface overlap energy intergral). The d-valence band is partially occupied.

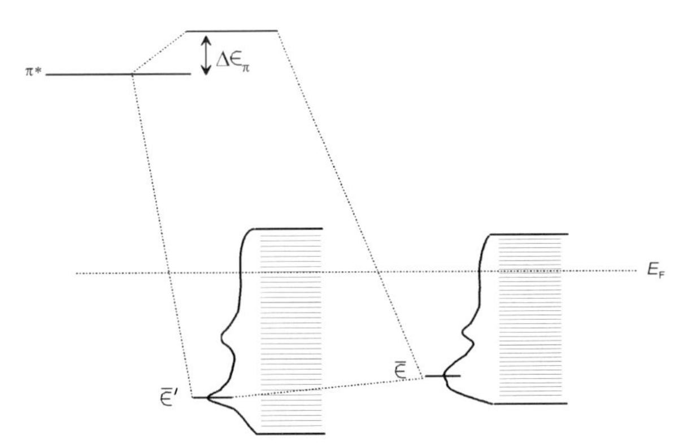

Figure 3.11. Scheme of back-donating interaction with the unoccupied adsorbate orbital π^*. The expression for ΔE_π is similar to ΔE_σ in Fig. 3.10.

and that the interaction in that regime is bonding. The back-donative orbital interaction scheme is shown in Fig. 3.11

The differences in symmetry between π and σ orbitals lead them to interact with different metal orbitals. The antisymmetric $2\pi^*$ orbital on CO prefers to interact with d_{xz} or d_{yz} atomic orbitals, whereas the 5σ orbital prefers to interact with the metal d_{z^2} orbital. The orbital interactions between the metal d–valence electrons and adsorbate orbital is now bonding. This bonding interaction stabilizes the energy by a downward shift of $\overline{\in}_{\mathrm{d}}$. Since $\overline{\in}_d$ moves upwards when the number of surface metal atoms decreases the interaction energy increases with a decrease in the number of metal surface–atom neighbors for the metal atoms involved in the chemisorptive bond.

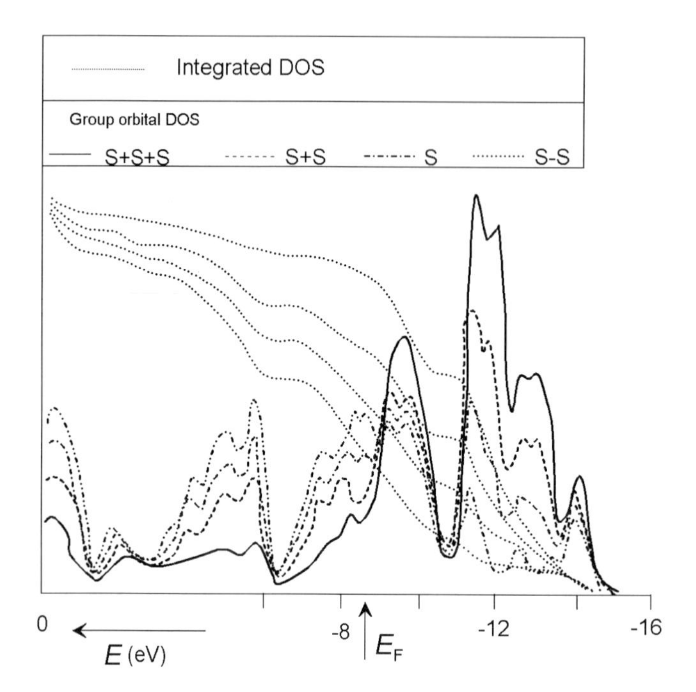

Figure 3.12. Group orbital Density of States as computed using Extended Hückel theory for the s atomic orbitals of the Ag(111) surface[10]. The metal electron energy increases from right to left.

The orbital interaction scheme in which the attractive contribution of the adsorbate surface bond is estimated from the donative and respective back-donative interactions is called the Blyholder model. It is the analogous to the Chatt–Dewar model which is used to describe chemical bonds in organometallic complexes.

Figure 3.12 illustrates the importance of the concept of surface group orbitals. A surface group orbital is defined as a symmetry combination of atomic orbitals on the metal atoms to which an adsorbed molecule or atom is attached. For example, the adsorption of a probe molecule such as CO at a three-fold coordination site results in an interaction between the 5σ lone-pair orbital on CO, which is σ symmetric, and, hence when we limit the example to the interaction with s-atomic metal orbitals with the symmetric combination of the s-atomic orbitals on the coordinating metal atom.

$$\psi_g^s = \frac{1}{\sqrt{3+6S}}\left[\varphi_1(s) + \varphi_2(s) + \varphi_3(s)\right] \tag{3.18}$$

The CO π-symmetric orbitals does not interact with the metal ψ_g^s state but instead with the two antisymmetric s-atomic orbital combinations:

$$\psi_g^a(1) = \frac{1}{\sqrt{2-2S}}\left[\varphi_1(s) - \varphi_2(s)\right] \tag{3.19}$$

$$\psi_g^a(2) = \frac{1}{\sqrt{6-6S}}\left[\varphi_1(s) + \varphi_2(s) - 2\varphi_3(s)\right] \tag{3.20}$$

Analogous surface group orbitals can be constructed from the d, or p-atomic orbitals. Figure 3.12 shows the local density of states of such group orbitals as a function of electron energy. The results are shown for extended Huckel calculations on Ag(111) indicate that the symmetric group orbitals have a maximum density at the lower energies. At somewhat higher energies, the single atomic orbital densities (relevant for atop adsorption) become the maximum. At the highest energies, the antisymmetric atomic orbital energies have the maximum in density. Hence, a low electron occupation of the metal-valence band provides a high electron density at the Fermi level for the high coordination s-symmetric interaction. A higher electron occupation of the valence band favors s-symmetric interactions in the atop configuration. It pushes adsorbates to atop positions. π-Type interactions are optimum at higher electron occupations and always tend to prefer high coordination.

The preference for CO to be adsorbed atop is typical for Co, Rh, Ru, Ir and Pt. On Ni and Pd, however, CO prefers to adsorb at the higher coordination sites. We have already discussed that the preference for atop adsorption is the result of the minimization of the repulsive interaction between doubly occupied π and σ states on CO and the occupied d-valence electron orbitals on the metal atom. Back-donation into unoccupied $2\pi^*$ molecular orbitals is maximum in high coordination sites. The back-donative interaction is enhanced by the additional interactions with antisymmetric combinations of surface s-atomic orbitals. The shift to higher coordination of CO on Ni and Pd indicates less involvement of the d-valence electrons in the surface chemical bond, consistent with the decrease in spatial extension of the d-atomic orbitals, when a metal changes position moving upward in a column or from left to right in a row of the periodic table.

For NH_3 adsorption to Cu in two-fold coordination (Fig. 3.4) an analysis of the surface chemical bonds has been made with the surface orbital densities decomposed according the corresponding group orbitals. One notes the density of states at lower energies as well as the lower energy position of bonding and antibonding interactions for two-fold coordination compared to one-fold coordination.

3.3.1 Bonding in Transition-Metal Complexes

The importance of hybridization (see Addendum 3.11), between d-, s- and p-metal orbitals in chemical bonding is easily understood for the molecular organometallic transition-metal complexes. The bonding in the tetrahedral $Ni(CO)_4$ complex, for instance, can best be understood by initially considering the 5d–atomic orbitals doubly occupied with 10 electrons.

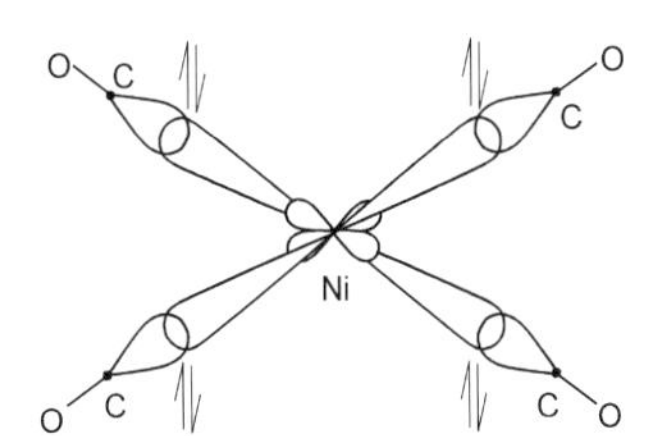

Figure 3.13. Interaction of four 5σ orbitals of CO with the four empty Ni sp^3 orbitals (schematic).

The 4s and three 2p orbitals of Ni are empty and can hybridize into four empty equivalent sp^3 orbitals. Combination with the four doubly occupied $CO_{5\sigma}$ orbitals leads to the formation of four bonding and four antibonding σ-type orbitals. The corresponding molecular orbital scheme is shown schematically in Fig. 3.14. The bonding σ as well as the non-bonding d-orbitals are occupied. The $Ni(CO)_4$ complex is further stabilized by

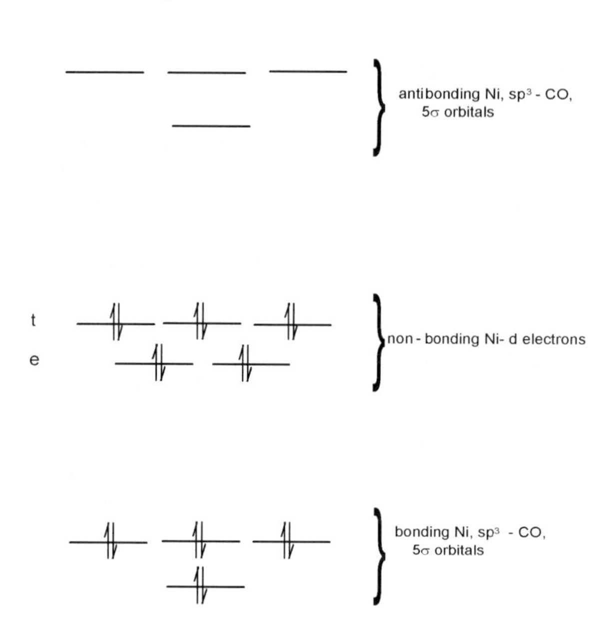

Figure 3.14. The σ electronic interaction scheme with Ni in $Ni(CO)_4$, including ligand field splitting of Ni d–atomic orbitals.

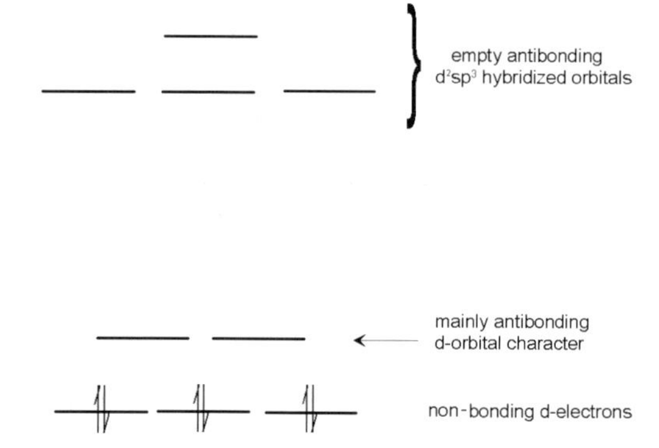

Figure 3.15. Electronic structure of CrL_6.

the additional back-donating interacting between the Ni-d orbitals and CO $2\pi^*$ orbitals. The bonding scheme nicely follows the 18-electron stability rule, which can be used to rationalize the stability of carbonyl complexes. In the $Ni(CO)_4$ complex, the Ni atom contributes ten valence electrons whereas the four CO molecules contribute eight valence electrons.

It explains, for instance the stability of isoelectronic $Co(CO)_4^-$ and the consequential acidity of $HCo(CO)_4$. It also predicts that the $Co_2(CO)_8$ complex is a stable dimer. There is a change in the ratio of number of CO ligands with metal-atom electron count. For Fe, which has 8 d-electrons, this becomes $Fe(CO)_5$. For chromium this is $Cr(CO)_6$, since Cr has 6 d-atomic electrons. In this way these complexes again satisfy the 18 electron rule. For CrL_6 the resulting molecular orbital scheme is given in Fig. 3.15. The change in ligand number implies a change in shape of the molecule and of the hybridization.

The $Ni(CO)_4$ sp^3 hybridization scheme is based on the formation of the four bonding sp^3 orbitals directed towards the corners of a tetrahedron. The prototype example for such bonding is CH_4. The two other hybridization schemes of importance are d^2sp^3, giving 6 directed orbitals for octahedral coordination, and dsp^2, given four directed orbitals for a square planar coordination. The latter is characteristic for complexes which contain 16 valence electrons, e.g. $PtCl_4{}^{2-}$. In $PtCl_4{}^{2-}$ two σ electrons are counted for each Cl^- ion.

The recombination of $Co(CO)_4$ to $Co_2(CO)_8$ can also be understood within the octahedral d^2sp^3 hybridization scheme. Let us put six of the nine Co d-electrons into the three nonbonding Co d-orbitals. Four of the six d^2sp^3 orbitals form bonding orbitals with the four 5σ CO orbitals, that each donate two electrons. Two lone–pair type d^2sp^3 orbitals are left. One orbital can be doubly occupied whereas the other contains the final electron that is left. The latter orbital can combine with an equivalent orbital of another $Co(CO)_4$ radical. A $Co(CO)_3$ fragment contains three dangling d^2sp^3 orbitals each of which is occupied with one electron. Hence it can combine with three other such fragments to form $Co_4(CO)_{12}$ (see also Chapter 5,pages 226, 227).

For Pt, the $Pt_4(CO)_{10}$ carbonyl complex is formed. In the Pt_4 framework, the Pt atoms have fully occupied d-atomic orbitals. Each Pt atom has one dangling free sp^3 orbital and three sp^3 orbitals that interact with the three neighboring Pt atoms. These form six σ bonds between the Pt atoms. Within this bonding scheme, all of these orbitals are initially empty. A total of 20 electrons have to be donated by the σ-lone pair orbitals of CO in order to fill the ten bonding σ-type orbitals. This is accomplished by the adsorption of four CO molecules at the apices and six CO molecules at the edges. In contrast to $Ni_4(CO)_{10}$ such multi–metal carbonyl Pt complexes are stable. Because the Pt-Pt distances are relatively long, Pauli repulsion between the doubly occupied d-atomic orbitals is reduced and thus easier to overcome than for the shorter metal–metal bonds that would appear in complexes formed from Ni. As we will see, the stabilizing interaction between Pt and the CO $2\pi^*$ orbitals is also more efficient because of the larger spatial extension of Pt d-atomic orbitals.

Interestingly, the negatively charged Chini complexes can be analyzed using these same principles. They are based on the triangular Pt_3 bonding motif. In line with our earlier discussion one predicts the neutral Pt trimer to be $Pt_3(CO)_9$, with a CO molecule coordinated between two Pt atoms and a CO molecules coordinated end-on the layered Chini complexes. The Pt s,p-atomic orbitals are sp^2 hybridized. The Pt p_z atomic orbital remains unoccupied. The layered Chini complexes are then constructed by stacking of negatively charged $Pt_3(CO)_6$ units. Whereas the σ framework has doubly occupied σ-type orbitals, the $Pt(p_z)$ orbitals are empty and hence can combine with a second

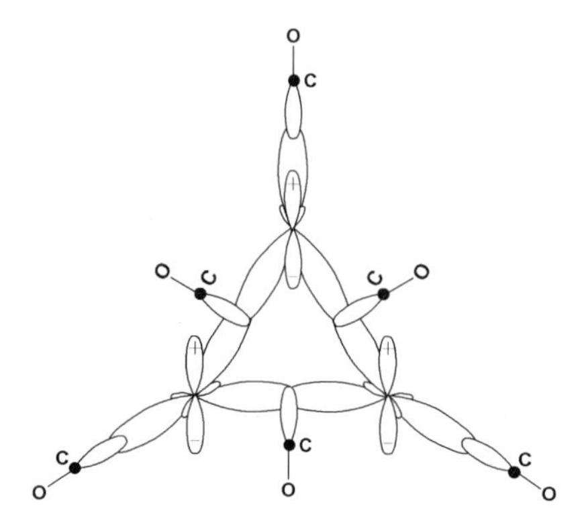

Figure 3.16. The layered Chini complexes are constructed from $Pt_3(CO)_6$ units.

$Pt_3(CO)_6$ unit, when electrons are available to occupy the bonding orbital combination that is formed between the $Pt(p_z)$ orbitals of the two trimers. Doubly charged layered Chini complexes are stable compounds. The corresponding orbital scheme of a $Pt_3(CO)_6$ unit is shown in Fig. 3.17. The partial density of states for the Chini complex is shown in Fig. 3.17 along with a representation of some of the corresponding orbitals. The d-valence region of the Pt atoms lies between −10 and −5 eV. In the energy range −16 to −14.8 eV and −13 to −11 eV the density of states is dominated by the bonding of the lone pair σ-CO interactions with the initially empty Ptσ-atomic orbitals. The lone pair interaction results in a rehybridization of 4σ and 5σ orbitals which is very similar to that for CO adsorbed on a metal surface. In the energy regime of -13 to -11 eV there is the interaction with the Pt d–atomic orbital and CO 1π orbitals.

The electron density range −9 to −6.6 eV is dominated by bonding, nonbonding and antibonding interactions between the d electrons and also the interactions with rehybridized 1π, $2\pi^*$, 4σ and 5σ orbitals. This leads to localization of charge on oxygen atoms as well as antibonding d-electron interactions with the hybridized 4σ and 5σ CO orbitals. The HOMO consists mainly of occupied nonbonding d_{xz} and d_{yz} orbitals. The LUMO is comprised of the bonding combination of 3Pt $5p_z$ orbitals and the unoccupied CO $2\pi^*$-orbital fragments. The electron density regime representing the non-occupied orbitals between −4 and −1 eV consists mainly of the antibonding interaction between the CO $2\pi^*$ orbital and the Pt atomic orbitals.

We note that in clusters as on the surfaces the contribution of metal d as well as s and p electrons to the formation of ligand or adsorbate bonds plays a very important role. Electron donation of the 5σ–ligand orbitals can "glue" together cluster-atom fragments that otherwise would not be stable. Comparison of Fig. 3.17 and 3.6 indicates that there are many similarities between the interaction of CO on a transition-metal surface and a carbonyl complex. A main difference is the low CO/metal atom ratio at the surface as compared to that in the carbonyl complex. Hence binding of CO to a transition-metal surface tends to weaken metal atom bonds, whereas in carbonyl complexes the binding to CO is essential to the stability of the complex.

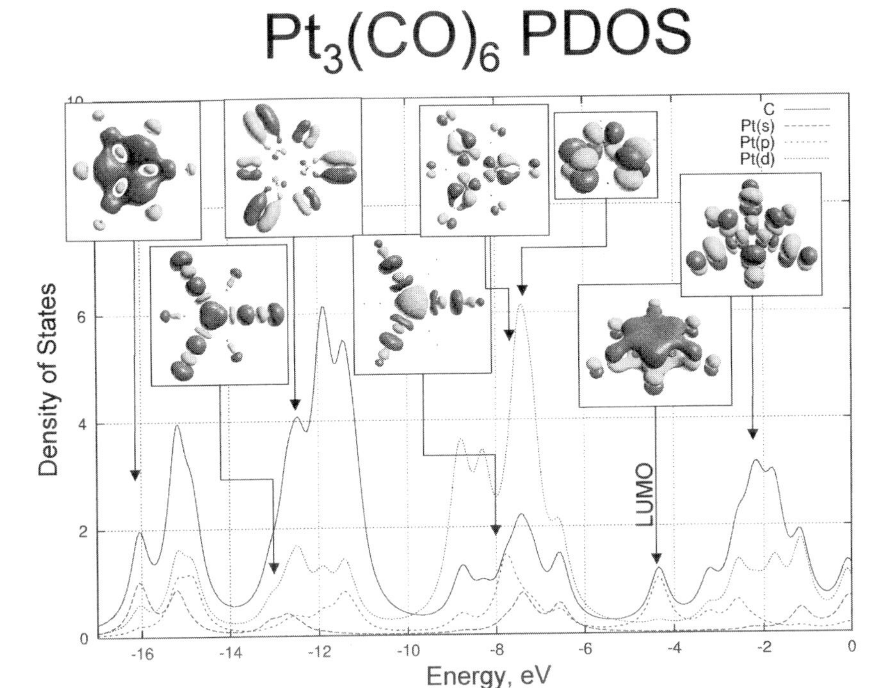

Figure 3.17. DFT calculated molecular orbitals of $Pt_3(CO)_6$ as well as their local density of state contributions on Pt and C.

3.4 Chemisorption of Atoms: Periodic Trends

The dissociation of diatomic molecules such as CO, NO, and N_2 leads to the formation of the strongly bound adatoms at the surface. We focus here initially on the binding of carbon and oxygen adatoms. The results, however, generally apply to other adatom bonding to transition-metal surfaces. The preferred adsorption site for atomic carbon and oxygen depends on the surface topology. They generally prefer binding to the higher coordination sites. On the dense (111) surfaces of face-centered cubic structures, this is typically the three-fold coordination site. The strong bond energy stems from the fact that both O and C atoms have empty, or partially occupied, low-energy 2p atomic orbitals, that can form strong bonding orbitals with available surface sites. The corresponding antibonding orbital fragments are only partially occupied. Second, the repulsive Pauli interaction between electrons in the doubly-occupied adatom atomic orbitals and doubly-occupied surface valence-electron orbitals is reduced, as result of the weak interaction with the doubly-occupied low-energy atomic 2s orbitals. Figure 3.18 compares the interaction energies for O bound to different Group VIII and IB metals. Figure 3.18(a) shows the O $2p_x$ local density of states for each of the different metals. The d-electron occupation increases from the left to right. The d-electron density of the transition-metal surface before adsorption is also shown. The spatial extension of the d-valence electron density decreases from left to right in the periodic table, and therefore also the corresponding overlap with adatom orbitals. As a consequence, the differences in energy between bonding and antibonding $2p_x$ transition-metal surface-orbital fragments also decrease. The antibonding orbital fragments progressively become more occupied, which is apparent from the local density of states plot in Fig. 3.18. The adatom bond energies then decrease with increasing d-valence electron occupation, as seen in Fig. 3.19. Figure 3.20 illustrates the general observation that along a row of the periodic table the adatom-metal surface interaction energy tends to decrease with increasing d-valence electron occupation.

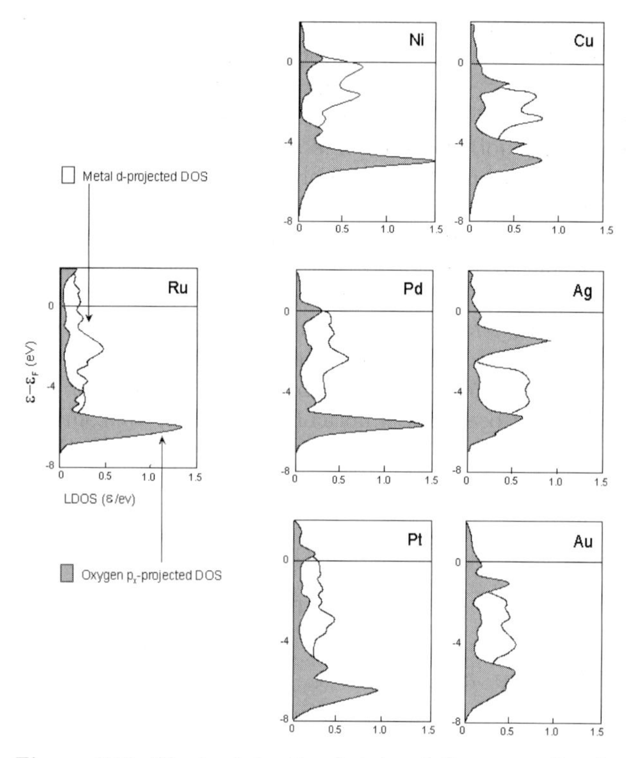

Figure 3.18. The local density of states of the oxygen $2p_x$ for oxygen adsorbed on different metal surfaces, adapted from Hammer and Nørskov[4].

Table 3.1 DFT-calculated binding energy for atomic C on the most dense surface for selected noble netals. Values are reported in kJ/mol.

	Fe*	Co	Ni	Cu
	-769	-668	-629	-476
	Ru	Rh	Pd	Ag
	-688	-690	-645	-338
Re	Os	Ir	Pt	Au
-713	-696	-675	-657	-411

* hcp

Table 3.2 DFT-calculated binding energy for atomic O on the most dense surface for selected noble netals. Values are reported in kJ/mol.

	Fe*	Co	Ni	Cu
	-714	-550	-496	-429
	Ru	Rh	Pd	Ag
	-557	-469	-382	-321
Re	Os	Ir	Pt	Au
-634	-533	-428	-354	-270

* hcp

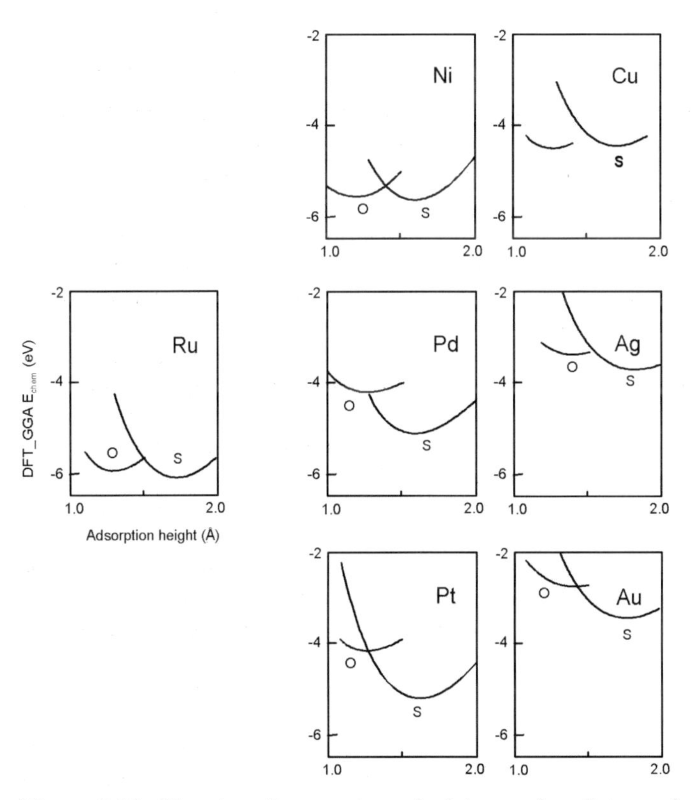

Figure 3.19. The adsorption energies and adatom surface distances for atomic O and S adsorbed on different transition metals, adapted from Hammer and Nørskov[4].

DFT-computed binding energies of C and O are summarized in Tables 3.1 and 3.2. One notes in both tables the large differences in the adsorption energies of the adatom. C is bound more strongly than O since fewer antibonding surface fragment orbitals are occupied. The adatom bond energies are seen to decrease across individual rows in the periodic table. Ru and Rh are exceptions, which may be the result of fact that Ru has the hcp structure whereas Rh has the fcc structure. As one proceeds down a column in the periodic table the interaction with the O atom decreases. The binding energy of the C adatom increases on moving down a column of the periodic table. This is mainly due to the increasing overlap with the more spatially extended d-valence orbitals. The trends for oxygen are different for two reasons. The first is that the occupation of antibonding surface orbitals is larger for oxygen than carbon because of the larger number of electrons on oxygen. This results in a larger Pauli repulsion when the overlap with the d-orbital electrons increases. The second factor relates to the fact that the work function increases down a column, which results in a decreased back-donation from the metal to the adatom. This will influence oxygen more than carbon due to the relatively lower position of the oxygen atomic orbitals over carbon with respect to the Fermi level of the metal. For the Group VIII metals, the bond energies of adatoms such as H varies much less when different metals are compared, since there is only interaction through a single σ-type bond. A characteristic value is 120 kJ/mol, which is just enough for the H_2 molecule to dissociate.

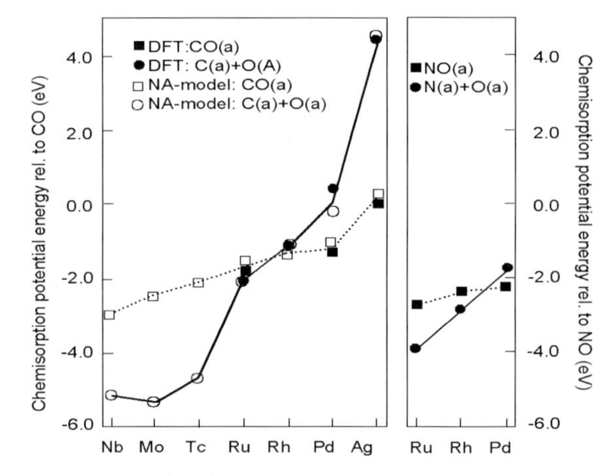

Figure 3.20. (Left) Calculated and model estimates of the variation in the adsorpion energy of molecular CO compared with atomically adsorbed C and O for the most close-packed surface of the 4d transition metals. (Right) Calculated molecular and dissociative chemisorption of NO. Solid symbols are DFT calculations; open symbols are Newns–Anderson model effective medium[3] calculations. For CO, dissociative chemisorption appears to the left of rhodium. For NO, dissociative chemisorption appears further to the right, i.e., also on rhodium, adapted from Hammer and Nørskov[4].

Figure 3.20 illustrates that there are much larger changes in bond energies for atoms as we span over different metals across the periodic table than for changes in the bond energy for molecules such as CO. The changes in the bond energies for CO across the periodic table are much less. This is due to the conflicting tendencies for the donative and back-donative interactions.

The other important parameter for the surface chemical bond is surface topology, i.e. the dependence of the surface bond energy of a surface adatom on the coordinative unsaturation of surface metal atoms. The changes in the adsorption properties of C on different Ru surfaces with different local coordination numbers have been studied in detail. A summary of the results is given in Table 3.3. Note that on the dense Ru(0001) surface atomic carbon prefers to bind to the three-fold coordination sites.

Table 3.3. DFT adsorption energies of C atoms adsorbed with different coordination to different Ru surfaces

Surface	Site	Adsorption energy (kJ/mol)	Number of Ru neighbors for C (and for Ru)
Ru(0001)	top	497	1 (9)
	bridge	631	2 (9)
	hollow hcp	688	3 (9)
	hollow fcc	648	3 (9)
Ru(1120)	top up	549	1 (7)
	top down	675	3 (7) + 1 (11)
	bridge short	666	2 (7)
	bridge long	579	2 (7) + 2 (11)
Ru(1010)	hollow	678	2 (11) + 1 (7)
Ru(1015)	hcp	714	2 (7) + 1 (9)

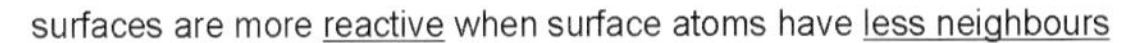

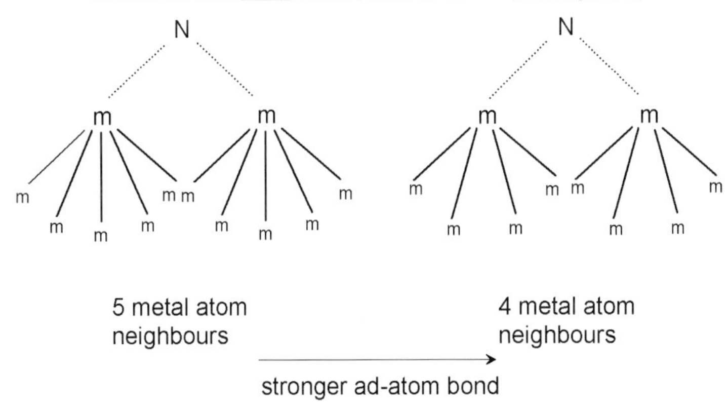

Figure 3.21. Adatom chemisorption. Surfaces bind more strongly when surface atoms have fewer neighbors.

However, for topological reasons, C prefers the two-fold bridge sites on the more open Ru surfaces. The bond energies are listed along with the site code, $x(y)$, which describes the adsorption site, where x refers the number of Ru neighbors coordinated to C, and y refers to the nearest number of metal Ru neighbors. One notes that at constant x, the interaction energy uniformly increases with decreasing y, in line with predictions according to the Bond Order Principle[11] (see also Section 3.5). The result is illustrated in a schematic fashion (Fig. 3.21). The chemical bonding between an adsorbate and the metal surface controls the adsorbate's potential reactivity. The metal surface provides binding sites that can stabilize active fragments critical to the overall catalysis, thus dramatically lowering the activation barrier from that found in the analogous vapor-phase chemistry. Strong chemical bonds between the adsorbate and the surface generally help to activate internal adsorbate bonds. Weaker adsorbate-surface bonds on the other hand generally help to enhance bond-making processes or association reactions. The nature of the metal–adsorbate bonding therefore is critical to the overall catalytic cycle since the cycle is nothing more than a complex array of bond breaking and making processes.

Bond breaking within the adsorbing molecule occurs when the metal can lower the energetics associated activating specific bonds in the adsorbate. For CO, this involves populating the antibonding CO $2\pi^*$ orbital which weakens the CO bond energy. The vibrational bond frequency typically decreases and the CO bond length tends to increase by a few tenths of an angstrom . Similarely bond weakening occurs between the metal atoms within the transition–metal surface upon adsorption. The stronger the metal–adsorbate bond, the greater is the degree of metal–metal bond weakening. This again is consistent with bond order conservation principles. An adlayer of atomic intermediates such as C, O or N can significantly weaken the metal–metal surface bonds. As a consequence, the metal–metal distances in the surface layer as well as the interlayer spacing of the outermost surface layer can lengthen[12]. The results of periodic DFT calculations for a 2 x 2 adlayer of nitrogen bound to Pt(111), for example, show that the Pt atoms directly bound to N expand outward whereas the neighboring Pt atoms (not bound to N) contract inwards (as seen in Fig. 3.22), thus roughening the surface.

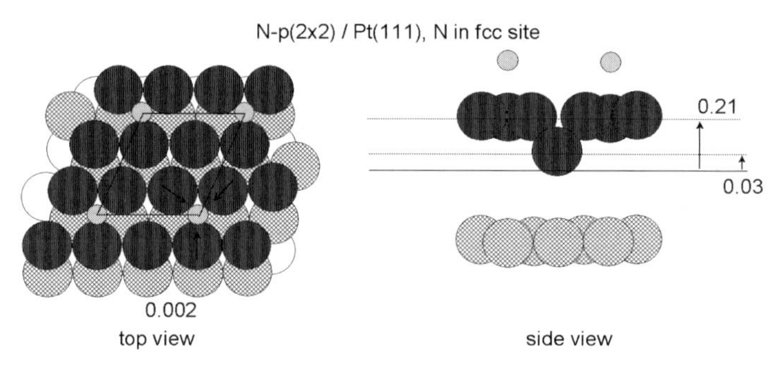

Figure 3.22. The computed DFT changes in bond distances between Pt atoms at the Pt(111) surface upon adsorption of nitrogen atoms.

Bond weakening within a metal cluster may be much more significant than that within a metal surface since the coordination numbers of the metal atoms involved in adsorbate bonding are typically lower in the cluster than those in the closed packed surface. For example, the metal–metal atom distances for O adsorbed to a Pd_6 cluster and a Pd_{18} cluster shown in Fig. 3.23 are significantly longer for the Pd atoms bonded to O in the small cluster as compared with that in the large cluster. The Pd coordination numbers on the bare Pd_6 cluster are all 4 whereas those in the center of the Pd_{18} cluster are 9, which matches that of the close-packed Pd(111) surface. The smaller Pd_6 cluster can accommodate the forces due to bond weakening more completely. Small clusters can even rearrange upon the adsorption of a single molecule. Pd clusters, for example, show a significant expansion in the Pd bond lengths upon the adsorption of ethylene, thus opening up these bonds and changing the geometry of the metal cluster.

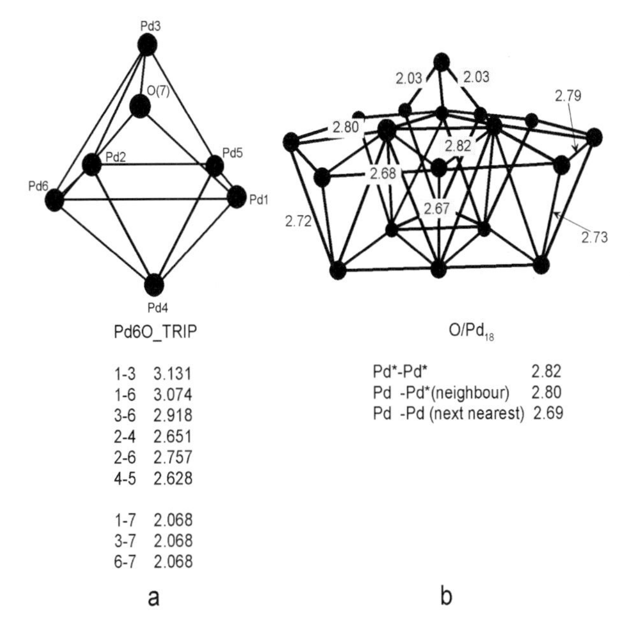

Figure 3.23. Adsorption-induced changes in bond length on a small and a large cluster[13].

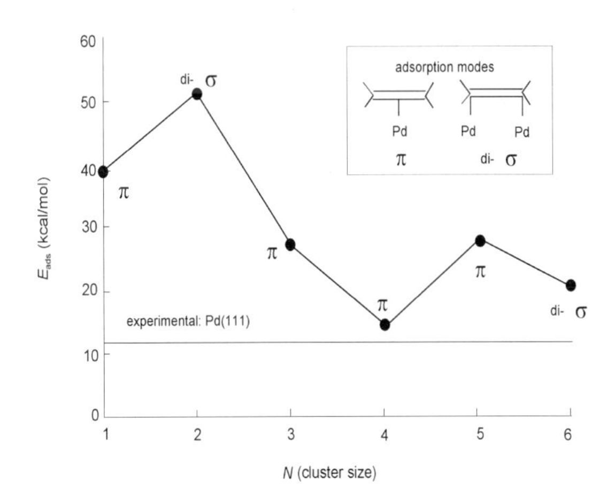

Figure 3.24. The calculated bond energy of ethylene as a function of cluster size[14].

Figure 3.24 shows the changes in adsorption energy for ethylene on Pd as a function of Pd_x cluster size. Ethylene prefers to adsorb in the π-bound state on the very small Pd clusters. Di-σ-adsorption in which two of the C atoms of ethylene interact with the two Pd atoms can result in a much stronger interaction which would weaken the Pd–Pd bonds and induce large deformations in the cluster shape (see Fig. 3.25a and b). The large change in the cluster structure weakens ethylene adsorbed in the di-σ mode, thus resulting in a preference for ethylene to adsorb in the π-adsorption mode on the cluster. Because of the high coordinative unsaturation of the Pd atoms in small clusters, ethylene binds strongly and the bond energy is substantially higher than that of Pd adsorbed

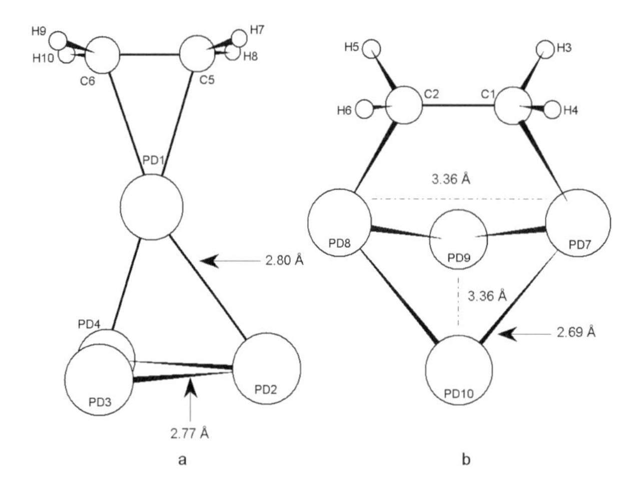

Figure 3.25. The adsorption of π (a) and di-σ (b) bonded ethylene to a Pd_4 cluster[14].

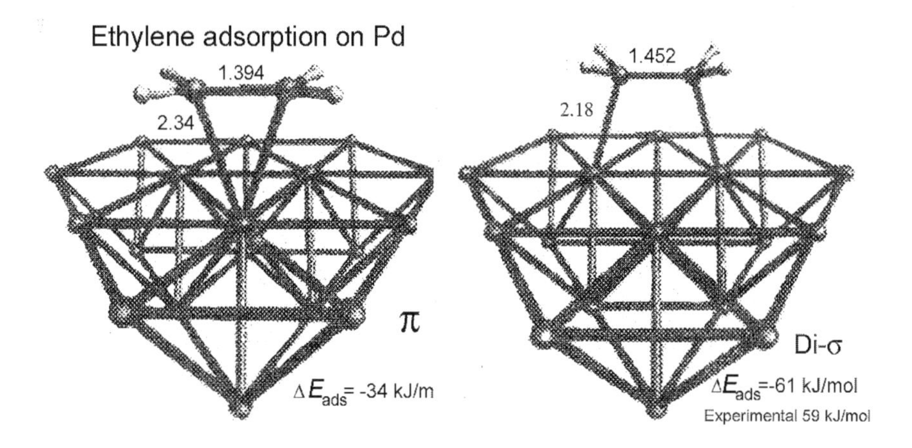

Figure 3.26. Adsorption of ethylene on Pd_{18} cluster[15].

on the Pd(111) surface. The clusters must typically approach around 20 atoms and in addition, the atoms to which the adsorbate binds must maintain the same number of nearest neighbor metal atoms as on a closed packed surface. The calculated interaction energies between the cluster and the bulk then become comparable. This is illustrated in Fig. 3.26.

The reactivity of the cluster edge atoms is, of course, higher than that for the higher coordinated sites for atoms found in the center. Therefore, comparison of different adsorption modes should be restricted to calculations performed on the same cluster, otherwise differences in cluster response may determine difference in adsorption energies, rather than differences in adsorption topology.

3.5 Elementary Quantum Chemistry of the Surface Chemical Bond

The interaction of an adatom with transition-metal surface atoms results in the broadening and shifting of the atomic orbital energies. There is not only a change in the electron distribution on the adatom, but also in the metal atoms involved in the surface chemical bond. Upon formation of the adsorbate surface chemical bond, the interaction between the surface atoms weakens.

In a very elementary fashion, this can be described with Bond Order Conservation theory[11]. This approximate theory assumes spherical electron densities around each of the atoms, and hence ignores hybridization. It is based on Pauling's observation that bond lengths between atoms in complexes depend logaritmically on the bond order x of each bond:

$$x = \mathrm{e}^{-\frac{r-r_0}{a}} \tag{3.21}$$

If the two center interaction that comprises the chemical bond is described by a Morse potential then a simple relation follows between the potential $Q(x)$ and the bond order:

$$Q(x) = Q_0 \, (2x - x^2) \tag{3.22}$$

where Q_0 is the bond strength for the bond at equilibrium. Bond Order Conservation implies:

1) If an atom has n neighbors rather than one, the total bond strength can be written as a sum of two-body interactions:

$$Q_n^t = \sum_{i=1}^{n} Q_i \tag{3.23}$$

2) The total bond order x_n^t of an atom is independent of number and type of neighboring atoms. The total bond order is then conserved:

$$x_n^t = \sum_{i=1}^{n} x_i \tag{3.24}$$

This enables one to relate the bond orders of chemical bonds with different numbers of neighbors. For instance, let us calculate the bond order for a chemical bond between two atoms a complex in which the central atom has n neighbors:

$$n x_i = x_0 = 1 \tag{3.25a}$$

$$x_i = \frac{1}{n} \tag{3.25b}$$

The bond order per bond is found to decrease with increasing number of bonds n. Since x_i depends exponentially on bond distances, the bond length is found to increase logarithmically with n. Proposition 1 then provides an expression for the bond strength as a function of coordination number:

$$Q_n^t = \sum_{i}^{n} Q_i \tag{3.26a}$$

$$= Q_0 \sum_{i}^{n} (2x_i - x_i^2) \tag{3.26b}$$

$$= Q_0 \, (2 - \frac{1}{n}) \tag{3.26c}$$

The total bond strength increases with decreasing coordination number, but less than proportionally. This analysis assumes that all of the atoms are equivalent. When the adatom is different from the metal atom, the changes in chemical bonding are more complex.

Within a molecular orbital scheme, the relevant parameters are the overlap energies β' of the atomic orbitals with surface atom orbitals and the overlap energies between the metal atomic orbitals on the different atoms β. The other important parameters are the relative energies of adatom orbitals and metal-atom orbital energies.

If one assumes a one-dimensional open-chain model for the adatom–metal system, with one atomic orbital per atom and all orbital energies equivalent, within tight binding theory, analytical solutions for the molecular orbital energies can be found[3].

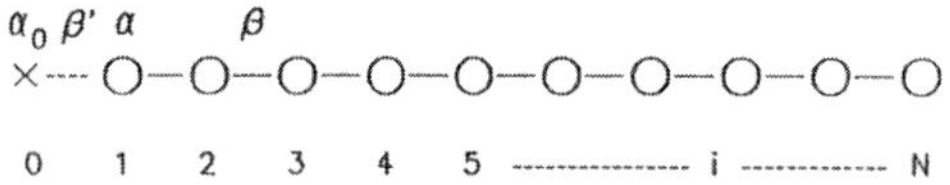

Figure 3.27. Open-chain model with the end atom being different from the other atoms ($\alpha_0 = \alpha$).

Without the adatom, the electrons in the chain are distributed over molecular orbitals between the metal orbital energies $\alpha_m + 2\beta$ and $\alpha_m - 2\beta$, α_m. In the chain each metal atom has two neighbors. When this number is eight or twelve as in a bulk metal, the electron distribution of the electrons within the metal would be found to vary approximately between $\alpha_m + 2\sqrt{\frac{n}{2}}\,\beta$ and $\alpha_m - 2\sqrt{\frac{n}{2}}\,\beta$. When the metal electron valence bond is half occupied and all bonding orbitals make an attractive contribution to the chemical bond, this interaction is proportional to $\sqrt{n}$. This is a very general result.

In Fig. 3.28 the molecular orbital scheme for a two-atom system is compared with that for a multi-atom cluster, with the central atom having n neighbors. In the latter case only the two molecular orbitals are shown that arise when the central atom interacts with its neighbors. The atomic orbital on the atom at the center is assumed to be spherical as an s-atomic orbital, and hence it interacts only with a symmetric combination of atomic orbitals on the neighboring atoms ψ_n:

$$\psi_n = \frac{1}{\sqrt{n}} \sum_{i=1}^{n} \varphi_i \tag{3.27}$$

where φ_i are the atomic orbitals on the neighboring atoms. As a consequence, the band gap between the bonding and antibonding orbitals formed by the interaction of φ_0 on the central atom and ψ_n is found to be proportional to $\sqrt{n}$.

The repulsive part to the interaction energy, which one finds when all the orbitals are doubly occupied, is proportional to n as follows from the expression

$$E_{\text{Pauli}} = -2\Delta(n) \cdot S(n) \approx -n\beta' S \tag{3.28}$$

The solution that one finds for the electron density distribution of the adatom in the open chain (Fig. 3.27) depends sensitively on the ratio $\mu = \frac{\beta'}{\beta}$, i.e. the relative interaction between adatom and surface and between the surface atoms[3]. Two different scenarios

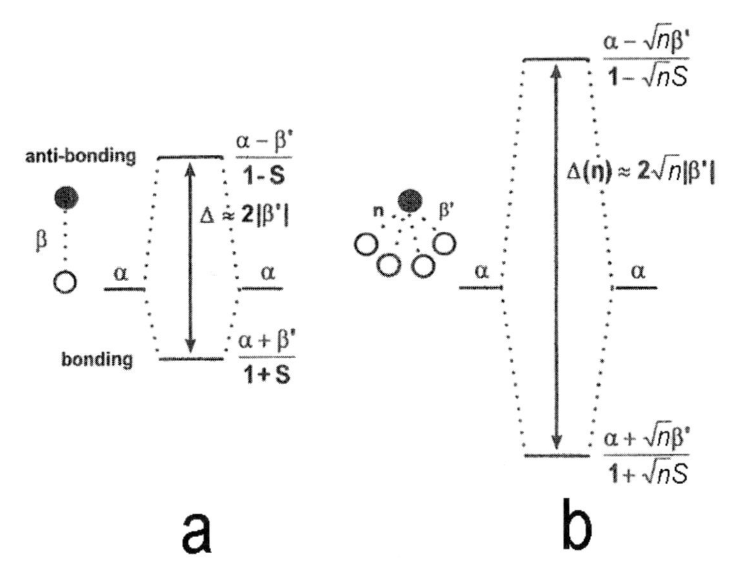

Figure 3.28. Interaction scheme between two atoms and one atom in a cluster[16], $E_{\mathrm{attr}} \sim \sqrt{n}\,\beta'$; $E_{\mathrm{rep}} = -n\beta' S$.

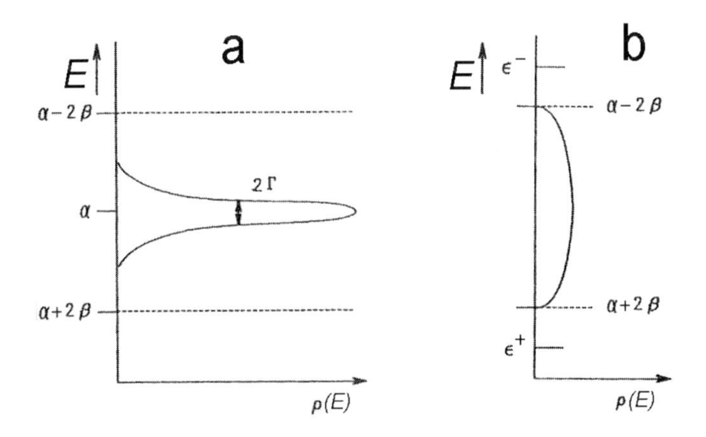

Figure 3.29. (a) Weak adsorption limit $\mu \ll 1$; (b) surface molecule limit $\mu \gg 1$. Local density of state $\varrho(\in)$ on adatom O (see Fig. 3.27). $\alpha + 2\beta$ and $\alpha - 2\beta$ are the boundaries of the metal electron density of states assumed to be a linear chain.

can result. The first when $\mu \ll 1$, which is known as the weak adsorption limit. In the second, $\mu \gg 1$, which is known as the surface molecule limit. These two situations are shown schematically in Fig. 3.29.
In the weak adsorption limit $\mu \ll 1$ (Fig. 3.29a), there is a small broadening of the electron density at the atoms. The broadening $2\Gamma_e$ equals

$$\Gamma_e = \frac{n'\beta'^2}{\sqrt{\frac{n_s}{2}}|\beta|} \tag{3.29}$$

For the one-dimensional chain, $n' = 1$ and $n = 2$. When the adatom orbitals have n' surface atom neighbors the interaction appears to depend linearly on n'. It decreases with the number of surface atom neighbors n_s.

When all of the bonding orbitals are occupied, the attractive contribution to the interaction energy is proportional to Γ. The inverse dependence on β stems from the need to localize an electron at a surface atom in order to bind with an adatom electron. The larger the bandwidth, the larger is the delocalization energy with a weakening of the interaction energy. The binding energy to a surface atom of increasing coordinative saturation increases with decreasing number of surface-atom neighbors as $(\sqrt{n_s})^{-1}$.

In the weak adsorption limit, the interaction energy is proportional to the metal local electronic density of states at the Fermi level $[\rho(E)_F]$:

$$\Delta E_{\text{ads}} \approx -\beta'^2 \rho(E_F) \tag{3.30}$$

An example of the local density of states at a Rh surface is given in Fig. 3.3b. The local DOS initially increases with the electron energy, then shows large fluctuations and finally decreases when the electron-valence band is fully occupied. Expression (3.30), however, is only valid for very weak interactions. For this reason, experimental studies have never obtained evidence that the adsorption energies relate strongly to the details of the local density of states at the Fermi level. For stronger interactions, the local density if states is sampled over an energy interval around the Fermi level that is proportional to the interaction energy. The general result is an increase in the interaction energy with electron occupation of the valence band as long as the bonding surface fragment orbitals are occupied and a decreasing interaction energy when the antibonding surface fragment orbitals also become occupied.

The limiting case of $\mu > 1$ corresponds to the surface molecule limit. This is usually close to the surface chemical bonding situation found in practice. The corresponding local electron density of states distribution is shown schematically in Fig. 3.29b. This electron distribution corresponds now to a bonding and antibonding orbital of the adatom–surface atom molecular complex with orbital energies $\in^+$ and $\in^-$, respectively, and a small contribution to the local density of states from electrons of the same energy of the metal valence electrons. In the one-dimensional system the orbital energies $\in^+$ and $\in^-$ are $\alpha + \beta'$ and $\alpha - \beta'$, respectively. This implies nearly complete localization of an electron at atom 1 and a very small bond order between surface metal atoms 1 and 2.

Bonding in the surface molecule limit is illustrated in a different way by the surface bond formation scheme presented in Fig. 3.30. Step 1 represents the localization of an electron on a surface atom, with complete rupture of the chemical bonds with neighboring atoms. Step 2 represents the formation of the molecule complex between the adatom and the surface atoms. Step 3 involves the formation of the adsorption complex by embedding of the surface molecule complex into the vacancy of the surface atom.

$$\Delta E_{\text{ads}} = E_{\text{MA}} + E_{\text{loc}} + E_{\text{emb}} \tag{3.31}$$

In the surface molecule limit, $|E_{\text{MA}}| \gg |E_{\text{loc}}|$ and $|E_{\text{emb}}| \ll E_{\text{loc}}$. This implies that the dominant correction to the surface–molecule complex energy E_{MA} is E_{loc}, which is proportional to the sublimation energy of a surface atom, which increases with increasing coordinative saturation as $\sqrt{n_s}$. The embedding energy is equal to the weakened metal

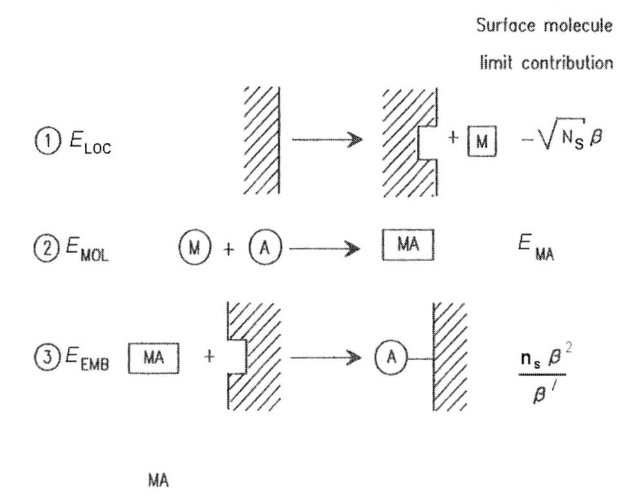

Figure 3.30. Chemisorption can be considered as the sum of three terms: $E = E_{\mathrm{MA}} + E_{\mathrm{loc}} + E_{\mathrm{emb}}$[3].

surface atom interaction energy:

$$E_{\mathrm{emb}} \approx \frac{n_s \beta^2}{\beta'} = \frac{n_s \beta}{\mu} \tag{3.32}$$

The stronger the interaction with the surface (increasing μ), the smaller is the embedding energy. The surface metal–metal bonds weaken owing to the interactions of the metal atoms with the adsorbate. The general result of the above analysis is that electronic effects that influence the adsorption bond are largest when the interaction energy is weakest. Therefore, changes in interaction energy due to coordinative unsaturation of surface atoms will be largest when the interaction energy is small compared with the metal–metal bond strength.

The analysis presented implies that in order to model appropriately the dynamics of surface processes the force field potentials used for MD simulations will require potentials that incorporate bond weakening (with the neighboring atoms) not directly involved in the adatom–surface metal atom complex. Changes in next-neighbor bonds also occur but decrease exponentially with distance. Interestingly, when the chemical bonds with neighboring metal atoms weaken, the bonds between atoms two and three then slightly increase (Fig. 3.27). This is a consequence of Bond Order Conservation. The decrease in bond order of one of the neighboring bonds increases the bond order of the other.

An important quantitative approach to determine force field parameters that satisfy the above criteria is the Embedded-Atom Method and modifications[17] such as the Modified Embedded Atom Method. According to the Embedded Atom Method, the energy per metal atom at the position $r = r_0$ equals

$$E_i = -\frac{n}{n_0} E_0 + F\left[n^i_{\mathrm{eff}}(r_0)\right] \tag{3.33}$$

where n denotes the number of nearest-neighbor atoms and n_0 the number of nearest neighbors in the bulk of the metal, E_0 is the sublimation energy and the function F is

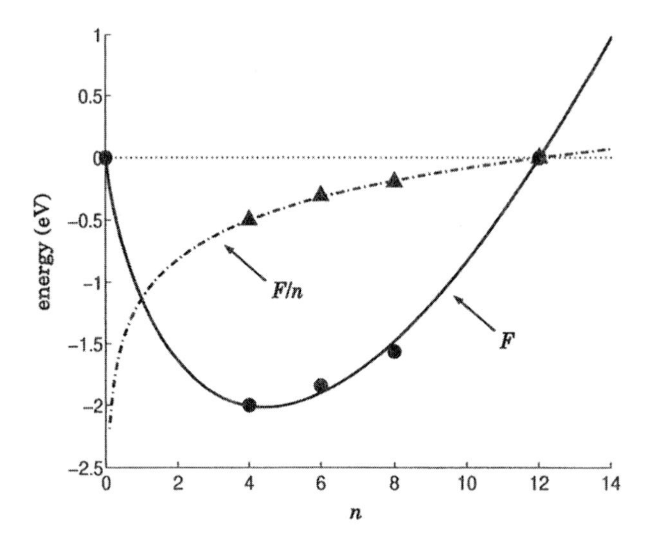

Figure 3.31. Bond-order function F and bond-order correction energy per bond F/n for Rh. Values indicated by the circles are obtained from the atomic energies for fcc, bcc, simple cubic (sc), and diamond cubic (dc) systems. In all systems the atoms are equally spaced at r_0, the equilibrium nearest-neighbor distance of Rh in an fcc lattice. $AE_0 = 5.48$ eV, adapted from P. van Beurden[18].

the Bond Order Function given by

$$F(n_{\text{eff}}) = AE_0 \frac{n_{\text{eff}}}{n_0} \ln\left(\frac{n_{\text{eff}}}{n_0}\right) \tag{3.34}$$

The functional form of F gives a similar logarithmic relationship between the bond length and the number of bonds as earlier used in the Bond Order Conservation formulation. n^i_{eff} in its most elementary definition equals to

$$n^i_{\text{eff}} = \sum_{j \neq i} \rho(r_{ij}) \tag{3.35}$$

where $\rho(r_{ij})$ is the spherically averaged atomic electron density of atom j. In the Modified Embedded Atom Method n_{eff} can be seen as an effective coordination number. Hence F can be considered a Bond Order correction term to the total energy.

An example of the bond-order function F is shown in Fig. 3.31. In Fig. 3.31 one notes the decreasing value of the bond order per bond $\frac{F}{n}$ with increasing coordination number, as predicted according to the Bond Order Conservation theory.

3.5.1 Molecular Orbital View of Chemisorption. A Summary

A schematic illustration that summarizes the essential electronic structure features that determine chemisorption was proposed by Hoffmann[1]and is reproduced here in Fig. 3.35. In the weak adsorption limit interactions (1) and (2) represent the attractive HOMO–LUMO interactions described in Section 3.3. Interaction (3) between the doubly occupied

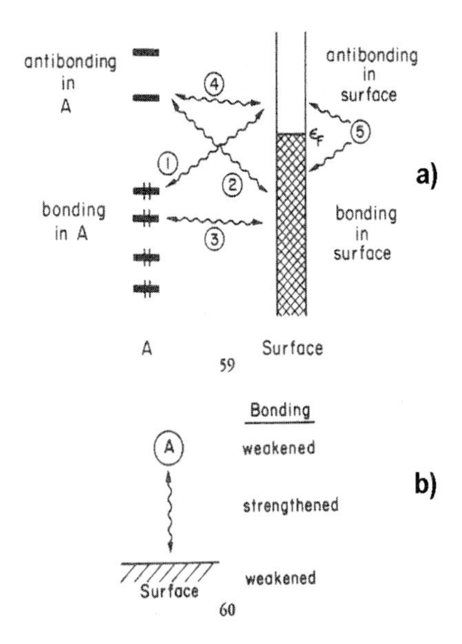

Figure 3.32. a) The electronic interactions of the surface chemical bond. b) Bondweakening and bond strengthening[1].

adsorbate orbitals and occupied surface orbitals is repulsive. The interaction between the adsorbate and the metal can lead to electron excitation within the adsorbate and the metal surface. In this process, surface–metal bonding orbitals become depleted and antibonding surface orbitals above the Fermi level become partially filled (process 5). This weakens the surface–metal bond energies. The relative position of the Fermi level of the metal with respect to the bonding and antibonding adsorbate–surface orbitals determines their electron occupation. Depletion of newly formed antibonding orbitals with energies higher than the Fermi level increases the adsorbate–surface energy (process 4). Figure 3.32b highlights the balance of bond weakening and strengthening interactions upon formation of the chemisorptive bond.

3.6 Elementary Reaction Steps on Transition-Metal Surfaces. Trends with Position of a Metal in the Periodic Table

3.6.1 General Considerations

In this section, we move from the elucidation of molecular and atomic adsorption to the fundamental features that control surface reactivity. We start by initially describing dissociative adsorption processes. We focus on elucidating surface chemistry as well as the understanding of how the metal substrate influences the intrinsic surface reactivity. We will also pay attention to geometric ensemble-size related requirements. The Brønsted–Evans–Polanyi relationship between transition-state energy and reaction energy discussed in Chapter 2 is particularly useful in understanding differences in reactivity between different metal surfaces.

$$E_{TST} = E^{\circ}_{TST} + \alpha \delta E_R \tag{3.36}$$

where E_{TST} is the transition-state energy and δE_R is the change in reaction energy compared to a reference state with transition-state energy E°_{TS}. When the reaction paths for elementary steps are similar, the change in activation energy can be predicted from the overall change in the reaction energy. For bond breaking reactions, this involves the energy difference between the adsorbed molecule state and the dissociated surface product fragments. We will use the Brønsted–Evans–Polanyi relationship quite extensively to predict trends in the activation energies with changes in the metal substrate.

We will continue to base our exposition to a significant extent on theoretically obtained results. We use mainly the results from periodic DFT slab calculations which represent the transition-metal surface. Ground-state properties and especially the transition-state energies may depend sensitively on many detailed aspects of the calculations. Important factors, for instance, are:

- the density functional used to compute the exchange correlation contribution to the energy
- accounting for spin polarization of the surface electrons
- the determination of local energy minima and maxima
- the size of the unit cell, which may contribute spurious or desired lateral interaction terms
- the number of metal atom layers used in the slab model and the distance between the slabs
- the particular optimization scheme used and the details of which metal atoms are optimized
- the inclusion of zero-point vibrational frequency corrections

The errors in computed energies may vary for each item by more than 10 kJ/mol. For this reason one has to be extremely careful when comparing energies obtained by different authors directly. Fortunately, the situation is much better when one compares spectroscopic data such as vibrational energies. The primary goal of theoretical analysis has to be a qualitative understanding supported by "semi"-quantative numerical correlations. The best approach is to compare systematically results obtained by the same method or similar model systems. Notwithstanding the great insights obtained by the use of current computational approaches, there remains a need for more accurate methods applicable to large systems.

As an illustration we will conclude this section with a short discussion on studies to establish theoretically the transition-state energy for CH_4 dissociation on an Ni(111) surface. Using a slab of Ni atoms of three layers, a 2 x 2 unit cell with the top layer and adsorbate fully relaxed, Watwe et al.[19] predicted an activation energy of CH_4 on Ni(111) of 127 kJ/mol. Increasing the size of the unit cell to 3 x 3, including spin polarization and increasing the number of layers to five decreases the transition state energy to 101 kJ/mol[20]. The zero-point vibrational correction decreases the barrier energy by at least further a 10 kJ/mol[21]. This reduces the transition-state energy computed with the most frequently used approach by approximately 30 kJ/mol, which amounts to 25%. Yang and Whitten[22], using a very different ab initio configuration interaction approach with embedded clusters obtained a value for the transition state energy of 71 kJ/mol. Whereas the latter values are close to experimental data, here one has to be careful also. As we have seen in Chapter 2, and we will note in the following section also, step edges or kinks may be present experimentally and act to reduce the measured activation energies significantly. Therefore, even when present in minute amounts they may dominate the outcome of experiments. Notwithstanding the difficulties mentioned, remarkable new insights have been obtained recently owing to the close comparison of theoretical results and experimental data. Keeping the above comments in mind we will now proceed with an analysis of computed results, not always mentioning the details of the models used as these can be found in the papers referred to.

3.6.2 Activation of CO and Other Diatomics

We start this subsection with reaction paths for the activation of CO and subsequently extend this to other diatomic molecules. In a subsequent subsection we then advance ideas learned from diatomic molecules to slightly more complex molecules such as methane and ethane to examine C–H and C–C activation.

In Fig. 3.20 we show that the change in metal affects the adsorption energy of adatoms much more so than that of adsorbed molecules. The thermodynamics for dissociative adsorption therefore become more unfavorable with increasing d-electron valence-bond occupation of the metal atoms in a row of the periodic table. This trend is a general result that is likely observed for all dissociation reactions.

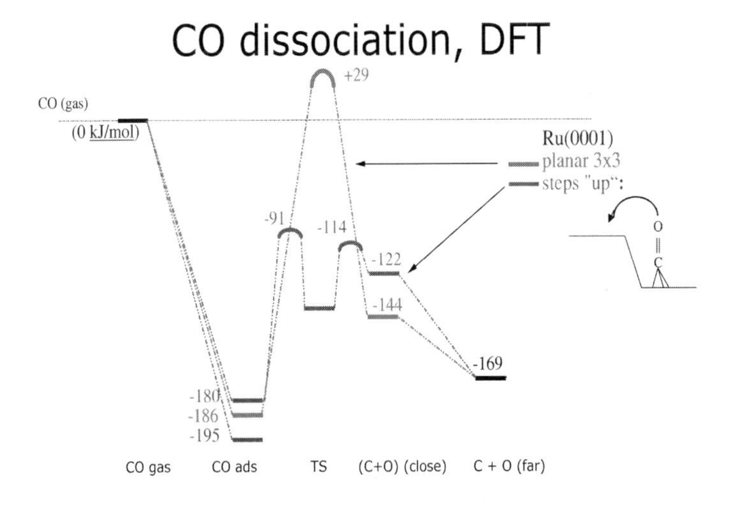

Figure 3.33. Reaction energy diagram of CO dissociation on the terrace or on the stepped surface of Ru(0001)[23]. The barrier energy $E = -195$ kJ/mol corresponds to the energy of adsorption of molecule atop; $E = -186$ kJ/mol, CO three-fold adsorbed; $E = -180$ kJ/mol, CO adsorbed in the hollow of the step.

The electronic energy and geometric changes for CO dissociation on a terrace and on a step of Ru(0001) are given in Fig. 3.33. On a terrace of the Ru(0001) surface, CO shifts from an atop to three-fold position at a cost of 12 kJ/mol. The CO bond, which is initially oriented perpendicular to the surface, bends towards the bridging position between two metal atoms as it stretches. The C–O bond in the transition state was found to be nearly parallel to the surface with a nearly zero bond order between the carbon and oxygen atom. The C atom is close to its final state above an hcp site. The oxygen atom is asymmetrically situated over a neighboring bridge site. Two of the three final O bonds are close to their final state. The activation barrier was calculated to be quite high at 224 kJ/mol. Additional energy is released when the neighboring C and O atoms diffuse away from each other.

The reaction proceeds by stretching the CO bond as well as tilting the CO axis toward the surface. This helps to lower the $2\pi^*$ state and enhances the transfer of electrons from the metal into this state. This charge transfer of electrons into the $2\pi^*$ state (back-donation) weakens the CO bond and thus aids CO activation.

The preferred reaction path for CO dissociation occurs over the bridge site, thus avoiding the closer atop site (see Fig. 3.34a). Activation over the atop site is actually higher in energy and in addition would form C and O atoms at hcp adsorption sites. The resulting hcp adsorption site for O would be 50 kJ/mol less favorable than the adsorption of oxygen at an fcc site. In Fig. 3.34b the stretched CO intermediate is shown for the dissociation of CO along a step edge.

Table 3.4. CO transition-state energies according to the Brønsted–Evans–Polanyi relation (kJ/mol). $\delta E_{TST} = 0.85\,\delta\Delta E_R$

	Fe * 166	Co 251	Ni 355	Cu 517
	Ru 227	Rh 315	Pd 424	Ag 592
Re 122	Os 227	Ir 336	Pt 416	Au 581

*based on 2 x 2 Ru(0001), fcc

Hammer and Nørskov[4] isolated the transition states for various diatomic molecules such as CO, N_2, and O_2 and nicely demonstrated that they all show very similar structures whereby the adsorbate–adsobate bond is significantly stretched and the product framents can form fairly strong bonds with the surface. The transition states are all considered late (for a definition of a late transition state, see Chapter 2) and have comparable values of α of approximately 0.9.

The Brønsted–Evans–Polanyi relation applied to the CO dissociation reaction results in Table 3.4. The quantum-chemical result upon which this is based is the dissociation of CO over Ru(0001) with 2 x 2 coverage.

The results show that there is an increase in the activation barrier for the transition state moving across a row from left to right in the periodic table. This increase corresponds with the increase in the d-valence electron occupation.

The high reactivity of the metals such as Fe, Co or Ru, as reflected in the relatively low activation energies, is essentially due to the increased stabilization of the adatom products such as C or O compared with CO on the metal surface. As we learned in previous sections, the heat of adsorption of a molecule varies much less with the adsorption site or the metal than do adsorbed atoms. Therefore, the relative interaction energies of the latter dominate the trends in reactivity.

Inspection of the changes in the local density of states of CO in the transition state and ground state show that there is a reduced interaction between C and O in the transition state. Essentially the C and O can be considered already separated in the transition state.

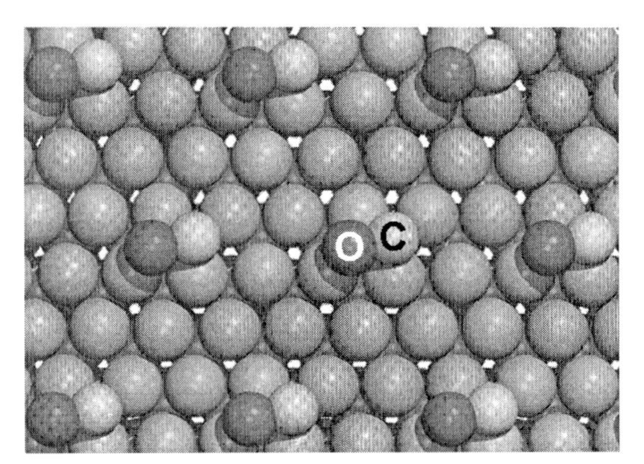

Figure 3.34a. The transition state for CO decomposition on a flat Ru(0001) surface[23].

Figure 3.34b. The transition state for CO decomposition on a stepped Ru(0001) surface, side and top views[23].

The local density of states on C and O are now more similar to those of the separated atoms than to those from an adsorbed molecule. A similar conclusion can be found for dissociation of many diatomic molecules.

Dissociation at stepped surfaces leads to significantly reduced activation barriers for many reaction systems. The activation energy and the corresponding transition state for CO dissociation on a stepped Ru(0001) surface are shown in Figs. 3.32 and 3.33b. The computed activation energy for CO has now become 104 kJ/mol and an intermediate state along the step can be identified.

The reduced activation energies for diatomic molecules at step edges appears to be quite general. Two factors tend to contribute. First, as Fig. 3.35 indicates, dissociation of the molecule over a step provides enhanced $2\pi^*$ back-donation into the bond weakening antibonding CO orbital.

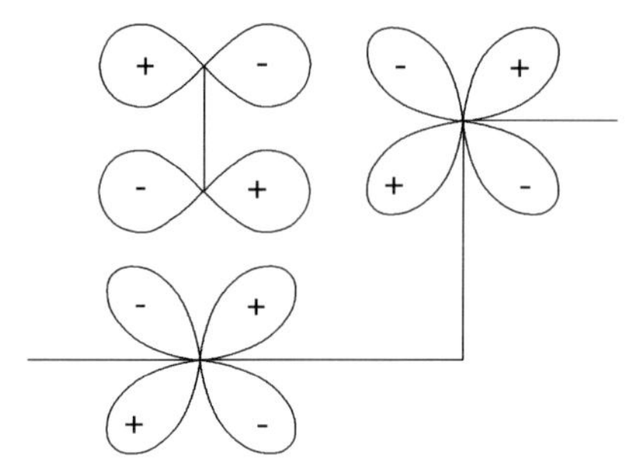

Figure 3.35. Orbital interaction between CO $2\pi^*$ and edge atom d-atomic orbitals (schematic).

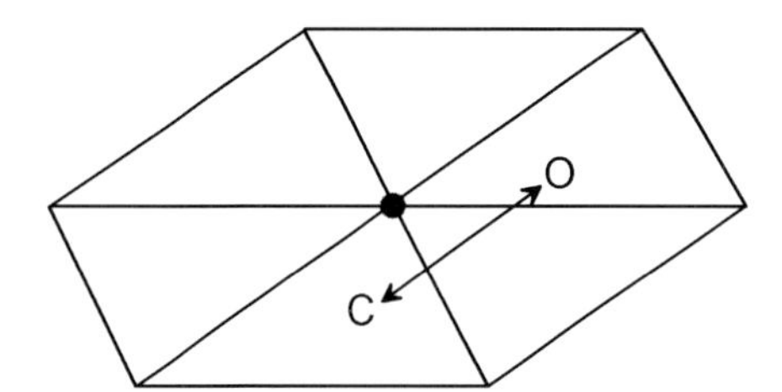

Figure 3.36. Reaction path for CO dissociation on the Ru(0001) surface (hcp→hcp) schematic).

Another important contribution that can result in a significant lowering of the activation energy is the specific configuration that the resulting product fragments adopt with respect to one another and the metal surface. For the dissociation of CO on a terrace, the C and O atoms that form tend to share a bond with a metal atom on the surface, as seen in Fig. 3.36. This ultimately destabilizes this state by 25 kJ/mol compared with the state where C and O are further removed from one another and do not share metal atoms. This raises the barrier for CO activation. When CO dissociates at a step, C and O atoms that form preferentially bind to the bottom and the edge of the step, respectively. The destabilization that results from a shared bonding with a metal surface atom is absent.

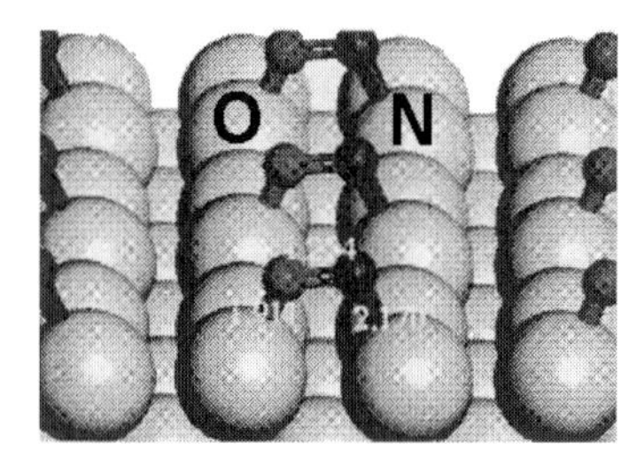

Figure 3.37. Structure of the "side–on" NO at the bridge sites on Pt(100)[24a].

The latter concept is basic to the specific dependence of the activation energies of elementary reaction steps on different surfaces. There is not always an immediate relation with the coordinative unsaturation of the metal surface atoms. Ge and Neurock[24a] noted an exceptionally low barrier for the dissociation of NO adsorbed on the non-reconstructed Pt(100) surface. The corresponding transition state is shown in Fig. 3.37. The calculated activation energies for NO dissociation over the (111) and (110) surfaces are 160 and 105 kJ/mol, respectively. The calculated barrier on the (100) surface is only 93 kJ/mol. These results are in very good agreement with those presented by Eichler and Hafner[24b]. In addition, since the dissociative reaction energy is 86 kJ/mol, it implies a barrier of only 7 kJ/mol for the recombinative association of N_{ads} and O_{ads} on this surface. We will return to this point in later sections. The low barrier for NO dissociation on the (100) surface relates to two factors. First, since NO dissociation proceeds over the valley created by the four neighboring Pt atoms, it allows for substantial electron back-donation. Second, the N_{ads} and O_{ads} product atoms that form do not share bonds with the same surface metal atom which significantly reduces the repulsive interaactions.

Nørskov et al.[25] elegantly demonstrated that the similarity in the transition state structures for N_2, O_2, CO, and NO could be used to establish a universal relationship between the activation energy and the heat of reaction for the dissociation of all of these molecules over different metal surfaces. They derived the following relationships for reactions that occur over the close-packed surfaces and step edges all energies in eV.

$$E_a = (2.07 \pm 0.07) + E_R(0.9 \pm 0.4) \text{ close-packed surfaces}$$
$$E_a = (1.34 \pm 0.09) + E_R(0.87 \pm 0.05) \textit{steps}:$$

The large value of α in these Brønsted–Evans–Polanyi relations is consistent with a late transition state for the dissociation reactions as discussed in Chapter 2. The only parameter in the universal relations is the reaction energy E_R, which can be easily calculated.

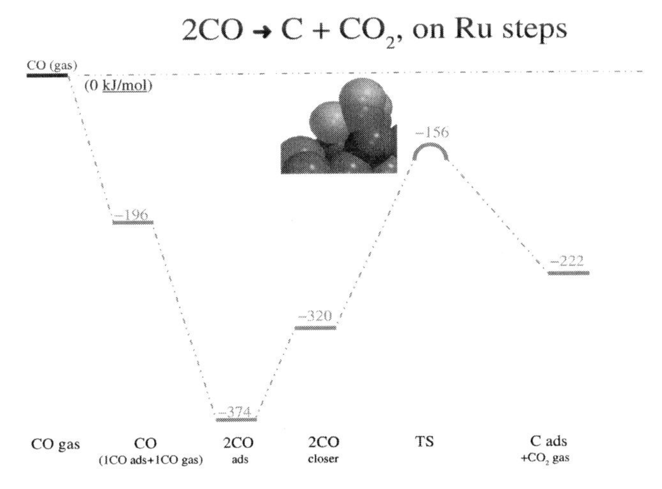

Figure 3.38. Transition-state energies for the Boudouard reaction $2CO \longrightarrow CO_2 + C_{ads}$ at a step on the Ru(0001) surface[23].

The dissociation of CO can also occur via the Boudouard disproportionation reaction where two CO molecules react together to form CO_2 and surface carbon: $2CO \rightarrow CO_2 + C$. The transition state for this reaction on a step of the Ru(0001) surface is given in Fig. 3.38. This reaction proceeds via an associative reaction in which a CO molecule at the bottom of a step recombines with a CO molecule adsorbed at the edge of the step. A C atom is then generated at the bottom of the step to form CO_2. The bent form of the CO_2-type transition state is indicative of its negative charge. The computed activation energy of 206 kJ/mol is higher than that for CO dissociation at a step edge.

3.6.3 Association Reactions; Carbon–Carbon Bond Formation

The activation of CO is one of the critical elementary steps that controls Fischer–Tropsch synthesis in the production of higher hydrocarbons from synthesis gas. It is well established that the Fischer–Tropsch reaction proceeds by activating CO to form surface carbon and oxygen[26]. The surface carbon subsequently hydrogenates to form various CH_x intermediates which can react further with hydrogen, couple with other CH_x fragments and ultimately desorb as different hydrocarbon products.

In addition to the CO dissociation paths discussed in the previous subsection, the presence of coadsorbed hydrogen offers a third potential reaction path for the activation of the CO bond. CO bond activation, which proceeds through interaction with the metal surface, becomes more difficult for Group VIII metals at the bottom-right corner of the periodic table. In contrast, the weaker M–CO bonds tend to help promote the hydrogenation of CO to form formaldehyde or methanol. This reaction proceeds through the formation of surface formyl (CHO) intermediates.

The calculated barrier for the reaction of CO and hydrogen to form the surface formyl intermediate over Ru is 143 kJ/mol (see Fig. 3.39). The subsequent CO bond activation of the the formyl intermediate on Ru is then only 30 kJ/mol. Hence, on the Ru terrace this scheme of first adding hydrogen before CO dissociation will be the preferred path to cleave the CO bond compared with the direct CO dissociation path. On the step, the barrier to formyl formation is similar to that on the terrace, hence higher than the barrier to direct CO dissociation. On surface steps the latter dissociation step path will be the preferred path for the formation of C_1 adsorbed species.

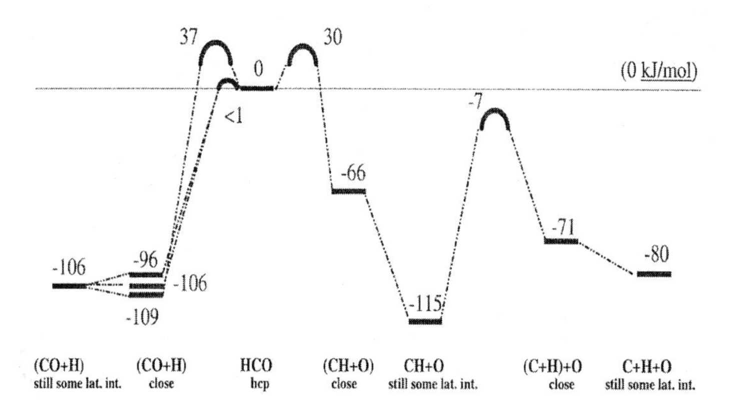

Figure 3.39. The reaction energy diagram for the C–O bond-cleavage reaction via intermediate formyl formation[23].

The trends for the insertion reaction are opposite of those for dissociation since association is the microscopic reverse of dissociation. The weaker the adsorption of CO and H, the easier their recombination should be. The calculated activation barrier for the association of surface CO and hydrogen to form the formyl intermediate on Pd(111) is 70 kJ/mol[27], which is much lower than the earlier value reported over Ru. This is directly in line with expectation, since CO and H bind more strongly to Ru than they do to Pd. The formation of the adsorbed formyl intermediate is lowered now to 40 kJ/mol endothermic. In the presence of hydrogen, the path to consecutive hydrogenation to methanol now competes with the CO cleavage reaction. Because of the weak metal adsorbate bonds the reactions steps towards methanol formation are also exothermic. In contrast, the CO dissociation reaction over Pd(111) has a fairly high activation barrier and is highly endothermic[27].

The comparison of the CO activation and CO hydrogenation barriers over Pd strongly indicates that CO hydrogenation to methanol is much more prevalent than CO activation subsequently to form CH_4. This discussion also illustrates why on a transition-metal surface (in the absence of any steric constraints) there can be a preference for a particular reaction channel. On Pd(111) the overall reaction barrier for methanol formation is significantly lower than that for the CO dissociation reaction.

3.7 Organometallic Chemistry of the Hydroformulation Reaction

Homogeneous catalytic reactions can typically provide considerable insight into the mechanisms that govern analogous heterogeneous catalyzed pathways[28]. Organometallic complexes are structurally well defined and can be characterized by a number of in-situ spectroscopic techniques. In the previous section, we discussed possible reaction paths and mechanisms for the hydrogenation of CO to methanol. This reaction is analogous to the homogeneous hydroformulation reaction. Hydroformulation involves the reaction of an alkene with hydrogen and CO over a homogeneous organometallic catalyst and is used in the production of aldehydes and alcohols. This reaction is thought to proceed by the hydrogenation of ethylene to form an ethyl fragment followed by the insertion of the ethyl intermediate into a metal–CO bond. We will discuss here the elementary reaction steps involved in the hydroformulation reaction as proposed for the cobalt carbonyl complex. Cationic Pd or Rh complexes with phosphine ligands are currently also widely used for this reaction. The structure of the $HCo(CO)_4$ complex that reacts with ethylene is given in Fig. 3.40. In order to react with ethylene, one of the CO ligands has to desorb so that alkyl formation can occur from the reaction of π-bonded ethylene with the H atom bound to CO. In complex II, which is shown in Fig. 3.41, Co has a formal charge of +1 and a d-electron occupation of d^8. The insertion reaction can be considered to occur on the $(OC)_3Co$ fragment we discussed before (Section 3.4), with three empty ligand positions. The three dangling bonds can form one symmetric and two antisymmetric combinations. In complex II, one of the antisymmetric d_{xz} symmetric orbitals is unoccupied. Complex II can subsequently react to form complex III.

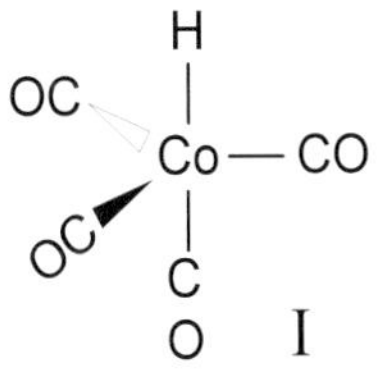

Figure 3.40. The structure of $HCo(CO)_4$ (complex I).

In the first step, a CO molecule inserts into the M–C bond of adsorbed ethyl. Subsequently, H_2 adsorbs on the vacant ligand position. In the final step, H_2 dissociates heterolytically and subsequently reacts with the $CH_3CH_2CO^*$ fragment to form the corresponding aldehyde along with regeneration of the catalytically active carbonyl complex.

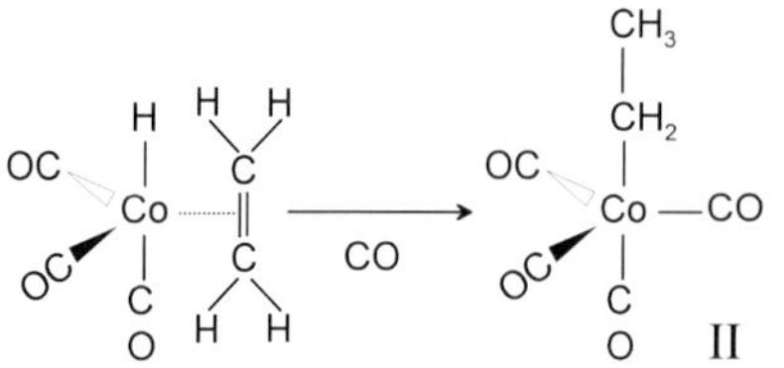

Figure 3.41. Formation of a Co alkyl–carbonyl complex.

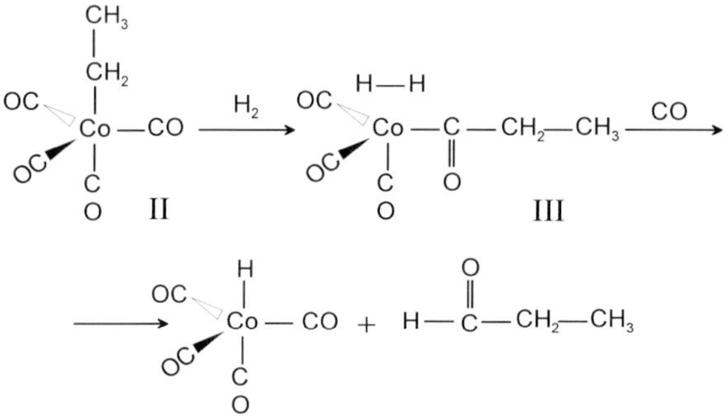

Figure 3.42. Insertion of CO and aldehyde formation.

The barrier necessary to activate CO for insertion stems from the Pauli repulsion between the doubly occupied σ-ethyl orbital and the doubly occupied 5σ-CO orbital of the inserting CO molecule[29]. When the ethyl carbon atom and the CO carbon atom approach each other, a bonding and antibonding orbital combination comprised of the two σ-type orbitals is formed. Both are doubly occupied, thus resulting in Pauli repulsion. This Pauli repulsion is reduced by the donative interaction of this doubly occupied antibonding σ orbital combination with the antisymmetric unoccupied Co d_{xz} atomic orbital. This is a very general process and has been investigated for several different metal systems[29]. This is schematically illustrated in Fig. 3.43 for the insertion of CO into the metal–methyl bond of a Pd^{2+} phosphine complex. The complex in Fig. 3.43 is planar. Donation of electrons from the antibonding 5σ–CH_3 orbital into the empty low energy $d_{x^2-y^2}$ metal orbital lowers the activation barrier for the insertion reaction.

3.8 Activation of CH_4, NH_3 and H_2O

The dissociation of CO, NO, and other diatomic molecules predominantly occurs through the back-donation of electrons from the metal surface into the low-energy π^* antibonding orbitals on the adsorbate, thus leading to a significant weakening of the adsorbate–adsorbate bond. In the transition state the intramolecular bond is typically nearly broken. The dissociating complex essentially exists as two fragments coordinated to the surface that tend to share bonding with a common metal atom. This metal atom sharing typically leads to repulsive interactions between the product fragments. The activation barrier with respect to the dissociated fragments is primarily determined by the degree of weakening of the bonds between the dissociating fragments and the metal surface. The barrier height for recombination is only weakly dependent upon the variation of the adatom metal surface interaction energy.

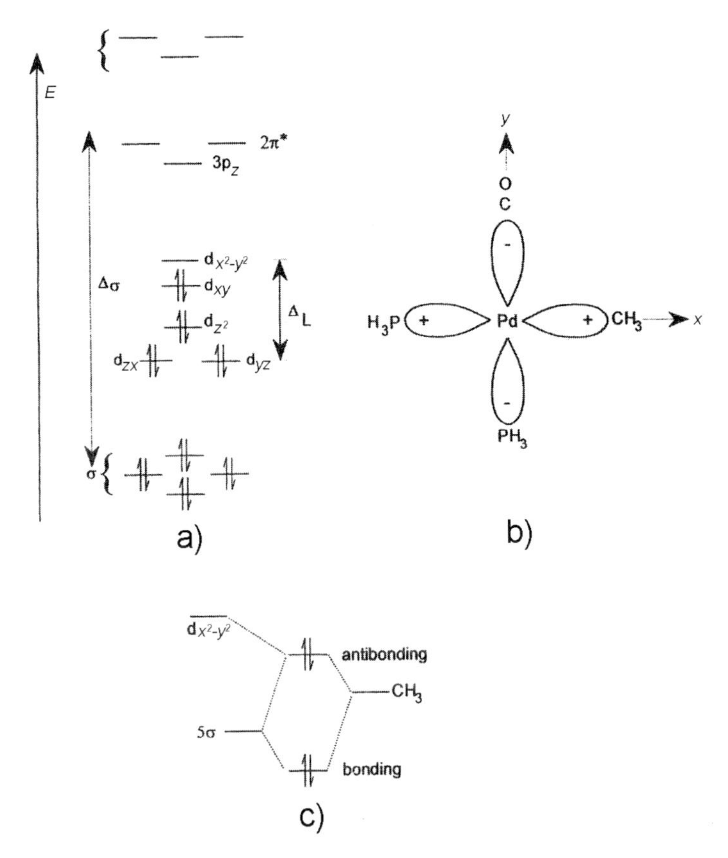

Figure 3.43. (a) Molecular orbital scheme for the $Pd(CO)(CH_3)(PH_3)_2$ complex. (b) The Pd $d_{x^2-y^2}$ orbital of the $Pd(CO)(CH_3)(PH_3)_2$ complex. c) Release of the repulsive interaction between CH_3 and CO by back-donation into the empty $d_{x^2-y^2}$ orbital.

The activation of C–H or N–H or alkane C–C σ bonds is much more difficult than for π bonds since the unoccupied antibonding orbitals of the former are much higher in energy. The bond that is to be broken has to be significantly stretched from its initial state as one proceeds along the minimum energy reaction path, but still maintain a significant bond order. In the transition state the CH bond is extensively stretched. The corresponding transition states are therefore best characterized as late transition states (see Chapter 2 for the definition).

The activation of the bond that is broken occurs via a similar mechanism to that for the oxidative addition over an organometallic substrate. Dissociation leads to the formation of negatively charged adsorbate-fragment orbitals with formally oxidized metal surface atoms. Dissociation typically occurs over the top of a surface atom. The critical point in the activation of the C–H or N–H bond occurs when it is stretched sufficiently such that the empty antibonding bond orbital lowers close enough to the Fermi level to allow for back-donation and electron transfer from metal into the antibonding state of the absorbate. This is illustrated in Fig. 3.44 for the dissociation of H_2 over different metal surfaces (see also ref. [4]). In the oxidative addition reaction, the reactant-molecular

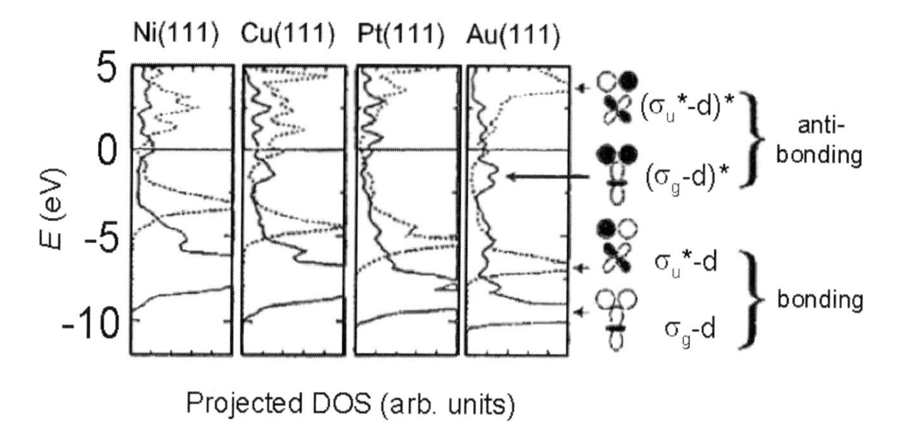

Figure 3.44. The DOS projected onto σ_g and σ_u* for H_2 in the dissociation transition state on Cu(111), Ni(111), and Au(111), and Pt(111) surfaces, adapted from Hammer and Nørskov[4].

bond distance has to be increased so that that empty antibonding molecular orbitals are low enough to become occupied by electron back-donation. In the reverse reaction, the reductive elimination, in order to form a chemical bond, the antibonding intramolecular orbital has to be pushed above the Fermi level so as to reduce the repulsive intramolecular interaction by electron donation to the metal, similarly as illustrated in Fig. 3.44. The activation of methane occurs as depicted in Fig. 3.45.

The activation of the CH bond can occur by stretching it over the top of an Ni atom on the surface or through a valley between Ni atoms. Methane activation occurs with slight preference for the path that proceeds through the valley between Ni atoms. In contrast, CH_3 activation occurs by stretching the CH bond over the top of an Ni atom. With the exception of Cu, Pd, Pt and possibly Ir, the CH_x fragments that are generated prefer adsorption in three-fold coordination sites on many of the close-packed surfaces. On the excepted metals, the interaction between adsorbate and the highly occupied, spatially extended d-valence electrons forces the CH_3 fragment to the atop position (CH_3 has one empty sp^3 lone-pair orbital that binds to the metal) and the CH_2 fragment to a two-fold position (CH_2 has two empty orbitals), as predicted according to bond hybridization arguments[3]. The difference in the coordination of CH_3 to Ni on which CH_3 prefers three-fold coordination and Pt where CH_3 binds atop derives from the small spatial extension of the d-atomic orbitals of Ni compared with Pt. The repulsive surface–adsorbate d–lectron interactions on Ni are significantly reduced.

Because of the particular geometry of the Ru(1120) surface, the adatoms now prefer two-fold coordination surface sites. The coordination number of the metal atoms on the Ru(1000) surface is nine, on the more open Ru(1120) surface the coordination number is seven. The lower coordinated Ru(1120) surface should therefore be more reactive. The activation energy for CH_4 dissociation is found to be lowered by over 30 kJ/mol moving from the Ru(0001) to the Ru(1120) surface, see Fig. 3.46 a and b. In this figure (moving from right to left) the activation energies are compared for subsequent elementary C-H bond activation steps in the decomposition of CH_4 to adsorbed carbon. These surfaces were studied at two different surface coverages (25% versus 10%). At higher surface occupations the repulsive interactions between adsorbates decrease bond energies and, hence

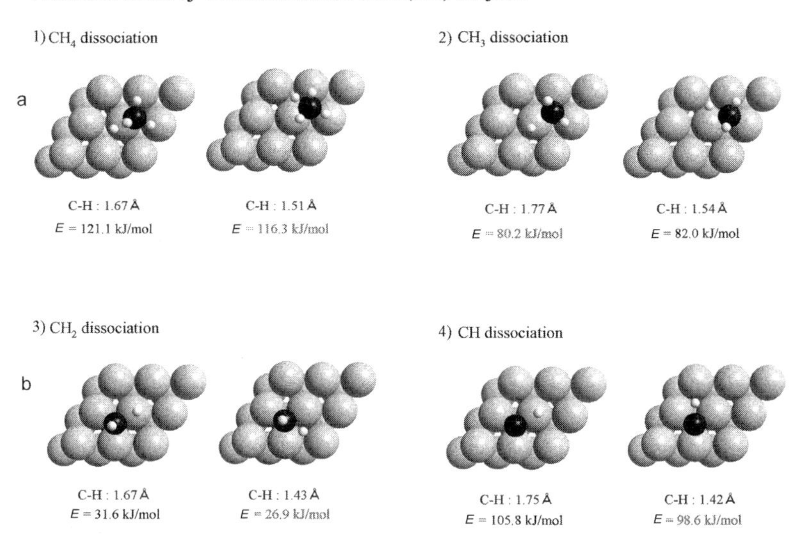

Figure 3.45. DFT-computed reaction paths for the dissociation of CH_4 and subsequent reaction intermediates on the Ni(111) surface[30].

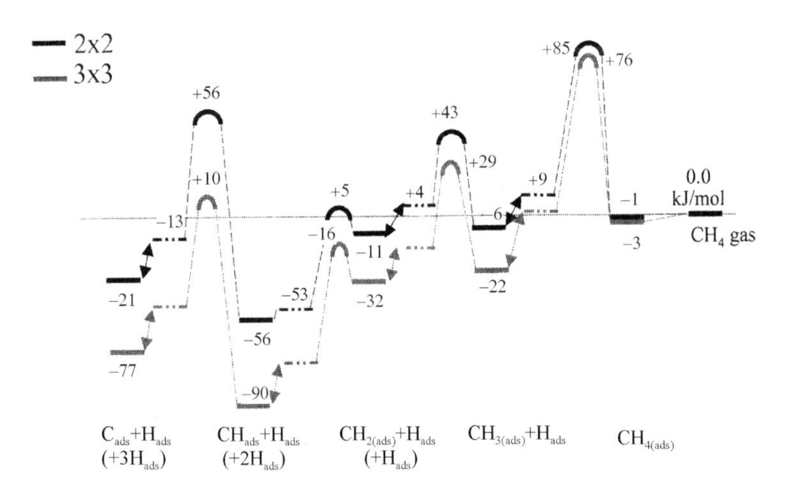

Figure 3.46a. Reaction energy diagram for CH_4 decomposition over the Ru(0001) surface at different coverages of intermediates[31a]. Reaction proceeds from right to the left. Top curves (2 x 2) unit cell, low curves (3 x 3) unit cell.

increase activation barriers. Of the dissociated fragments, the relative energies are compared immediately after dissociation and when they are at an infinite distance. Whereas on the Ru(0001) surface the CH adsorbed fragment is most stable, CH_2 and CH have comparable energies on the more open surface. CH prefers high coordination sites on the (0001) surface, but is limited on the (1120) surface to two-fold coordination sites. Trends in reactivity for different metals are illustrated in Table 3.5.

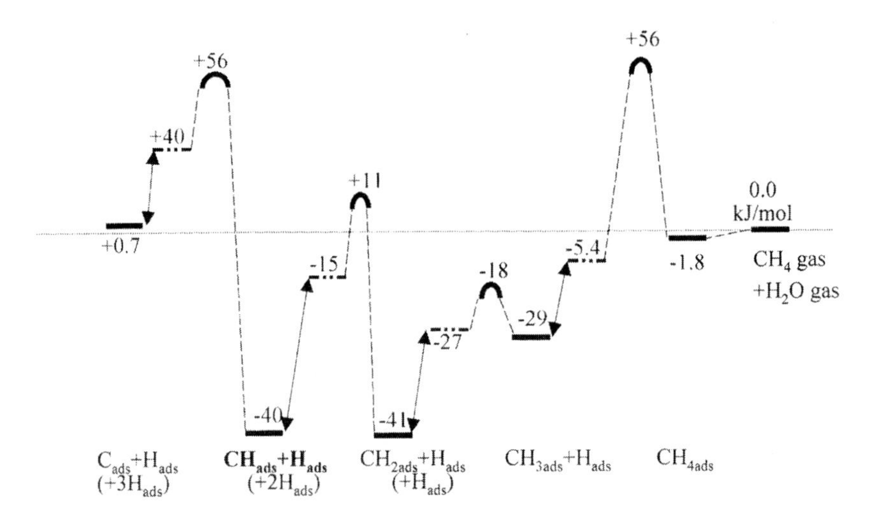

Figure 3.46b. DFT-calculated reaction diagram of CH_4 decomposition on the Ru (1120) surface; (2 x 2) unit cell[31b].

Activation barriers for row 4 metals appear to be the highest, those for the row 5 metals are lower and those for row 6 metals tend to be lowest. Surprising is the low value found for Pd, which may be an artifact of the calculation.

Table 3.5. DFT-computed activation energies in kJ/mol for CH_4 decomposition to CH_3 computed in a (2 x 2) unit cell[21,32–34,41]

	100	118
	Co(0001)[21]	Ni(111)[21]
77	67	66
Ru(0001)[33]	Rh(111)[32]	Pd(111)[32,41]
	≈40	75
	Ir(111)[34]	Pt(111)[33,41]

There is an important difference between the trends found here for methane activation and those reported earlier for CO dissociation. The CO dissociation energy trend is determined by the O_{ads} adsorption energy. Note that whereas CO dissociation on Pt(111) is more difficult than on the Ni(111) surface, it is the reverse for CH_4 activation.

Also the activation of C–C bonds and the formation of C–C bonds for partially hydrogenated intermediates do not have to behave similar to that for C–H activation. This follows, for example, from inspection of Fig. 3.51 which will be discussed more extensively later. In this figure, the C–C coupling reactions on Co and Ru are compared. While the barrier for the C–H cleavage of CH_4 is higher on Co than on Ru, the barriers for C–C bond formation and C–C bond dissociation are lower on Co than on Ru. On Ru the stronger

M–C interaction in the product as well as reactant state results in weaker interaction between reacting hydrocarbon fragments, so that barriers for the reaction in both directions increase. This in line with the observed lower rate of hydrocarbon hydrogenolysis observed for Pt as compared with Ni. On Ni the rate of C–H activation is lowered more than that for the CH_x–CH_y bond cleavage reaction. Because of the stronger metal–carbon bond of Pt than Ni, the rate of methane formation by recombination of adsorbed hydrogen with adsorbed CH_3 will be lower on the former.

Now let us compare the dissociation of water with methane. On Ni, the activation energy for H_2O dissociation is lower than that of CH_4 because of the substantially stronger metal–OH interaction on Ni, $E_{\text{ads}}(\text{OH}) = 326$ kJ/mol, as compared with the metal–CH_3 interaction. In Fig. 3.47 the reaction energy diagrams are shown for decomposition of H_2O to adsorbed oxygen and hydrogen. The dissociation of water over Ni(111) proceeds via a path through the valley of three Ni atoms. The dissociative adsorption of water also strongly depends on coordinative unsaturation of the surface metal atoms. Bengaard et al.[20] predict that the activation energy for H_2O dissociation is decreased by 50 kJ/mol on comparing the open (211) surface with the close-packed less reactive Ni(111) surface.

Returning to our discussion of methane activation, as has been pointed out by Liu and Hu[32] and by Abbott and Harrison[35], the experimentally measured activation energies for CH_4 differ by more than 30 kJ/mol for the same surfaces. Whereas this may be partially due to an overestimate of the activation energies due to the inaccuracy of density functional theory itself, a more likely explanation is crystal imperfection on the experimental single crystal surfaces. For Rh and Pd the large reductions in activation energies for CH_4 and CO dissociation on surface kinks and steps are compared with those on terraces in Table 3.6.

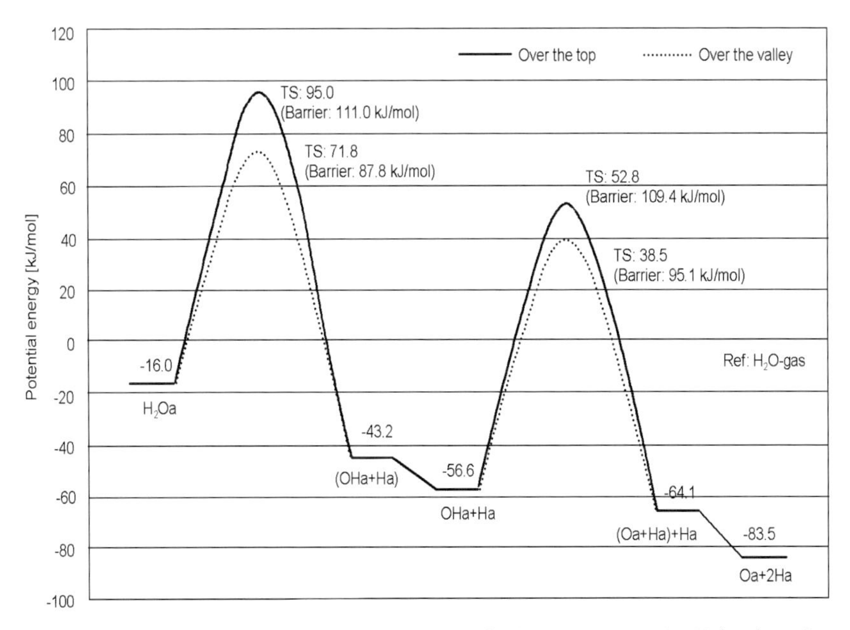

Figure 3.47a. Reaction energy diagrams of H_2O dissociation on the Ni(111) surface; energies with respect to adsorbed hydrogen.

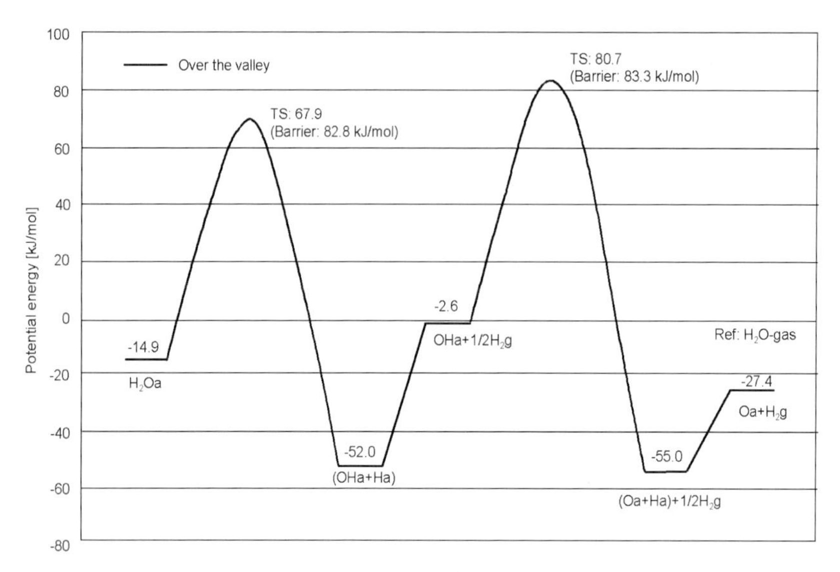

Figure 3.47b. Reaction energy diagram of H_2O dissociation on the Ru(0001) surface, energies with respect to H_2 gas phase.

Table 3.6. The calculated dissociation barriers[33] (E_a^{dis}) for $CH_4(g) \rightarrow CH_3^* + H^*$ and $CO^* \rightarrow C + O^*$ and the barriers (E_a^{as}) for their reverse reactions on different Rh and Pd surfaces[a]

	$CH_4(g) \leftrightarrow CH_3^* + H^*$		$CO^* \leftrightarrow C + O^*$		
	E_a^{dis}	E_a^{as}	E_a^{dis}	E_a^{as}	CN
Rh(111)	0.67	0.65	1.17	1.84	9
Rh-step	0.32	0.59	0.30	1.18	7
Rh-kink	0.20	0.49	0.21	1.09	6
Pd(111)	0.66	0.68	1.87	1.98	9
Pd-step	0.38	0.63	0.57	0.68	7
Pd-kink	0.41	0.53	0.38	0.49	6

[a]The least coordination number (CN) of the metal atoms involved in the TS_s on flat surfaces, steps and kinks are also listed for comparison. The unit of the barriers is in eV.

One notes the large decreases in activation energies for the activation of CH_4 on the more coordinatively unsaturated edge and kink atoms. The reaction path proceeds over the top of a single metal atom. Interestingly, for the recombination reaction of H_{ads} and CH_{3ads}, this difference in activation energies largely disappears. For CO dissociation, on the other hand, the activation energy for recombination also shows a large reduction.

The behavior of CH_4 activation follows that which is expected for late transitions as was discussed in Chapter 2. This only applies, however, for systems that have similar reaction pathways. This is not the case, however, to CO. There is a significantly large decrease in the activation energies for CO, where the C and O atom that form in the transition state do not share the same metal atoms, whereas they do on a terrace. Since the reaction paths are very similar on the step and kink sites, the differences between the activation energies in the forward and backward directions are relatively small.

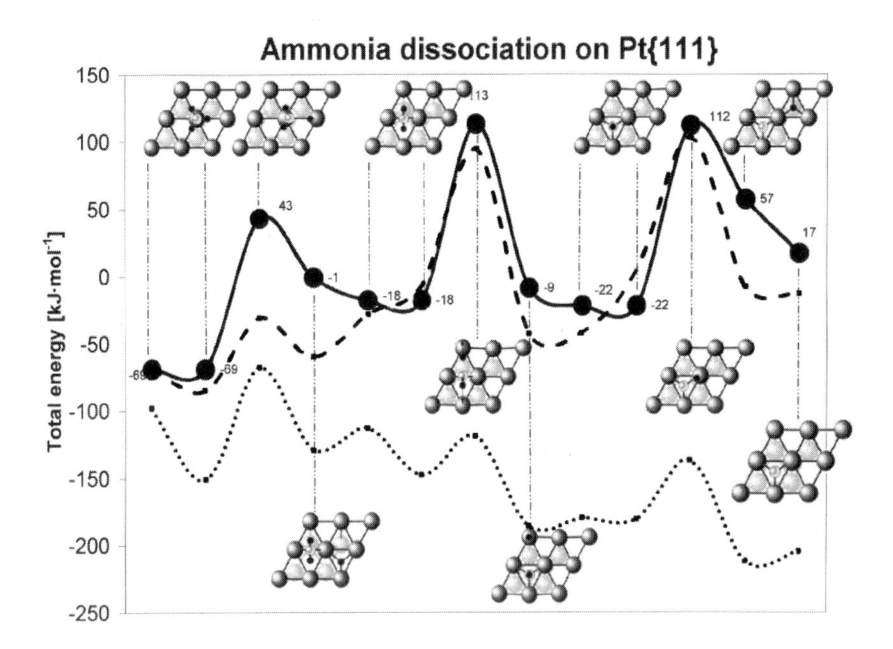

Figure 3.48. The activation of NH_3 by Pt(111). Reaction energy paths and structures of reaction intermediates and their corresponding transition states. Reaction energy diagram for the transformation of NH_3 to N_{ad}. — — — NH_3; — · —· NH_3+O_{ads}; ··· NH_3+OH_{ads}[36].

Two classes of reactions can be distinguished: structure sensitive and structure insensitive. The cleavage of the C–H bond that occurs for the oxidative addition is structure sensitive. The reverse reaction for the reductive elimination, however, is not ($\alpha = 1$!). This is due to the reaction path, that involves C–H activation via a transition state that involves a metal atom that shares bonding to the molecule CH_3 and H fragments. More coordinatively unsaturated metal atoms bind the reaction fragments more strongly and this acts to help lower the barrier for C–H activation. In contrast, CO dissociation, as well as CO formation, are both structure-sensitive reactions. The interaction with a large ensemble of atoms with the right topology (a geometric effect) lowers the energy of the transition state. Low activation barriers are found for the forward and reverse reactions when the fragments do not share metal surface atoms in the transition state. This concept of metal-atom sharing can also be used to help understand the promotion of adsorbate bond activation by coadsorption. Ammonia decomposition, for example, can be activated in the presence of surface oxygen. Figure 3.48 illustrates the reaction energy paths for NH_3 activation as found on a Pt(111) surface[36]. On clean Pt, the dissociation of adsorbed NH_3 is thermodynamically unfavorable. Experiments indicate that at low temperature NH_3 does not decompose on the close-packed Pt surfaces. The presence of oxygen, however, can help to promote this reaction. The activation of NH_3 over the clean Pt surface is compared here with activation of NH_3 in the presence of coadsorbed atomic oxygen and OH. Nitrogen is the main product at low temperature. The competitive product is NO. Reaction energies and reaction paths for the recombination of N and O adatoms are shown in Fig. 3.49. The recombination of nitrogen adatoms to form N_2 is quite similar.

The coadsorption of atomic oxygen enhances the adsorption energy of ammonia by 18 kJ/mol. This enhancement of the adsorption energy occurs as long as adsorbed NH_3 and atomic oxygen do not share a bond with the same metal atom. According to the Bond Order conservation principle, the weakened metal–metal bonds next to adsorbed oxygen enhance the reactivity of the surface metal atom to which NH_3 adsorbs. The reaction of ammonia with coadsorbed oxygen reduces the activation barrier by over 68 kJ/mol. In the transition state, the oxygen atom moves to a two-fold position and the hydrogen–oxygen bond has already been partially formed.

The barrier heights for the subsequent elementary reaction steps with adsorbed O increase. This is due to the fact that O_{ad} and NH_{2ad} have to share metal surface atoms in the transition state. This results in strong repulsive interactions. The overall result is that the rate of the initial dissociative NH_3 adsorption with coadsorbed O is increased, but the reactivity of adsorbed NH_2 and NH are reduced.

On Pt, OH adsorbs on an atop site. Therefore, its interactions with NH_3, NH_2 and NH with OH in the transition state do not share binding to the same metal atom. Reactions with adsorbed OH to form H_2O lower the activation energies for all three of these cases.

The effect of surface steps on the activation of the NH bond over Pt was found to be negligible. Nevertheless, the experimental rate of ammonia activation will be higher near the step than on the terrace. This is due to an increase in the surface concentration of ammonia at the step over the concentration on the terrace rather than the intrinsic activation barriers. The higher coverage of ammonia at the step is the result of a high adsorption energy of ammonia at the step. The heat of adsorption of NH_3 is 20 kJ/mol higher on a terrace than on the step. Hence the apparent activation energy for the dissociative adsorption, at low coverage, which is equal to $E_{TST} + E_{\text{ads}}$, is decreased by the same amount. The overall effect will be an increased rate of dissociative adsorption.

The apparent activation energy for the dissociative adsorption of NH_3 with preadsorbed O is significantly lower. It then follows that the effect of coadsorbed oxygen is much larger than that of the activation by steps.

The low-temperature oxidation reaction of NH_3 is extensively discussed in Chapter 6. The experimental evidence indicates that ammonia will only dissociatively adsorb when coadsorbed oxygen is present on the Pt surface. N_2 is the initial product that forms in the presence of oxygen in a flow system operating under mild conditions. N_2O is formed as a co-product as the catalyst begins to deactivate.

The recombination of nitrogen adatoms occurs at at step edges, with a low barrier ($E_{\text{act}} = 70$ kJ/mol) and, hence, at low temperature. The formation of N_2 at low temperature, however, can also occur via the reaction of NO with NH_3. This reaction requires the formation of surface NH_2 to proceed (see Chapter 6). Because of its high adsorption energy ($E_{\text{ads}} = 187$ kJ/mol), NO once formed will not desorb, especially at low temperatures. NO, however, can be indirectly detected if it reacts to form N_2O. N_2O readily desorbs and is a co-product that indeed is observed.

The activation energies and reaction paths for the recombination of N and O to give NO on a terrace and a stepped Pt(111) surface are shown in Fig. 3.49. The activation energy for N + O recombination ($\text{E}_{\text{act}} = 223$ kJ/mol)is slightly lower than that for the N + N recombination reaction ($\text{E}_{\text{act}} = 234$ kJ/mol). The decrease of the activation energy for the formation of NO on the terrace step ($E_{\text{act}} = 65$ kJ/mol is so large that it occurs readily at low temperatures. However, the desorption energy of NO from a terrace is 187 kJ/mol and from the hollow position on the step is 133 kJ/mol, hence NO will not desorb at low temperatures. The low reactivity of the Pt(111) surface, without steps, is clearly

Total energy [kJ/mol]
100
50
0
-50
-100
-150
-200
59
-68
-103
-138
-135
-164
-133
-187
+

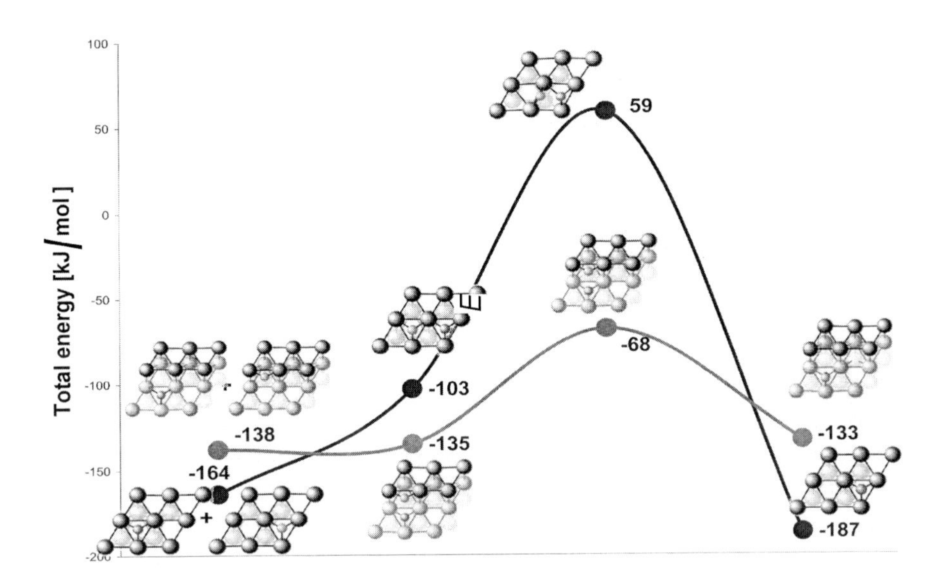

Figure 3.49. The recombination reaction path for nitrogen and oxygen on a stepped Pt(111) surface. Reactions at terraces and steps are compared.

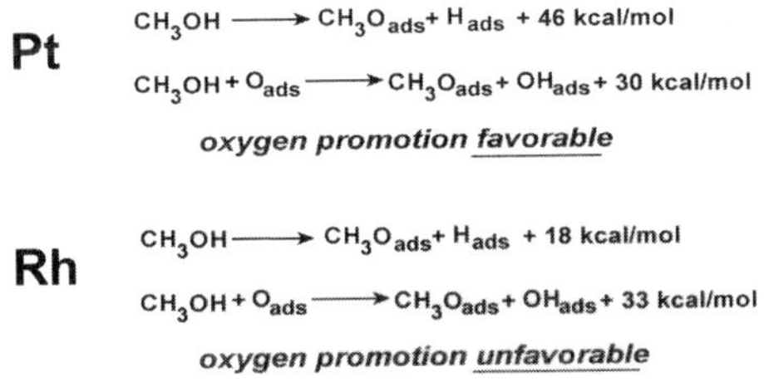

Pt
CH3OH → CH3Oads+ Hads + 46 kcal/mol
CH3OH + Oads → CH3Oads+ OHads+ 30 kcal/mol
oxygen promotion favorable
Rh
CH3OH → CH3Oads+ Hads + 18 kcal/mol
CH3OH + Oads → CH3Oads+ OHads+ 33 kcal/mol
oxygen promotion unfavorable

Figure 3.50. Coadsorbed oxygen atoms are not always promoters[45].

due to the high barriers for the recombination reaction of $N_{ads} + O_{ads}$. The presence of steps, however, provide low activation barrier paths for these recombination reactions. As discussed in the previous subsection, NO has a unique reactivity on the Pt(100) surface. Qe and Neurock[24a] showed that NO dissociation is strongly enhanced, but the N and O adsorption energies are less increased. A low barrier (~7 kJ/mol) for N_{ads} and O_{ads} recombination is found owing to the unique transition-state configuration over the four-fold hollow site made up of squarely arranged Pt atoms. NO recombination therefore also proceeds over the hollow site of squarely arranged Pt atoms. The N and O atoms do not share binding to the same metal atom in the transition state.

The reaction of NH_3 with adsorbed O to form hydroxyl intermediates on Pt decreases the reaction energy for the dissociative adsorption of NH_3. In contrast, such a reaction on Rh acts to increase the reaction energy. This relates to the much stronger Rh–O_{ads} bond (–110 kcal/mol) as compared with the Pt–O_{ads} (–66 kcal/mol) bond. For dissociative adsorption of methanol, this is illustrated in Fig. 3.50. The stronger oxygen adatom bond to Rh than to Pt may have a beneficial selectivity effect in reactions where hydrogen formation competes with water formation. Hickman and co-workers[39] studied methane oxidation on Rh and Pt monoliths. The main products from this reaction were CO, CO_2, H_2 and H_2O. On Rh they found for reaction conditions which enhanced the selectivity to CO production over that of CO_2 a substantially higher selectivity to H_2 compared with H_2O. The activation energy for H_2O formation from $2H_{ads}$ and O_{ads} over Rh equals 120 kJ/mol. On Pt the corresponding value is only 60 kJ/mol. The bond energy of hydrogen with Rh or Pt is quite similar and, hence, the activation energy for H_2 recombination also will be similar. The improved H_2 selectivity on Rh is due to the suppressed rate of H_2O formation on this surface caused by the stronger metal–oxygen bond energy.

A final remark has to be made on the nature of the transition states. The transition states for the first bond cleavage reaction of CH_4, NH_3 and H_2O to produce H_{ads} and adsorbed CH_3, NH_2 or OH, respectively, have slightly higher entropies than the bond cleavage of the second or successive X–H bonds. In the transition state for the NH cleavage of NH_3, the entropy remains relatively high because the HNH part of the molecule remains nearly freely rotating. The lower transition-state entropies for NH_2 and NH cleavage support the view that the bond cleavage reactions proceed through tight transition states[40]. The higher transition-state entropy for the cleavage of the first X–H bond implies a slightly looser transition state and, hence, a smaller dependence on the degree of coordinative unsaturation of the surface atoms. On the other hand, surface reactions tend to proceed through activation barriers of low activation entropy and, hence, bond cleavage reactions show a strong dependence on overall reaction enthalpy. This will be reflected in BEP transition state with α values close to 1. For reactions involving CH_x and OH_{ads} surface species this has been confirmed by Michaelides et al.[41].

3.9 Carbon–Carbon Bond Cleavage and Formation Reactions, a Comparison with CO Oxidation

Watwe et al.[42] describe the C–C bond cleavage over Pt(111) and Pt(211) surfaces. Despite the differences between the two surfaces, the structures for the adsorbates on the two surfaces are similar. A common feature of the transition states is the significant extension of the C–C bond compared with that in the ground state. In the transition state, the C–C bond was found to be at least 25% longer. The transition states are late. As follows from Table 3.7, the barrier for breaking the C–C bond increases with increasing bond order, and is lower for those bonds which generate fragments that are more stabilized by the metal surface.

There is a substantial lowering of the barrier when the C–C bond is activated over a step site. In the gas phase, the activation of σ-C–C bonds is thermodynamically preferred over that of alkane C–H bonds. On the surface, however, C–H activation is found to be easier than C–C bond activation. This relates to some extent to the differences between the M–CH_3 and M–H bond energies. The C–H bond is more readily activated on the surface since there is initially less steric hinderance than that for H_xC–CH_y activation. In the adsorbed state, the hydrogen atoms of an alkyl intermediate directly interact with the surface through weak van der Waals interactions. The carbon atoms, on the other hand, are not in direct contact.

Carbon–carbon bond formation reactions are of critical importance to the Fischer–Tropsch synthesis of linear hydrocarbons over transition metals such as Co, Fe or Ru. Fischer–Tropsch synthesis involves the activation of CO and H_2 over the metal to form adsorbed carbon, oxygen and hydrogen. The carbon atoms that form hydrogenate to form different CH_x intermediates that can subsequently couple to form longer hydrocarbon chain intermediates and products. We present here theoretical results obtained out over well-defined transition-metal surfaces for some of the critical steps in the mechanism. Zheng et al.[43] demonstrated that on terraces of the transition metals the CH_x–CH_y recombination reaction requires low values of x and y otherwise repulsive interactions between the hydrogen atoms on the CH_2 and CH_3 groups will prevent recombination. This conclusion is in line with the high barrier for the ethyl C–C cleavage reaction found on the Pt(111) surface reported in Table 3.7. The barrier is due to the large repulsive interaction that the hydrogen atoms experience with the surface.

Table 3.7. Transition-state energies for C–C bond cleavage on two surfaces of Pt[42]

Fragment	Surface	E_{TST} (kJ/mol)
C_2H_5	111	173
	211	102
C_2H_4	211	193
$CHCH_3$	111	106
$CHCH_2$ (parallel to step)	111	160
	211	160
$CHCH_2$ (over the step)	211	100

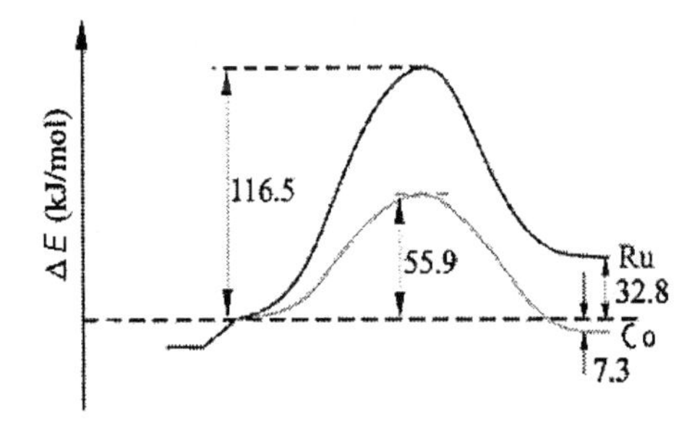

Figure 3.51. Schematic potential energy surfaces (relative energies) on Co and Ru for C–C coupling of CH with CH_2[45].

A similar argument holds for the recombination of CH_3 and CO on the terrace of a transition metal. The strong repulsive interactions between the CH and the metal surface in the transition state with the CO surface species has to be overcome. Reaction paths that cross over a step edge or occur directly at a step edge are expected to experience far less repulsion interactions between these adsorbates. Calculated barriers for CH_x–CH_y recombination on stepped and unstepped Ru surfaces are given in Table 3.8. Ethylene formation is a preferred reaction on the steps. Also C + CH and C + CH_2 recombination are preferred recombination steps.

The low barrier for the recombination of CH and CH_3 surface species on terrace sites is quite striking. Work by Ge et al.[45] allows a comparison of the CH + CH_2 recombination reaction on the less reactive Co(0001) compared with that on the Ru(0001) surface. The recombination reaction occurs with a substantially lower barrier on Co than Ru (see Fig. 3.51). This is due to the weaker M–CH_x bonds on Co. The proposition that the C_1 adsorbed intermediates, namely CH_{ads}, are the most abundant reaction intermediates (MARI) in Fischer–Tropsch synthesis is consistent with the fact that the the CH_{ads} species is calculated to be more stable on the surface than the C_{ads} species. Chain growth would then require the following reaction steps:

$$CH_{ads} + CH_{3_{ads}} \longrightarrow CHCH_{3_{ads}} \quad (1)$$

$$CHCH_{3_{ads}} + H_{ads} \longrightarrow CH_2CH_{3_{ads}} \quad (2)$$

$$CH_{ads} + CH_2CH_{3_{ads}} \longrightarrow CHCH_2CH_{3_{as}} \quad (3)$$

The calculated transition state of 77 kJ/mol for step (2) implies that in this cycle of steps, that hydrogen addition may be the rate-limiting step. While the results from theoretical predictions suggest that we can speculate as to the MARI and the rate-limiting step, we can not be sure unless we carry out full simulations of the elementary processes occurring simultaneously to establish the actual kinetic outcome.

Table 3.8. Calculated reaction barriers (E_a) for C–C coupling and some hydrogenation addition reactions on the Ru surface (eV)[44]

	Ru step	Ru(0001)
C + C	1.05	1.51
C + CH	0.43	1.01
C + CH_2	0.56	1.08
CH + CH	0.95	0.87
CH + CH_2	1.20	0.97
CH_2 + CH_2	0.59	1.23
CH_2 + CH_3	1.40	1.80
CH + CH_3		48
$CHCH_3$ + H→CH_2CH_3		77
$CHCH_2$ + H→$CHCH_3$		40

As long as there are enough CH species available on the surface, the chain growth reaction may be faster than the chain termination reaction. On steps and surfaces where CH_2 species are stabilized [e.g. the Ru(1121) surface] ethylene formation will compete strongly with the chain growth reaction. Alkene formation can also proceed by the β-CH cleavage of adsorbed alkyl species. For Pd the activation barrier for this reaction is 70 kJ/mol, which competes with the activation energy for hydrogenation of ethyl. On Ru the activation barrier of the β-CH reaction is even lower. Therefore, alkene formation is preferred over alkane formation in the chain termination step.

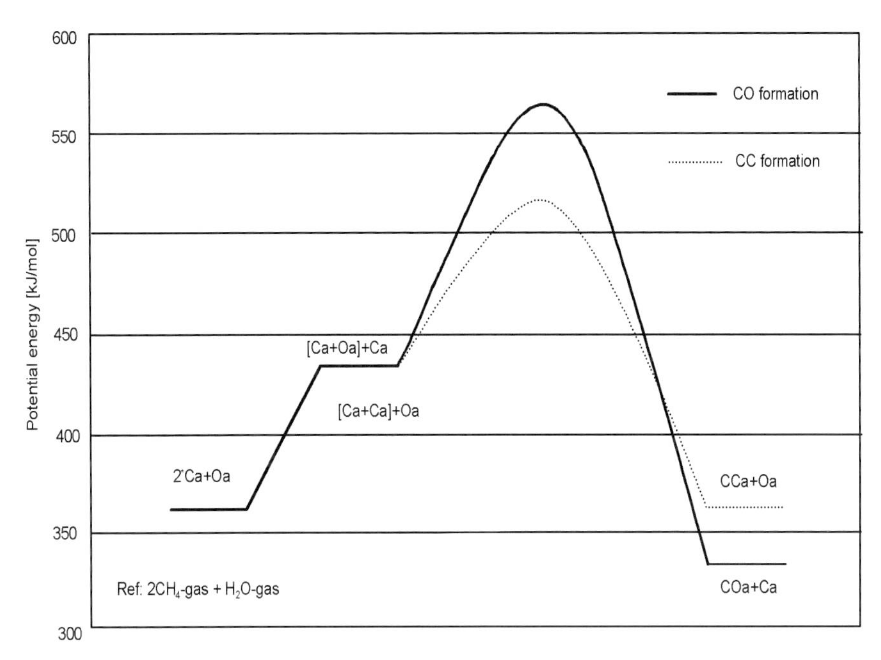

Figure 3.52a. A comparison of reaction paths for the C–C and C–O recombination reactions on Ru(0001). (DFT slab calculations).

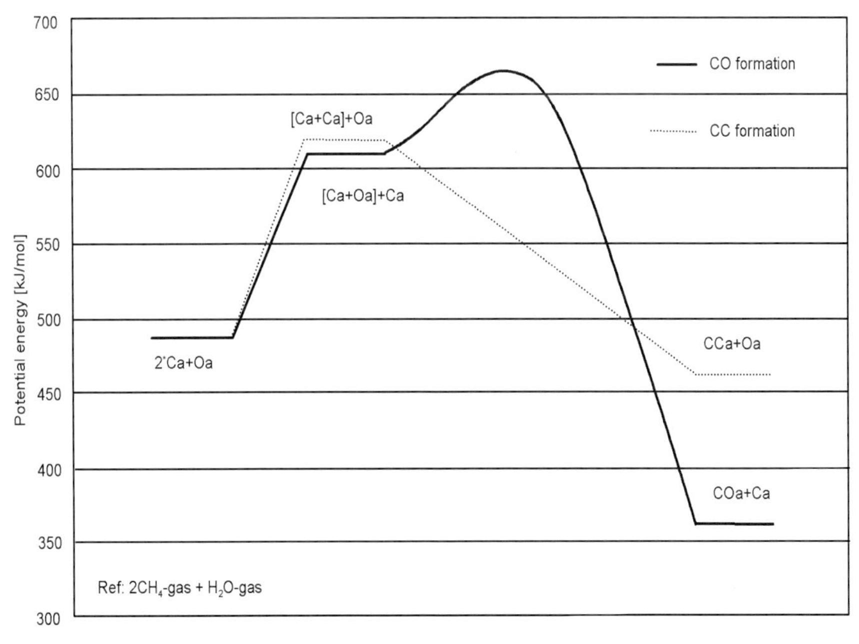

Figure 3.52b. A comparison of the reaction paths for the C–C and C–O recombination reactions on Ni(111). (DFT slab calculations).

Fe	Co	Ni	Cu
0	25	7.3	0
Ru	**Rh**	**Pd**	**Ag**
19	4.1	0	0

Figure 3.53. Fraction of adsorbed C_1 species formed via CH_4 decomposition at 400 °C[48], cooled to room temperature and subsequently converted to ethane and propane upon hydrogenation at room temperature[47].

The third alternative termination path is by direct insertion of CO into the metal–alkyl bond. This insertion reaction requires the presence of coadsorbed CO and occurs on surfaces where CO bond cleavage competes with C–C bond formation as shown in Section 3.7. CO insertion or hydroformulation requires cationic metal centers and is therefore an unfavorable reaction at a reduced metal surface.

There is ample experimental evidence[26] that suggests that the Fischer–Tropsch reaction proceeds via the formation of C_1 species, as proposed above. Experimental evidence indicates that the C_1 species are favored as surface CH_x intermediates[46].

In order to dissociate CO, the barrier for CO activation has to be low. As illustrated in Table 3.4, the rate of dissociative CO adsorption appears to be fastest over Ru, Co and Fe, the preferred transition metals for the Fischer–Tropsch reaction. The barrier for CO dissociation, however, is very high compared with that of the chain growth or termination reactions, especially on the close-packed terraces. While the presence of steps helps to lower the CO activation barrier, it is still considered to be rate limiting.

The requirement for a low activation energy for CO dissociation is opposite to that for efficient carbon–carbon bond formation, which favors weak metal–carbon bonds. This is nicely illustrated in Fig. 3.52a and b where we compare the C–C bond formation and C–O bond formation energies on Ru(111) and Ni(111) surfaces, respectively. Table 3.4 summarizes the transition-state energies for CO dissociation, and indicates that Ni is less reactive than Rh. The M–C bonds are weaker on Ni than on Ru as reflected by the higher endothermicity for carbon deposition on Ni. Whereas C–C bond formation on Ru has a substantial barrier, such a barrier is absent on Ni. In the presence of hydrogen, the C–C bond formation reaction, which is the precursor to graphite formation, is suppressed. Once CH species are present on the surface, the graphitization reaction becomes much more difficult and typically does not occur. In the presence of hydrogen, the CH_x-CH_y bond formation reaction begins to compete with the methanation reaction. In the methanation reaction, the metal–carbon bonds become completely broken. The C–C bond formation reaction, however, only depends weakly on variation in the M–C bond energies. Hence methanation will be more strongly dependent on the M–C bond energy than the chain growth reaction is. This prediction is confirmed by experiments[47] in which CH_{ads} species were generated by dissociative adsorption of CH_4 on the surface. In the presence of hydrogen, C–C coupling occurs and the alkanes desorb. The competitive reaction here is CH_4 formation. The selectivity towards higher hydrocarbons is shown in Fig. 3.53. The stronger the metal–carbon bond, the larger is the selectivity towards higher hydrocarbons. There is no reaction on metals such as Cu, or Ag where H_2 cannot dissociate. There is also no reaction when the M–C bond is too strong, as is the case for Fe.

3.10 Lateral Interactions

3.10.1 Introduction

Up to this point, we have focused on modeling the intrinsic reactivity of individual molecules on a surface which can be used for comparison with experiments carried out at low surface coverages under UHV conditions. Reactions that are carried out under more industrially relevant conditions, however, are typically run at pressures which are many orders of magnitude greater. This is known as the pressure-gap problem in surface science. These higher pressures can lead to significantly higher surface coverages, which can subsequently change the surface composition as well as the catalytic performance. At higher surface coverages, the intermolecular interactions between adsorbed intermediates become important in dictating the bond strength of the adsorbate to the surface and also its reactivity. The interactions between adsorbed species are known as lateral interactions and can be defined in terms of *through-space* or *through-surface* interactions. The *through-space* interactions are typically due to local steric or electrostatic interactions between two or more species. These interactions are based solely on the position of the adsorbate with respect to other molecules or intermediates on the surface. They exist even in the absence of the metal surface. Through-surface interactions are the result of electronic interactions between two or more adsorbates that are mediated through changes in the electronic structure of the surface metal atoms. We discuss the chemical bonding aspects of lateral interactions in more detail in Section 3.10.2.

The interactions between coadsorbed molecules or atoms can be either attractive or repulsive depending upon the local positioning of the adsorbates with respect to one another. If two or more adsorbates form bonds with the same metal atom, the interactions are typically repulsive, as would be predicted by Bond Order Conservation (BOC) principles. The interactions between adsorbates that are distant from one another by a single metal bond are typically attractive, which again would follow BOC principles. At higher surface coverages, the interactions between adsorbates are typically, but not always, repulsive. Repulsive interactions weaken the adsorption energy and thus lead to a decrease in the desorption energy with coverage. The metal-mediated electronic changes can lead to overlayer ordering and, in some cases, even drive surface reconstruction, as was discussed in Chapter 2. The substantial weakening of the adatom or adsorbed molecule interaction occurs when two or more adsorbates bind to the same surface metal atom. The reduction in the binding energy with coverage implies that such adsorption states are likely populated only at low temperature and high pressure and not likely accessible under UHV conditions.

The adsorbates which are more weakly bound to the surface are more likely to interact with other surface species through bond-making processes. An example of this situation will be discussed in Section 3.10.3 where we examine the ethylene hydrogenation mechanism as a function of surface coverage. We specifically analyze the elementary reaction steps for both π- and σ-bonded ethylene intermediates.

Lateral interactions influence the reactants, products, intermediates and even transition states for a reaction. Reactant molecules likely adsorb in different local environments and are therefore exposed to different lateral interactions depending upon the relative number, type and position of neighboring adsorbates. Stochastic kinetic methods provide the best hope of capturing these molecular differences. Traditional deterministic modeling of catalytic systems average over the surface coverage and thus provide only a mean field description. Individual surface sites, as well as intermolecular interactions, however, can be treated by adopting kinetic or dynamic Monte Carlo simulations which are mathematically rigorous solutions. We describe kinetic Monte Carlo simulation methodology in more detail in the Appendix. Various different applications are describe later in this chapter and in other chapters.

3.10.2 Lateral Interaction Models

First-principle quantum chemical calculations, have proven to be instrumental in quantifying the energetics for spatially explicit interactions between species at specific surface sites. For example, strongly adsorbed species such as oxygen and nitrogen are typically repulsive when they share either one or more metal atoms. The repulsive interactions increase as the number of shared metal atom neighbors increases. On copper, the repulsive interactions between atomic oxygen and nitrogen bound to fcc sites which share a single metal atom are repulsive by 30 kJ/mol. The repulsive interaction energies increases to 130 kJ/mol when these same two species sit at sites which share two metal atoms. Part of this strong repulsion is also likely due to the significantly shorter distance between the two oxygen atoms which sit at adjacent fcc and hcp sites and share two metal atoms. This distance is on the order of 1.5 Å, thus leading to strong repulsive effects. For adsorbates which lie at nearest-neighbor sites but do not share metal atoms, the interactions can actually be attractive. For example, the ethylene–ethylene interaction is attractive by 11 kJ/mol per ethylene pair when the two ethylene molecules bind to bridge sites that are separated by one vacant bridge site.

Attractive or repulsive through-surface interactions are readily understood in terms of the Bond Order Conservation principles. When an adatom binds to a neighboring surface metal atom, the metal–metal bonds that form to the surface metal atom of interest are weakened. This increases the potential reactivity of the neighboring metal atoms since less of its electron density is tied to the metal atom involved in the surface–adatom bond. Thus, another adatom bound to the neighboring surface metal atom would have an increased interaction energy. Through-surface interactions are repulsive when two or more adsorbates share a metal atom, but attractive when the adsorbates sit at neighboring metal atom sites. These effects are illustrated in Fig. 3.54.

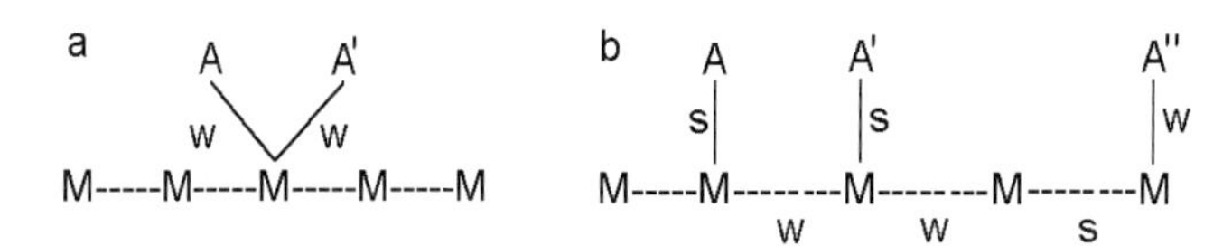

Figure 3.54. Schematic illustration of the use of Bond Order Conservation to predict attractive or repulsive interaction between adsorbates. (a) Adsorbates which bond to the same surface metal atom are weakened by the presence of one another as the result of competition for electron density from the same metal atom. These interactions are repulsive. (b) Adsorbates that are bound to metal atoms which are neighbors have an effective attractive interaction, because of the weakening of the metal–metal bond due to their coadsorption. Bond order conservation indicates that attractive and repulsive interactions alternate through bonds. Binding to a next-nearest metal atom neighbor such as A'' versus A' has a weaker interaction and, hence, this through-surface interaction is repulsive.

Although first-principle calculations offer quantitative estimates for specific configurations, the shear number of different scenarios which arise in any kinetic or dynamic Monte Carlo simulation make it impossible to compute all of the possible configurations that might arise from first-principle calculations. Currently the only practical way to include lateral interactions would be to develop a simpler coarse-grained model that can be used to calculate the interactions as they arise within the simulation.

Various models have been proposed in the literature to model adsorbate–adsorbate interaction[49]. At the simplest level, the interactions can be described by a single parameter ω_{AA} which treats the repulsion ($\omega > 0$) and attractive ($\omega < 0$) interactions between two species labeled A:

$$Q_{\mathrm{A}} = Q_0 - \omega_{\mathrm{AA}} \tag{3.37}$$

This approach has been adapted for systems whereby the lateral interactions are lumped into a single parameter. More advanced treatments add a second parameter ($\omega'_{\mathrm{AA'}}$) in order to describe next-nearest neighbor interactions. Q_0 here refers to the binding energy of the isolated A molecule. The simplest molecular level treatment would view these as pairwise additive interactions. The change in the binding energies and activation barriers can easily be computed by simply adding (or subtracting) the effect of all pairwise interactions that result from direct nearest (ω) and next-nearest neighbors (ω'), This is shown for the adsorption of A in the following expression:

$$Q_{\mathrm{A}} = Q_0 - \sum_{A_i} \omega_{\mathrm{AA_i}} - \sum_{j} \omega'_{\mathrm{AA_j}} \tag{3.38}$$

A much more sophisticated treatment for modeling all interaction pairs as well as trimer interactions was developed by Kreuzer and co-workers[50]. They subsequently extended the approach to examine oxygen on Ru using first principle DFT calculations[51]. They calculated a large number of different possible configurations for oxygen on Ru, and then regressed the coefficients for both pair and ternary interactions in the model to these constants. This is an elegant study whereby a theory was used to establish a coarse-grained interaction model. The results follow the experimental TPD curves for oxygen on Ru(0001) very well. The number ab initio calculations necessary to establish a ternary model for catalytic systems with multiple different adsorbates would be prohibitive.

A second approach which may be attractive for more complex surface systems involves the application of the Bond Order Conservation model that was developed by Shustorovich and co-workers[52–65]. The BOC model treats the interaction between the adsorbate and the surface atom through the use of a Morse potential. The total heat of adsorption is then described by summing all interactions. The BOC model is based on the concept that the bonding potential for every atom in the system is conserved. The heat of adsorption for an atomic species A is described by the following expression:

$$Q_{\mathrm{A}n} = Q_{0\mathrm{A}}(2 - 1/n) \tag{3.39}$$

where A and n refer to the adsorbate A and the number of metal atoms, respectively.

Shustorovich and Sellers [61,65] developed a systematic set of equations which can be used to estimate activation barriers for adsorption, surface reaction and desorption processes based upon the strength of atomic interactions Although this approach may not yield quantitative predictions for all systems, it has been very effective in estimating the barriers and adsorption energies for a number of systems.

Hansen and Neurock[66] showed how such an approach can be coupled with first-principle DFT calculations in order to model quantitatively the interactions of O/Ru(100).

The resulting model was used internal in a kinetic Monte Carlo simulation in order to provide predictions of the lateral interactions along with site specificity as the reaction progressed.

A third approach to treat lateral interactions internal in the atomistic (or molecular) simulation involves the use of semiempirical quantum chemical methods. Lateral interactions can be described in this method by extracting smaller grids from the surface and using an extended Hückel molecular orbital theory or a tight-binding method to calculate lateral interactions within each grid[67]. After thousands of EHT interactions, a substantial database can be built up by which the interactions can then be directly embedded into the simulation. A model for the interactions between ethylene, hydrogen, and ethyl intermediates was developed to describe their surface interactions by using a series of first-principle calculations along with over 1700 EHT calculations of adsorbate–adsorbate interactions by which to parameterize a simple coarse-grained empirical model based on a radial distance between adsorbates as well as the molecular size of each adsorbate[67]. This radial function model was subsequently used internal to a Monte Carlo simulation of ethylene TPD and hydrogen kinetics.

3.10.3 Hydrogenation of Ethylene; the Importance of Lateral Interactions

Ethylene adsorbs on metal surface in two predominant adsorption modes: di-σ and π. In the di-σ adsorption mode, the orbitals on the carbon atoms can rehybridize to form two direct σ bonds with two different metal surface atoms. This is the preferred adsorption mode on a number of different surfaces, especially at lower coverages. In the π-adsorption mode, ethylene coordinates to a single metal atom, similar to that found for ethylene binding to an organometallic complex (see Fig. 3.55). The chemical bonding is best described by donative-back-donative interactions. At low coverage, the adsorption energy of ethylene in the di-σ mode was calculated to be –60 kJ/mol at a coverage of 0.33 monolayer (ML). The adsorption energy in the π-adsorption mode at the same coverage was calculated to be –30 kJ/mol. When the ethylene coverage increases, the interactions between di-σ-adsorbed ethylene is initially attractive. DFT calculations indicate that there is an attractive interaction of –20 kJ/mol between two coadsorbed ethylene species bound to neighboring Pd-dimer pairs, that do not share any metal bonds on the Pd(100) surface. As expected from the Bond Order Conservation principle, the weakened Pd–Pd bonds enhance the adsorption of ethylene at these sites[65]. As the coverage within the overlayer increases, the ethylene molecules begin to interact more directly with one another through metal-atom sharing. This decreases the energy of adsorption.

Hydrogenation is thought to occur via a Horiuta–Polanyi mechanism which involves the sequential addition of hydrogen atoms to the adsorbed ethylene. Molecular hydrogen dissociates to form two atomic hydrogen atoms on the surface which ultimately react with coadsorbed ethylene [67,68]. At low surface coverage, the reactivity with the π-adsorbed ethylene is low. The interaction of ethylene with adsorbed hydrogen instead pushes the π-bonded ethylene to the more favorable di-σ-bonded state. At higher coverages, however, hydrogenation can occur through the addition to either the di-σ- or π-adsorbed ethylene. As Fig. 3.55 illustrates, the barriers, as well as the two transition-state structures for the two hydrogen additions, are quite similar.

The addition of a single hydrogen atom to ethylene leads to the formation of an atop adsorbed ethyl intermediate, which occurs with an overall activation energy of 88 kJ/mol. (The activation barrier for the reverse reaction to activate the C–H bond in ethane is 106 kJ/mol, which is the same order or magnitude as the values found for the activation of

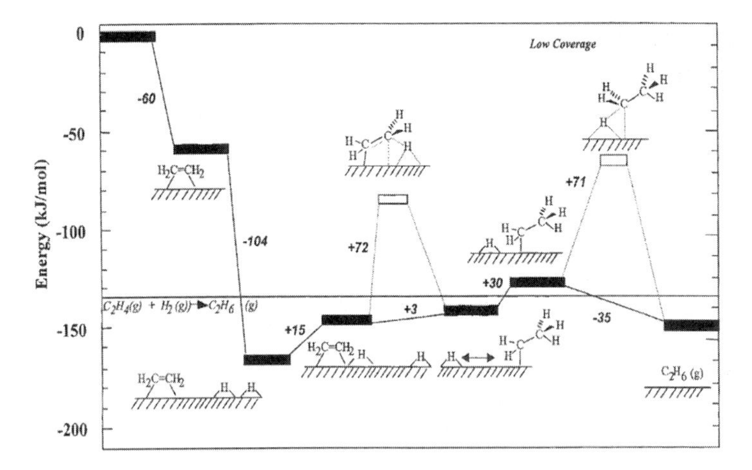

Figure 3.55a. Reaction energy diagram of ethylene hydrogenation on a Pd(111) surface at low ethylene surface concentration[68].

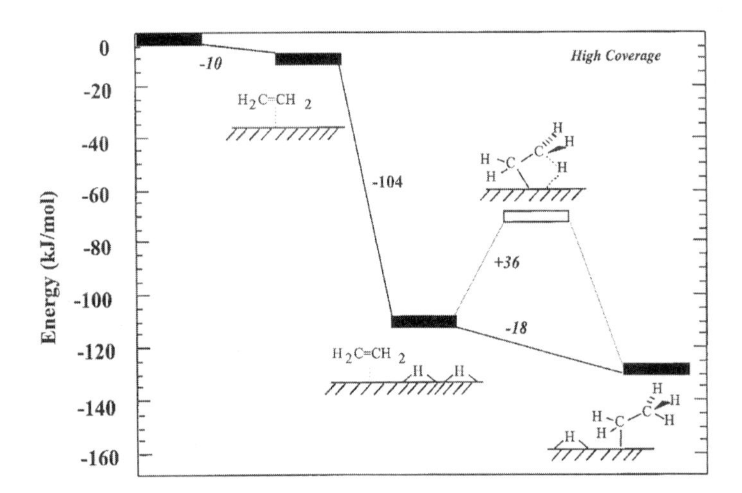

Figure 3.55b. Reaction energy diagram of ethylene hydrogenation on a Pd(111) surface at high ethylene surface concentration[68].

methane). The barriers found for hydrogen addition to ethylene are significantly lower than those for the competitive reaction paths in which ethylene C–H bonds are cleaved with formation of ethylidene or CH_x species. At higher coverages of ethylene, or when the metal surface becomes covered with hydrogenolysis products, the di-σ-adsorption of ethylene becomes suppressed and π-bonded ethylene becomes more favorable. The activation barrier for hydrogen addition to a π-bonded ethylene surface intermediate as depicted in Fig. 3.55b was 36 kJ/mol, which is significantly lower than that for the low-coverage di-σ intermediate. This barrier is also significantly lower than the barrier computed for insertion of hydrogen into ethylene adsorbed within an organometallic complex of Pd^{2+}. Siegbahn[48] studied the insertion of the hydride ion into the metal–carbon bond. In the process, the ethylene molecule slips from its π-bonded coordination mode along the Pd^{2+} cation, to form the ethyl ligand.

The barrier for the hydrogenation reaction is largely due to the formation of antibonding occupied orbitals between hydride and reacting carbon atom. This is very similar to that discussed previously for the insertion reaction of ethyl and CO discussed in section 3.7. This repulsive interaction is reduced by the donation of electrons from the C–H bond into empty transition-metal cation orbitals. The predicted barrier for the hydrogenation of CO on the Pd complex was 103 kJ/mol. This is significantly higher than the barrier which was calculated on the surface. When the density on the metal-ion center decreases the barrier height is reduced. For instance, the barrier height for this reaction is reduced to 75 kJ/mol when it is carried out over Mo^{5+}. This difference in barrier energies between the surface and organometallic complex illustrates the importance and the benefit of the unique topologies that surfaces offer due to the large ensemble size of the metal atoms involved in the reaction.

3.10.4 Lateral Interactions; the Simulation of Overall Surface Reaction Rates

Reactant or product states of surface reactions are often (de-)stabilized by the presence of other adsorbates. This implies a change in the reaction energy as a function of overlayer composition. The Brønsted–Evans–Polanyi relation again provides an elegant procedure to estimate the effect of lateral interactions on changes in the activation energies.

$$\delta\Delta E_{\mathrm{act}} = \alpha\,\delta\Delta E_R \tag{3.40}$$

The change in the activation energy due to lateral interactions is seen to be simply proportional to the difference of the lateral interaction energies of the reaction intermediates before and after reaction.

A nice example of such an effect is given by the recombination of CO with adsorbed O on the Pd(111) surface as studied by Zhang and Hu[69]. With respect to the reverse reaction, which involves the cleavage of CO_2, the transition state is considered to form late along the reaction channel, thus forming a tight transition-state complex. In the transition state, CO_2 has an angle of $\sim 100°$, which is indicative of ${CO_2}^{\delta-}$. The barrier of recombination should be quite sensitive to the interaction of atomic oxygen with the transition-metal surface. As in all related transition states, the three-fold adsorbed oxygen moves towards the two-fold adsorption site. CO has a small barrier to move from its preferred adsorption site at the three-fold position towards an atop position on a third metal atom. This helps to lower the activation barrier for the reaction of CO with O, which is weakly bound to the two-fold bridge site, thus allowing the recombination to occur. The barrier to activation is dominated by the need to reduce the metal–oxygen and metal–CO bond energies. Zhang and Hu calculated the activation energy for CO–O recombination for surfaces that are 25% covered with oxygen and 16% with oxygen. The barrier was found to increase from 93 to 150 kJ/mol as the coverage was reduced. The oxygen adsorption energies changed from 370 and 420 kJ/mol, respectively, and that of CO from 160 and 195 kJ/mol, respectively, as the coverage was reduced.

By properly including the lateral interactions between adsorbed species into a dynamic or kinetic Monte Carlo algorithm, one can simulate the response of different experimental protocols including the simulation of temperature-programmed desorption (TPD) and temperature-programmed reaction (TPR) spectroscopy, steady-state and transient kinetics. The application of dynamic or kinetic Monte Carlo simulation thus offers a more accurate treatment surface kinetics for higher coverage systems since it allows the kinetic and equilibrium properties to be coverage dependent. In such simulations, the surface is typically represented by some form of a lattice. As such, each atop, bridge and hollow site along with the specific ad-species on the surface can explicitly be followed as a function of time and reaction conditions. The state system is then defined by the specific atomic surface structure, along with the specific location and configuration of all of the intermediates in the adlayer. The evolution of the system with time or processing conditions requires tracking the changes of the system state. The state of the surface changes via individual elementary physicochemical kinetic processes such as reaction, diffusion, adsorption or desorption. The evolution of the states of the system as a function of (real) time can be described by means of the chemical Master Equation:

$$\frac{\mathrm{d}P(c,t)}{\mathrm{d}t} = \sum_{c' \neq c} \left[k_{cc'} P\left(c', t\right) - k_{c'c} P(c,t) \right] \tag{3.41}$$

where $P(c,t)$ denotes the probability of finding the system in a specific state or configuration c at time t; $k_{c',c}$ is the transition probability per unit time of the elementary process that changes the system from state C to C$'$. The transition probability can be interpreted as a microscopic rate constant, that can be described by the Arrhenius equation:

$$k_{c'c} = \nu_{c'c}\, exp\left(\frac{E_{c',c}}{kT} \right) \tag{3.42}$$

An analytical solution to the master equation is only possible for very simple systems. The master equation, however, can readily be simulated by using stochastic kinetics or more specifically kinetic Monte Carlo simulation. Several Monte Carlo algorithms exist. More details on kinetic Monte Carlo simulation can be found in the Appendix.

Since we will also discuss in later chapters (Chapter 8 and 9) the use of cellular automata to study surface reactions, it is important to compare kinetic Monte Carlo with cellular automata methods. The main characteristic of cellular automata is that each cell, which corresponds to a grid point of a surface model, is updated simultaneously. The realism of such an assumption is questionable since reaction appears to be a random process. Randomness can be incorporated by using probabilistic cellular automata, in which updates are done with some probability. Probabilistic cellular automata simulations can be developed that are equivalent to the Random Selection Method.

As a first example, we discuss the simulation of ethylene desorption from the Pd(100) surface[67]. As mentioned in Section 3.10.2, a set of lateral interaction parameters was developed by regressing over 1700 extended Hückel calculations at different ethylene coverages. The simulation was then used to simulate both temperature-programmed desorption of adsorbed of ethylene and the kinetics for the high-pressure hydrogenation of ethylene. The simulation of the TPD spectra for ethylene on Pd(100) demonstrated that the initial high coverage state of the surface was comprised of a disordered array of ethylene molecules bound in random orientations, with repulsive interactions between the adsorbed ethylene molecules. Two desorption peaks were observed. The first peak, which appeared at low temperatures, was due to the desorption of weakly bonded ethylene. This peak was the direct result of lateral repulsive interactions between ethylene intermediates that occcurred at the higher coverages which formed at low temperature. As the surface was heated, ethylene desorbed, thus freeing up surface sites. At higher temperatures, a stable well-ordered (2 x 2) ethylene overlayer began to appear. This was the result of the attractive interactions between ethyene molecules when they adsorb in a (2 x 2) arrangement, as was confirmed by DFT calculations. The higher temperature ethylene TPD peak is due to the desorption of ethylene from this ordered overlayer. The simulated adsorption peaks agreed reasonably well with those measured experimentally.

The other example is for NO dissociation on Rh(111) to produce N_2[70]. A lattice model of atop, fcc and hcp sites was used to model the surface. The interactions between next-nearest and next-next-nearest neighbors were explicitly included. NO adsorption was considered on both the three-fold and atop adsorption sites. Nitrogen and oxygen atoms were only allowed to adsorb at three-fold coordination sites. The kinetic parameters for NO, N and O diffusion and NO desorption, N_2 formation and NO desorption were defined by comparison with experiment. The simulated TPD curves were found to be in very good agreement with the experimental curves, as is shown in Fig. 3.56.

The simulations were able to reproduce the ordered structures found experimentally and the lateral interaction parameters regressed from this system were consistent with the values derived from DFT calculations. The low-coverage region (< 0.2 ML) was characterized by the dissociation of all NO. NO dissociation was found to be complete at 300 K. N_2 then desorbs around 550 K. The medium coverage region was characterized by the partial decomposition of the three-fold bound NO. N and O form islands separated from NO islands; dissociation stops when the NO islands are compressed into an ordered structure of 0.50 ML coverage. NO can be compressed more easily than N and O that are constrained to three-fold adsorption sites. Only when part of the NO desorbs above 400 K, can more NO dissociate. This is the reason why the N_2 desorption peak appears at the temperature where NO desorbs. At coverages higher than saturation (above 0.5 ML), dissociation is completely inhibited. NO adsorbs atop and initially can only desorb. NO dissociation is not yet possible dowing to the lack of available sites. Dissociation remains suppressed until some of the three–fold adsorbed NO starts to desorb.

Ab initio-based kinetic Monte Carlo studies have been implemented by Neurock and co-workers[66,67,72] to follow a fairly comprehensive set of adsorption, surface diffusion, desorption and surface reaction processes in order to monitor the surface kinetics for various different reaction systems including NO decomposition[72e,f], ethylene hydrogenation [67,72a–d], and vinyl acetate synthesis[72g]. We briefly described some of the simulation results for the influence of alloying Pd with Au on the kinetics for ethylene hydrogenation and vinyl acetate synthesis in Chapter 2. In order to present the utility of the approach to follow the influence of lateral interactions and model high-coverage conditions, we discuss here the hydrogenation of ethylene over Pd.

This example provides a natural extension to show how ab initio Monte Carlo simulations can be extended from the TPD studies under UHV reported in the previous two examples described above to catalytic kinetics over surfaces under more realistic reaction conditions. Neurock and co-workers[67,72a–d] extended the ab initio-based DFT formalism discussed above to steady-state and transient ethylene hydrogenation catalytic kinetics at higher pressures. The DFT-calculated potential energy profiles for the ethylene hydrogenation presented in Fig. 3.55 were used together with the ethylene, hydrogen, ethyl lateral interaction model, which was described earlier, to follow the elementary adsorption, desorption, surface diffusion and surface reaction processes along with the lateral interactions using ab initio-based kinetic Monte Carlo simulations. More specifically, the simulations tracked the fate of individual molecules on the surface as a function of reac-

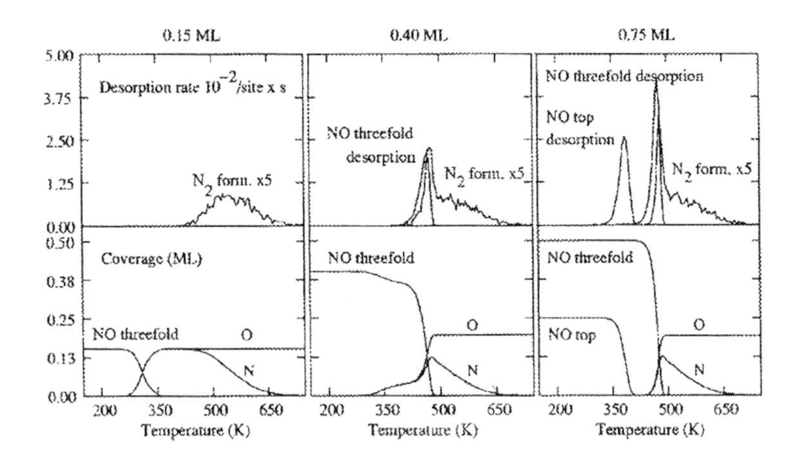

Figure 3.56a. NO and N_2 desorption rates (top), and NO, nitrogen and oxygen coverages (bottom), during temperature-programmed desorption. Starting coverages are (from left panel to right) 0.15, 0.40 and 0.75 ML. N_2 desorption rates have been multiplied by 5; the heating rate was 10 K/s[71].

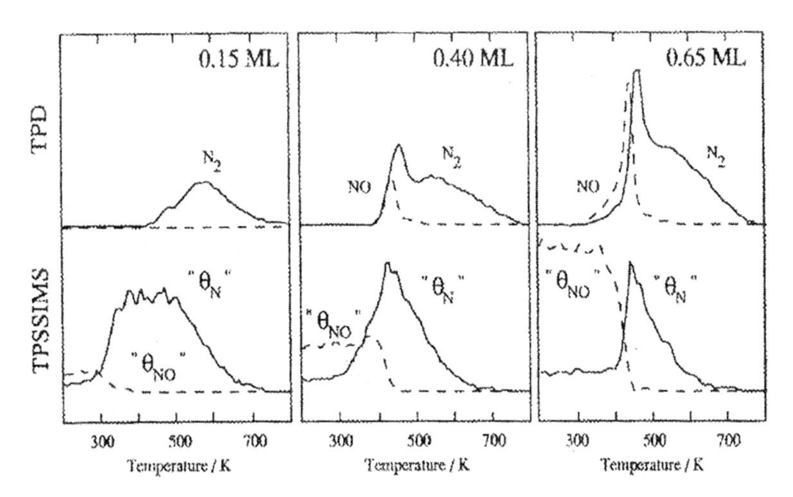

Figure 3.56b. The NO (- - - -) and N_2 (—) TPD rates (top) and $\sum_{n=1,2} Rh_n\ NO_n^+$ (- - - -) and Rh_2N^+/Rh_2^+ (—) TPSSIMS ion intensity ratios (bottom), during the temperature-programmed reaction of NO on Rh(111) for low (left panel), medium (central panel) and high (right panel) initial NO coverages. The NO TPD spectra have been divided by a factor of 4 with respect to the N_2 TPD spectra. The adsorption temperature was 100 K: the heating rate was 10 K/s[71].

tion conditions to determine the number of ethane product molecules that form as a function of the number of active sites. This is the catalytic turnover frequency. The reaction conditions such as temperature and pressure and also the surface composition can all be varied to establish their influence on the overall catalytic performance. As such, the simulations can be used as "virtual experiments" in order to predict macroscopic features such as the turnover frequency, selectivity, apparent activation energies and reaction orders for more direct comparison with experiment. In addition, the simulation also provides for a full molecular-level description, thus explicitly tracking the occupancy of individual atop, bridge, and three-fold fcc and hcp hollow sites throughout the simulation.

In the previous TPD simulations, the surface was allowed to equilbrate at low temperature before starting the simulation runs. The reagents were thus no longer allowed to adsorb on the surface. In simulating the kinetics, the background partial pressure of ethylene and hydrogen, and also the temperature, were set to the conditions of interest and then held constant in order to simulate the steady state. Molecules were allowed to adsorb and desorb continuously, thus providing the ability to repopulate the surface as the reaction proceeds. This is governed by the kinetics for adsorption and desorption. The higher pressures of ethylene ($P_{C_2H_4} = 25$ torr) and hydrogen ($P_{H_2} = 100$ torr) used in these simulations resulted in surface coverages of ethylene of about 0.20 ML and hydrogen of about 0.42 ML. The total surface coverage was significantly higher than that reported for UHV conditions. The higher coverages led to lateral interactions on the surface that were predominantly repulsive, which ultimately lowers the activation energy.

The simulations allowed for the formation and reaction of both π- and di-σ-bound ethylene surface intermediates. At higher coverages, the reaction proceeds through both intermediates. While the di-σ intermediate is present in higher surface concentrations, the π-bound intermediates form and can rapidly hydrogenate. The results agree with the experimental studies of Cremer and Somorja [73] on ethylene hydrogenation on Pt that distinguish the π-bound ethylene as the reactive intermediate. The simulations here on Pd(111), however, show that hydrogenation can still proceed through the di-σ route also. Repulsive interactions ultimately weaken both the π and di-σ species, making them both reactive channels.

Simulations were run at a series of different temperatures in order to determine surface-averaged activation barriers. The intrinsic barriers for low-coverage ethylene hydrogenation discussed earlier were 15 kcal/mol. The simulations of the apparent activation barriers, however, were found to be significantly lower at 9.2 kcal/mol, which agrees very well with reported experimental values of 9–12 kcal/mol[74]. The dramatic reduction in the barrier between the zero coverage limit and actual surface conditions is due to the lateral repulsive interactions that exist between surface adsorbates, as shown for the DFT calculations for ethylene hydrogenation at higher coverages presented in Fig. 3.55. The simulations were run at various partial pressures of ethylene and hydrogen in order to determine the reaction orders for both ethylene and hydrogen. The reaction orders for ethylene and hydrogen were calculated to be 0.65–0.85 and 0.16–0.03, respectively. This agrees quite well with the known experimental literature values (0.5–1.0 for hydrogen and –0.5 to 0.0 for ethylene)[74]. The concluding message is that while the adsorbate surface bond strength is critical in determining reactivity, the extrinsic factors such as coverage effects can be just as important owing to the changes that they can impart on the metal–adsorbate bonding.

In Chapter 2, we decribed the simulation results for vinyl acetate synthesis over Pd and Pd/Au alloys. The results showed that acid and oxygen preferentially dissociate on Pd sites. The addition of Au was found to decrease the Pd ensemble size and, hence, the surface coverage. This reduces the blocking of the Pd sites by acetate anions. The addition of Au opened up sites for ethylene to adsorb and to coexist with both acetate and oxygen. The adsorption of ethylene on Pd(111) without Au was significantly suppressed because of the strong acetate adsorption. This system is an interesting example of a heterogeneous catalyst where the mixture of two metals creates two different adsorption sites to permit the reaction between two different molecules that otherwise would not react or would react with difficulty. Alloying in this example provides a synergy the coadsorption of reactants that would otherwise be quite difficult. This appears to be a general effect of alloying, also observed in very different reaction systems.

For example, the unique properties of the Pt–Ru alloy in electrochemical CO oxidation appears to relate to OH generation on Ru, which is difficult on Pt since the reaction is suppressed by CO[75] poisoning on Pt. Kinetic Monte Carlo calculations can also be used to simulate voltammograms and adsorption in electrochemical experiments.

We will briefly describe here the so-called butterfly voltammogram found for adsorption of anions on single-crystal electrode surfaces[76]. As illustrated in Fig. 3.57, the formation of ordered adlayers of anions is often accompanied by a characteristic sharply peaked current response in the cyclic voltammograms.

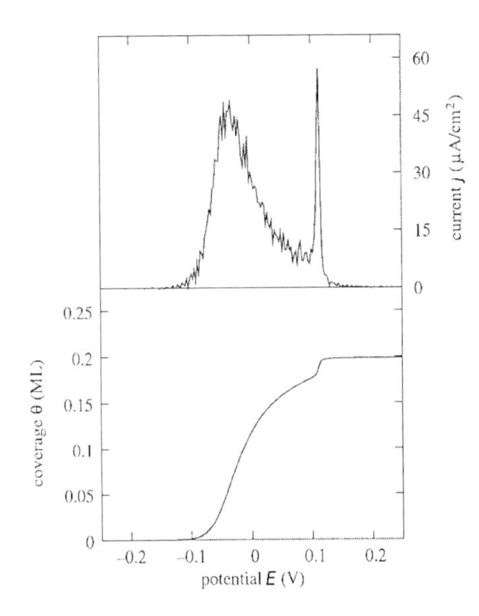

Figure 3.57. Simulated voltammogram (top) and adsorption isotherm (bottom) for (bi)sulfate adsorption on an fcc(111) surface.

To produce this voltammogram, adsorbate adsorption was modeled by Monte Carlo simulation employing the lattice-gas model for the adsorbate (eg sulfate anion), $A^{2-} + 2{*} \rightleftharpoons A_{ads} + 2e^-$, where * denotes an empty site. The interaction between adsorbates can be included in several ways. In the example in Fig. 3.57, a shell of purely hard sphere interactions is considered, in which the simultaneous bonding of two anions to neighboring sites is excluded. The isotherm can be calculated by including adsorption, desorption and surface diffusion steps and scanning potential E. The rate constants for adsorption and desorption are of the form

$$k_{ads} = k^0 \exp\left(\frac{-\alpha_{ads}\,\gamma e\,E}{k_B T}\right) \tag{3.43}$$

where α is the BEP coefficient for adsorption, taken as $\frac{1}{2}$, γ is the electrosorption valency (taken as -2), e the elementary charge, and E the electrode potential. The current j

follows from the expression

$$j = -e\gamma\Gamma_m \nu \frac{\mathrm{d}\theta}{\mathrm{d}E} \tag{3.44}$$

where Γ_m is the number of surface sites per unit surface area and ν the potential sweep rate. Going from more negative to more positive potential, the anion adsorbs between -0.1 and +0.1 V in a disordered phase (see Fig. 3.57; $k^0 = 10^3\,\mathrm{s}^{-1}$, $k_{\mathrm{diff}} = 10^5\,\mathrm{s}^{-1}$, $\nu = 50$ mV/s).

The simulation results show a broad adsorption peak in the voltammogram. A disorder–order transition occurs at 0.11 eV, which is distinguished by the sharp peak in the voltammogram. At this voltage, the anion coverage rapidly increases to saturation coverage. The onset of the disorder–order transition is shown in Fig. 3.58c and the ordered state in Fig. 3.58d.

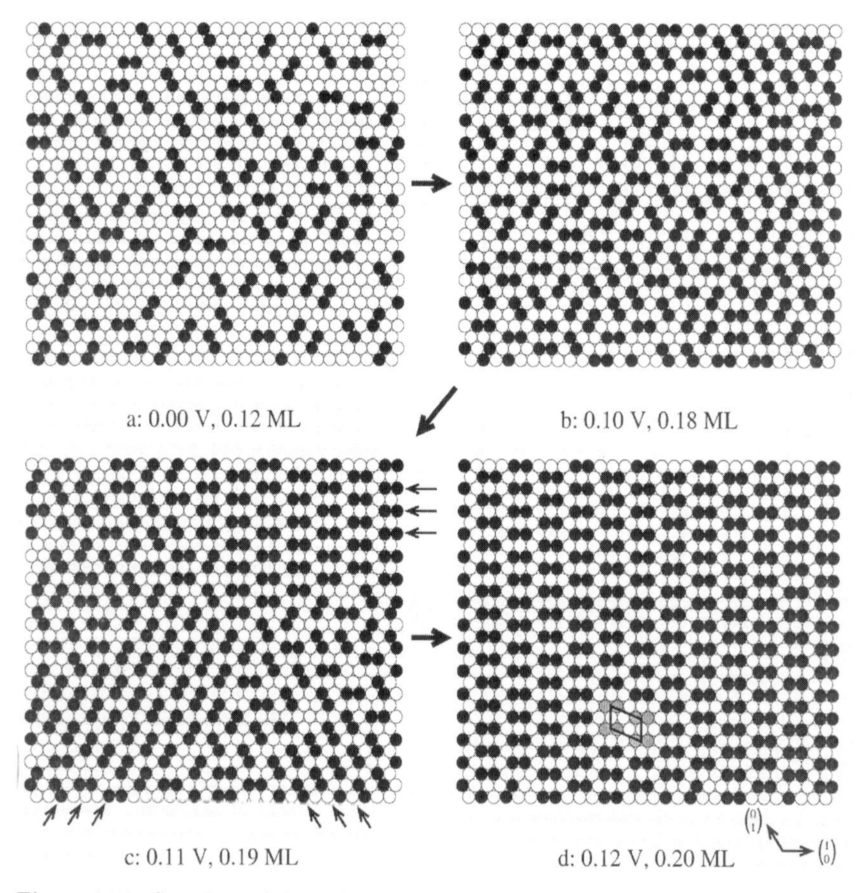

Figure 3.58. Snapshots of the surface during anion adsorption for the model with first neighbor shell exclusion. Before the disorder–order transition (a) and (b) there is no ordering; during the disorder–order transition ($\sqrt{3}$ x $\sqrt{7}$) islands grow [(c); the three different domain orientations are indicated by the small arrows]; after the disorder–order transition large islands dominate (d). The ($\sqrt{3}$ x $\sqrt{7}$) unit cell is indicated in (d).

Whereas the accuracy of computed adsorption and activation energies is usually not better than 10 kJ/mol, remarkably, simulated kinetics often compares much better with experiment than one might expect based on the accuracy of quantum chemically obtained data. When this inaccuracy is based on systematic errors, the explanation is provided by a compensation effect[77]. It intimately relates to the Brønsted–Eyring–Polanyi relation and also to the Sabatier principle. For example, according to the BEP relation a decrease in the activation energy of an elementary reaction step is proportional to an increase of adsorption energies. Hence it will increase the equilibrium concentration of surface adsorbates. Typically, the overall rate of an important surface reaction such as a dissociation reaction is given by Eq. 2.21 (page 42), which indicates a strong sensitivety to surface concentration. Beyond the Sabatier maximum the overall rate decreases with surface concentration. The decrease in rate due to the loss in surface vacancies is compensated by the higher rate of the elementary dissociation step owing to its lower activation energy. Near the Sabatier maximum the overall rate is maximum as a function of adsorption energies. Therefore, at this optimum value of the adsorbate interaction energy, by definition, the computed rate is least sensitive to variation of computed adsorption energies. Hence predictions of catalytic turnover near the Sabatier maximum will have the smallest error.

3.11 Addendum; Hybridization

We will first introduce the hybridization concept by discussing the electronic structure of the linear BeH_2 molecule.

BeH_2

The valence atomic orbitals are: for H_2 two φ_{1s} and for Be: φ_{2s}, φ_{2p_x}, φ_{2p_y} and φ_{2p_z}.

If we choose the z-axis to be oriented along the BeH_2 axis, only the σ-type orbitals interact: $2\varphi_{1s}(H)$, $\varphi_{2s}(Be)$, $\varphi_{2p_z}(Be)$. The $\varphi_{2s}(Be)$ atomic orbital interacts only with the symmetric combination of the two hydrogen atomic φ_{1s} orbitals and the $\varphi_{2p_z}(Be)$ orbital interacts only with the antisymmetric combination of the two hydrogen $\varphi_{1s}(H)$ orbitals.

The resulting molecular orbital diagram is given in Fig. 3.59.

The orbital splitting due to the respective symmetric and antisymmetric orbital interactions result in two σ and σ_2 subsystems. The $2p_x$ and $2p_y$ orbitals on Be remain non-interacting and are the two $\epsilon_5(\pi)$ and $\epsilon_6(\pi)$ BeH_2 orbitals. Within the Hückel approximation the differences in energy of the respective bonding and antibonding σ orbital sets are given by

$$2\Delta_1 = \sqrt{\Big[\in_{1s}(\mathrm{H}) - \in_{2\mathrm{s}}(\mathrm{Be})\Big]^2 + 8\beta_{2\mathrm{s},1\mathrm{s}}{}^2} \tag{3.44a}$$

$$2\Delta_2 = \sqrt{\Big[\in_{1s}(\mathrm{H}) - \in_{2\mathrm{p_z}}(\mathrm{Be})\Big]^2 + 8\beta_{2\mathrm{p_z},1\mathrm{s}}{}^2} \tag{3.44b}$$

where $\in_i$ are the energies of the atomic orbitals and β the corresponding overlap energies.

The bonding and antibonding orbitals are degenerate when

$$\in_{2s}(\mathrm{Be}) = \in_{2p_z}(\mathrm{Be}) \tag{3.45a}$$

$$\beta_{2s,1s} = \beta_{2p_z,1s} \tag{3.45b}$$

This defines the condition for ideal hybridization, with $\delta = 0$.

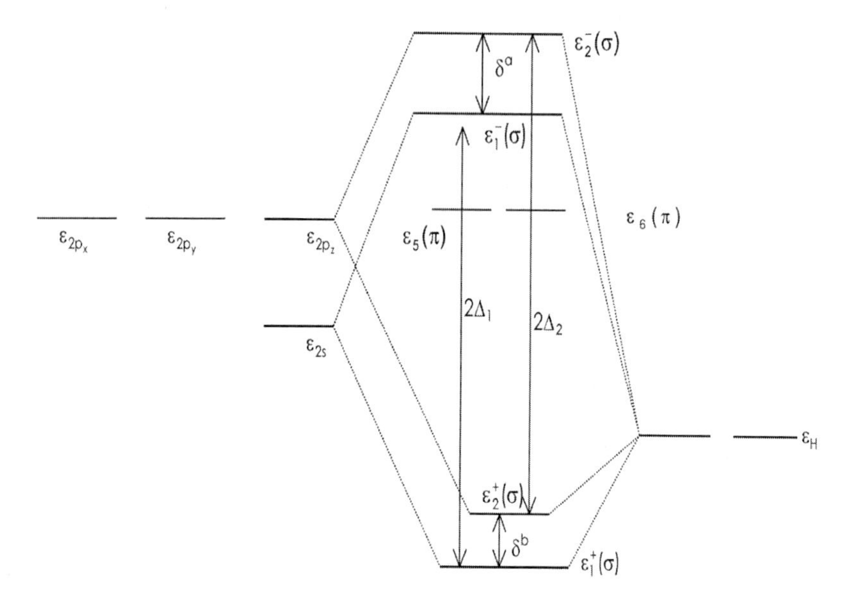

Figure 3.59. The molecular orbital energies for the BeH_2 molecule. 2δ is the difference in energy of the bonding and antibonding σ-orbital pairs; δ is the difference in energy of the two ϵ_{2p_z}, and ϵ_{2s}, based bonding and antibonding orbitals, respectively.

One constructs hybridized atomic orbitals on Be by the combination of atomic orbitals along the symmetry axis of a molecule. In our example for Be:

$$\varphi^+ = \frac{1}{\sqrt{2}}(2s + 2p_z) \tag{3.46a}$$

$$\varphi^- = \frac{1}{\sqrt{2}}(2s - 2p_z) \tag{3.46b}$$

They are illustrated in Fig. 3.60., where φ^+ is an orbital oriented to the right and φ^- is an orbital directed to the left. The energies of those orbitals are:

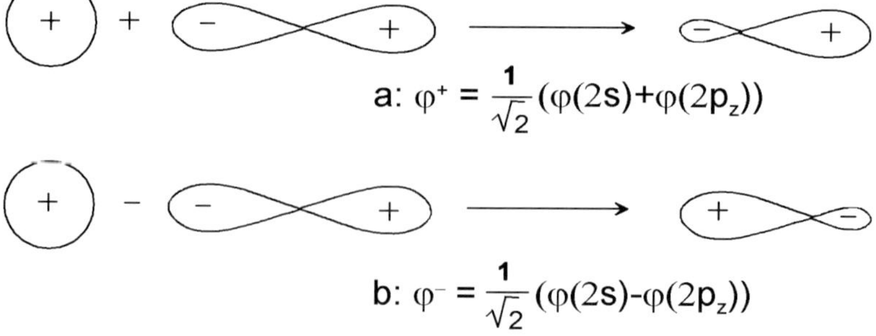

$$\text{a: } \varphi^+ = \frac{1}{\sqrt{2}}(\varphi(2s)+\varphi(2p_z))$$

$$\text{b: } \varphi^- = \frac{1}{\sqrt{2}}(\varphi(2s)-\varphi(2p_z))$$

Figure 3.60. As can be seen, φ^+ is oriented to the right and φ^- to the left. The hybridized orbitals φ^+ and φ^- are orthogonal.

$$\in^+ = \in^- = \frac{1}{2}(\in_{2s} + \in_{2p_z}) \tag{3.48}$$

the overlap energy of orbitals φ^+(Be) and φ^-(Be) is

$$\beta^{+-} = \langle\varphi^+|H|\varphi^-\rangle = \frac{1}{2}(\in_{2s} - \in_{2p_z}) \tag{3.49}$$

When the ideal hybridization condition is satisfied, $\beta^{+-} = 0$. The hybridized Beφ^+ orbital has the following overlap energies with hydrogen atoms a and b, located respectively right and left from Be in BeH_2:

$$\beta_{1s(a),\varphi^+} =< \varphi_{1s}(a)|H|\varphi^+ >= \frac{1}{\sqrt{2}}(\beta_{1s,2s} + \beta_{2s,2p_z}) \tag{3.50}$$

$$\beta_{1s(b),\varphi^-} =< \varphi_{1s}(a)|H|\varphi^- >= \frac{1}{\sqrt{2}}(\beta_{1s,2s} - \beta_{2s,2p_z}) \tag{3.51}$$

In the ideal hybridization limit:

$$\beta_{1s(a),\varphi^+} = \sqrt{2}\,\beta_{1s,2s} \tag{3.52}$$

$$\beta_{1s(a),\varphi^-} = 0 \tag{3.53}$$

A similar relation then holds between hybridized orbital φ^- and the hydrogen atomic orbital on the atom at the left hand position of Be.

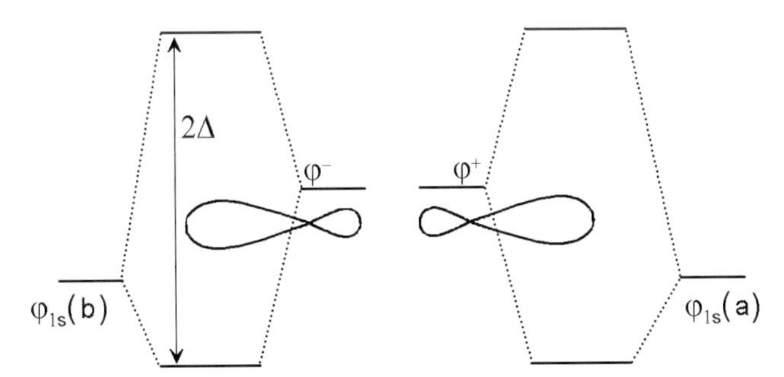

Figure 3.61. Ideal hybridization molecular orbital interaction scheme in BeH_2.

The φ^+ orbital interacts with the hydrogen atom to the right of the Be atom, the φ^- orbital interacts with the hydrogen atom at the left of the Be atom. The difference between the resulting degenerate bonding and antibonding orbitals (see Fig. 3.61) becomes in the ideal hybridization limit

$$\Delta = \Delta_1 = \Delta_2 \tag{3.54}$$

As long as $\frac{\delta}{\Delta} \ll 1$ (see Fig. 3.59), the hybridization model can be considered a satisfactory description of the chemical bond.

The condition for the approximate validity of this description is that the overlap energies of the 2s and 2p atomic orbitals with their neighboring atoms are approximately the same and these overlap energies are large compared with the Be atomic orbital energy difference of $\in_{2s}$ and $\in_{2p_z}$.

References

1. R. Hoffmann, *Solids and Surfaces*, VCH, Weinheim (1988)
2. F. Fréchard, R.A. van Santen, A. Siokou, J.W. Niemantsverdriet, J. Hafner, *J. Chem. Phys.* 111, 8124 (1999)
3. R.A. van Santen, *Theoretical Heterogeneous Catalysis*, World Scientific, Singapore (1991)
4. B. Hammer, J.K. Nørskov, *Adv. Catal.* 45, 71 (2000)
5. W. Biemolt, G.J.C.S. van de Kerkhof, P.R. Davies, A.P.J. Jansen, R.A. van Santen, *Chem. Phys. Lett.* 188, 477 (1992)
6. G. Blyholder, *J. Phys. Chem.* 68, 2772 (1964)
7. A. Föhlisch, M. Nyberg, P. Bennich, L. Triguero, J. Hasselström, O. Karis, L.G.M. Pettersson, A. Nilsson, *J. Chem. Phys.* 112, 1946 (2000)
8. M.T.M. Koper, R.A. van Santen, *J. Electroanal. Chem.* 476, 64 (1999)
9. W. Biemolt, *Thesis*, Eindhoven (1995)
10. M.C. Zonnevylle, R. Hoffmann, P.J. van den Hoek, R.A. van Santen, *Surf. Sci.* 223, 233 (1989)
11. E.M. Shustorovich, *Surf. Sci. Rep.* 6, 1 (1986)
12. G.A. Somorjai, M.A. van Hove, *Prog. Surf. Sci.* 30, 201 (1989)
13. R.A. van Santen, M. Neurock, *Catal. Rev. Sci. Eng.* 37, 557 (1995)
14. A. Fahmi, R.A. van Santen, *J. Phys. Chem.* 100, 5676 (1996)
15. R.A. van Santen *Chem. Eng. Sci.* 37, 557 (1995);
    M. Neurock, personal communication
16. R.A. van Santen, P.W.N.M. van Leeuwen, J.A. Moulijn, B.A. Averill, *Catalysis, an Integrated Approach*, Elsevier, Amsterdam (1996)
17. M.S. Daw, S.M. Foiles, M.I. Baskes, *Mater. Sci. Rep.* 9, 251 (1993)
18. P. van Beurden, *Thesis*, Eindhoven (2003)
19. R.M. Watwe, H.S. Bengaard, J.R. Rostrup–Nielsen, J.A. Dumesic, J.K. Nørskov, *J. Catal.* 189, 16 (2000)
20. H.S. Bengaard, J.K. Nørskov, J. Selested, B.S. Clausen, L.P. Nielsen, A.M. Molenbroek, J.R. Rostrup–Nielsen, *J. Catal.* 209, 365 (2002)
21. P. Kratzer, B. Hammer, J.K. Nørskov, *J. Chem. Phys.* 105, 5595 (1996)
22. H. Yang, J.L. Whitten, *J. Chem Phys.* 96, 5529 (1991)
23. I.M. Ciobica, R.A. van Santen, *J. Phys. Chem. B*, 107, 3808 (2003)
24. (a) Q. Ge, M. Neurock, *J. Am. Chem. Soc.* 126, 1551 (2004);
    (b) A. Eichler, J. Hafner, *J. Chem. Phys. Lett.* 343, 383 (2004)
25. J.K. Nørskov, T. Bligaard, A. Logadottir, S. Bahn, L.B. Hansen, M. Bollinger, H. Bengaard, B. Hammer, Z. Sljivancanin, M. Mavrikakis, Y. Xu, S. Dahl, C.J.H. Jacobsen, *J. Catal.* 209, 275 (2002)
26. P. Biloen, W.M.H. Sachtler, *Adv. Catal.* 30, 165 (1981)
27. M. Neurock, *Top. in Catal.* 9, 135 (1999)
28. P.W.N.M. van Leeuwen, *Homogeneous Catalysis*, Kluwer, Dordrecht (2004)
29. D.L. Thorn, R.J. Hoffmann, *J. Am. Chem. Soc.* 97, 4445 (1978)
30. I. Ciobica, R.A. van Santen, personal communications
31. (a) I.M. Ciobica, F. Fréchard, R.A. van Santen, A.W. Kleijn, J. Hafner, *J. Phys. Chem. B* 104, 3364 (2000);
    (b) I.M. Ciobica, R.A. van Santen, *J. Phys. Chem. B*, 106, 6200 (2002)
32. Z.-P. Liu, P. Hu, *J. Am. Chem. Soc.* 125, 1958 (2003)

33. A. Michaelides, P. Hu, *J. Am. Chem. Soc.* 123, 4235 (2001)
34. G. Henkelman, H. Jonsson *Phys. Rev. Lett.* 86, 664 (2001)
35. H.L.Abbott, I. Harrison, to be published
36. W. Offermans, R.A. van Santen, *Thesis,* Eindhoven (2005)
37. J.M. Bradley, A. Hopkinson, D.A. King, *Surf. Sci.* 371, 225 (1997);
M.F.H. van Tol, J. Siera, P.D. Cobden, B.E. Nieuwenhuys, *Surf.Sci.* 274, 63 (1992);
S.A.C. Carabineiro, B.E. Nieuwenhuys, *Surf. Sci.* 532, 87 (2003);
T.S. Amorelli, A.F. Carley, M.K. Rajumon, M.W. Roberts, P.B. Wells, *Surf. Sci.* 315, L990 (1994)
38. M. Chen, S.P. Bates, C.M. Friend, *J. Phys. Chem. B*, 101, 10051 (1997)
39. D.A. Hickman, L.D. Schmidt, *J. Catal.* 138, 267 (1992):
D.A. Dickman, E.A. Hanfoer, L.D. Schmidt, *Catal. Lett.* 17, 223 (1993)
40. W. Offermans, R.A. van Santen, in preparation
41. A. Michaelides, Z.P. Liu, C.J. Zhang, A. Alavi, D.P. King, P. Hu,
J. Am. Chem. Soc. 125, 3704 (2003)
42. R.M. Watwe, R.D. Cortright, J.K. Nørskov, J.A. Dumesic, *J. Phys. Chem. B*, 104, 2299 (2000)
43. C. Zheng, Y. Apeloig, R. Hoffmann, *J. Am. Chem. Soc.* 110, 749 (1988)
44. Z.P. Liu, P. Hu, *J. Chem. Phys.* 115, 4977 (2001)
45. Q. Ge, M. Neurock, H.A. Wright, N. Srinivasan, *J. Phys. Chem. B*, 106, 2826 (2002)
46. M. Cheng, A.W. Goodman, G.W. Zajac, *Catal. Lett.* 24, 23 (1954);
M. Cheng, D.W. Goodman, *J. Am. Chem. Soc.* 116, 1364 (1994)
47. T. Koerts, M.J.A.G. Deelen, R.A. van Santen, *J. Catal.* 138, 101 (1992)
48. P.E. Siegbahn, *J. Am. Chem. Soc.* 115, 5803 (1993)
49. R.I. Masel, *Principles of Adsorption and Reaction on Solid Surfaces,* Wiley, New York (1990)
50. (a) H.J. Kreuzer, *Surf. Sci.* 238, 305 (1990);
(b) H.J. Kreuzer, *Appl. Phys. A*, 51, 491 (1990);
(c) H.J. Kreuzer, *J. Zhang, Appl. Phys. A*, 51, 183 (1990);
(d) S.H. Payne, H.J. Kreuzer, *Surf. Sci.* 222, 404 (1989);
(e) S.H. Payne, H.J. Kreuzer, *Surf. Sci.* 338, 261 (1995)
51. C. Stampfl, H.J. Kreuzer, S.H. Payne, H. Pfnur, M. Scheffler, *Phys. Rev. Lett.* 83, 2993 (1999)
52. E. Shustorovich, *Surf. Sci.* 150, L115 (1985)
53. E. Shustorovich, *Surf. Sci.* 175, 561(1986)
54. E. Shustorovich, *Surf. Sci. Rep.* 6, 1 (1986)
55. E. Shustorovich, *Acc. Chem. Res.* 21, 183 (1988)
56. E. Shustorovich, *Surf. Sci.* 205,336 (1988)
57. E. Shustorovich, *Adv. Catal.* 37, 101 (1990)
58. E. Shustorovich, A.T. Bell, *Surf. Sci.* 253, 386 (1991)
59. E. Shustorovich, A.T. Bell, *Surf. Sci.* 248, 359 (1991)
60. E. Shustorovich, A.T. Bell, *Surf. Sci.* 289, 127 (1993)
61. E. Shustorovich, *Surf. Sci. Rep.* 31, 1 (1998)
62. A.T. Bell, E. Shustorovich, *Surf. Sci.* 235, 343 (1990)
63. H.L. Sellers, *Surf. Sci.* 294, 650 (1993)
64. H.L. Sellers, *Surf. Sci.* 310, 281 (1994)
65. H.L. Sellers, E. Shustorovich, *Surf. Sci.* 356, 209 (1996)

66. (a) E.W. Hansen, M. Neurock, *Surf. Sci.* 464, 91 (2000);
(b) E.W. Hansen, *Surf. Sci.* 441, 410 (1999).
67. (a) E.W. Hansen, M. Neurock, *J. Catal.* 196, 241 (2001);
(b) E.W. Hansen, M. Neurock, *Chem. Eng. Sci.* 54, 3411
68. M. Neurock, R.A. van Santen, *J. Phys. Chem. B*, 104, 11127 (2000)
69. C.J. Zhang, P. Hu, *J. Am. Chem. Soc.* 123, 1166 (2001)
70. R.J. Gelten, A.P.J. Jansen, R.A. van Santen, in *Molecular Dynamics; from Classical to Quantum Methods*, P.B. Balbuena, J.M. Seminara (eds.), Elsevier, Amsterdam , pages 737–784 (1999)
71. C.G.M. Hermse, F. Frechard, A.P. van Bavel, J.J. Lukien, J.W. Niemantsverdriet, R.A. van Santen, A.P.J. Jansen, *J. Chem. Phys.* 118, 7081 (2003)
72. (a) M. Neurock, E.W. Hansen, and D. Mei, *Stud. Surf. Sci. and Catal.* 133, 1 (2001);
(b) M. Neurock, E. Hansen, D. Mei, P. S. Venkataraman, *Stud. Surf. Sci. Catal.* 133, 19 (2001);
(c) M. Neurock, and D. Mei, *Top. Catal.* 20, 1, 5 (2002);
(d) D. Mei, E.W. Hansen, and M. Neurock, *J. Phys. Chem. B*, 107, 798 (2003);
(e) D. Mei, Q. Ge, M. Neurock, L. Kieken, and J. Lerou, *Mol. Phys.* 102, 361 (2004);
(f) L. Kieken, M. Neurock, D. Mei, *J. Phys. Chem. B*, 109, 2234 (2005);
(g) M. Neurock, *J. Catal.* 216, 73 (2003)
73. (a) P. Cremer, G. A. Somorjai, *J. Chem. Soc. Faraday Trans.* 91, 3671 (1995);
(b) P. Cremer, G.A Somorjai, *Catal. Lett.* 40, 143 (1996)
74. (a) A.N.R. Bos, E.S. Bootsma, F. Foeth, H.W.J. Sleyster, K.R. Westerterp, *Chem. Eng. Pro.* 32, 53 (1993);
(b) G.C.A. Schuit, L.V. Reijen, *Adv. Catal.*, 10, 242 (1958);
(c) R. J. Davis, M. Boudart, *Catal. Sci. Tech.* 1, 129 (1991); Y. Takasu, T. Sakuma, Y. Matsuda, *Chem. Lett.* 48, 1179 (1985);
(d) W. Tysoe, G. Nyberg, R. Lambert. *J. Phys. Chem.* 88, 1960 (1984)
75. M.T.M. Koper, J.J. Lukien, A.P.J. Jansen, R.A. van Santen, *J. Phys. Chem.* 103, 552 (1999)
76. J. Clavier, in *Interfacial Electrochemistry, Theory, Experiment and Applications*, J. Clavier (ed.), Marcel Dekker, New York, p. 231 (1999)
77. K. Honkala, A, Hellman, I.N. Remediakis, A. Logadottir, A. Carlsson, S. Dahl, C.H. Christensen, J.K. Nørskov, *Science*, 307, 555 (2005)

CHAPTER 4
Shape-Selective Microporous Catalysts, the Zeolites

4.1 Zeolite Catalysis, an Introduction

4.1.1 Zeolite Structural Features

Zeolites are crystalline aluminosilicates which assemble into well-defined three-dimensional structures comprised of microporous channels that interconnnect cavities which approach molecular scale dimensions ranging from 2 to 12 Å. Mesoporous silica materials with pore sizes over 40 nm have also been made. They are relatively easy to synthesize and offer outstanding control over the pore size as well as its three-dimensional pore architecture, which makes them ideal for gas separations and shape-selective catalysis. Zeolites are perhaps the most widely recognized catalytic material. They are found in nature in over 50 distinct mineral forms. In addition, over 140 man-made zeolites have also been synthesized, with a total of 165 different framework structures[1].

In addition to their microporous environment, zeolites offer a range of other important chemical properties. They are highly acidic materials. One can tune both the number of acid sites and the strength of these sites and thus control their overall acidity. They have high thermal stabilities. Metal ions can be readily exchanged, impregnated or added into the synthesis to open up many other types of reactions.

The primary building units of the zeolite are tetrahedral (MO_4) structures typically comprised of a silicon or aluminum atom that sits at the center of the tetrahedron and four oxygen atoms which sit at the vertices of the tetrahedron. These oxygen atoms interconnect tetrahedra forming an interconnected network of three-dimensional channels. Each tetrahedron is bound to the others through the oxygen atoms at their vertices. The tetrahedra can subsequently assemble and form different secondary building units (16 SBUs are currently known) or chain-like structures which can go on to form 6, 8, 10 and 12 rings. These rings can be considered to assemble into two-dimensional structures that go on to form three-dimensional cavities or cages bounded by well-defined 6-, 8-, 10- and 12-ring apertures. The cages are connected via different SBUs to comprise the tertiary zeolite framework. The porous network can form either straight or zig-zag pore structures depending on how the tertiary structure assembles. Two well-known, catalytically important zeolite structures mordenite and chabazite are shown in Fig. 4.1.

Zeolites have wide industrial uses covering a range of different commercial processes including: catalytic cracking of gas oil to gasoline, hydrocracking gas oil to kerosene, dewaxing middle distillates to lubricants, benzene alkylation to styrene, toluene disproportionation to xylenes, xylene isomerization, methanol-to-gasoline conversion, methanol-to-alkenes, halogenations and nitrations of arenes, isomerization of different hydrocarbons, hydration and dehydrations of alcohols and acids, hydrogenation and dehydrogenations of hydrocarbons, hydroformylation, and oxidation, to name just a few.

In this chapter, we focus solely on the catalysis of zeolitic systems. Zeolite synthesis is discussed in Chapter 8 and mesoporous systems are discussed in sections dealing with biomineralization in Chapter 9.

The purely silica framework (SiO_2) alone is charge neutral. The substitution of Si^{4+}, however, with Fe^{3+}, B^{3+} or Al^{3+} will impart a negative charge on the framework oxygen atoms. Protons attached to the framework or positively charged cations located in the cavities or zeolite channels subsequently act to maintain overall charge neutrality.

This framework charge is compensated by positively charged cations positioned in

Molecular Heterogeneous Catalysis. Rutger Anthony van Santen and Matthew Neurock

ISBN: 3-527-29662-X

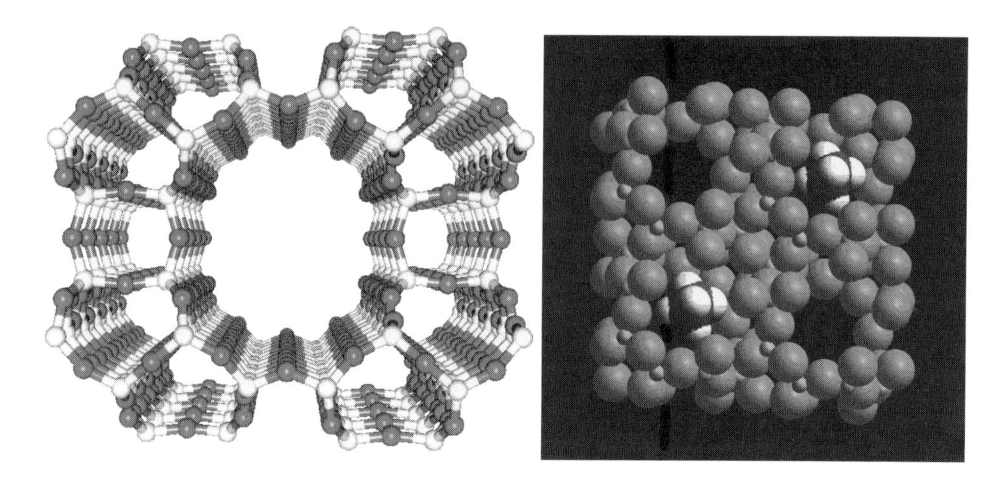

Figure 4.1a. (Lefthand side) Ball-and-stick representation of mordenite. (Righthand side) Space filling model of chabazite. Note the difference in size of oxygen and silicon atoms. Ethane is adsorbed in two of the cavities.

the zeolite channels and cavities. Ammonium ions can be used to compensate for the negative charge on the framework lattice. Ammonium ions, however, thermally decompose at higher temperatures, leaving behind a proton which binds to the framework oxygen. This subsequently leads to high Brønsted acidity at the framework oxygen sites. The concentration, type and size of the cations contained in the micropores ultimately control the zeolite's catalytic activity. For example, Zn^{2+} or Fe^{2+} can act as Lewis acid or redox centers when used as charge-compensating cations of negatively charged zeolite lattices. Another class of zeolites contain reducible cations such as Co^{2+} (as in $AlPO_4$), or Fe^{3+} and Ti^{4+} in the center of the lattice tetrahedra. As such, these materials are useful for oxidation catalysis.

The $AlPO_4$ polymorphs are quite similar to the zeolites. The framework for the standard $AlPO_4$ is charge neutral. The substitution of Al^{3+} with Zn^{2+}, Co^{2+} or Ni^{2+}, on the other hand, can lead to a negatively charged framework. Much of what we have described for the zeolites holds for the $AlPO_4$ polymorph systems also.

In this section, we discuss the general aspects of chemical bonding in zeolites and the zeolite O–H bond. Brønsted and Lewis acid catalysis by zeolites is presented in Section 4.2. Section 4.3 covers redox catalysis by zeolites. The final three sections describe the catalytic cycle and the role of adsorption and diffusion on catalytic performance. An important question that arises in each of these sections is the relation between the micropore structure of the zeolite and its activity and selectivity.

A fundamental understanding of the nature of chemical bonds within the zeolite and between the zeolite and adsorbate molecules is necessary in order to provide a more comprehensive picture of the molecular aspects that control zeolite catalysis. We describe both the structure and the electronic features that control bonding. We start by first describing the purely siliceous systems and then compare these with the results from cation-exchanged systems, exploring proton-induced acidity and the addition of metal cations for Lewis acidity and redox chemistry. The chemical bonding in the siliceous zeolitic polymorphs of silica is largely covalent. The electrostatic contribution to the chemical bond energy is only about 10%[2].

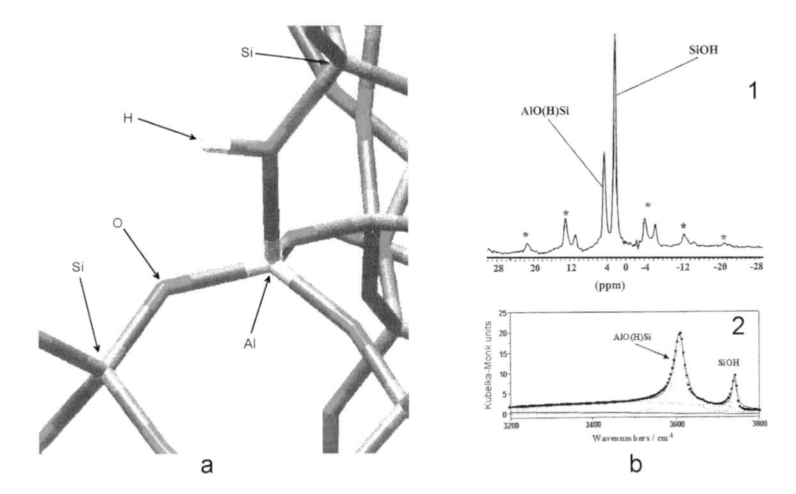

Figure 4.2. The structure of zeolites;(a) Brønsted acidic site. (b) HZSM-5 zeolite with Si/Al = 45; (1) ^{1}H MAS NMR spectrum of OH groups; (2) DRIFT spectrum, adapted from Kazansky et al.[5a].

The dielectric constant of these materials is not sensitive to the micropore size. The siliceous part of the zeolite framework is hydrophobic and creates an apolar environment for adsorbed reactants. The chemical reactivity within the zeolite more closely resembles the reactivity in a vacuum than in a solution. Interestingly, the hydrophobic part of enzymes, as discussed in Chapter 7, has a dielectric constant comparable to that of the zeolite framework suggesting that zeolites, in principle, may be able to carry out similar reaction processes to those in the enzyme.

As one would expect, the local structure and bonding within zeolite are similar to those of silica. The potential energy surface of Si–O–Si angle bond deformation is rather flat. For instance, a 10° change in angle changes the energy only by a few kJ/mol. The energy needed to deform the tetrahedral configuration around Si is substantially greater since the rehybridization energy of the covalent SiO tetrahedral bond is quite large. In addition, a significant amount of energy is required to stretch the Si–O bond.

Structural changes of zeolites can fairly easily be accommodated by small changes in the Si–O–Si bond angles. This is the reason for the minor differences in the heats of formation of different siliceous polymorphs of the zeolites. The difference in energy between high-density and low-density zeolites usually does not exceed 15 kJ/mol per site. This is another indicator that the long-range electrostatic interactions result in only minor contributions to the chemical bonding in these systems.

In contrast to the purely siliceous framework bonds, the interaction between the negatively charged framework and the charge-compensating cations, such as Na^+ and Zn^{2+}, is primarily electrostatic and therefore quite strong[3]. Protons are covalently bound to the oxygen atoms that bridge the lattice Si and Al atoms (Fig. 4.2a). The charge on the proton is essentially zero. The proton–oxygen bond is therefore quite strong and predominantly covalent. Bonding between the proton and the oxygen changes the hybridization of the oxygen atom to which it is bond.

In classical chemical bonding theory, the hybridization on oxygen would change from approximately sp, when the Si–O–Al angle is nearly linear, to approximately sp^2, when oxygen is three coordinated. This is in line with the decrease in the average Si–O–Si angle of 144° in quartz to a smaller Si–O–Al angle of about 120° in the protonated system. This leads to the formation of a strong O–H bond. The energy for the heterolytic cleavage of the O–H bond is about 1250 kJ/mol typically.

The bond between the zeolite and the proton can be considered an internal silanol (SiOH) group which interacts with a framework Al^{3+} cation through the oxygen atom. The SiO–H bond weakens when the silanol oxygen atom connects with the Al atom. The change in hybridization of the silanol oxygen from approximately sp to sp_2 implies a decrease in the s-character of the chemical bond. The corresponding weakening of the OH bonds arises from the fact that the 2s states are lower in energy than the 2p states. The OH bond of the [Si–OH–Al] unit is therefore weakened compared with that of the silanol. The increased polarizability of the zeolitic proton compared with that for silanol is nicely illustrated by the greater infrared adsorption intensities of the zeolitic proton compared with that for a free surface silanol group.

For instance, the results from NMR studies on HZSM-5, as seen in Fig. 4.2b(1), show a dominance of surface silanol groups compared with acidic protons. The results shown in Fig. 4.2b(2), however, show a much larger infrared adsorption intensity, which should be assigned to the acidic protons the low-frequency peak in Fig. 4.2b(2). Normalizing these intensities per proton results in an 8-fold increase in the infrared intensity of the zeolitic proton compared with that of silanol. Since the infrared intensity is proportional to the polarizability, one concludes that there is a larger polarizability of the zeolitic proton. Hence it is easier to induce a positive charge on the zeolitic bonded proton than that on silanol when it interacts with a Lewis basic molecule. The H–O bond energy, which is approximately 1250 kJ/mol, is rather strong.

When the oxygen–proton bond is broken within a catalytic reaction, the metal–oxygen bond gains more s character, the Si–O and Al–O bond energies therefore increase and the irrespective distances decrease. Consequently, the effective volume of the [Al–O–Si] unit is smaller than that of the [Si–OH–Al] unit. The resulting stress on the neighboring atoms is partially reduced by changes in their bond distances and angles. This results in small differences in local acidity in a zeolite due to differences in the local compressibility of the structure[4].

The differences in proton bond cleavage energies in zeolites are also related to the framework composition. The OH bond energies at Al/Si ratios that are greater than 0.1 are 10–40 kJ/mol higher than those at Al/Si ratios that are smaller than 0.1. The differences in energy relate to differences in the local lattice-relaxation energies and also electronic relaxation effects when the proton–oxygen bond is cleaved. At high Al concentrations, the effective negative charge excess increases and, thus, the proton interaction energy increases. This increased OH bond energy corresponds to a weaker acid site for catalysis.

The strong electrostatic interaction with micropore cations will induce significant local structural changes. This local lattice relaxation in which the Si–O–Si and Si–O–Al angles are altered at small energy cost controls significantly the differences in the relative energies of cations adsorbed in different exchange locations [5b].

For non-diffusion-limited reactions carried out in low Al/Si ratio systems, the overall rate for a proton-catalyzed reaction increases linearly with the proton concentration, as illustrated schematically in Fig. 4.4[6]. The rate, when normalized against the framework proton concentration, however, is a constant. When the lattice Al/Si exceeds 10%, the proton–zeolite interaction energy increases. This increase in the proton–zeolite interaction decreases the intrinsic Brønsted acidity of the zeolite. At this concentration, the tetrahedra containing Al start to share a silicon tetrahedron. This increases the effective negative

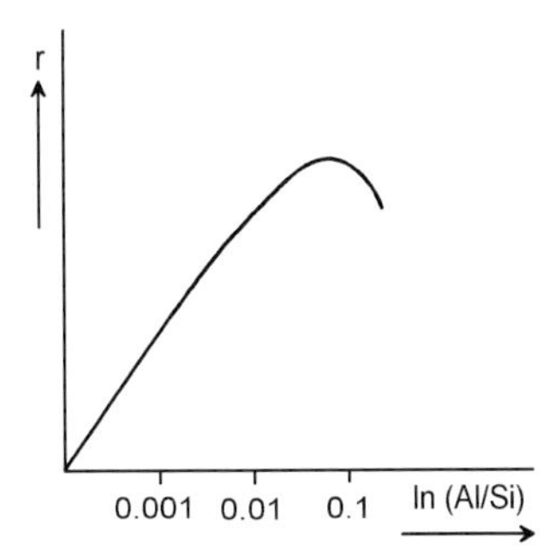

Figure 4.3. Dependence of the rate of a zeolite-catalyzed reaction on Al/Si ratio (schematic).

charge on the oxygen atoms. As a consequence, the zeolitic protons can no longer be considered to be independent. The effect of having a negative charge on two or more tetrahedra containing Al prevents them from being in nearest-neighbor tetrahedra. This is the so-called Löwenstein rule.

4.2 Activation of Reactant Molecules

4.2.1 Proton-Activated Reactivity

When a molecule adsorbs on the siliceous part of the micropore of a zeolite, the main interaction it experiences is a dispersive van der Waals-type interaction. This is due to the dominant interaction with the large polarizable oxygen atoms that make up the zeolite framework. For example, the interaction between a hydrocarbon CH_3 or CH_2 group and the siliceous framework typically results in an interaction energy that is on the order of 5–10 kJ/mol. These electrostatic interactions are small.

When an organic molecule approaches the zeolitic proton, in addition to the van der Waals dispersion forces, there is a weak additional interaction which is on the order of 5 kJ/mol. The large interaction energy of hydrocarbons with the siliceous zeolite channel [e.g. hexane in silicalite (ZSM-5), 60 kJ/mol] is due to the fact that there are multiple contacts between the hydrocarbon and the zeolite channel which are additive in nature. This helps to illustrates the fact that the siliceous zeolite channel is quite hydrophobic[7].

The interaction of a zeolitic proton with a polar adsorbate is much stronger than with an apolar hydrocarbon. The heat of adsorption of CH_3OH, for example, to the zeolitic proton is on the order of 80 kJ/mol. This is largely due to the strong interaction between the OH group on methanol and the zeolite proton. Its bonding features are well understood and sketched in Fig. 4.4[8].

```
      CH3
       |
      /O·..
     H    H
     :    |
 \ /O\  /O\  /O\
  Si   Al   Si
  /\   /\   /\
```

Figure 4.4. Coordination of CH_3OH to zeolitic proton. Solid lines indicate covalent bonds whereas dotted lines represent hydrogen bonds.

The proton–oxygen bond is covalent, but also highly polarizable. The zeolite proton does not transfer completely to the adsorbing molecule upon adsorption. The interaction is largely comprised of the two hydrogen bonds that form between methanol and the zeolite. The first is the bond between the now weakened zeolite proton and the basic oxygen atom of methanol, the second is between the acidic methanol proton and a second basic oxygen atom of the Al tetrahedron. The large interaction between the polar groups on the molecule and proton site in the zeolite is indicative of the hydrophilic nature of the protonic sites.

In order to activate reactant C–O or C–C bonds, the zeolitic Brønsted OH bond must dissociate. To dissociate this bond into an H^+ and a negatively charged zeolite costs ~ 1250 kJ/mol. This energy cost is lowered by the energy gain of binding the proton to the reactant hydrocarbon. The low dielectric constant of the hydrophobic zeolite framework ($\varepsilon_{\text{silicalite}} = 2$, compared with $\varepsilon_{H_2O} = 80$) disfavors charge separation. The fragments of opposite charge therefore tend to attract one another. Positively charged intermediates are, therefore, rarely the lowest energy ground state. Protonated molecular fragments will more readily adsorb to form as an alkoxy intermediate. This is illustrated in Fig. 4.5.

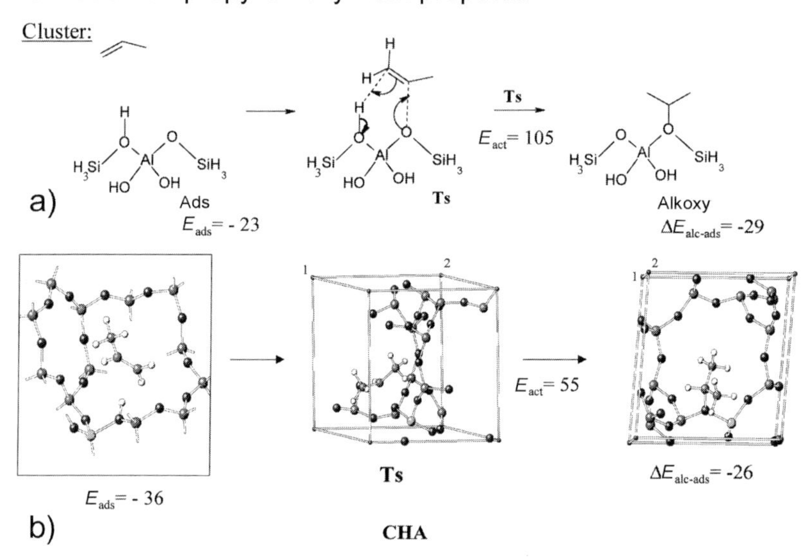

Figure 4.5. Comparison of the DFT-calculated structures of the reactant, transition and product states of the protonation reaction of propylene by a zeolitic proton. (a) Results of calculations using zeolite clusters. (b)Results from periodic DFT calculations on the structure and the resulting energy for the protonation of propylene by the protonated form of chabazite. Al values are in kJ/mol.

Figure 4.5 shows the energies of the initial weak hydrogen-bonded adsorbed state of propylene, the proton-activated transition state and the final alkoxy product state of the protonated propylene. The structures and energies are established from DFT cluster calculations using the model structure shown in Fig. 4.5a and periodic DFT calculations using the unit cell of chabazite and the zeolitic protons (Fig. 4.5b). The cluster used in Fig.

4.5a was created by "cutting" it out from the periodic crystalline framework and capping the terminating oxygen atoms with hydrogen atoms so as to neutralize the cluster. One notes the relatively small difference in energies of the adsorbed propylene before proton transfer (initial state) and the protonated propylene bound to the zeolite as an alkoxy species (product state). The transition state for this reaction requires a significant stretch (activation) of the OH bond to give the near formation of a protonated propyl cation and this is high in energy. This high barrier is due to the large energy cost required to cleave the zeolite OH bond heterolytically . This is only partially compensated by the formation of a new C–H bond and the electrostatic stabilization between the protonated propylene and negatively charged zeolite. The comparison of the results from cluster and periodic DFT calculations shows only small differences for the neutral-bonded state and the alkoxy state. The difference in energy between the relative energies for the protonated carbenium-like transition states, however, is significantly greater. A considerable degree of charge separation occurs in the transition state, which is partially screened by the polarization of the large oxygen atoms of the zeolite cavity. This screening of the dipole between the positively charged transition state and negatively charged zeolite framework reduces the activation barrier for protonation from 105 kJ/mol in the cluster model to 55 kJ/mol[8] in the periodic approach.

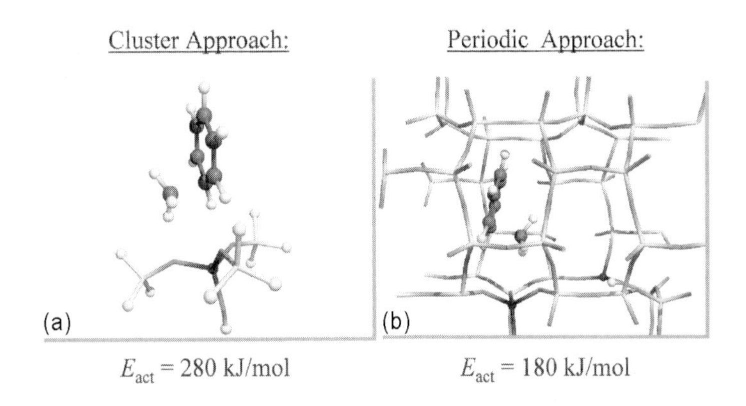

Figure 4.6. Transition states and their energies with respect to the reactant state of adsorbed toluene. (a) Toluene activated by cluster; (b) toluene activated by a proton in the mordenite structure.

This concept of electrostatic screening of the charge separation in the transition state is illustrated in detail in Fig. 4.6 for the proton-activated methyl carbon bond cleavage of toluene. The calculated transition-state structure and the energies for the protonated cleavage of the C–C bond of toluene in the pore of the mordenite channel that result from periodic DFT calculations, are shown in Fig. 4.6b for comparison with the cluster results shown in Fig. 4.6a. The similarity in predicted transition-state structures for both the cluster and periodic results is noteworthy. In the transition state, a proton attaches to the aromatic ring. The $CH_3{}^+$ group that forms subsequently tilts out of the plane of the molecule and binds at an angle which is close to 90°. The transition state structures are quite similar to the free protonated toluene cation and do not experience a steric constraint due to the occlusion in the zeolite channel. The energetic difference is due

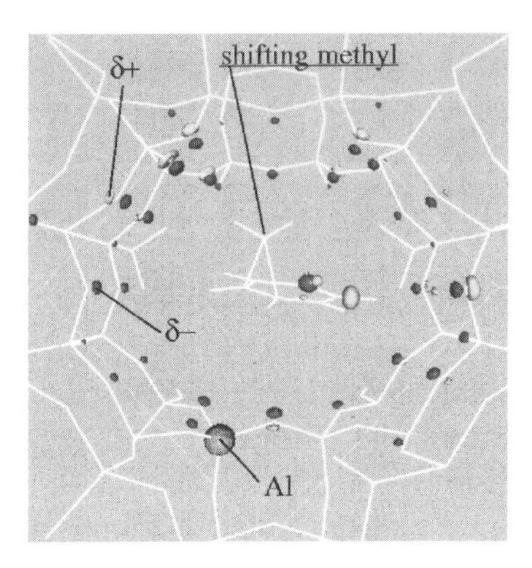

Figure 4.6c. Polarization of the electronic density in the shift isomerization transition state of toluene catalyzed by an acidic mordenite.

to the electrostatic interaction of the protonated toluene transition state with the negatively charged zeolite framework which is more realistically represented in the periodic simulations.

The charges that form on the transition-state complex are screened by the oxygen atoms. The changes in charges that occur on the oxygen atoms in the zeolite channel are shown in Fig. 4.6c. This screening of charged intermediates results in a significant reduction in the energy of charge separation. This is a consequence of the match between the size and shape of the positively charged reaction intermediates and the size and shape of the cavity. The protonated intermediates in the transition states shown in Figs. 4.5 and 4.6 can be considered to be the analogues of a classical carbonium ion formed in superacid solutions[11] (see also Chapter 5, page 237). Carbonium-ion and carbenium chemistry is well understood in solution media. Carbonium ions contain protonated saturated C–C or C–H bonds, whereas carbenium ions result from the protonation of alkenes. The carbonium ion is classically defined as containing carbon atoms with a coordination number of five. The carbenium ion, on the other hand, contains carbon atoms with classical valencies with a coordination number of three. The carbenium ion typically takes on a planar configuration.

In zeolite catalysis, carbenium- or carbonium-ion intermediates are energetically located at the top of the reaction energy barriers. In contrast, in superacid solutions, these protonated intermediates are ground-state reactants. The zeolite carbonium- and carbenium-ion transition state concepts are illustrated for C–C activation and olefin isomerization reactions below.

The transition states for the proton-activated cleavage of propane and butane are shown in Fig. 4.7. These transition-state intermediates can be considered as protonated propane and butane and hence are the analogues of classical carbonium ions. The structures shown in Fig. 4.7 have been computed in the chabazite cavity [12]. The calculated activation energies for proton activation are quite high, between 170 and 200 kJ/mol. The n-butane molecule does not fit well in the small pore of the chabazite framework. Therefore, it must adopt a *gauche* conformation.

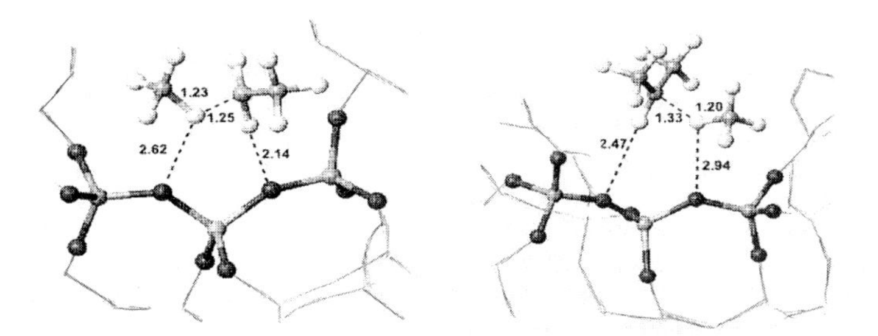

Figure 4.7. Transition-state structures of propane and n-butane cracking in the chabazite framework. Only the lattice atoms in contact with the substrate molecules are clearly visible, adapted from Angyan et al.[12].

Reaction routes for alkane transformation through carbonium ions can lead to C–H or C–C bond cleavage reactions as illustrated for isobutane in Fig. 4.8.

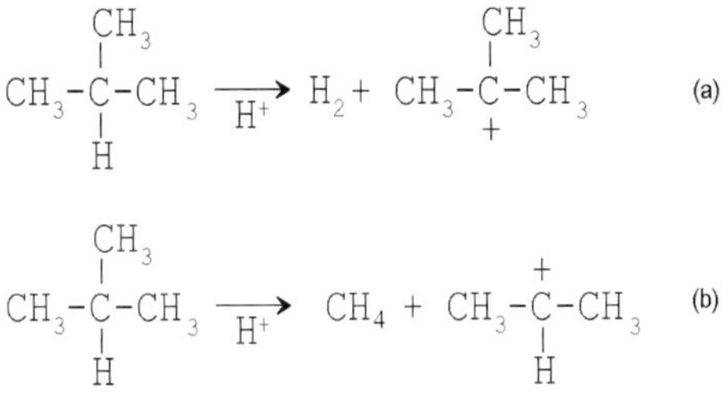

$$CH_3-\underset{\substack{|\\H}}{\overset{\substack{CH_3\\|}}{C}}-CH_3 \xrightarrow[H^+]{} H_2 + CH_3-\underset{+}{\overset{\substack{CH_3\\|}}{C}}-CH_3 \qquad \text{(a)}$$

$$CH_3-\underset{\substack{|\\H}}{\overset{\substack{CH_3\\|}}{C}}-CH_3 \xrightarrow[H^+]{} CH_4 + CH_3-\underset{\substack{|\\H}}{\overset{+}{C}}-CH_3 \qquad \text{(b)}$$

Figure 4.8. Carbonium ion intermediated reactions of isobutane (schematic). (a) Dehydrogenation; (b) hydrogenolysis.

For reactions that can proceed through different reaction channels, each channel implies the formation of a different transition state with different shape, energy and entropy. In the CH cleavage reaction (Fig. 4.8a), the protonated carbonium ion transition state (with five-fold coordination on the protonated carbon) has the approximate geometry

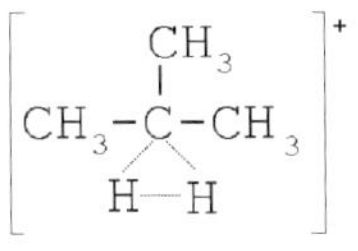

$$\left[\begin{array}{c} CH_3 \\ | \\ CH_3-C-CH_3 \\ \diagup\ \diagdown \\ H-H \end{array}\right]^+$$

The transition–state geometry for the carbon–carbon cleavage reaction is similar to that shown in Fig. 4.7:

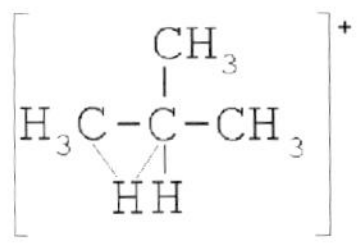

$$\left[\begin{array}{c} CH_3 \\ | \\ H_3C-C-CH_3 \\ \diagdown\ \diagup\ | \\ HH \end{array}\right]^+$$

The charged carbocations produced after the C–H and C–C bond cleavage can be considered protonated alkenes or carbenium ions, in which the hybridization around the positively charged carbon is sp^2. The carbenium ions subsequently adsorb to form alkoxy intermediates on the zeolite lattice, as illustrated in Fig. 4.5.

Reaction steps that proceed through the formation of primary carbenium ions, i.e. with the positive charge developing on the terminal carbon atom of the hydrocarbon chain, are energetically unfavorable and considered to be forbidden in classical liquid-phase carbenium ion chemistry. In zeolites, however, the formation of a primary terminal carbenium ion can be stabilized by its interaction with the negatively charged zeolite framework. For instance, the isomerization of n-butene to isobutene is catalyzed by the zeolite ferrierite and occurs initially as a monomolecular reaction. This isomerization reaction has to proceed through the formation of a primary carbenium ion intermediate. Butene adsorbs and subsequently isomerizes in the zeolite, thus leading to the formation of the cyclopropyl cationic intermediate (I) sketched in Fig. 4.9. The ring opening of the cyclopropyl intermediate that follows leads to the formation of the primary carbenium ion (II). Intermediate (I), is formed in the transition state. Intermediate (II), on the other hand, is stabilized as an alkoxy intermediate.

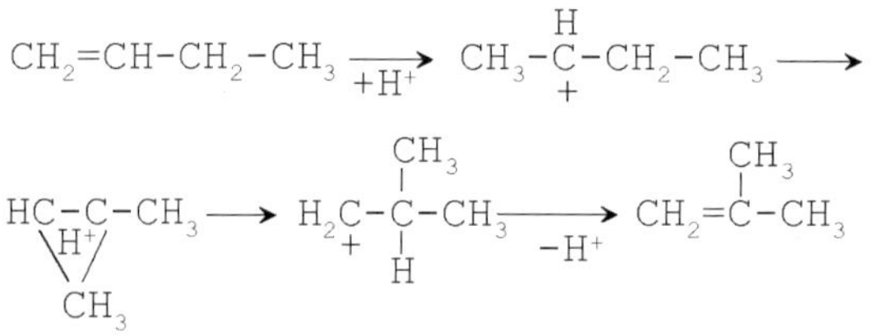

$$CH_2{=}CH{-}CH_2{-}CH_3 \xrightarrow[+H^+]{} CH_3{-}\overset{H}{\underset{+}{C}}{-}CH_2{-}CH_3 \longrightarrow$$

$$\underset{\text{I}}{HC{-}C{-}CH_3\ (\text{bridged by } H^+ \text{ to } CH_3)} \longrightarrow \underset{\text{II}}{H_2\underset{+}{C}{-}\overset{CH_3}{\underset{H}{C}}{-}CH_3} \xrightarrow[-H^+]{} CH_2{=}\overset{CH_3}{C}{-}CH_3$$

Figure 4.9. Isomerization of n-butene to isobutene via the formation of a primary carbenium ion intermediate (schematic).

The analogous states involved in pentene isomerization in mordenite [15] are illustrated in Fig. 4.10a and b. Pentene isomerization occurs via the secondary rather than the primary carbenium ion.

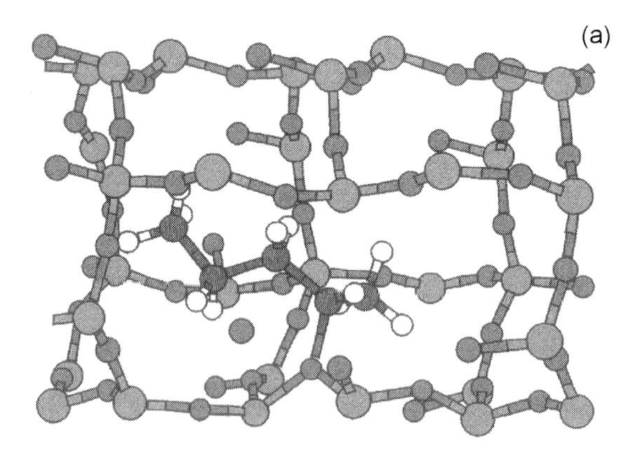

Figure 4.10a. Protonated pentene adsorbed as a secondary alkoxy species to mordenite zeolite framework[15].

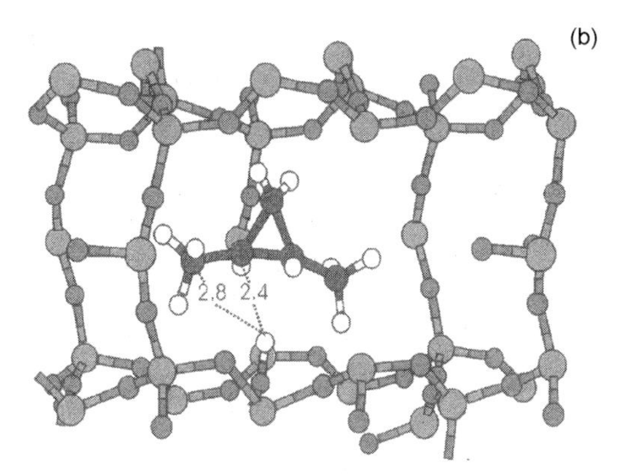

Figure 4.10b. The calculated structure of the adsorbed dimethylcyclopropyl intermediate[15].

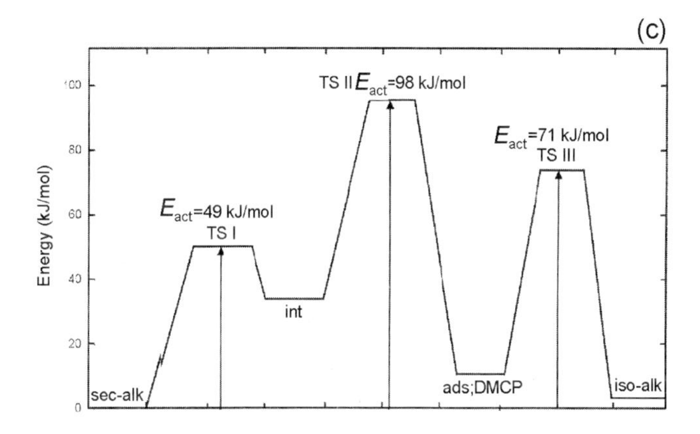

Figure 4.10c. The calculated energy diagram for the isomerization of pentene to adsorbed isobutene[15].

The corresponding computed reaction energy diagram which is shown in Fig. 4.10c[15] proceeds through the formation of the three adsorbed intermediates shown in Fig. 4.10a and b. The first transition state (TSI) is a secondary n-carbenium ion-like state that forms as the result of protonation of pentane (Fig. 4.10a) and leads to the formation of the adsorbed n-pentyl intermediate. The second transition state (TSII) corresponds to the state that leads to cyclopentyl formation (structure 4.10b). The third transition state (TSIII) leads to the formation of isobutyl through C–C cleavage of the cyclopentyl ring. Let us return now to the question of whether the stabilization by a zeolite makes protonation reactions via primary carbenium ions possible. We analyze this here for the two protonation options of isobutene shown in Fig. 4.11.

The computed reaction energy diagrams for the two different reaction paths in two different zeolites which have cylindrical micropores are calculated here. The mordenite zeolite has a one-dimensional 12-ring channel and ferrierite (TON; TON is the nomenclature according to the International Zeolite Association) a one-dimensional 10-ring channel. Hence there are differences in the curvature of the channels between the two zeolites. The reaction energy changes are shown for the reaction paths proceeding through a primary carbenium or tertiary carbenium ion in Fig. 4.12[16a].

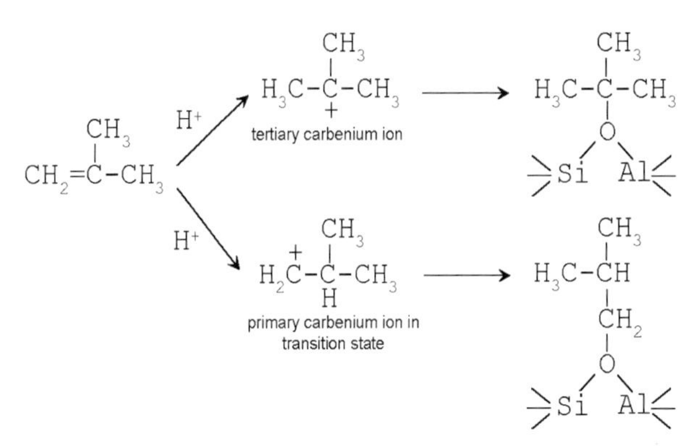

Figure 4.11. Protonation options for isobutane (schematic).

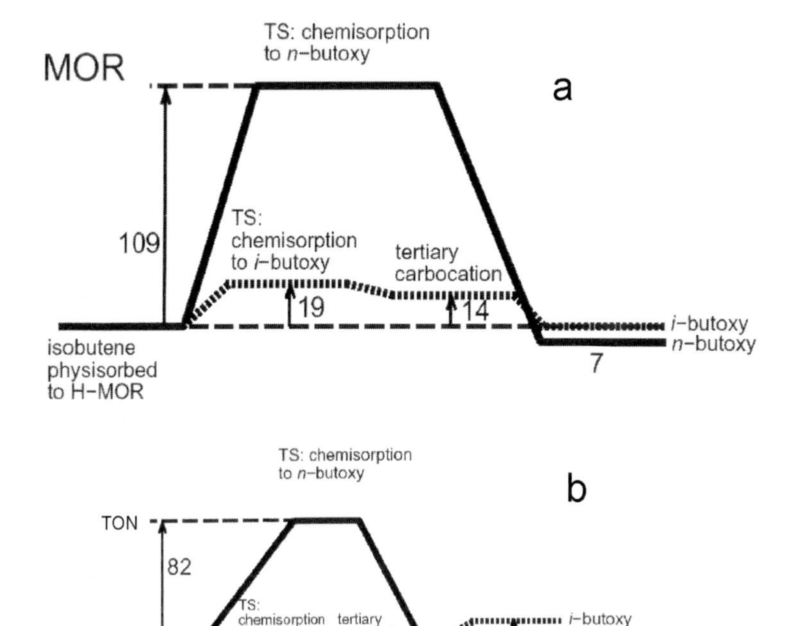

Figure 4.12. Protonation reaction energies of isobutene. A comparison of the formation of primary and tertiary carbenium ions. (DFT–VASP calculations). (a) Mordenite. (b) Ferrierite (DFT–VASP calculations) (TON)[16b.

First it is important to note the small difference in energy of the protonated ground-state primary (n-butoxy) and tertiary (isobutoxy) alkoxy species in mordenite. The transition-state energies of the two corresponding intermediate carbenium ions, however, demonstrate a much larger energy difference.

The activation energy for protonation through a primary carbenium ion is 60–80 kJ/mol higher than that through the tertiary carbenium ion. The height of this barrier is of the same order of magnitude as that for the isomerization of pentene, a reaction that readily occurs in a zeolite at 550 K. Hence it can be concluded that isomerization and other hydrocarbon conversion reactions via primary carbenium ions are possible at reasonable temperatures because of the stabilization of the carbenium ion transition state by the zeolite framework. The protonation via the tertiary carbenium ion, however, is substantially more favorable, which is in agreement with physical organic theory.

In the narrow pore ferrierite (TON), the energies of the two alkoxyspecies are quite different. For the reaction that proceeds via the tertiary carbenium ion, the free protonated isobutyl cation is even more stable than covalently bonded isobutoxy intermediate. The curvature of the ferrierite channel prevents the close approach of methyl groups on the isobutyl intermediate to oxygen atoms in the zeolite wall, hence the tertiary carbenium ion is a stable but freely moving intermediate.

These examples help to illustrate the importance of the shape and dimensions of the micropore and their influence on the electrostatic screening of the charges generated in the transition state. The previous discussion illustrates that the formation of the carbon–oxygen bond between protonated species and zeolitic oxygen atom is counteracted by repulsive interactions arising from the bulkiness of the protonated intermediate and the zeolite micropores. When the curvature of the cavity becomes significantly large, the bulkiness of the protonated intermediate prevents the formation of the corresponding alkoxy species, hence the free carbenium ion becomes a stable intermediate. If the reaction intermediates become larger than the micropore cavity, they will not be formed. The same holds for reactant molecules that are too large to enter a micropore. The suppression of the formation of intermediates larger than the micropore cavity can lead to a reduction in coke formation. For this reason, solid acid reactions carried out in zeolitic micropores are less susceptible to coke formation.

In the first part of this section we have shown for zeolite solid acids that carbenium or carbonium ion intermediates are typically present as transition states or unstable intermediates. The activation energies depend on the deprotonation energy of the zeolite, the stabilization of the charged cationic intermediates by screening effects and by their interaction with the negative charge left on the zeolite lattice.

Three additional mechanistic aspects that are also essential to zeolite catalysis include:

- pre-transition state orientation
- associative versus alkoxy intermediate reactions
- scaffolding effects of coadsorbed polar molecules

We will illustrate the effects of pre-transition state orientation for the dissociation of methanol. This reaction is essential for the formation of dimethyl ether, which is described below. This is followed by a discussion of the alkylation mechanisms.

Let us consider dimethyl ether formation from methanol, which proceeds through a consecutive reaction mechanism[17]. Figure 4.13a illustrates the reaction intermediates for the first reaction step in which the C–O bond in methanol is cleaved. The calculated reaction energy diagram for this reaction is shown in Fig. 4.13b. The reaction products that form are water and adsorbed methoxy.

C–O bond cleavage by protonation only occurs when methanol is rotated from its most stable adsorption mode (end-on), which has two hydrogen bonds, to the methanol side-on adsorbed mode which has only one hydrogen bridge between the zeolite proton and the

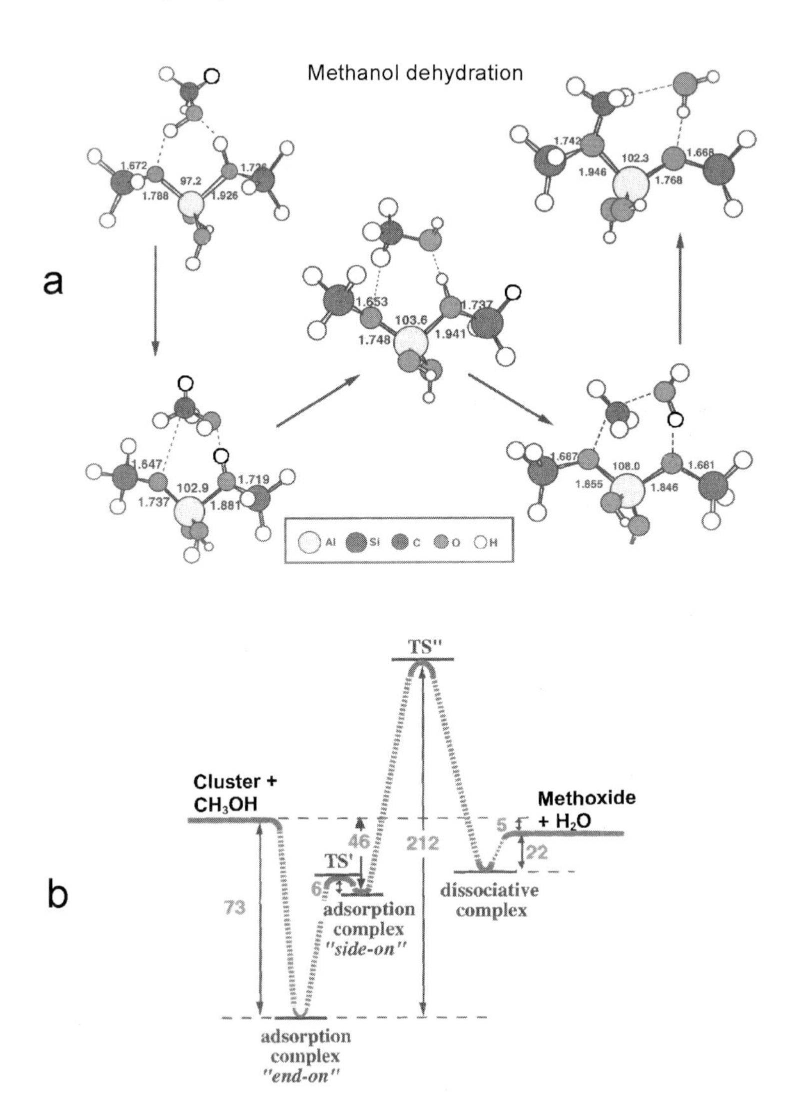

Figure 4.13. (a) Reaction intermediates for the proton activated dissociation of CH_3OH, and in addition, (b) the corresponding reaction energy diagram. The structures and energies reported are from DFT calculations on small zeolitic cluster models[17].

adsorbed molecule and along with weak interaction between the framework oxygen atoms and the methyl group. The unsaturated fragment that is generated upon dissociation is stabilized by bonding to a surface oxygen atom. The side-on mode is now the only mode that enables the $CH_3{}^+$ cation to adsorb on the negatively charged oxygen atom attached to an aluminum atom in the lattice. This oxygen atom is different from the one to which the proton was initially bonded. The reaction energies shown in Fig. 4.13b suggest that under reaction conditions only a very small minority of the adsorbed molecules will be adsorbed in this so-called pre-transition state mode. The reaction energy diagram shown

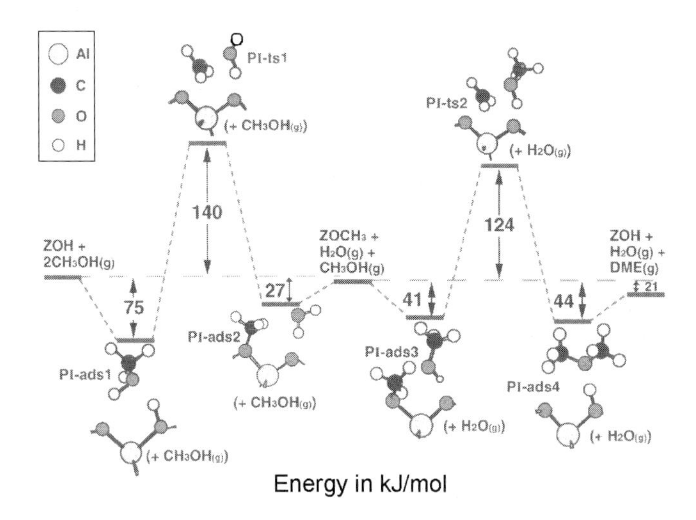

Figure 4.14a. Reaction energy scheme for the consecutive direct reaction path towards dimethyl ether formation[16].

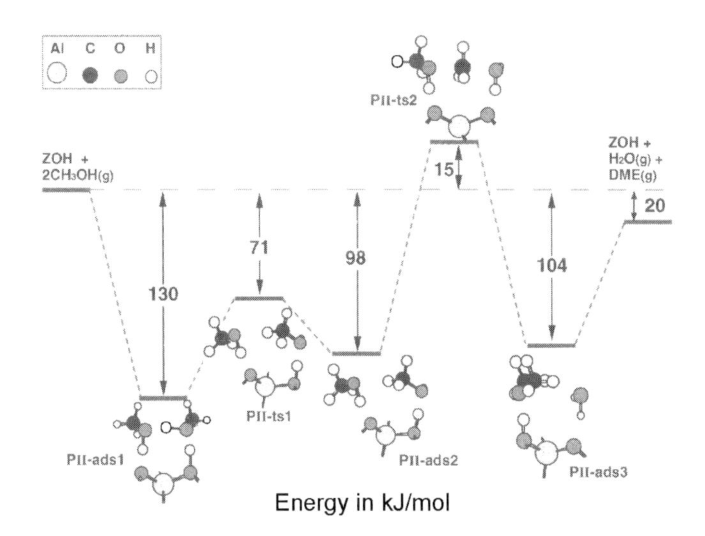

Figure 4.14b. Associative reaction path towards formation of dimethyl ether from methanol[16].

in Fig. 4.13b, however, has been computed for a cluster simulating the zeolitic proton. The transition-state value should be lowered by approximately 50 kJ/mol to correct for the absence of screening effects by the zeolite lattice.

The methoxy species formed by the dissociation of methanol can react with a second methanol molecule to give dimethyl ether and a proton. This is termed the "alkoxy" intermediate for the consecutive direct reaction mechanism. Alternative reaction paths exist which can be described as "associative" reaction mechanisms in which the product of an association reaction is formed without the formation of an alkoxy intermediate species[16]. For two coadsorbed methanol molecules, the most stable adsorption mode does not correspond to the pre-transition state adsorption mode from which the reaction occurs. In the preferred associative reaction path, the $CH_3{}^+$ cation generated in the transition state is directly transferred to the oxygen atom of methanol without the formation of the methoxy intermediate. The reaction energies for the two reaction mechanisms involved in dimethyl ether formation are compared in Fig. 4.14. Note that with respect to the energies of the molecules in the gas phase, the activation barriers for the elementary reaction steps that correspond to the association reaction path are substantially lower than the barriers for the consecutive reaction path.

We conclude this section by presenting the scaffolding effect, whereby adsorbed polar molecules such as H_2O or H_2S can significantly lower the activation energies for different hydrocarbon conversion reactions. In order to illustrate this effect, we will focus on the trans–alkylation reaction[17]. In this reaction, a methyl group from one aromatic mole–cule is transferred to another aromatic molecule. In the "alkoxy"-mediated reaction path the methoxy species are formed by the C–C bond cleavage of the methyl–phenyl bond. The transition state for this C–C bond cleavage is shown in Fig. 4.6. Figure 4.15 shows the transition state for the same reaction but now in the presence of water[18]. The activation energy for this C–C bond formation reaction from the adsorbed methoxy is lowered by 50 kJ/mol in the presence of water. The presence of water significantly stabilizes the charged carbenium ion that forms, thus significantly lowering the activation barrier. The presence of water alters the structure of the carbenium ion by increasing the angle between the $CH_3{}^+$ group and the benzene ring. The angle is nearly perpendicular when water is present (see Fig. 4.15) thus optimizing the interaction with the benzene π-electrons. The effect of coadsorbing non-reacting polar molecules is, in this case, purely geometric. The water molecule acts as a scaffold. Similar assistance effects have been found in several other reactions such as the C–C bond formation reaction from methanol which is discussed later.

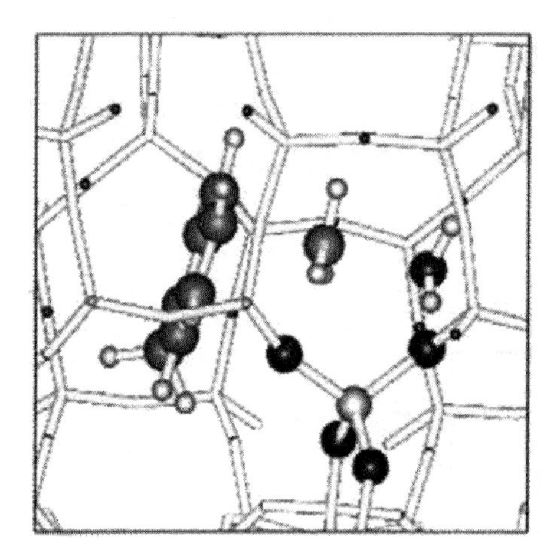

Figure 4.15. Transition state for alkylation of toluene by methoxy species in the presence of water[18]. Within lattice negatively charged oxygen atoms around Al are indicated.

4.2.2 Transition-State Selectivity. Alkylation of Toluene by Methanol Catalyzed by Mordenite

In the previous section, we explained the steric and energetic consequences of the zeolite micropore shape for the activation energy of an elementary zeolite-catalyzed reaction step. In this section we discuss transition-state selectivity whereby differences in selectivity are ascribed to a more or less optimum match of the reaction transition state with the micropore cavity. We will demonstrate that the difference in the selectivity for the reaction is determined by the probability that a preferred pre-transition state orientation can form rather than by differences in the activation barrier for the reaction step in which protonation occurs. This result is analogous to the finding that the preferred reaction channel for enantioselective homogeneous catalysts proceeds through the adsorption complex with the most favorable free energy (see Section 2.4.4). Similarly, we will discuss the importance of the pre-transition-state complex in enzyme catalysis in Chapter 7.

To illustrate the influence of pre-transition-state control for a zeolite-catalyzed reaction, we describe the results of a quantum-chemical study of the alkylation of toluene by methanol[18]. The selectivities to produce *ortho-*, *meta-* or *para-*xylene in the channel of the Mordenite zeolite are studied. The study illustrates the effect of the spatial constraint induced by the one-dimensional 12-ring micropore channel dimension on the selectivity of the zeolite catalyzed-reaction.

The reaction mechanism for the alkylation of toluene is well understood and is illustrated in Fig. 4.16 for the production of *para-*xylene.

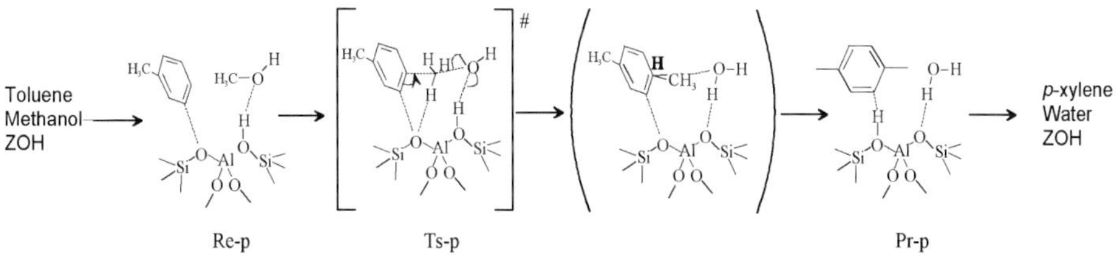

Toluene Methanol ZOH → Re-p → Ts-p → Pr-p → *p*-xylene Water ZOH

Figure 4.16. Mechanism of the alkylation reaction of toluene by methanol catalyzed by an acidic zeolite.

Methanol is first activated in the zeolite to form the $CH_3{}^+$ intermediate as was described in Section 4.2.1. The $CH_3{}^+$ intermediate subsequently adds to the *para* position on the adsorbed toluene to form *para-*xylene. The $CH_3{}^+$ intermediate can similarly attack the adsorbed toluene at the ortho and meta positions to give *ortho-* and *meta-*xylene, respectively.

The reaction energy diagram for the alkylation of toluene by methanol in mordenite is presented in Fig. 4.17. In this reaction energy diagram the energy changes for the first two steps refer to the heat release that occurs from the adsorption of methanol and toluene, respectively. The interaction of toluene with the zeolite channel atoms is primarily controlled by the van der Waals interactions between the toluene atoms and channel oxygen atoms. DFT calculations tend to describe these interactions poorly. The van der Waals-type interactions have therefore been empirically estimated and added to the quantum-chemical interaction energies.

The key to understanding of the energy differences of this zeolite-catalyzed reaction is an appreciation of the differences in energy cost to reorient adsorbed CH_3OH towards the *para*, *meta* or *ortho*-carbon atom of toluene that becomes alkylated. In mordenite, the energy differences of these pre-transition-state structures are controlled by the repulsive interactions that arise due to discrepancies of intermediate shapes for *ortho*-, *meta*- and *para*-orientated pre-alkylation complex with the channel shape and size. With respect to the ground-state energies of the pre-transition-state intermediates, methylation of toluene occurs with nearly similar energies to the *ortho*-, *meta*- and *para*-positions. Since the *para* pre-transition-state structure is most stabilized, this reaction channel gives the lower overall activation energy for formation of *para*-xylene compared with the reaction channels that give *ortho*- and *meta*-xylene.

The concept of the pre-transition states relates zeolite catalysis to enzyme catalysis. As we will see in Chapter 7, a major difference between the zeolite and the enzyme is the limited stabilization of the intermediate in the zeolites which is due to the high rigidity of the zeolite framework. In contrast to enzymes, the zeolite lattice is rather inflexible. It will not adjust to the shape of the desired transition-state structure in previous figures, hence the barriers for proton activated reactions will remain high as compared with those of enzyme-activated reactions.

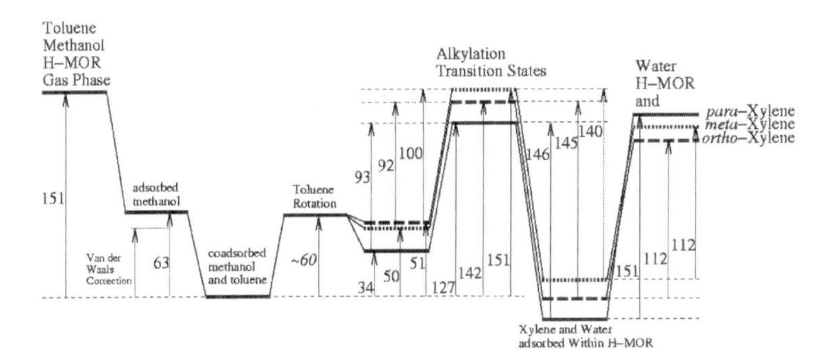

Figure 4.17. Zeolite transition-state selectivity. Toluene alkylation with methanol catalyzed by H-MOR showing the energies of the key reaction intermediates[18]. Reaction energy diagram for *ortho*-, *meta*- and *para*-xylene are compared.

4.2.3 Lewis Acid Catalysis

4.2.3.1 Lewis Acidity in Zeolites; Cations Compared with Oxy-Cations

The reactivity of a zeolite activated by ion exchange with a soft Lewis acid cation will be examined in detail by following C–H bond activation over a Zn^{2+} ion. We compare the results with those for the reactivity of the $ZnOZn^{2+}$ oxycation, often also formed in experimental systems during the ion exchange-reaction of Zn^{2+} into zeolites. As an introduction to Chapter 7 on biocatalysis, we will also discuss the hydrolysis of acetonitrile by a Zn^{2+} ion exchanged into the micropore of a zeolite. A comparison will be made with the reactivity of Ga^{+} and polarization effects due to a hard Lewis acid such as Mg^{2+}.

Let us first analyze the interaction of probe molecules such as CO with the Zn^{2+} and related cations in some detail. The adsorption of CO can be used to help understand the differences in the electrostatic polarizing properties of cations in zeolites. For non-reducible cations, there is an upwards shift of the CO frequency which is proportional to $\frac{q}{r}$, where q and r are the charge and the radius of the cation, respectively. Rehybridization of predominantly the σ-type orbitals in CO leads to a depopulation of the antibonding C–O orbital. This results in a strengthening of the CO bond and, hence, an upward shift of the CO vibrational frequency.

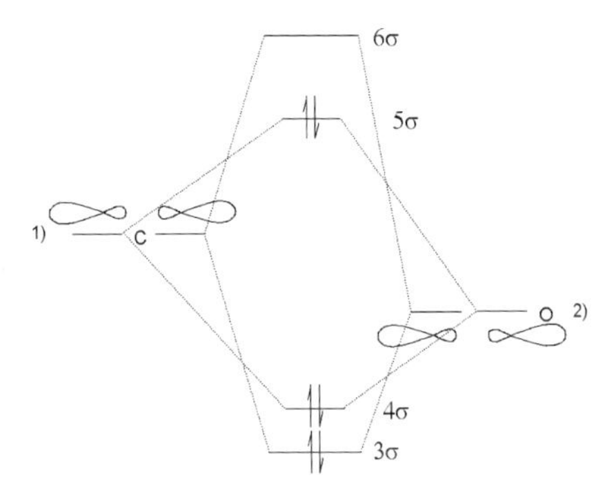

Figure 4.18. Hybridization model of CO σ-type valence orbitals. The relation with the molecular orbitals as computed for CO (see Fig. 3.4) is indicated. 1) Hybridized s,p_z atomic orbitals on C, 2) Hybridized s,p_z atomic orbitals on O.

As illustrated in Fig. 4.18, the antibonding nature of the CO 5σ orbital can be readily deduced from a CO chemical bonding picture based on hybridization of the C and O, 2s and $2p_z$ atomic orbitals. In Chapter 3 we presented calculated CO molecular orbitals and energies (page 93) and in the Addendum to that chapter we gave a short introduction to hybridization

The linear combination of the 2s and $2p_z$ atomic orbitals on each of the atoms leads to the formation of four molecular orbitals. A bonding and antibonding pair of molecular orbitals with a large difference in energy is generated from the hybridized atomic orbitals oriented toward each other, called 3σ and 6σ orbitals. The hybridized atomic orbitals involved strongly overlap and, hence, occupation of the bonding orbital strongly contributes to the strength of the C–O bond. The energy difference between the other pair of bonding and antibonding orbitals is much smaller since they are formed from the hybridized atomic orbitals that are not directed towards one another.

The results in Fig. 4.18 also show the higher occupied 5σ orbital as the lone-pair orbital localized on C, which is antibonding with respect to the C–O bond. The corresponding bonding combination is the 4σ orbital, that is, the lone pair orbital localized mainly on the oxygen atom.

Table 4.1. CO orbital energies and their relative shifts (eV)

	CO	Z–Sr–CO	Z–Mg–CO	Z–Zn–CO
4σ	–14.1	–15.9	–16.2	–16.4
1π	–11.6	–13.4	–13.8	–13.8
5σ	–9.0	–11.5	–12.5	–12.1
Δ_σ	5.1	4.4	3.7	4.3
Δ_π	9.6	9.4	9.4	9.4
$\Delta_{\sigma\pi}$	2.6	1.9	1.3	1.7

Table 4.1 and Fig. 4.19 compare DFT-computed orbitals of CO when adsorbed on Mg^{2+}, Sr^{2+} and Zn^{2+} cations adsorbed to a four-ring structure of Si- and Al-containing tetrahedra.

$\Delta_\sigma = E_{5\sigma} - E_{4\sigma}$, $\Delta_\pi = E_{2\pi^*} - E_{1\pi}$, $\Delta_{\sigma\pi} = E_{5\sigma} - E_{1\pi}$

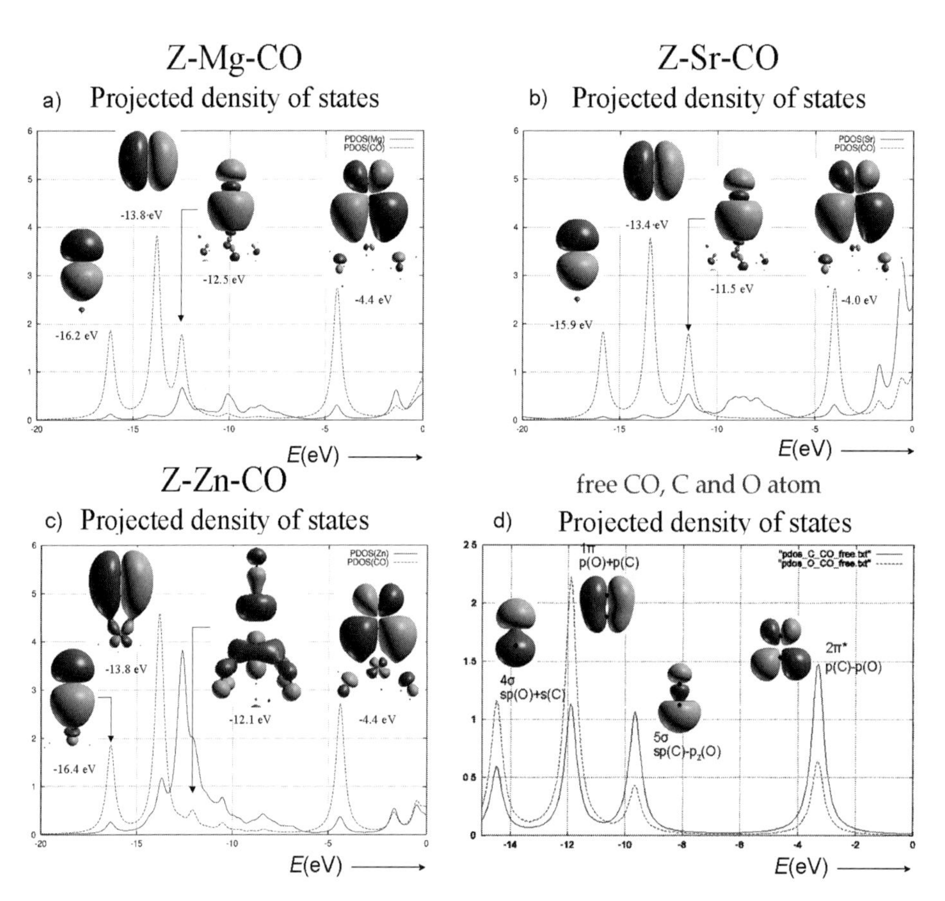

Figure 4.19. The interaction of CO with Mg^{2+}, Sr^{2+} and Zn^{2+} coordinated to a four-ring cluster $Si_2Al_2O_4(OH)_4{}^{2-}$: Local densities of states of CO orbitals are shown as a function of orbital energy:
(a) Mg^{2+}: E_{ads}=–39 kJ/mol; r_{CO}=1.135 Å; r_{C-Mg}=2.357 Å
(b) Sr^{2+}: E_{ads}=–4 kJ/mol; r_{CO}=1.136 Å; r_{C-Sr}=3.158 Å
(c) Zn^{2+}: E_{ads}=–42 kJ/mol; r_{CO}=1.136 Å; r_{C-Zn}=2.138 Å
(d) CO molecular orbitals; r_{CO}=1.144 kcal/mol. The local densities of states projected on C and O are shown. The exact LDOS are delta functions. They have been artificially broadened for ease of visualization.

Table 4.1 compares the relative energy positions of the CO molecular orbitals and their respective energy differences.

The interaction between the cation and CO is seen to lower all of the CO orbital energies. The reduction of the difference in energy between CO 4σ and 5σ molecular orbitals is due to the stronger lone-pair CO 5σ interaction with the cation. This results in a small rehybridization of the CO σ orbital. The increased CO bond strength relates to a reduction of the antibonding character of the 5σ orbital. The increased interaction with Zn^{2+} is clearly seen to be due to the additional interaction with the Zn d_{z^2} orbitals.

One should also note the decreased difference in energy between CO 1π and 5σ molecular orbitals for CO adsorbed to Sr^{2+} and Mg^{2+}. The difference in energy for CO adsorbed on Mg^{2+} is lower than that for Sr^{2+}. The 5σ orbital directed towards the cation experiences a larger electrostatic attraction than the 1π orbital perpendicular to the CO–cation interaction axis. The decrease in 5σ–1π interaction is nearly proportional to the difference in cation–carbon distance.

The bond energies decrease rapidly with bond distance. One expects for polarizable systems a dependence with radius of r^{-4}. Inspection of the orbital pictures in Fig. 4.19 immediately indicates the difference in the interaction between hard Lewis acid cations such as Mg^{2+} and Sr^{2+} and soft cations such as Zn^{2+}.

The presence of cations in zeolites can significantly affects the reactivity of coadsorbed small metal particles[19]. The influence of cations on transition-metal clusters is mainly electrostatic. The electrostatic field generated by the cation polarizes the metal particle. This polarization dramatically affects the reactivity of the small metal cluster. Calculations analyzing the interaction of an H_2 molecule with a Ir_4 cluster in the presence and absence of an Mg^{2+} cation illustrates the importance of induced polarization effects.[20].

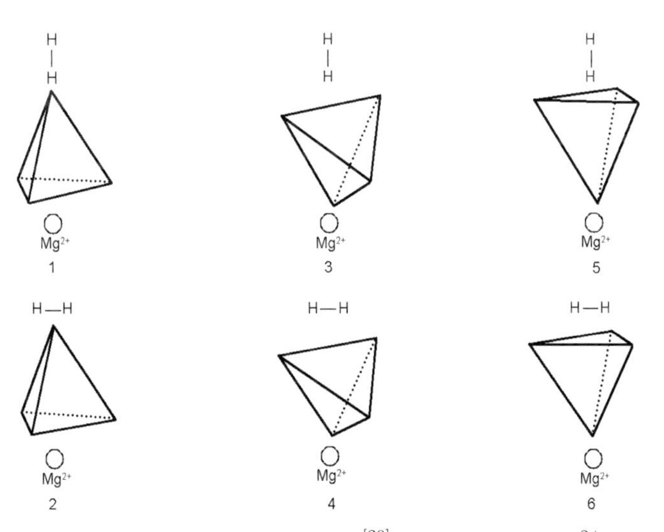

Figure 4.20. Adsorption geometries[20] for hydrogen on Mg^{2+}-promoted Ir_4 clusters.

H_2 weakly interacts with an isolated Ir_4 tetrahedron in an end-on perpendicular adsorption mode. The adsorption of H_2 to the same cluster is largely enhanced, however, if the Ir_4 cluster interacts with an Mg^{2+} cation. Hydrogen adsorption is enhanced at positions where the electron density is most reduced due to the polarization of the Ir_4 atoms by Mg^{2+}. Most striking is the increase in electron density of the antibonding H_2 orbital that results from structure 6 in Fig. 4.20 (see Table 4.2).

In the absence of Mg^{2+}, the interaction between H_2 and Ir_4 is dominated by the strong Pauli repulsive interaction between the doubly occupied orbitals of H_2 and Ir_4. In the presence of Mg^{2+}, however, the Ir_4 is polarized by Mg^{2+}, which reduces the electron density between H_2 and Ir_4. The resulting reduction of the Pauli repulsion allows the H_2 molecule to approach the Ir_4 plane in the parallel adsorption mode with strong orbital overlap. The result is a strong activation and weakening of the H_2 bond. This study illustrates how polarization of metal particles can dramatically alter their chemical reactivity, especially with respect to closed shell systems such as H_2 or CH bonds.

Table 4.2. The molecular orbital population and the interaction energy (ΔE kJ/mol) for hydrogen chemisorbed at different positions on the Mg^{2+} promoted Ir_4 clusters shown in the structures of Fig. 4.20[20]

	Ir_4–H_2					
	onefold		twofold		threefold	
	end-on	side-on	end-on	side-on	end-on	side-on
	1	2	3	4	5	6
H_2:σ_g	1.861	1.888	1.883	1.960	1.944	1.988
H_2:σ_u^*	0.045	0.060	0.082	0.012	0.055	0.003
ΔE (kJ/mol)	–22	–28	–16	–3	–7.5	–2
	Mg^{2+}–Ir_4–H_2					
	onefold		twofold		threefold	
	end-on	side-on	end-on	side-on	end-on	side-on
	1	2	3	4	5	6
H_2:σ_g	1.826	1.859	1.818		1.730	1.596
H_2:σ_u^*	0.008	0.008	0.038		0.059	0.194
ΔE (kJ/mol)	–36	–48	–40		–61	–88

The interaction of a reacting molecule with a metal cation in the zeolite is not only determined by cationic chemical-bonding properties, but also by the negative lattice charge distribution that compensates for the positive charge of the cation. This is elegantly demonstrated by infrared measurements of the absorption intensities for methane adsorbed on Zn^{2+} cations in low Al-ZSM-5 and high framework Al-zeolite Y[21]. In the latter, the framework negative charge is the highest and hence the effective positive charge on Zn^{2+} is the smallest. For this reason, the infrared intensity for the vibrational excitation of methane that is only spectroscopically allowed in the presence of the electrostatic field of the zeolite is highest in the low Al-ZSM-5. This is shown in Fig. 4.21.

In Fig. 4.21 one notes the large difference in the intensity of the vibrational excitation around 2800 cm^{-1} in methane adsorbed on Zn^{2+} in ZSM-5 and zeolite Y. This excitation is not observed in the gas phase, where by symmetry it is vibrationally forbidden. The effective charge of the Zn^{2+} cation polarizes the molecule, which results in symmetry breaking, thus allowing for the excitation of this symmetric CH_4 mode. The lower effective charge of Zn^{2+} in the faujasite structure zeolite Y with higher Al concentration results in a smaller polarization and hence a decreased relative intensity.

Cations prefer particular sites in the zeolite. The divalent Zn^{2+} cations prefer adsorption in a ring of framework tetrahedra with at least two Al framework cations, so that an overall charge neutral site is generated. There is also a relationship between the size

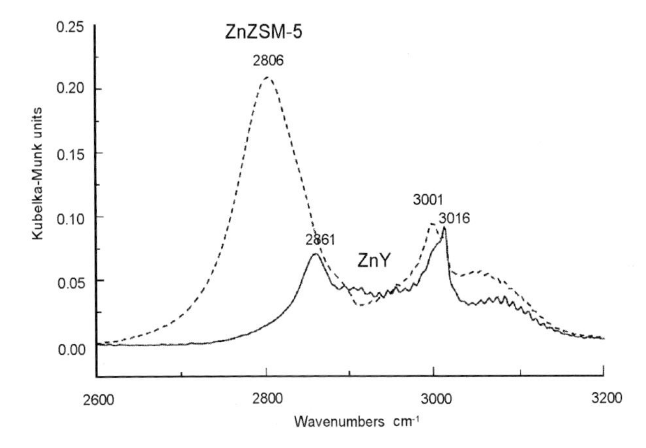

Figure 4.21. Methane adsorption by zinc-modified zeolite. Comparison between a low Al/Si framework ratio in ZSM-5 and a high Al/Si ratio in zeolite Y[21].

of the cation and the size of the $(Si_xAl_{(1-x)}O_2)_n$ ring system, which determines which adsorption is the most preferred. For Zn^{2+} coordination, the six-ring of $(Si_{2/3}Al_{2/3}O_2)_6$ is preferred. In some zeolites such six-rings have different local zeolite framework environments. The tendency to accommodate lattice deformations, due to Zn^{2+} attachment, may then vary, which will also affect the preferred siting. Zn^{2+}, for example, prefers the six-ring lattice position whereas the larger $ZnOZn^{2+}$ oxycation prefers the larger zeolite 8-ring[22] as its optimal site.

The activation of C–H bonds for different hydrocarbons can occur both at Zn^{2+} and $ZnOZn^{2+}$ sites. We will first discuss hydrocarbon activation by Zn^{2+}. The results presented here are based on quantum-chemical cluster calculations. The reaction energies involved in the overall catalytic cycle for the activation of ethane over a Zn^{2+} cation and a $ZnOZn^{2+}$ oxycation adsorbed on a representative cluster chosen to model the ZSM-5 adsorption site are compared in Fig. 4.22.

The activation of an alkane by Zn^{2+} occurs through formation of a Zn–alkyl species and a zeolitic H^+. The proton adsorbs on a basic lattice oxygen atom that connects a silicon with an aluminum lattice cation[24].

When the alkane molecule reacts with the $ZnOZn^{2+}$ oxycation the proton binds to the oxycationic oxygen atom, which has a much higher reactivity than the zeolite lattice oxygen atom. As a consquence, the initial activation of the C–H bond at the $ZnOZn^{2+}$ site is highly preferred over Zn^{2+}.

In a subsequent reaction step, the alkene is generated from the Zn–alkyl complex, by a β-C–H cleavage reaction, which is endothermic. The catalytic cycle closes by recombination of the two hydrogen atoms to give H_2. This final step is much more difficult for the $ZnOZn^{2+}$ center than for Zn^{2+}, because of the much larger [ZnOHZn] hydroxyl-bond energy compared with that of the zeolitic proton.

In the case of Zn^{2+}, the basic zeolite oxygen atoms can be considered to act as a reactive ligand to Zn^{2+} assisting heterolytic cleavage reactions. This is quite common for reactive charged cations in zeolites and can be considered as a spillover effect. The reduction of cations will not always lead to the generation of zeolitic protons.

The activation of H_2 by Ga^+ is proposed to proceed differently since no protons are observed experimentally when H_2 or alkane adsorbs. Homolytic hydrogen dissociation

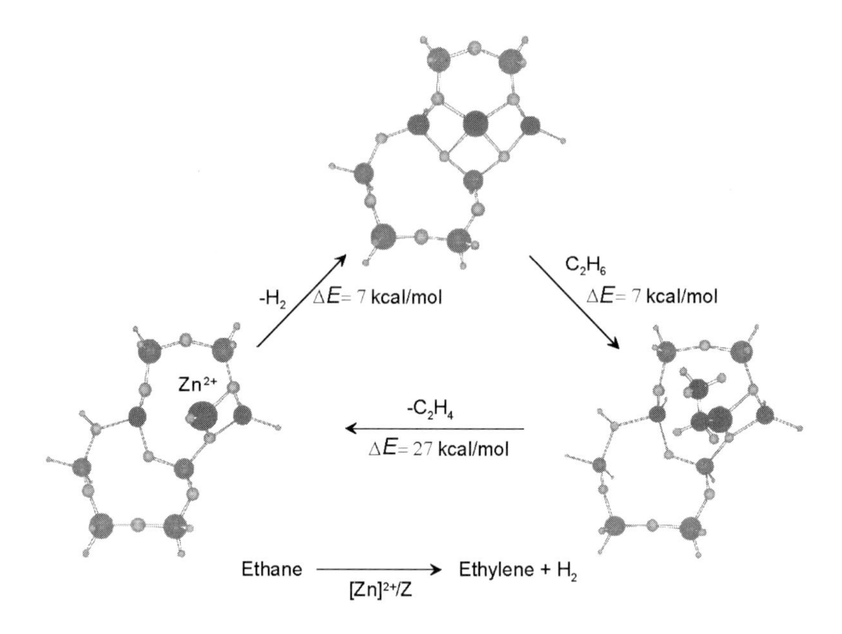

Figure 4.22a. The structures and energies involved in the catalytic activation of ethane by the Zn^{2+} exchanged at a ZSM-5 adsorption site[23].

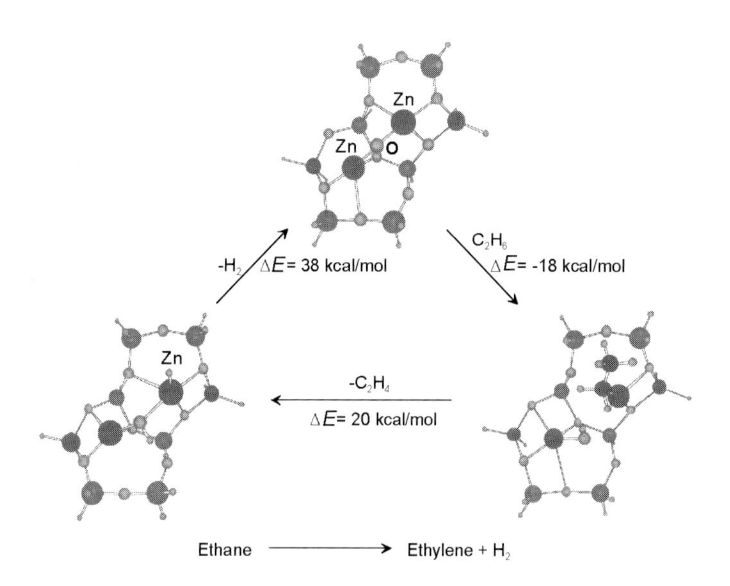

Figure 4.22b. Activation of ethane by an ion-exchanged $ZnOZn^{2+}$ cluster[23].

is concluded to occur because no zeolite protons are generated by H_2 reduction. This reaction sequence has been studied theoretically by Gonzales[25b] et al.

Homolytic oxidative addition of Ga^+ with H_2 is a slow reaction ($E_{act} = 240$ kJ/mol, because the back-donative interaction with a Ga^+ d-atomic orbital is very weak (see Chapter 3, page 129). This reaction is quite different from heterolytic dissociation with an activation barrier of only 60 kJ/mol for H_2 on Zn^{2+} to form ZnH^+ and a zeolite proton. Catalysis by Ga is complex in its chemistry. The GaO^+ species can activate the C–H bond, but H_2 recombination is slow. Also, GaO^+ may reduce during reaction and ${GaH_2}^+$ can be formed. Both Ga^+ and ${GaH_2}^+$ can activate the C–H bonds of alkanes. The chemistry is clarified by theoretical results that show that actually both systems for hydrocarbons heterolytic bond cleavage is the preferred reaction path[24b]. However, the intermediate $GaHR^+$ is thermodynamically preferred over ZOH–GaR. Therefore, after initial heterolytic dissociation by Ga^+ the intermediate $GaHR^+$ is rapidly formed. The consecutive reaction to alkenes proceeds via $GaHR^+$, explaining the absence of a proton signal in infrared experiments[24b]. According to Kazansky et al.[25], the reduction of GaO^+ results in the formation of low-coordinated gallium or gallium hydrides;

$$ZO^- \cdot\cdot GaO^+ + H_2 \longrightarrow ZO^- \cdot\cdot HGaOH^+ \quad \text{(a)}$$

$$ZO^- \cdot\cdot HGaOH^+ \longrightarrow ZO^- \cdot\cdot Ga^+ + H_2O \quad \text{(b)}$$

$$ZO^- \cdot\cdot Ga^+ + H_2 \longrightarrow ZO^- \cdot\cdot (GaH_2)^+ \quad \text{(c)}$$

They predicted for reaction (a) a reaction energy of −243 kJ/mol, reaction (b) a reaction energy of +130 kJ/mol and for the oxidative addition reaction (c) a reaction energy of −57 kJ/mol. For the last reaction an activation energy of at least 240 kJ/mol has been computed, to be compared with only 60 kJ/mol for reaction (a).

The reduction of cations such as Pd^{2+} and other transition metals proceed analogously to the C–H activation events discussed above for Zn^{2+}. Hydrogen dissociatively adsorbs to form $[PdH]^+$ and a zeolite proton. This is followed by the subsequent activation of a second H_2 molecule to form PdH_2 and another zeolitic proton. Hydrogen readily desorbs from Pd, leaving a reduced metal atom next to two zeolitic protons[19].

Cations such as Zn^{2+} or Ga^+ behave as soft Lewis acids in the reactions discussed above with the formation of intermediate metal–alkyl or metal–hydride species. This implies an electron transfer between the ligand and cation. To illustrate further the Lewis acid nature of Zn^{2+}, we analyze the mechanism for the hydrolysis of CH_3CN in which there is no change in formal valency of Zn^{2+}, and compare the energetics for this ion-exchanged Zn^{2+} reaction with that for the zeolitic proton[26]. The overall reaction scheme is

$$CH_3CN + H_2O \longrightarrow CH_3C(O)NH_2$$

Acetonitrile adsorbs strongly on Zn^{2+}. Its calculated interaction energy with the Zn^{2+} site in the zeolite model is −126 kJ/mol. In contrast, it interacts much more weakly with the zeolitic proton: $E_{ads} = -46$ kJ/mol. The product molecule of the hydrolysis reaction, acetamide [$CH_3C(O)NH_2$], however, has an adsorption energy of −73 kJ/mol with Zn^{2+} as it only coordinates through the basic nitrogen atom of acetamide. Acetamide binds to the proton in the protonated zeolite at −90 kJ/mol. It adsorbs in a bidentate configuration where the carbonyl oxygen atom binds to the H^+ and via the amide N–H group with the basic zeolite oxygen atom. The rate of the overall hydrolysis reaction of acetonitrile appears to be product inhibited. The inhibition is stronger for the protonic zeolite system where the acetamide adsorption is stronger. The acetamide desorption, however, can be assisted by a concerted adsorption step with the reactant nitrile. The overall thermodynamics for product desorption over Zn^{2+} is exothermic. The analogous reaction, however, still remains endothermic for the protonic case. Zn^{2+} is preferred for this reaction because there is no product inhibition.

Two of the key reaction steps are the cleavage of H_2O (see Fig. 4.23a) to produce an OH^- intermediate that will attach to the C atom of the nitrile, and the subsequent proton transfer from C–OH to form the first NH bond (se Fig. 4.23b). Figures 4.23 compares the activation energies for the activation of water to produce OH^- in both the absence and the presence of coadsorbed H_2O.

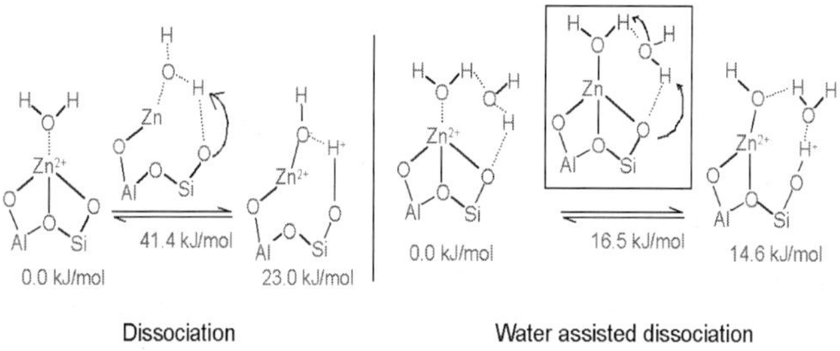

Figure 4.23a. Comparison of the effect of coadsorbed water on the generation of $ZnOH^+$ and the zeolitic H^+ by dissociation of water H_2O.[26].

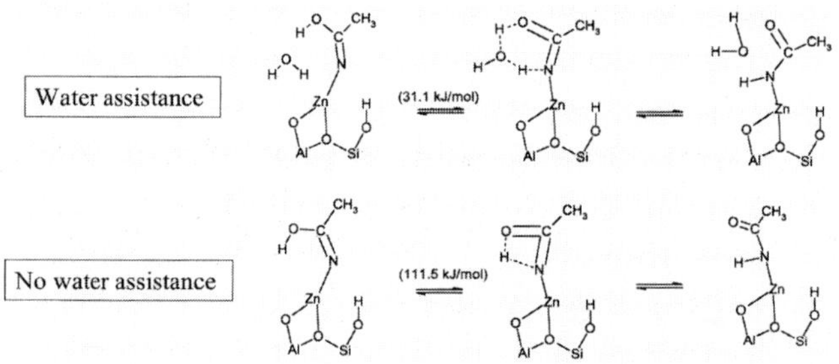

Figure 4.23b. The most difficult step in initial hydrolysis is the proton transfer within the intermediate (iminol) to form the keto group. This reaction step is made more facile by synergetic effect of a coadsorbed water molecule, which catalyzes the proton transfer from the OH to the NH group[26].

Figure 4.23b illustrates that the proton migration from the COH to form NH is assisted by coadsorbed water, which provides a low-energy path for proton transfer. The addition of a second water molecule here significantly lowers the activation barrier. Water directly participates in the transition state, providing a low-energy conduit for proton transfer by the rearrangement of the proton oxygen bonds around the water molecules. Proton transfer paths that proceed via consecutive H_2O proton bond formation and breaking reactions are well known in bulk water, aqueous and hydroxylated metal surfaces (see Chapter 6) and in enzyme catalysis (see Chapter 7).

The presence of coadsorbed water enhances this proton transfer path by providing a more optimal transition-state structure that does not require the dramatic distortion of metal–adsorbate bond angles. In addition, water stabilizes charge transfer, thus lowering the activation energies.

The promotional effect of protic polar molecules for proton transfer is very general. As an example, we mention the influence of methanol on the epoxidation of alkenes by Ti substituted in the zeolite lattice (see Scheme 4.1). The reaction is promoted by methanol, because it stabilizes the reactant structure and provides for a direct proton transfer path (see also Chapter 8 and ref. [27]). Alcohol or water formation restores the catalyst. Coadsorbed methanol assists proton transfer from the zeolite to the peroxide to produce the alcohol (ROH). The proton from (TiOHSi) is transferred to methanol and the proton from methanol assists the cleavage of the peroxide O–O bond.

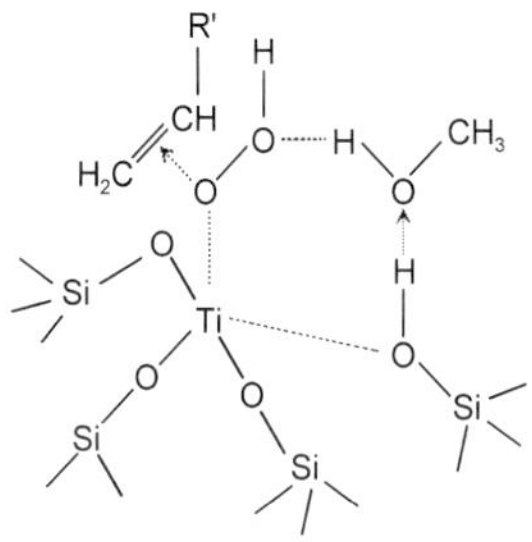

Scheme 4.1 Enhancement of the epoxidation activity of framework Ti by coadsorbed methanol. The methanol proton is transferred to form ROH (schematic).

4.3 Redox Catalysis

Redox reactions can be catalyzed by reducible cations substituted into the framework of zeolitic systems as well as polymorphic $AlPO_4$ systems or by cations not located in the framework but in the micropores. In Chapter 8 we will discuss more extensively catalysis by $Ti_xSi_{(1-x)}O_2$ systems using peroxides. Here we will initiate the discussion on redox catalysis with $Co_xAl_{(1-x)}PO_4$ oxidation catalysts where reducible ions such as Co^{3+} substitute for Al^{3+}. Catalytic oxidation carried out with oxygen provides an opportunity to discuss radical-type chemistry. A second system that we will discuss is photochemical oxidation induced by the strong electrostatic field of ion-exchanged cations. We will subsequently discuss catalysis by Fe^{3+} and Fe^{2+} ion exchanged zeolites with comparisons to Zn^{2+} systems and the important role of the corresponding oxycation.

For the iron system we first discuss N_2O decomposition and then describe the selective oxidation with N_2O to produce benzene from phenol. The N_2O decomposition reaction will be an example that illustrates additional complexity of catalytic systems, with self-organizing features.

4.3.1 Selective Oxidation of Alkanes Using the Reducible $M_xAl_{1-x}PO_4$ Zeolitic Polymorphs

We will follow closely the analysis given by Labinger[28]. The selective oxidation of cyclic and linear alkanes with O_2 over reducible $M_xAl_{1-x}PO_4$ catalytic materials has been reported under mild conditions. Dugal et al.[29] designed Co- and Mn-containing aluminophosphates, where the Co or Mn substitute for Al framework positions, with microporous structures similar to those of zeolites. They discovered that the catalysts with the smallest pore diameters show the highest selectivities towards oxidation of the terminal carbon atoms (see Table 4.3).

In radical chain oxidation reactions, the relative rates of the termination steps differentiate oxidation steps and, hence, affect the selectivity. The termination of tertiary peroxides is much faster than that of primary peroxides. The constraints of the zeolite micropore dimensions limit the geometry of the bimolecular encounter of alkane and oxyradical.

In zeolites with small dimensions, the initial interaction will involve the terminal methyl group of the alkane. The main products that form are the diacids. Radical-chain autoxidation proceeds by a sequence of initiation, chain propagation and chain termination steps. The initiation steps involve the formation of an OOH (OOH •) and a hydrocarbon free radical (R •). The formation of the reactive ROO • species occurs by reaction of the alkyl radical with O_2. A reaction chain propagation step is

$$ROO\bullet + RH \longrightarrow ROOH + R\bullet$$

A chain termination step is

$$2R(H)OO\bullet \longrightarrow ROH + (R\text{–}H){=}O + O_2$$

The metals participate in redox steps such as

$$ROOH + M^{3+} \longrightarrow ROO + H^+ + M^{2+} \qquad \text{(reduction)}$$

or

$$ROOH + M^{2+} \longrightarrow RO + OH^- + M^{3+} \qquad \text{(oxidation)}$$

Table 4.3. Oxidation of n-alkanes over MAPOs: primary selectivity

Framework	Pore dimensions (nm).	Substrate	Metal	Primary sel. (%)
$AIPO_4$-18	0.38 x 0.38 nm	n-pentane	Co	33
			Mn	39
		n-hexane	Co	61
			Mn	66
		n-octane	Co	60
			Mn	62
$AIPO_4$-11	0.39 x 0.63 nm	n-hexane	Co	19
$AIPO_4$-36	0.65 x 0.75 nm	n-pentane	Co	5
			Mn	0
		n-hexane	Co	23
			Mn	0
		n-octane	Co	12
			Mn	7
$AIPO_4$-5	0.73 x 0.73 nm	n-hexane	Co	9

In the zeolitic micropore, chain termination will also be suppressed because of the small likelihood that two peroxy radicals are present in the same small cavity. This favors the chain propagation reaction that depends on the CH bond strength, which in turn favors the primary oxidation reactions. While the primary carbenium ions are high in energy as compared with secondary or tertiary carbenium ions, the opposite order of stability is found for the alkyl radicals! Interestingly, oxidation reactions that take place in enzymes, which have been proposed to proceed through radical intermediates, are thought to be selective because the reactions are constrained by rebound within the enzyme cavity (see Chapter 7, page 328).

Radical reactions can be initiated also by radical centers generated in zeolites by heat treatment. Brønsted acidic sites, on the Si–OH–Al sites can be converted to highly reactive Lewis acid sites by high temperature with the elimination of water. The nature of these sites is still a matter of debate (see Kühl[31]). With hydrocarbons ESR-active radical cations are generated from such sites, which are part of catalytic reaction cycles. Such radicals may play a role in coke deposition and have been proposed also to play a role in catalytic cracking, Orchilles[32] and Corma[33] have studied the catalytic oxidation of hydrocarbons by such systems in detail, and have sustained catalysis over more than 4000 cycles. Dehydroxylation of the Si–OH–Al site creates three-fold coordinated Si and Al as well as an Si–O–Al center. In the dehydroxylation process also Lewis acidic AlO^+ sites may have been generated that adsorb to the negatively charged Si–O–Al sites. Interaction with 2,5-dimethylhexa-2,4-diene (DMHD) produces ESR signals that can be readily followed. Leu and Rodriner[34] concluded the following reaction sequence, in which single-electron transfer sites (SETS) are regenerated:

$$\begin{aligned} \mathrm{DMHD} + \mathrm{SETS} &\longrightarrow \mathrm{DMHD}^+ + \mathrm{SETS}^- \\ \mathrm{SETS}^- + \mathrm{O_2} &\longrightarrow \mathrm{SETS} + \mathrm{O_2}^- \\ \mathrm{DMHD}^+ + \mathrm{O_2}^- &\longrightarrow \text{reaction products} \end{aligned}$$

Reaction products have not been analyzed, because product molecules remained adsorbed in the zeolite. The chemistry of this dark oxidation reaction is related to the photochemical reaction steps discussed above.

4.3.2 Photo Catalytic Oxidation

Cations such as Na^+, Ba^{2+} or Ca^{2+} ion-exchanged into zeolites have been shown by Blatter et al.[30 to play an important role in the selective photo-oxidation of alkenes and alkanes. We learned earlier in Section 4.2.3 that the electrostatic field-induced polarization of an adsorbed molecule changes the adsorption intensity of vibrational transitions in the infrared spectroscopic regime. When an organic molecule and O_2 adsorb on the cations, the energy of electron transfer between the organic molecule and the oxygen molecule is lowered, with important consequences for oxidation catalysis initiated by this electron excitation event. In the gas or liquid phase, charge transfer between O_2 and the hydrocarbon to give $O_2{}^-$ and a positively charged hydrocarbon occurs via an electron excitation induced by UV or visible light. Subsequent oxidation steps occur through radical chain reaction pathways that result from OOH and R. They tend to have low selectivity.

Longer wavelength visible light can be used instead of UV photons needed in the gas phase. This has the important advantage that the radical-generating reactions which compete with the desired oxidation radical chain reaction are now suppressed. For example, light with $\lambda \ll 600$ nm will induce a selective oxidation reaction chain between O_2 and toluene towards benzaldehyde, with no consecutive oxidation of benzaldehyde . The lower energy photons are not able to overcome the higher ionization potential of benzaldehyde (9.5 eV) compared with 8.8 eV for toluene, necessary for electron transfer towards oxygen.

Reactions are proposed to proceed through radical chain reactions and intermediate formation of the corresponding hydroperoxides that have been trapped at low temperature. These hydroperoxides can be produced with high selectivity because of the constraints on the recombination between molecules or radicals from the cavity. Thirdly, dehydration of peroxides occurs readily in the ionic zeolite environment via heterolytic mechanisms leading to carbonyl products without side reactions. The use of low-energy photons and a low-temperature environment also precludes homolysis of the peroxide bond, which minimizes non-selective gas-phase oxidation reactions. A representative reaction scheme for propane is shown in Fig. 4.24. The origin of the low-energy electron transfer reaction through cation interactions appears to be only partially the direct consequence of the high electric fields near the cations (order of magnitude is 0.6 Å^{-1}), but can be considered to be the result of a confinement effect[24b] as has been found theoretically for dimethylbutene oxidation. Reaction only occurs when cations with a specific size are properly located with respect to each other. One of the cations adsorbs O_2 and the other one adsorbs the alkene. Their heats of adsoption have to be comparable.

Light-activated charge transfer takes place with low energy between O_2 and the alkene oriented by interaction with the reaction such that there is overlap between the respective HOMO and LUMO orbitals. Adsorption on the cations overcomes the repulsive interaction between molecules when they approach so close that van der Waals radii overlap, with the result that a (photo) chemical reaction occurs. This can be considered again as an example of pre-transition state stabilization, that was discussed earlier in Section 4.2.1.

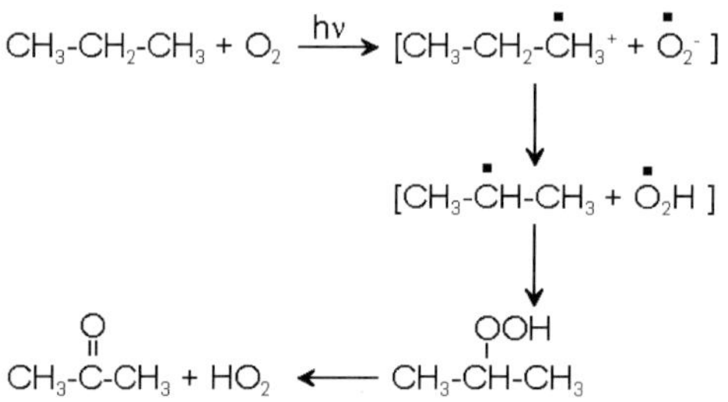

$$CH_3\text{-}CH_2\text{-}CH_3 + O_2 \xrightarrow{h\nu} [CH_3\text{-}CH_2\text{-}\dot{C}H_3^{+} + \dot{O}_2^{-}]$$

$$\downarrow$$

$$[CH_3\text{-}\dot{C}H\text{-}CH_3 + \dot{O}_2H]$$

$$\downarrow$$

$$CH_3\text{-}CH(OOH)\text{-}CH_3$$

$$CH_3\text{-}C(=O)\text{-}CH_3 + HO_2 \longleftarrow CH_3\text{-}CH(OOH)\text{-}CH_3$$

Figure 4.24. Proposed mechanism for photo-oxidation of alkanes.

4.3.3 The N_2O Decomposition Reaction; Self-Organization in Zeolite Catalysis

We will initially examine N_2O decomposition over single cationic metal atom centers and then subsequently continue the discussion on N_2O decomposition over dimer oxycation species. The reaction energy diagrams for N_2O decomposition over single Fe^{3+}, Co^{3+} and Rh^{3+} centers established from ab initio density functional theory calculations are presented in Fig. 4.25 a and b[35]. The initial adsorption of N_2O appears to be strongest on Co^{3+}, but the metal–oxygen bond generated as the result of N_2O decomposition is strongest for $[FeO]^{3+}$. Therefore, the reaction with the second N_2O molecule to form O_2

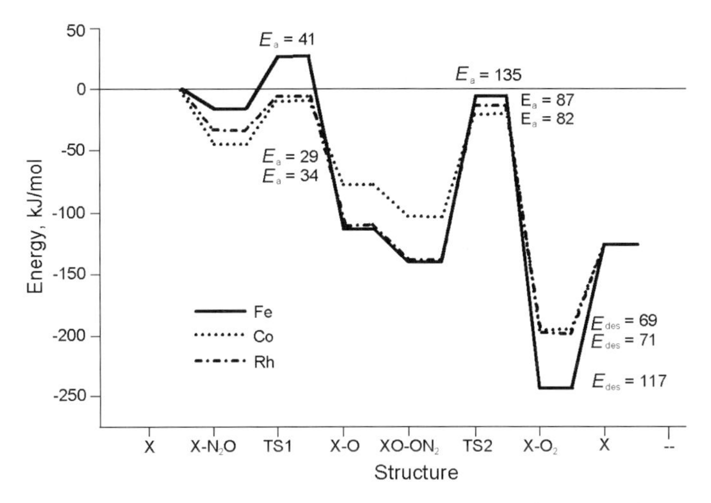

Figure 4.25a. Comparison of the reaction energy profiles for N_2O decomposition over Co, Rh and Fe sites on a cluster model of HZSM-5.

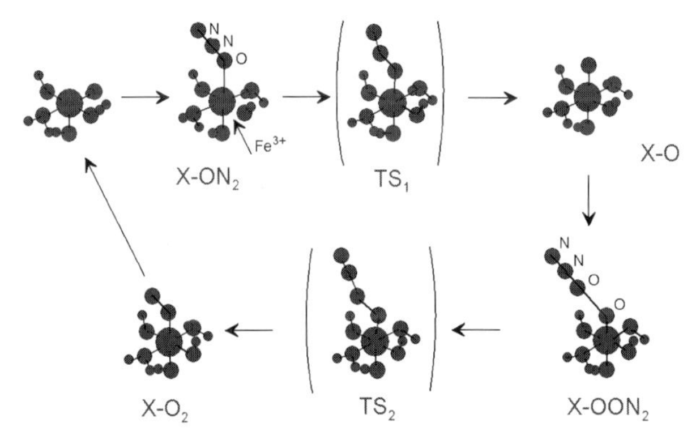

Figure 4.25b. N_2O decomposition on a single Fe^{3+} center. The structures that correspond to the energies in the reaction-energy diagram of Fig. 4.25a.

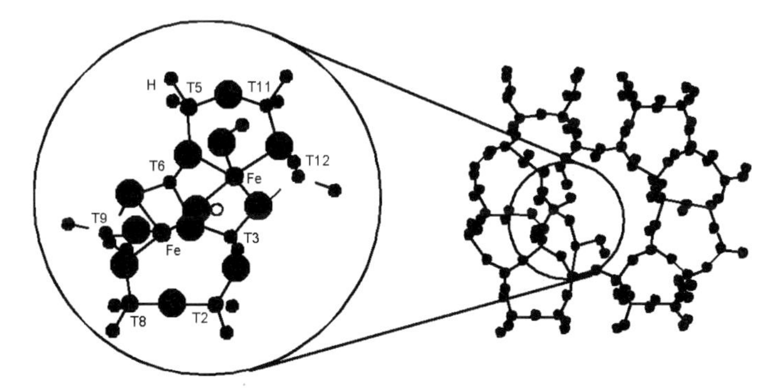

Figure 4.26. N_2O decomposition catalyzed by Fe/ZSM-5. Cluster model of the binuclear iron oxide-hydroxide site[36].

is most difficult on the Fe^{3+} center but easier on Rh^{3+} or Co^{3+} centers to which atomic oxygen is more weakly bound. The small angle of N_2O in the transition state (see Fig. 4.25b) implies that there is a small amount of electron donation towards N_2O (remember that N_2O^- is isoelectronic with NO_2). Figure 4.26 shows the $[HOFe^{3+}–O–Fe^{3+}–OH]^{2+}$ dimer iron oxyhydroxy cation in contact with the negatively charged α-site of ZSM-5. Such dimers have been proposed as dominant species for N_2O decomposition in iron-exchanged zeolites. The hydroxylated dimer cation is thermodynamically stable for N_2O decomposition in the presence of gas-phase water[36].

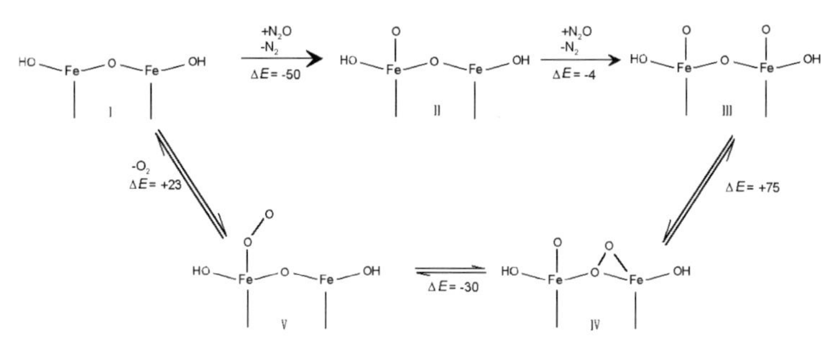

Figure 4.27. The intermediate structures and energies in the N_2O decomposition reaction of binuclear iron(III+)oxydehydroxide[36]. Energies are reported in kJ/mol. The energies reported all refer to the reaction direction from the lower numbers to the higher number. For example, $\Delta E_(I- > V)$ =+23 kJ/mol.

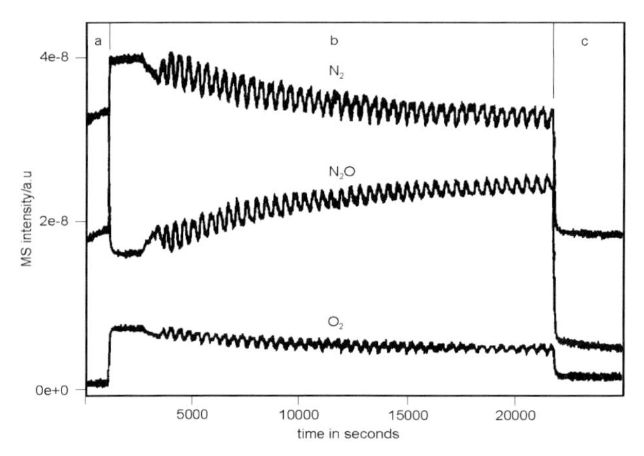

Figure 4.28. N_2O decomposition over Fe/ZSM-5 shows oscillating behavior in the presence of water[37].

The energies for the N_2O decomposition reaction on this iron-dimer cluster are shown in Fig. 4.27.

The N_2O decomposition reaction is especially interesting because under particular conditions the reaction can be induced to oscillate (see Fig. 4.28[37]). If non-isothermal effects can be excluded, this implies the presence of an auto-catalytic elementary reaction step in the overall catalytic reaction cycle (see Chapter 8). In this case, the auto-catalysis results from N_2O decomposition catalyzed by both mono-center and bi-center iron complexes.

There is ample experimental evidence for the presence of partial hydroxylated monomer- and dimer-iron complexes in zeolites that decompose N_2O. At high temperatures N_2O easily decomposes on a single-center $Fe^{3+}(OH)^-$ cationic complex to form N_2 and the $Fe^{3+}{=}O$ complex

$$N_2O + Fe^{3+}(OH)^- \longrightarrow (OH)^-Fe^{3+}{=}O + N_2$$

The formation of O_2 by a consecutive reaction of this oxidized center with N_2O requires a high activation energy (see Fig. 4.25a).

As we can deduce from Fig. 4.27, the recombination of two oxygen atoms on the bi-center iron complex is easy. The oscillatory time-dependent behavior of the overall reaction is consistent with the following auto-catalytic reaction sequence:

$$(OH)^-Fe^{3+} + (OH)^-Fe^{3+}O \longrightarrow \left[(HO)Fe^{3+}OFe^{3+}(OH)\right]^{4+} \tag{d}$$

$$N_2O + \left[(HO)Fe^{3+}OFe^{3+}(OH)\right]^{4+} \longrightarrow O_2\uparrow + N_2\uparrow + 2(OH)^-Fe^{3+} \tag{e}$$

The overall result of reaction (d) and (e) is that one $(OH)^-Fe^{3+}$ species generates, in an inorganic reaction with $(OH)^-Fe^{3+}{=}O$, two $(OH)^-Fe^{3+}$ intermediates. Such overall stoichiometry defines an auto-catalytic reaction. An important conclusion from this analysis is that the catalytically reactive phase only establishes itself during the course of reaction. The catalytic system is therefore considered dynamic. Monomers and dimers are formed and disappear in reactions with N_2O and desorption of O_2. The dynamic patterns arise when these events are synchronized. This is mathematically similar to the Turing patterns discussed in Chapter 8. The decomposition reaction is the driving force for the self-organization of the inorganic system. As a corollary, the inorganic chemistry of the reactive phase cannot be established independently of the catalytic reaction.

Oscillating phenomena have also been observed for N_2O decomposition by Cu-exchanged ZSM-5 zeolites[36]. A dynamic state consisting of monomeric Cu^+ and dimeric $Cu^{2+}OCu^{2+}$ is also proposed here.

Heyden et al.[39] suggested that hydrated and dehydrated monomolecular iron sites in Fe-ZSM-5 are responsible for N_2O decomposition. They proposed that $Z^-[FeO]^+$ is a key intermediate. Furthermore, water strongly adsorbs to give $Z^-Fe(OH)_2{}^{+1}$. This deactivates the $Z^-[FeO]^+$ site. The activation energy for N_2O decomposition in the presence of water increases steeply compared with the anhydrous situation, because water has to desorb from $Z^-Fe(OH)_2{}^{+1}$ in order for N_2O reduction to occur. Hydration and subsequent dehydration of the oxy-iron complex may provide an alternative explanation for the oscillatory reaction found by El-Malki et al. shown in Fig. 4.28. If the reaction is not isothermal, the temperature fluctuations arising from the exothermic N_2O decomposition reaction may lead to fluctuation in the water adsorption. This may provide an alternative explanation of the oscillatory kinetic behavior in the Fe^{3+}-ZSM-5 system.

4.3.4 Oxidation of Benzene by N_2O; the Panov Reaction

The catalytic oxidation of benzene to phenol in iron-containing zeolites is known as the Panov reaction[40] The ZSM-5 zeolitic system is the preferred matrix. There are several ways in which the catalyst can be activated for this reaction.

It is now well established that the active component of the catalytic reaction is monomeric Fe^{2+}. The Panov reaction consists of two reaction steps:

N$_2$O decomposition : $$N_2O \underset{Fe^{2+}}{\longrightarrow} N_2\uparrow + (FeO)^{2+}$$

selective oxidation : $$\text{benzene} + FeO \longrightarrow Fe^{2+} + \text{phenol}$$

The uniqueness of the ZSM-5 catalyst relates to its stabilization of Fe^{2+} cations in the selected (α) sites of the zeolite micropores. There is increasing evidence that non-lattice alumina plays a promoting role, by potentially enhancing the relative stability of isolated Fe^{2+} centers[41].

As a preliminary to our later comparison with biochemical systems (see Chapter 7), it is relevant to note here that the enzyme cytochrome P-450 also contains a single Fe center attached to a porphyrin system. Cytochrome P-450 catalyzes the reaction of methane to methanol. There is also an enzyme that contains a two-iron cationic center that catalyzes the same reaction. In the methane monooxygenase enzyme, two Fe cations are bridged by oxygen and charge compensated by glutamate and histidine groups[42].

An important difference between the benzene oxidation reaction and methane activation, is the absence of an isotope effect in the benzene oxidation reaction. In the enzyme, CH_4 activation is initiated by hydrogen abstraction. This initiates a radical-type reaction. Benzene oxidation in the Panov system, however, follows a very different reaction path.

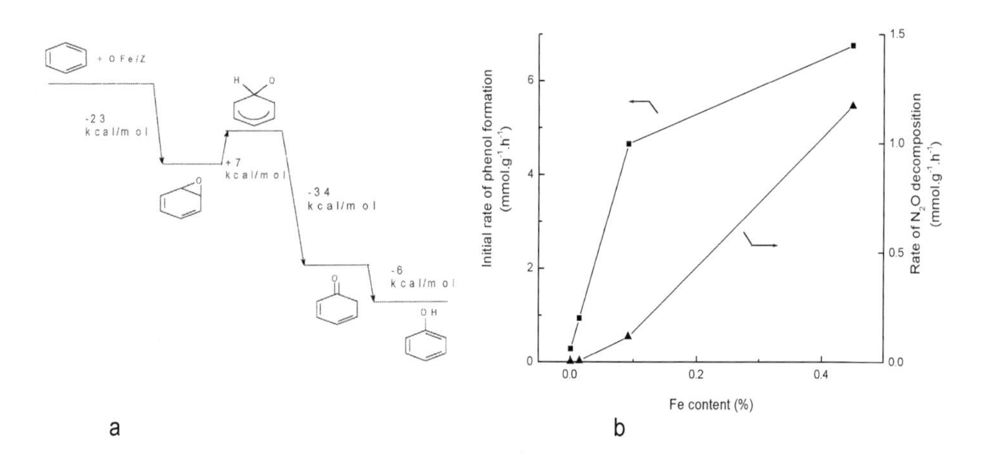

Figure 4.29. (a) The speculated reaction path for the oxidative transformation of benzene to phenol in the Panov reaction[43]. (b) Rates of phenol formation and N$_2$O decomposition as a function of Fe content in ZSM-5[44].

The unique feature of the $Fe^{2+}{=}O$ bond is its rather high bond energy of 250–290 kJ/mol. This is in contrast to the low bond energy in $Fe^{3+}{=}O$ (90 kJ/mol). This is found when Fe^{3+} is part of the oxycationic complex as well as the monomer.

Interestingly the $[FeO]^{2+}$ cation has been proposed[65] as the active oxidation species in the Fenton reagent[66] that, amongst others, hydroxylates aromatic substrates in the waterphase. Molecular orbital analysis of hydrated $[FeO]^{2+}$ shows that the HOMO of $[FeO]^{2+}$ has $2\pi^*$ character and the σ-type lone pair orbital is unoccupied. The reactant has, therefore, a strong electron donative capacity into the antibonding orbitals of the substrate CH orbitals.

Comparing the Fe^{3+} with the Fe^{2+} system, there is a difference of the order of 400 kJ/mol for the recombination energy of two oxygen atoms to form O_2. This would result in significantly slower O_2 evolution at the Fe^{2+} centers than from the Fe^{3+}-containing dimeric centers. This may also explain the uniqueness of Fe^{2+} for the benzene oxidation reaction. Oxygen generated by N_2O decomposition has to react with benzene and cannot recombine to O_2. The recombination reaction to give O_2 has to be suppressed without, however, suppressing the steps involved in the oxidation of benzene. The subsequent steps involved in the oxidation of benzene are shown in Fig. 4.29a.

The only step in the overall oxidation reaction cycle which is endothermic is step 2, which involves the direct insertion of oxygen into the C–H bond of benzene. This is costly since it requires the loss of aromaticity in the benzene ring. All other steps in the cycle are exothermic. Furthermore, matrix effects are absent in this reaction. The main role of the lattice appears to be to stabilize Fe^{2+} and prevent over-oxidation of N_2O decomposing Fe^{3+} oxyhydroxy dicationic clusters. The overall result is that the rate-limiting step for phenol formation is the rate of desorption of phenol. The relative concentration of the different sites varies with Fe loading, as illustrated in Fig. 4.29b. Whereas the rate of phenol formation increases steeply with the Fe content, when the Fe concentration is low, at higher Fe content N_2O decomposition increases, but phenol production is constant.

Whereas in the zeolite system the $[Fe^{2+}O]$ species is produced from N_2O, in the biochemical system the biochemical oxidation step of the enzyme with O_2 is coupled to an electrochemical reduction: the overall reaction for cytochrome P-450 that converts the alkane to an alcohol is

$$RCH + O_2 + 2H^+ + 2e \longrightarrow RCOH + H_2O$$

A similar stoichiometry holds for the reaction with the two-Fe center methane–monooxygenase enzyme.

The difference in the chemical environment and the valency of Fe may explain why the zeolite system does not produce methanol from CH_4 whereas the enzymes do.

4.4 The Zeolite Catalytic Cycle. Adsorption and Catalysis in Zeolites; the Principle of Least Optimum Fit

The dependence of the overall rate of the catalytic reaction on adsorption is extremely important in analyzing the kinetics for the overall rate of a zeolite-catalyzed reaction. We have already met this subject in Chapter 2 when analyzing the basis of the Sabatier principle. A proper understanding of adsorption effects is essential for establishing a theory of zeolite catalysis that predicts the dependence of kinetics on zeolite-micropore shape and connectivity.

Catalysis consists of a reaction cycle made up of several elementary reaction steps. For a zeolite, the catalytic cycle contains at least the following four elementary steps: adsorption, diffusion, substrate activation and desorption.

We will discuss in Section 4.6 explicitly the role of diffusion. In principle, diffusion effects can always be experimentally excluded by selecting crystallites so small that chemical

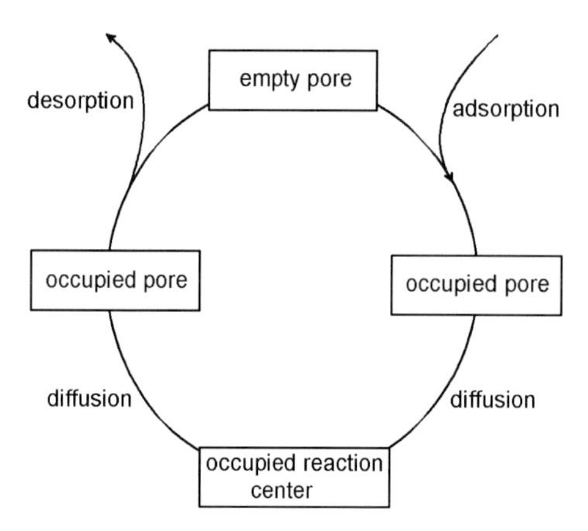

Figure 4.30. The catalytic cycle of a zeolite-catalyzed reaction.

activation or desorption is rate limiting. In the absence of diffusional constraints, the rate of product desorption competes with that of reactant activation. The elementary reaction steps of the latter processes proceed through steps with transition-state barriers that are typically between 50 and 150 kJ/mol.

As mentioned earlier, the interaction between hydrocarbons and the zeolite micropore depends upon a weak van der Waals attraction between the polarizable zeolite framework-oxygen atoms and the hydrocarbon units such as aromatic, CH_2 or CH_3 groups. These interactions are too weak to be described effectively by density functional theory. Although higher level CI calculations can more accurately model van der Waals interactions, they typically cannot model a large enough portion of the zeolite to provide an approriate model for a mesoporeous system. Empirical force fields, however, are effective in accurately describing these van der Waals interactions. Adsorption properties are, therefore, typically computed from force field calculations rather than from quantum mechanics. The use of such force fields in combination with the application of statistical mechanical techniques in Monte Carlo methods[45] allows the prediction of the chemical potential of adsorbed molecules as a function of temperature and pressure. This is very important since adsorption appears to be controlled not only by the adsorption enthalpy, but also by entropy effects that may depend strongly on the micropore filling with adsorbed molecules. For alkanes, the adsorption enthalpy is strongly dependent on their chain length which is incremental in the number of CH_2 units. The van der Waals interaction energy may vary between 5 and 15 kJ/mol per CH_x unit. The interaction energy increases when the pore dimension and the hydrocarbon chain dimension better match one another[46].

The implication is that the adsorption energies of hydrocarbons can be of the same order of magnitude or even larger than the activation barriers of the intrinsic rate constants. In this section we will discuss expressions for the overall rate when the intrinsic reaction rate (e.g. proton activation) is rate determining. In the next section we will discuss the consequences especially for selectivities of a reaction when the desorption rates are rate limiting. To introduce the subject, we first present some simple expressions used in modeling monomolecular reactions.

For a monomolecular reaction, the rate of the reaction normalized per reaction center is:

$$r = \frac{R_{\text{overall}}}{N_{\text{act.site}}} = k_{\text{act}} \cdot \theta_{\text{ads}} \tag{4.1}$$

where θ_{ads} is the equilibrium concentration of reactant adsorbed at the reaction center and k_{act} is the elementary rate constant for the conversion of the molecule adsorbed at the reaction center. In general, the expression for θ_{ads} is complex. It not only depends upon the many equilibrated intermediate reaction steps, but may also depend upon the micropore concentrations of adsorbed reactants and products. On the other hand, expression (4.1) is very general and, as we will discuss more extensively in Chapter 7, is used in heterogeneous, homogeneous and biocatalysis. In the last two cases, θ_{ads} is replaced by the concentration of the reaction complex formed by the interaction of the reactant with the organometallic complex or the biocatalyst (enzyme). Expression (4.1) is the equivalent of the Michaelis–Menten expression for the catalytic reaction rate in biochemistry.

As we will discuss in the next section, it is of interest to analyze the meaning of the expression when the Langmuir adsorption expression for θ is used. For a monomolecular gas-phase reaction, the expression for the catalytic reaction rate is

$$r = k_{act} \cdot \frac{K_{eq}^{ads} \cdot p}{1 + K_{eq}^{ads} \cdot p} \tag{4.2}$$

where K_{eq}^{ads} is the adsorption equilibrium constant for the reactant at the reaction center and p is the gas phase pressure of the reactant. Note that expression (4.2) is also simplified since the adsorption of the product molecule is ignored.

Expression (4.2) illustrates an important kinetic feature: the order of a catalytic reaction depends strongly on the reaction concentration. At low pressure, the rate is first order in reactant and at high pressure the rate is zero order in reactant. Second, the overall rate depends on the intrinsic rate constant of an elementary reaction step, k_{act}, and also on the adsorption constants. Expression (4.2) is valid only under the ideal conditions that all catalytic centers are similar and there are no interactions between reactant and (or) product molecules. These conditions are rarely satisfied and, for this reason, practical rate-expressions are often more complicated than Eq. (4.2). Expression (4.2) illustrates, however, that the interplay between surface coverage and elementary rate constants is very important, so that for an overall prediction of the reaction rate one needs to integrate intrinsic reaction rate predictions with surface state predictions. As mentioned earlier, the equilibrium constants for adsorption can be calculated using either statistical or dynamical Monte Carlo methods[45,46].

To illustrate the use of Eq. (4.1) in zeolite catalysis, we will discuss the results of theoretical and experimental studies of the hydroisomerization reaction of hexane[47]. In order to exclude diffusion limitation, two conditions have to be met. First, the Biot condition must be satisfied when the rate of desorption is limiting:

$$k_{des} \ll \frac{D}{R^2} \tag{4.3}$$

where R is the radius of a crystallite particle and D the rate of diffusion. If k_{des} is rate limiting,both diffusion and the elementary reaction rates within the zeolite micropore are fast compared with desorption.

More generally, the rate of diffusion should be much faster than the intrinsic rate of a chemical reaction:

$$R\sqrt{\frac{R_{overall}}{D}} \ll 1 \tag{4.4}$$

where R_{overall} is the rate constant defined in Eq. (4.1), D the rate of diffusion and R the radius of the crystallite. This expression is derived from the classical Thiele modulus expression where the intrinsic rate is thought to be controlling.

Zeolites used in the hydroisomerization of alkanes typically contain small transition-metal clusters such as Pt or Pd[47]. The hydrogenhydrocarbon ratio in these systems is typically 30 and the reaction occurs at around 250 °C. Hydroisomerization typically proceeds by alkane dehydrogenation over the metal. The alkene intermediate that forms can subsequently undergo isomerization at an acid site before it is rehydrogenated over the metal. The transition metal is necessary here to establish the alkane–alkene equilibrium which implies a very low hexene equilibrium concentration. At this low concentration of alkene, catalyst deactivation reactions which are the result of acid-catalyzed oligomerization are suppressed. The elementary isomerization step that is rate limiting involves the activation of the alkene. The isomerization of the branched alkene proceeds through a cyclopropyl-type transition state (see Fig. 4.9). When the alkene adsorbs to the proton, only the π bond interacts with the proton. The CH_2 and CH_3 groups of the molecule experience van der Waals interactions with the siliceous part of the zeolite lattice. The interaction between the zeolite micropore and the hydrocarbon is hydrophobic. For a molecule such as hexene, this implies that the adsorption energy and the activation energy for desorption may vary between 30 and 90 kJ/mol for different zeolites. This variation in the interaction energy arises from the difference in match between the radius of the alkyl portion of the molecule and the dimensions of the micropore, which results in a significant difference in the van der Waals interactions. The interaction of the π-bond of the hydrocarbon with the zeolitic proton is of the order of 30 kJ/mol. For Mordenite, a medium-sized pore zeolite, this results in the overall reaction energy scheme presented in Fig. 4.31.

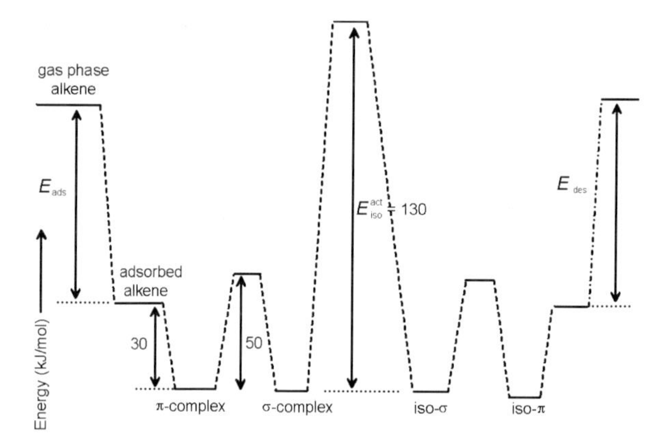

Figure 4.31. Reaction energy diagram for the isomerization of hexene in mordenite. The π-complex denotes hydrogen-bonded alkene whereas the σ-complex denotes the corresponding alkoxy intermediate (schematic).

The solvation of ionic complexes by the zeolite cavity is very limited, hence the protonation of an alkene and also the alkene isomerization require relatively high barriers. The shape of the protonated intermediate, its positive charge distribution, the zeolite microcavity size and shape and the negative charge distribution on the zeolite lattice determine the relative energy of the protonated intermediate. The zeolite lattice has very limited flexibility and, hence, the main additional stabilization derives from the polarization of the oxygen atoms by the positive charge of the protonated reaction intermediate. Clearly, the activation energy for isomerization and the desorption energy are close to being competitive and would be even more competitive for hydrocarbons with longer hydrocarbon chain lengths.

The importance of the entropy of adsorption is illustrated by experimental and calculated adsorption free energies for hexane in the 12-ring one-dimensional channel mordenite (MOR) and 10-ring one-dimensional channel of ferrierite (TON). Table 4.4 compares the simulated values for the heats of adsorption from configurationally biased Monte Carlo calculations valid at low micropore filling. The corresponding adsorption equilibrium constants are also compared in Table 4.4. One notes the increase in the energy of adsorption for the narrow-pore zeolite. However, at the temperature of reaction, the equilibrium adsorption constant is also a factor 10 lower for the narrow-pore zeolite.

Table 4.4. Calculated heats of adsorption and adsorption constants for various hydrocarbons in zeolites with different channel dimensions

	ΔH_{ads} kJ/mol simulation	$K_{ads}(T = 513K)$ mmol/g Pa simulation
n-pentane/TON	–63.6	4.8 x 10^{-6}
n-pentane/MOR	–61.5	4.8 x 10^{-5}
n-hexane/TON	–76.3	1.25 x 10^{-5}
n-hexane/MOR	–69.5	1.25 x 10^{-4}

The overall result at this temperature is a higher micropore filling in the wide pore material, notwithstanding its lower heat of adsorption. The loss in entropy that the molecules experience when their motion becomes more constrained in the micropore compared to their motion in the gas phase is less in the wide-pore zeolite than in the narrow-pore zeolite. This effect dominates, and as a result the coverage, θ, is largest in the zeolite in which the reacting molecule experiences the smallest entropy loss compared to the gas phase. Comparing the overall rates of reaction for different zeolites, the question arises, which of the two parameters is most sensitive to the pore size dimensions, the free energies of adsorption or the activation free energies of the elementary rate constant of isomerization.

Table 4.5. Results of activity measurements of the hydroisomerization of hexane (temp. = 240 °C, p_{nC_6} = 779 Pa)

	TOF (sec^{-1})	K_{ads,nC_6} (Pa^{-1})	$k_{isom}(sec^{-1})$
H-Beta	5.0 x 10^{-3}	6.4 x 10^{-5}	1.7 x 10^{-2}
H-Mor	1.1 x 10^{-2}	3.3 x 10^{-4}	2.7 x 10^{-2}
H-ZSM-5	4.1 x 10^{-3}	3.3 x 10^{-5}	2.8 x 10^{-2}
H-ZSM-22	1.6 x 10^{-3}	1.4 x 10^{-5}	1.7 x 10^{-2}

The results of experimental measurements[48] which were analyzed with expressions analogous to, but more sophisticated than, Eq. (4.2), are presented in Table 4.5. Zeolite β and ZSM-5 contain three-dimensional 10-ring channels. Mordenite has a one-dimensional 12-ring channel. H-ZSM22 has the ferrierite structure with a one-dimensional 10-ring channel. By comparing the overall rate normalized per protonic site (turn over frequency, TOF), one notes the large difference in overall rate with changes in the pore dimensions. In contrast, the elementary rate constants, k_{iso}, are basically the same.

The large difference in reactivity of the zeolites arises primarily from the large difference in micropore occupancy in the different zeolites. The large variation in micropore occupancy is deduced from the large differences in adsorption equilibrium constants shown in Table 4.5. The similarity of the elementary rate constants for isomerization implies that there is little variation in the corresponding activation energies. The overall rate of the reaction is found to be a maximum for the zeolite in which the adsorption concentration of the reactant is a maximum.

For a medium-sized pore zeolite such as mordenite, the match of reactant size and shape with the micropore cavity size and shape is not an optimum since smaller pore sizes would increase the enthalpy of the adsorbed state. The entropy of the adsorbed state in medium-pore zeolite, however, is larger than that found in the smaller pore zeolite. For wider pore zeolites, heats of adsorption are decreased so much that the entropy gain no longer compensates for its decrease. The maximum concentrations of molecules therefore tend to be found in the medium-micropore zeolites.

We have illustrated that for a catalytic reaction in a zeolite to have a maximum rate, the adsorption free energy should be a maximum. Zeolites with medium-sized cavities are preferred over zeolites with small cavities because in the latter entropy loss dominates the gain in enthalpy. This compares with the anti-lock-and-key behavior of some enantiomeric catalytic systems discussed in the final section of Chapter 2. The catalytic systems that have an optimal misfit with their cavity perform the best, again demonstrating that the occupation is maximized while minimizing the entropy loss.

For a monomolecular reaction one deduces from Eq. (4.2) that when $\theta = 1$, the apparent activation energy of the reaction is equal to its intrinsic activation barrier of the elementary reaction step ($E_{\text{act}}^{\text{iso}}$). Another important result that can be derived from Eq. (4.2), is the behavior of this system at low micropore occupation ($\theta \ll 1$) where the overall rate is linear in pressure. The apparent activation energy under these conditions is

$$E_{\text{act}}^{\text{app}} = E_{\text{activation}} + E_{\text{ads}} \tag{4.5}$$

At low site occupancy, the measured apparent activation energy is linear in the heat of adsorption of the hydrocarbon. This agrees with experimental observations. Narbeshuber et al.[49], studied the conversion of different linear alkanes all catalyzed by H-ZSM-5. This cracking reaction proceeds via an intermediate carbonium ion, as was discussed in Section 4.2.1. The site occupancy is low and, therefore, Eq. (4.1) applies. The intrinsic activation energy controlled by the proton donation is essentially independent of chain length. Narbeshuber et al.[49] found a linear decrease in the apparent activation barrier with an increase in the chain length of the hydrocarbon. Since the heat of n-alkane adsorption is linear in the chain length, this implies that the apparent activation energy is linear in the adsorption energy. At comparable conversions, the temperature of the reaction decreases with increasing hydrocarbon chain length.

The relative independence of the elementary rate constant for proton-activated hydrocarbon conversion reactions carried out in different zeolite structures would indicate that the differences in the overall rate for different zeolites are not determined by the small differences in the reactivity of the protons (intrinsic acidity) but rather by a variation in the concentration of reactant molecules adsorbed on the zeolitic reaction centers. The higher apparent reactivity of the longer hydrocarbon chains in zeolite-catalyzed hydrocarbon conversion reactions can be likely ascribed to the increased concentration of hydrocarbon adsorbed at the reactive centers and not to a higher intrinsic reactivity of the hydrocarbon molecules.

A very important concept to consider in designing zeolites for specific activity and/or selectivity is that the zeolite cavities can be considered as nano-size reactors. This view we discussed earlier when we presented the Labinger interpretation[28] of selective oxidation in zeolitic micropores. Hydrocarbon conversion processes also include oligomerization, cracking and aromatization reactions. As a consequence, carbonaceous residues will form even when the reactions are carried out in the presence of hydrogen. The size of the micropore cavities will limit the size of the residue molecules that can form and in this way will suppress coke formation in high-temperature reactions such as catalytic cracking. This dramatically increases the yield of desired alkenens or aromatic molecules and decreases the rate of catalyst deactivation. Interestingly, the precursors for coke formation can catalyze carbon–carbon bond-forming reactions. Such reactions may even overcome the sometimes more difficult direct initial activation reactions catalyzed by the zeolitic protons.

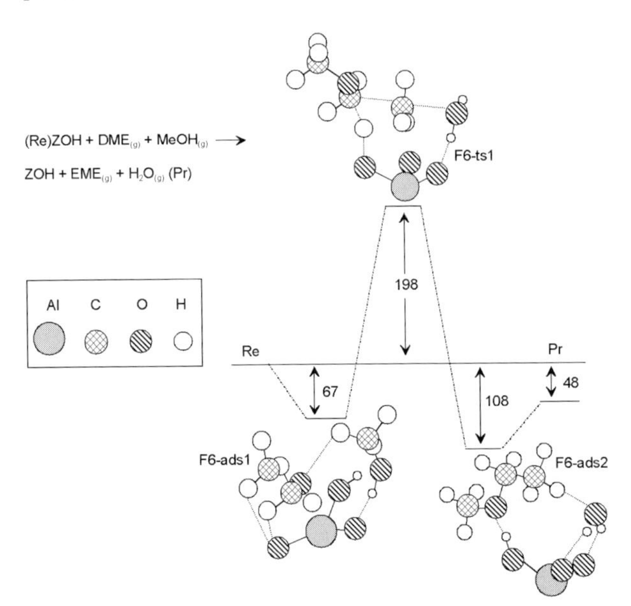

Figure 4.32. Ethyl ethyl ether formation via associative mechanism. Energies in kJ/mol[51].

An interesting example is the conversion of methanol to hydrocarbons catalyzed by the zeolitic SAPO-34 material[50]. This microporous $AlPO_4$-based material has the structure of chabazite. The protonic sites are similar to those in the zeolite, but of slightly less intrinsic acidity because of their embedding in the $AlPO_4$ structure. The formation of a C–C bond directly from two or three methanol molecules activated at the zeolitic proton site is possible, but requires higher temperatures to overcome the larger activation barriers. The most preferred reaction path was predicted based on cluster calculations to involve an associative reaction of a methanol molecule with dimethyl ether (see Fig. 4.32). An activation energy of 270 kJ/mol was predicted for this reaction over small clusters. The formation of ethylene by the decomposition of ethyl methyl ether requires an activation barrier of 140 kJ/mol. Ethylene can then easily oligomerize towards aromatics. The cavities of SAPO-34 are connected through cavity openings too small for aromatic molecules. Hence, when these aromatics are formed by consecutive reaction of ethylene, they will not leave the cavities. When such aromatic intermediates are present in the cavities, the direct ethylene-forming reaction from methanol is replaced by reaction sequences in which hexamethylbenzene and other related methyl-substituted benzenes play important roles. Propylene formation is favored by methylbenzenes with four to six methyl groups but ethylene is the predominant product from those with two or three methyl groups[46]. Smaller micropores favor the latter reaction. The reaction sequence given in Fig. 4.33 is proposed when hexamethylbenzene is the major reaction intermediate.

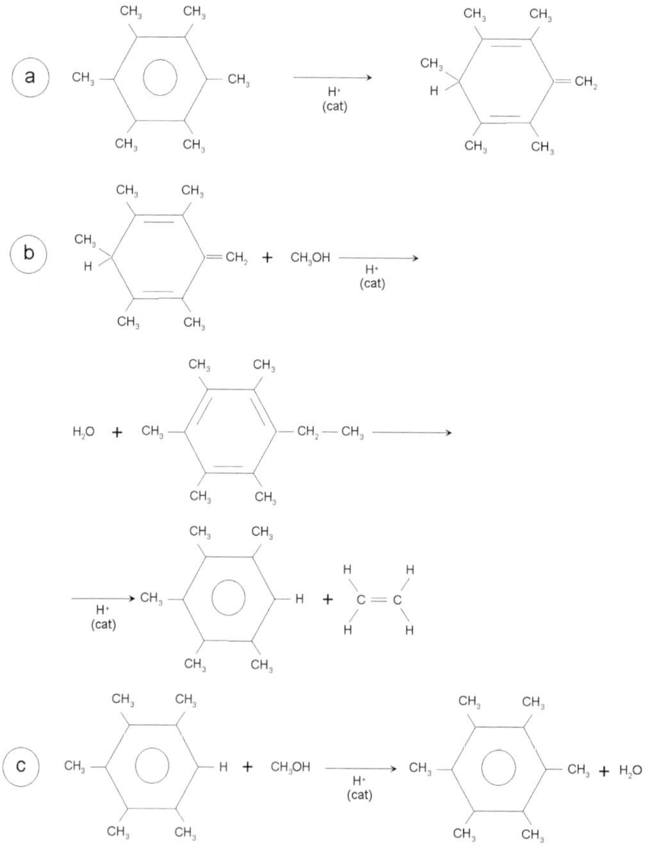

Figure 4.33. A detailed side-chain reaction route to ethylene.

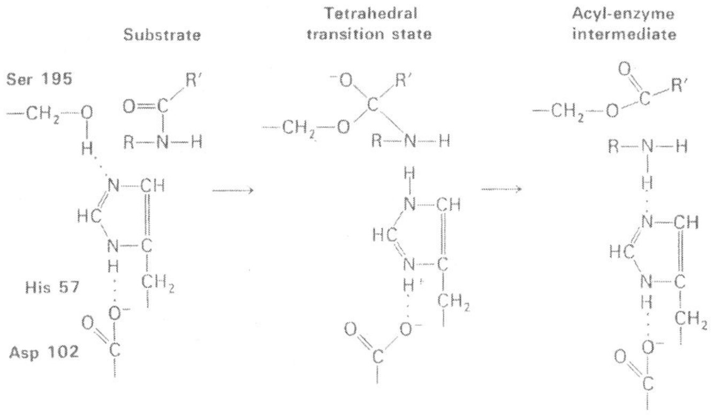

Substrate

Tetrahedral transition state

Acyl-enzyme intermediate

Ser 195

His 57

Asp 102

Figure 4.34a. First stage in the hydrolysis of a peptide by chymotrypsin: *acylation*. A tetrahedral transition state is formed, in which the peptide bond is cleaved. The amine component then rapidly diffuses away, leaving an acyl–enzyme intermediate, adapted from L. Stryer[53].

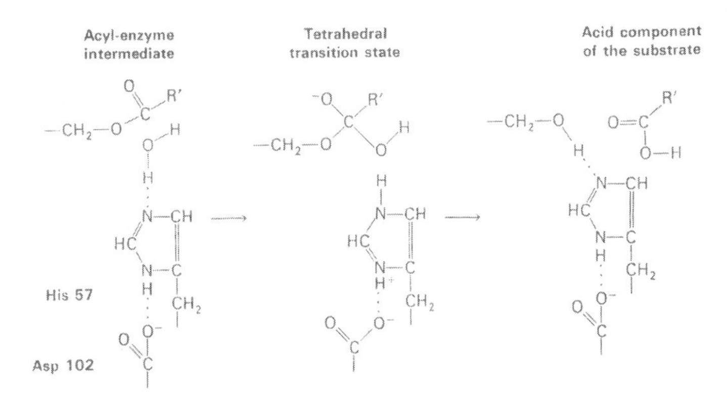

Figure 4.34b. Second stage in the hydrolysis of a peptide by chymotrypsin: *deacylation*. The acyl–enzyme intermediate is hydrolyzed by water. Note that deacylation is essentially the reverse of acylation, with water in the role of the amine component of the original substrate, adapted from L. Stryer[53].

The reaction path shown in Fig. 4.33 involves a fascinating sequence of reaction steps, with some similarity to enzyme catalytic steps, but is much less controlled. For example, in the hydrolysis of peptides catalyzed by the chymotrypsin enzyme, the peptide bond ruptures via the intermediate attachment of the acyl functional group to the activating $-CH_2-OH$ group of a serine peptide unit, near the catalytically reactive site that is part of the enzyme protein framework. The reaction steps that occur are sketched in Fig. 4.34. The OH group of serine is activated by a nearby imidizole group and proton transfer is stabilized by a negative charge on the carbonyl group of its next-neighboring aspartase unit. The free energies of activation of this reaction sequence are low because of the optimized locations of interacting acid and base groups around the reaction center.

This refinement is absent in the chabazite cavity where the location of the protonic center, negative charge on the zeolite wall and methylated benzene molecule all interact with one annother. Nonetheless, the combination of protonation, transition-state formation and cleavage and formation of covalent bonds has some similarity to the enzyme-type events.

The presence of reactive aromatic intermediates has also been invoked as catalysts for other hydrocarbon conversion reactions, such as the isomerization of n-butene to isobutene[54].

A positive effect of the presence of carbonaceous deposits on the skeletal isomerization of n-butene over a protonic ferrierite zeolite could also be related to a possible reaction mechanism in which the tertiary carbenium ions, which are located in the zeolite pores, act as catalytically active sites. Catalysis was found initially to proceed via oligomerization and cracking reactions.

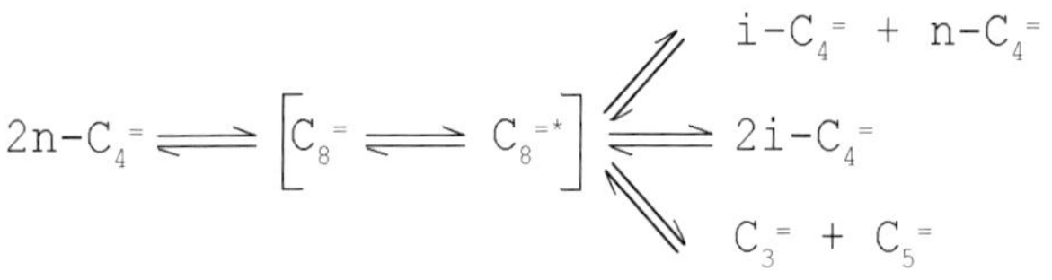

$$2n\text{-}C_4^{=} \rightleftharpoons \left[C_8^{=} \rightleftharpoons C_8^{=*}\right] \rightleftharpoons \begin{cases} i\text{-}C_4^{=} + n\text{-}C_4^{=} \\ 2i\text{-}C_4^{=} \\ C_3^{=} + C_5^{=} \end{cases}$$

Figure 4.35. Bimolecular reaction scheme for isobutene formation from n-butene. Formation of intermediate primary carbenium ions is circumvented.

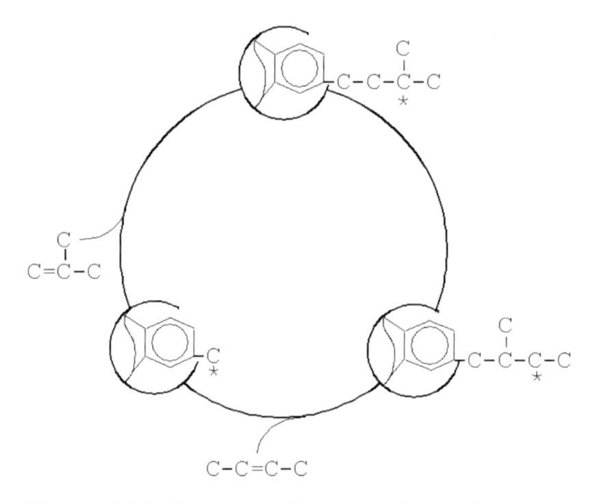

Figure 4.36. Formation of isobutene from n-butene through a tertiary carbon on coke aromatics[54].

As we discussed in Section 4.2.1, monomolecular n-butene to isobutene isomerization requires a relatively high activation energy because isomerization via the propyl cationic intermediate generates a primary carbenium ion in the transition state.

Isomerization initially proceeds through an intermediate bimolecular reaction as shown in Fig. 4.35, where the formation of carbonaceous deposits occurs in parallel. Over the course of time, isobutene results as the only product, but is now produced solely through the monomolecular path. A reaction sequence for the formation of isobutene as shown in Fig. 4.36 occurs at the tertiary carbon of a substituent alkyl group of the aromatic coke. The size of the microcavity inhibits further oligomerization reactions.

4.5 Adsorption Equilibria and Catalytic Selectivity

Figure 4.37 illustrates the sensitivity dependence of the molecular product distribution on zeolite micropore dimensions for the cracking of n-C_{16}. The product ratio of branched dimethylbutane (DMB) versus n-C_6 is taken as a measure of the selectivity. A maximum in selectivity towards the bulky branched molecule is found for the intermediate pore-size zeolite AFI. This result is curious since one would have expected the wide-pore zeolite Fau to lead to the maximum yield for the more bulky molecule. The differences from expectation appear to be related again to the adsorption properties of hydrocarbons.

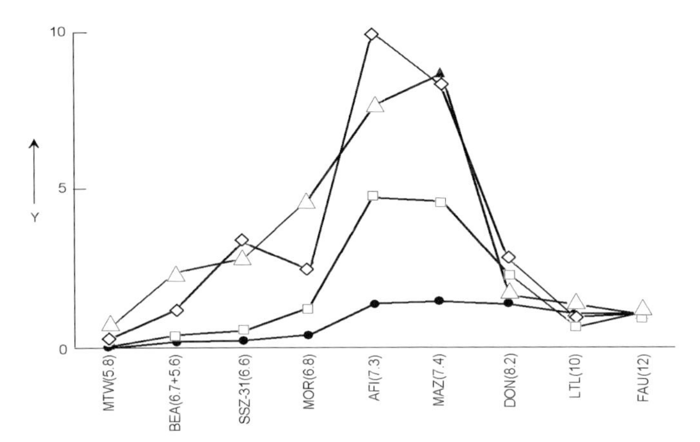

Figure 4.37. Experimental and simulated normalized DMB/n-C_6 yield ratios (y) for various zeolite structures at $T = 577$ K and $P = 3000$ kPa. The ratios were normalized by setting the value for FAU-type zeolite at one. The catalytic ratios △ were determined from n–C_6 hydroconversion experiments, the calculated ratios were taken from simulated adsorption isotherms of 2,2–DMB/n–C_6 (□) and 2,3–DMB/n–C_6 (◇) or from Henry coefficients (•). The numbers in parentheses are the average pore sizes in ångstrom[55].

The adsorption energy of an n-C_{16} molecule is at least 150 kJ/mol for a wide-pore zeolite and even greater for the smaller pore zeolites. This implies nearly complete filling of the zeolite micropore at the temperature and pressure of the cracking process. At such a high micropore filling, the diffusion of molecules becomes inhibited and, hence, becomes the rate-limiting step. Molecules in the micropore have a long residence time, thus allowing the product molecules to equilibrate in the micropore. The product distribution therefore reflects the differences in free energies of the adsorbed product molecules at high micropore filling. Three effects appear to play a role[55]:

- At high pore fillings the adsorption of small molecules is preferred over large molecules owing to entropic considerations[56].
- Linear molecules tend to stretch in narrow pores. This makes them larger and, hence at high pore filling their adsorption is suppressed compared with adsorption of smaller molecules.
- In small micropores the adsorption of branched molecules is suppressed.

The observed optimum in selectivity for the branched product relates strongly to the change in shape of the linear molecule and its larger effective volume at intermediate micropore sizes. In an analogous fashion, the different responses observed between the MFI zeolite and the MEL zeolite with respect to the conversion of C_{10} and C_7 hydrocarbons, is related to the very different pore filling in these two systems. The MFI zeolite has two different channel systems whereas the MEL zeolite contains only one type of channel. These different channels subsequently lead to the large differences in the heats of adsorption for hydrocarbons with different chain lengths. The computed adsorption isotherms are shown in Figs. 4.38 and 4.39[57].

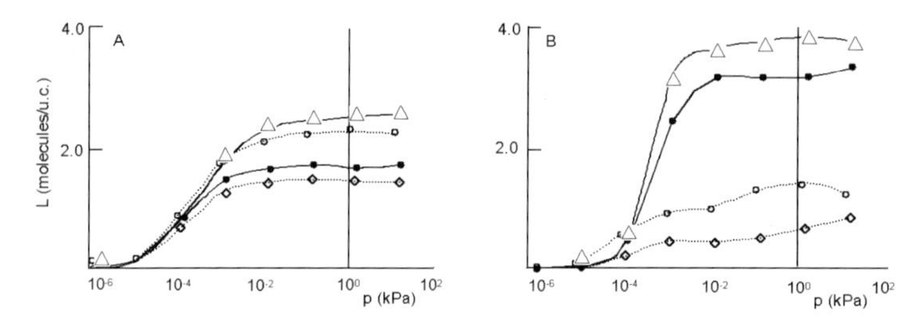

Figure 4.38. The adsorption isotherm at 415 K as calculated by CBMC calculations for binary mixture of 50% 2-methylnonane (–△–) and 50% n-decane (· · · ◇ · · ·), and of 50% 5-methyl nonane (· · · ○ · · ·) and 50% n-decane (–•–). (A) MFI-type and (B) MEL-type silica[57].

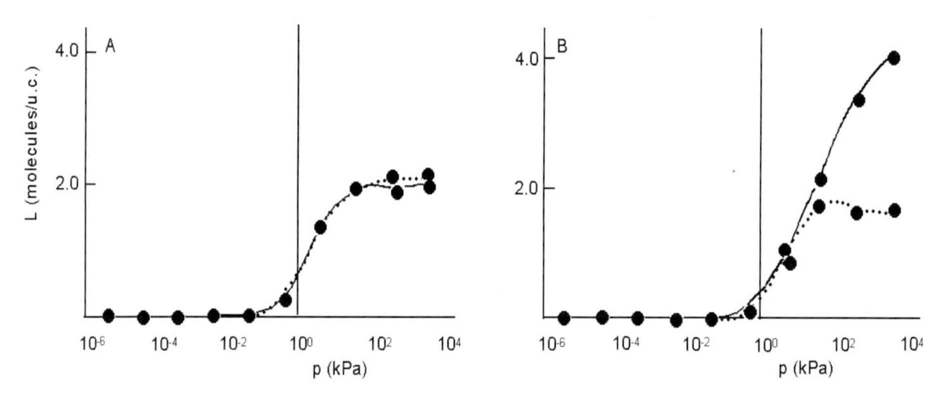

Figure 4.39. The adsorption isotherm at 523 K as calculated by CBMC calculations for binary mixture of 50% 2-methyl hexane. (· · · • · · ·) and 50% n-heptane (–•–). (A) MFI-type and (B) MEL-type silica[57].

At 1 kPa, for strongly adsorbing C_{10} alkanes in MEL-type zeolite, there is a large preference for adsorption of the linear alkane. This preference is much less than for the MFI zeolite. Differences appear at high micropore occupation. Competitive adsorption suppresses the formation of i-C_{10} in MEL owing to the difference in the channel cross-section geometry, where branched alkanes prefer to adsorb. As a consequence, the rate of n-C_{10} conversion is low towards i-C_{10}. The MFI zeolite, therefore has the superior rate since the rate of iC_{10} formation is higher. The reaction products are the result of consecutive reactions of i-C_{10}. In contrast, as one notes from Fig. 4.39, in MEL at 10 kPa for C_7 there is no such preference in adsorption for the n-C_7 versus i-C_7 molecule since under these conditions the adsorption concentration is still too low.

An approximate expression to compute the rate of productP_i formation is

$$\frac{\mathrm{d}P_i}{\mathrm{d}t} = \frac{1}{R^2}\sqrt{D_i R_i}$$

When desorption is rate limiting, this expression reduces to:

$$= \frac{1}{R^2}\sqrt{D_i.N_i\, k_{\text{des},i}.\theta_i}$$

At the high pore fillings conventionally used, product molecules can be assumed to be equilibrated in the micropores. This implies, as we have seen, that their relative concentration is determined by the molecular chemical potential in the micropores. Since the diffusion constants can also be a strong function of concentration, one has to apply these expressions with care, or even better one should use expressions that include the concentration dependence of diffusion constants such as the Maxwell–Stefan expression[58].

This section illustrates the very important result that in zeolites the conventionally used Langmuir–Hinshelwood rate expression for reactions does not apply very well in their present form. The Langmuir–Hinshelwood rate expression assumes reactant activation to be rate limiting:

$$r_i = k_i \frac{K_{\text{eq}}^{\text{ads}}(A)P_A}{1 + K_{\text{eq}}^{\text{ads}}(A)P_A + \sum\limits_{i \neq A} K_{\text{eq}}^{\text{ads}}(i)P_i}$$

Differences in product distribution depend on the elementary rate of conversion of reactant A to product i, rather than the micropore equilibrium concentration of product i. In this expression the adsorption equilibria are considered to be independent of the concentration of other components, a supposition clearly unacceptable for zeolites. Reaction mixtures of different components, however, behave strongly nonlinearly as a function of concentration. Entropic effects play a dominant role.

Another key result is that for reactions that occur at high pressure and with high occupation of the micropores, there is no equilibrium between the zeolite interior and exterior. Instead, product equilibrium is then established within the zeolite micropores and determined by the corresponding free energies of adsorption.

4.6 Diffusion in Zeolites

As noted in the previous section, in practical zeolite catalysis, diffusion controls the selectivity and activity for many reactions. When the zeolite micropore channel or cavity is significantly larger than the molecular dimensions, diffusion is of the Knudsen type. This implies that it scales with $m^{-1/2}$, as illustrated in Fig. 4.40.

Molecular dynamics simulation can be applied in such cases, since the activation energy for diffusion is low and hence simulations over a period of a few picoseconds are adequate. When the dimensions of the zeolite micropore decrease, the molecular residence near the surface increases and now diffusion becomes dominated by its motion along the zeolite wall. Micropore diffusion becomes fast compared with Knudsen-type diffusion when the dimensonal matches are such that the diffusing molecule will not leave the surface potential minimum regime except when desorbing from the micropore out of the zeolite. Typical surface-potential dominated diffusion (creeping motion) is illustrated in Fig. 4.41.

In contrast to the behavior in a wide-pore zeolite such as faujasite, one observes in Fig. 4.41 that, in the one-dimensional mordenite micropores with a diameter of 6.5 Å, the rate of diffusion is independent of the length of the hydrocarbon.

Diffusional constants may depend strongly on micropore filling. This, in essence, is due to site blocking effects. It explains the often observed relationship between overall experimentally measured diffusional rate constant activation energies and heats of adsorption.

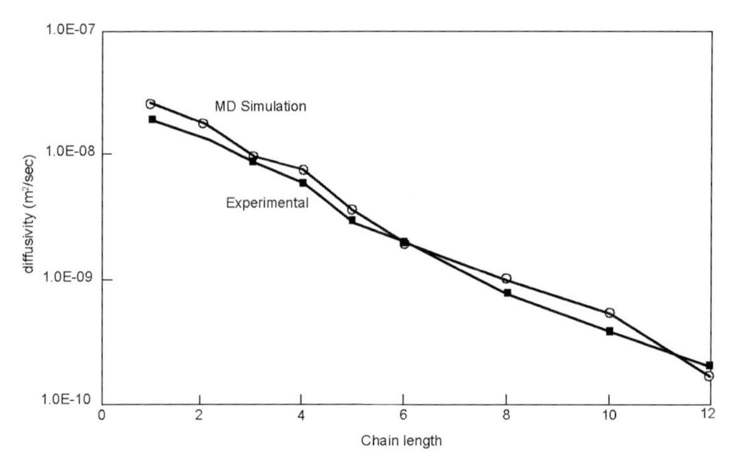

Figure 4.40. Simulated diffusion constants in faujasite as a function of hydrocarbon chain length[59].

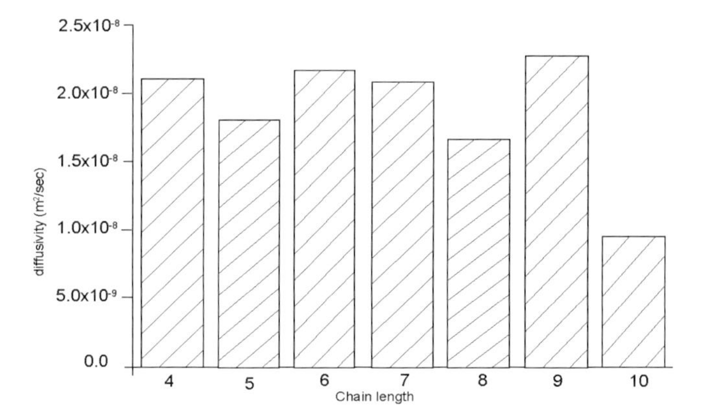

Figure 4.41. Simulated diffusion constants in mordenite as a function of hydrocarbon chain length.

In gas chromatographic diffusion measurement experiments, the molecular concentration in the zeolite micropore will vary with temperature. Hence the measured differential activation energy at constant pressure will contain a parameter that relates to the heat of adsorption. At higher temperatures, site blocking is decreased. This partially explains the difference from NMR or inelastic neutron scattering data. In the latter experiments, measurements are done at short time-scales, therefore, intra-pore elementary reaction steps are mainly controlled by the free diffusional pathlength in the micropore over a short distance, whereas data obtained macroscopically concern longer time-scales and, hence, larger diffusional paths, so that imperfections in the zeolite structure or impurities occupying the micropores may also have an influence.

In zeolites, diffusion constants will depend strongly on molecular shape. For example, in silicalite branched alkanes prefer to absorb in channel cross-sections, but linear alkanes prefer adsorption in the channels themselves. This has important consequences for differences in diffusion rates within mixtures. At high concentration, the rate of the diffusion of the branched alkane will control the rate of diffusion of the other alkanes[61].

A unique correlation in diffusional motion can occur in zeolites with one-dimensional micropores such as mordenite or ZSM-22 (TON). Single-file diffusion is defined as restricted mobility in one-dimensional pores, where molecules cannot pass each other. Indeed, dynamic Monte Carlo calculations that compute diffusion rates by considering them to be the result of the hopping of molecules between defined sites show a very steep decrease in the diffusion rate for a one-dimensional system compared with decreases found for three-dimensional systems.

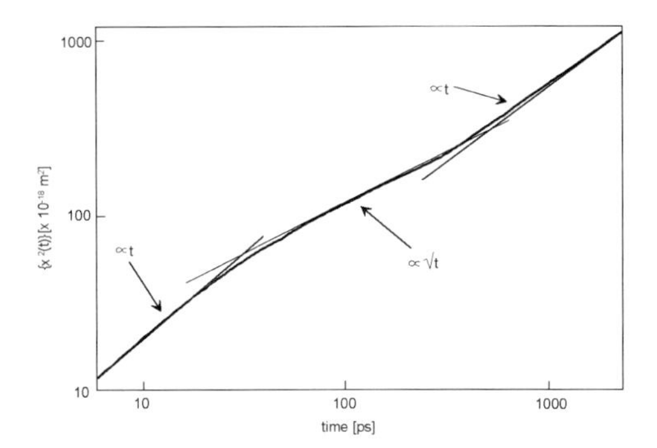

Figure 4.42. Dynamic Monte Carlo simulations of single-file diffusion. The different time regimes. Initial non–correlated balistic regime: $x^2 \sim t$ followed by single file-diffusion and finally collective center of mass diffusion[62].

However, the decrease in diffusion rate with increasing micropore filling is much faster than found experimentally or with molecular dynamics studies. A molecular dynamics model study, using an idealization of the one-dimensional zeolite micropore, shows that in open channels single-file diffusion behavior occurs only for particular time regimes (see Fig. 4.42).

Ballistic motion, or motion not controlled by the interaction between the reactant and the zeolite wall, but through hard sphere collisions, will never lead to single-file molecular motion. The direction of motion has to be partially randomized by collisional interaction with the corrugated micropore wall. This corrugation may be due to thermal mobility of the zeolite wall oxygen atoms. The characteristic relationship between the diffusion length and time for single-file diffusion is observed only when the motion between a large number of adsorbed molecules becomes correlated.

$$x^2 \approx Dt^{1/2} \tag{3}$$

wher x, D and t are the diffusion length, diffusion time, and diffusivity, respectively. Since molecular motion in a zeolite is only partially randomized and motion remains also partially ballistic, dynamic Monte Carlo hopping models of diffusion will overestimate the concentration regime in which single-file diffusion will occur.

After a longer time, single-file diffusional time dependence will disappear because diffusion will be controlled by the center of mass motion of the correlated molecules. The characteristic of single-file diffusion that remains is the concentration dependence and an increased effective mass, which now has to be taken as the center of mass of the collectively moving particles ($m_{c.m.}$), proportional to the length of the zeolite pore[64]:

$$D_{singlefile} \approx \frac{1-\theta}{\theta} \cdot \frac{1}{\sqrt{m_{c.m.}}}$$

A consequence of single-file diffusion is that catalysis in one-dimensional microporous systems becomes diffusion limited for crystal sizes much smaller than for three-dimensional systems. Hence equilibration of different reaction products within the zeolite micropores may be expected to occur most rapidly in one-dimensional microporous systems.

An important consequence of single-file diffusion to the overall rate of a catalytic reaction is that the overall rate of conversion may show a maximum as a function of micropore filling (increasing reactant pressure)[64]. Whereas at low micropore filling a reaction may show no diffusional limitation, with increasing micropore filling the rate of diffusion decreases strongly because of collective motion. This is illustrated in Fig. 4.43.

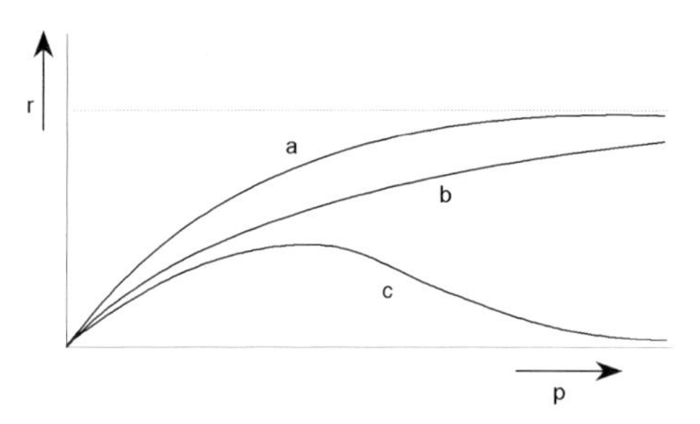

Figure 4.43. (a) The rate of a monomolecular zeolite-catalyzed reaction as a function of pressure (no diffusional limitation). (b) The rate of the monomolecular reaction that is diffusional limited (no single-file diffusion). (c) The rate of the monomolecular reaction that is diffusional limited (single-file diffusion).

References

1. C. Baerlocher, W.M. Meier, D.H. Olsson, *Atlas of Zeolite Structures Types*, 5th ed. Elsevier, London (2001)
2. A.J.M. de Man, B.W.M. van Beest, M. Leslie, R.A. van Santen, *J. Phys. Chem. B*, 94, 2524 (1990);
   A.J.M. de Man, R.A. van Santen, *Zeolites*, 12, 269 (1992)
3. L.A.M.M. Barbosa, R.A. van Santen, J. Hafner, *J. Am. Chem. Soc.* 123, 4530 (2002)
4. G.J. Kramer, R.A. van Santen, *J. Am. Chem. Soc.* 117, 1766 (1995)
5. (a) V.B. Kazansky, A.I. Serykh, V. Semmer-Herledam, J. Fraissard, *Phys. Chem. Chem. Phys.* 5, 966 (2003);
   (b) K.D. Hammonds, M. Deng, V. Heine, M.T. Dove, *Phys.Rev. Lett.* 78, 3701 (1997)
6. W. Haag, in *Stud. Surf. Sci. Catal.* 84,1375 (1994);
   F.C. Jentoft, B.C. Gates, *Top. Catal.* 4, 1 (1997);
   S. Kotrel, H, Knötzinger, B.C. Gates, *Micropor. Mesopor. Mater.* 11, 35 (2000)
7. H. Toulhoat, P. Raybaud, E. Benazzi, *J. Catal.* 221, 500 (2004)
8. R.A. van Santen, G.J. Kramer, *Chem. Rev.* 95, 637 (1995)
9. X. Rozanska, Th. Demuth, F. Hutschka, J. Hafner, R.A. van Santen, *J. Phys. Chem. B*, 106, 3248 (2002)
10. X. Rozanska, X. Saintigny, R.A. van Santen, F. Hutschka, *J. Catal.* 141 (2001);
    X.Rozanska, R.A. van Santen, F. Hutschka, *J. Phys. Chem. B*, 106, 202, 141 (2002)
11. G.A. Olah, G.K.S. Prakah, J. Sommer, *Super Acids*, Wiley, New York (1985)
12. J.G. Angyan, D. Parsons, Y, Jeanvoine, in *Theoretical aspects of Heterogeneous Catalysis*, M.A.C. Nascimento (ed.), Kluwer, Dordrecht p. 101 (2001)
13. Th. Demuth, X. Rozanska, L. Benco, J. Hafner, R.A. van Santen, H. Toulhoat, *J. Catal.* 214, 68 (2003)

14. X. Rozanska, R.A. van Santen, T. Demuth, F. Hutschka, J. Hafner *J. Phys. Chem. B*, 107, 1309 (2003)
15. S.R. Blaszkowski, R.A. van Santen, *J. Phys. Chem.* 99, 11728 (1995)
16. S.R. Blaszkowski, R.A. van Santen, *J. Am. Chem. Soc.* 118, 5152 (1996); S.R. Blaszkowski, R.A. van Santen, *J. Phys. Chem. B*, 101, 2292 (1997)
17. S.R. Blaszkowski, R.A. van Santen, in *Transition State Modelling for Catalysis,* L.G. Truhlar, K. Morokumada (eds.), ACS Symposium Series 721, 307. American Chemical Society, Washington, DC (1999); S.R. Blaszkowski, R.A. van Santen, *J. Am. Chem. Soc.* 119, 5020 (1997)
18. A.M. Vos, X. Rozanska, R.A. Schoonheydt, R.A. van Santen, F. Hutschka, J. Hafner, *J. Am. Chem. Soc.* 123, 2799 (2001)
19. W.M.H. Sachtler, Z. Zhang, *Adv. Catal.* 39, 139 (1993)
20. E. Sanchez Marcos, A.P.J. Jansen, R.A. van Santen *Chem. Phys. Lett.* 167, 399 (1990)
21. V.B. Kazansky, A.I. Serykh, E.A. Pidko, *J. Catal.* 225, 369 (2004)
22. L.A.M.M. Barbosa, R.A. van Santen, *J. Phys. Chem. B*, 107, 4532 (2003)
23. A.Yakovlev, A.A. Shubin, G.M. Zhidomirov, R.A. van Santen, *Catal. Lett.* 70, 175 (2000)
24. (a)M.V. Frash, R.A. van Santen, *Phys. Chem. Chem. Phys.* 2, 1085 (2000); M.V. Frash, R.A. van Santen, *J. Phys. Chem. A*, 104, 2468 (2000); (b) Y. Pidko, R.A. van Santen, in preparation
25. (a) V.B. Kazansky, I.R. Subbotina, R.A. van Santen, E.J.M. Hensen, *J. Catal.* 227, 263 (1994); b) N.O. Gonzales, A.R. Chakraborty, A.T. Bell, *Top. Catal.* 9, 207 (1999)
26. L.A.M.M. Barbosa, R.A. van Santen, *J. Mol. Catal. A*, 166, 101 (2001)
27. G, Sankar, J.M. Thomas, C.R.A. Catlow, C.M. Barker, D. Gleeson, N. Kaltsoyannis, *J. Phys. Chem. B*, 105, 9028 (2001); M. Neurock, L.E. Manzer, *Chem. Commun.* 1133, (1996)
28. J.A. Labinger, *CaTTech.* 5, 18 (1999)
29. M. Dugal, G. Sankar, R. Raja, J.M. Thomas, *Angew. Chem. Int. Ed.* 39, 2310 (2000).
30. F. Blatter, M. Sun, S. Vasenkov, H. Frei, *Catal. Today,* 41, 297 (1998)
31. G.H. Kühl, in *Catalysis and Zeolites, Fundamentals and Applications,* L. Puppe, J. Weitkamp (eds.), Springer, Berlin (1999)
32. A.V. Orchilles, *Micropor. Mesopor. Mater.* 35, 21 (2000)
33. A. Corma, M. Gareia, *Chem. Rev.* 102, 3837 (2002)
34. T.M. Leu, E. Rodriner, *J. Catal.* 228, 397 (2004)
35. A.L. Yakovlev, G.M. Zhidomirov, R.A. van Santen, *Catal. Lett.* 75, 45 (2001)
36. A.L. Yakovlev, G.M. Zhidomirov, R.A. van Santen, *J. Phys Chem. B*, 105, 12297 (2001)
37. El. M. El-Malki, R.A. van Santen, W.M.H. Sachtler, *Micropor. Mesopor. Mater.* 35–36, 235 (2000)
38. P.Ciambelli, E. Garufi, R. Pirone, G. Russo, F. Santagata, *Appl. Catal. B*, 8, 333 (1990)
39. A. Heyden, B. Peters, A.T. Bell, F.J. Keil, *J. Phys. Chem. B*, 109, 4801 (2005)
40. G.I. Panov, *CaTTech* 4, 18 (2000)
41. E.J.M. Hensen, Q. Zhu, M.M.R.M. Hendrix, A.R. Overweg, P.J. Kooyman, M.V. Sychev, R.A. van Santen, *J. Catal.* 221, 560 (2004);

Q. Zhu, R.M. van Teeffelen, R.A. van Santen, E.J.M. Hensen, *J. Catal.* 221, 575 (2004)
42. P.E.M. Siegbahn, R.H. Crabtree, *J. Am. Chem. Soc.* 119, 3103 (1997).
43. N.A. Kachurovskaya, G.M. Zhidomirov, E.M.J. Hensen, R.A. van Santen, *Catal. Lett.* 86, 25 (2003);
N.A. Kachurovskaya, G.M. Zhidomirov, R.A. van Santen, *J. Phys. Chem. B*, 108, 5944 (2004)
44. E.J.M. Hensen, Q. Zhu, R.A. van Santen, *J. Catal.* 233, 136 (2005)
45. D. Frenkel, B. Smit, *Understanding Molecular Simulation* Academic Press (1996);
B. Smit, in *Computer Modelling of Microporous Materials*, C.R.A. Catlow, R.A. van Santen, B. Smit (eds.), Elsevier, Amsterdam, p. 25 (1994)
46. S.P. Bates, R.A. van Santen, *Adv. Catal.* 42, 1 (1998);
S.P. Bates, W.J.M. van Well, R.A. van Santen, B. Smit, *J. Am. Chem. Soc.* 118, 6753 (1996)
47. P.B. Weisz, *Adv. Catal.* 13, 137 (1962;
M.L. Coonradt, W.E. Garwood, *Ind. Eng. Chem. Prod. Res. Dev.* 3, 38 (1964)
48. F.J.M.M. de Gauw, J. van Grondelle, R.A. van Santen, *J. Catal.* 206, 295 (2002)
49. F. Narbeshuber, H. Vinek, J.A. Lercher, *J. Catal.* 157, 388 (1995)
50. J.F. Haw, W. Song. D.M. Marcus, J.B. Nicholas, *Acc. Chem. Res.* 36, 317 (2003).
51. S.R. Blaszkowski, R.A. van Santen, *J. Am. Chem. Soc.* 119, 5020 (1997)
52. W.Song, H. Fu, J.F. Haw, *J. Am. Chem. Soc.* 123, 4749 (2001))
53. L. Stryer, *Biochemistry* Freeman, p. 226 (1995)
54. M. Guisnet, P. Andy, N.S. Gaep, C. Travers, E. Benazzi, *J. Chem. Soc. Chem. Commun.*, 1685 (1995).
55. M. Schenk, S. Calero, T.L.M. Maesen, L.L. van Benthem, M.G. Verbeek, B. Smit, *Angew. Chem. Int. Ed.* 41, 2499 (2002)
56. J. Talbot, *AIChE J.* 43, 2471 (1997)
57. Th. L. Maesen, M. Schenk, T.J.H. Vlugt, B. Smit, *J. Catal.* 203, 281 (2001)
58. F.J. Keil, R. Krishna, M.O. Coppens, *Rev. Chem. Eng.* 16, 71 (2000)
59. R.Q. Snurr, J. Kärger, *J. Phys. Chem. B*, 101, 6469 (1997)
60. D. Schuring, A.P.J. Jansen, R.A. van Santen, *J. Phys. Chem. B*, 104, 941 (2000)
61. D. Schuring, A.O. Koryabkina, A.M. de Jong, B. Smit, R.A. van Santen, *J. Phys. Chem. B*, 105, 7690 (2001)
62. P.H. Nelson, S.M. Auerbach, *J. Chem. Phys.* 110, 9235 (1999)
63. J. Kärcher, *Diffusion Under Confinement*, Sitzungs bericht der Sächsischen Akademie der Wissenschaften zu Leipzig – Mathemathisch – Naturwissenschaftliche Klasse – Band 128 – Heft 6 (2003)
64. F.J.M.M. de Gauw, J. van Grondelle, R.A. van Santen, *J. Catal.* 204, 53 (2001
65. B. Ensing, F. Buda, P. Blöchl, E.J. Baerends, *Ang. Chem. Int. Ed.* 40, 2893 (2001)
66. H.J.M. Fenton, *Chem. News.* 190 (1876)

CHAPTER 5
Catalysis by Oxides and Sulfides

5.1 General Introduction

This chapter generalizes the concepts of chemical bonding that we introduced in Chapters 3 and 4 on the reactivity of metals and zeolites in order to gain a fundamental understanding of the reactivity of metal oxides and sulfides. Metal oxides and sulfides are used to catalyze a wide range of industrial oxidation, hydrogenation and desulfurization reactions. Many of the metal and mixed-metal oxides used as catalysts take on both ionic characteristics which are important for acid or basic reactions as well as covalent characteristics which govern oxidation reactions. Metal sulfides reveal some of the same characteristics as the oxides and are therefore also discussed in this chapter. The first part of this chapter covers the general concepts associated with the bonding and the chemistry of metal oxide surfaces. The latter half of the chapter examines the reactivity of metal sulfide surfaces.

We start with a discussion in Section 5.2 on the general chemical reactivity of non-reducible oxides from the point of view that chemical bonding is considered to be completely electrostatic. This turns out to be a very useful framework for estimating the differences in reactivity of different metal oxide surfaces with varying surface topologies. Both metal oxides and metal sulfide surfaces demonstrate Lewis acid and base properties, which are subsequently converted into corresponding Brønsted acid and base forms upon reaction. The concepts associated with the electrostatic-surface interactions are subsequently extended to include the contributions to the chemical reactivity of the surface cations as the result of covalent bonding. We show how semiempirical chemical bonding schemes designed to predict the reactivity of coordination complexes can also be used to describe the reactivity of solid oxide surfaces.

We return to the elementary theory of Brønsted acidity which we introduced in Chapter 4 on zeolites in order to provide a more complete understanding of solid acid acidity and its comparison with strong liquid acids. The acidity is strongly tied to the ability of the medium to stabilize the anionic charge that forms yet destabilize the bonding between the proton and the specific medium. We develop the general ideas on what controls acidity by examining simple solution-phase acid–base systems. We use some of the general ideas on liquid-phase acidity in order to understand solid acidity. The differences between the liquid and solid acids become quite clear upon detailed comparisons and are highlighted herein. We conclude the discussion on acidity with a section which describes in some detail the acidity of heterpolyacids and their reactivity as a way to extend our understanding of other solid acids.

In Section 5.6, we turn our focus to oxidation catalysis. We begin by first examining the oxidation over reducible metal oxides. We describe some of the general mechanistic features for carrying out selective oxidation catalysis and then subsequently outline some general features that control the reactivity of reducible metal oxides. We demonstrate the importance of these features by presenting various different selective oxidation reaction systems. These examples are not inclusive and were chosen only to highlight some of the important concepts. We conclude this section on selective oxidation by describing non-reducible oxides and highlight some of their potential controlling features via different example systems.

Molecular Heterogeneous Catalysis. Rutger Anthony van Santen and Matthew Neurock

ISBN: 3-527-29662-X

The chemical bonding features of metal sulfide surfaces are related to those of the reducible oxides. We discuss in detail a key question for supported metal sulfide catalysts: "What is the state of the sulfide edge sites under catalytic conditions and the shape of the sulfide particles?" In addition, we attempt to help elucidate the role of ions such as Ni^{2+} or Co^{2+} in promoting reactivity for sulfide systems such as MoS_2.

5.2 Elementary Theory of Reactivity and Stability of Ionic Surfaces

In this section we present surface reactivity concepts based on the extreme view that the surface can be represented solely by a sequence of point charges. For most oxidic or sulfidic components, this provides a very approximate view of their chemical bond. We will build upon these ideas, however, in subsequent chapters, ultimately providing a more accurate and complete picture of the surface-chemical reactivity of oxides and sulfides.

In general, chemical bonds have to be formed or broken in order to generate a metal oxide or metal sulfide surface. There are, however, some oxides or sulfides where surface formation can occur without the cleavage of chemical bonds. An example of the latter system is found in the formation of the basal surface of layered clay materials or the basal plane of layered sulfides as, for instance, MoS_2. On such surfaces the covalent bonds between the atoms present in the bulk remain intact on the surface.

The solid basal planes of layered materials such as MoS_2 are held together by weak van der Waals interactions between S–Mo–S layers when they contact each other. These layers can also be held together via weak ionic interactions between the basal planes of layered materials, such as those found in V_2O_5. In the smectite clays, the layers are kept together by the electrostatic interactions with intercalated cations that compensate for the negative charges of the clay doubly layer. Interestingly, the internal surface of the microporous zeolitic systems, discussed in Chapter 4, is similar in that the micropore channel atoms are completely coordinatively saturated. This is characteristically different from silica or alumina surfaces that will be discussed later where the surface atoms are coordinatively unsaturated. The formation of surfaces by the cleavage of bonds perpendicular to the basal planes of layered compounds occurs with the loss of coordinating cations or anions at the surface. Such surfaces are, therefore, highly reactive. This is in contrast with the surfaces discussed above, in which no covalent bond has been broken. The basal MoS_2 surfaces expose coordinatively saturated S atoms which are chemically unreactive. Similarly, the basal surfaces of neutral clay layers are also unreactive. On the broken bond surfaces, the loss in the coordination of cations or anions leads to local charge fluctuations. These charge fluctuations can lead to enhanced surface reactivity. The splitting of charged surfaces requires a significant energy in order to separate the strong attractive charges. Such surfaces tend to be unstable. The only surfaces that have a low surface energy are those that are electrostatically neutral. In these systems the electrostatic interaction becomes zero at very long distances away from the surface.

For polar, charged surfaces, this situation is completely different. For polar surfaces such as ZnO, one surface is terminated by oxygen anions while the other terminated with Zn cations. One surface has an overall negative charge, while the other has an overall excess positive charge. As a result, a dipole moment builds proportional to the dimension of a particle. In non-layered compounds, polar surfaces have to reconstruct such that the external surface charge is reduced. Freund[1] describes three ways in which the system can adapt in order to reduce the overall charge in the first layer of the surface:

1. by a reconstruction of the surface affecting several layers.
2. by a terracing with a long-range periodicity.
3. by providing the surface with an adsorbed layer with only half the charge.

As we will learn later in this section, these processes are practically very important. The sites of highest catalytic reactivity are often the edged corner positions. The concentration of such sites is enhanced by Freund's adaption processes. The effective charges on the broken-bond surfaces are such that they induce Lewis acid- or Lewis basic-type reactivity features. The local charge excesses on an ionic surface, considered to exist as a series of point charges, can be estimated using Pauling's valency definitions[2].

The Pauling valency or the strength of an electrostatic bond with a cation or anion is defined as

$$S^{\pm} = \left| \frac{\text{formal ion charge}}{\text{number of nearest neighbor ions}} \right|$$

By way of example we use the concepts of Pauling valency to describe the reactivity of MgO. MgO takes on a rock-salt structure. In the bulk each cation or anion has six neighbors. For the Mg^{2+} ion, the Pauling bond strength equals

$$S^{+} = \frac{2}{6} = \frac{1}{3}$$

and for the O^{2-} ion, the Pauling bond strength becomes

$$S^{-} = \left| \frac{-2}{6} \right| = \frac{1}{3}$$

S^{+} and S^{-} are equal, as they should be since we are simply describing the same chemical bond. In NaCl, the structure is the same, but the ions are lower in charge, thus S^{+} and S^{-} are now reduced to a value of $\frac{1}{6}$.

Let us now consider the MgO (100) surface, in which we have an equal number of cations and anions and each has lost one neighbor, see Fig. 5.1. The charge excess of an ion, e, is defined as the formal charge contribution Q of the ion, compensated by the bond charge contribution of the nearest neighboring ions

$$e^{+} = Q^{+} - \sum_i S_i^{-}$$
$$e^{-} = Q^{-} + \sum_i S_i^{+}$$

On the MgO (100) surface, the excess ionic charges that result on the surface cation and anion respectively are

$$e^{\pm} = \pm \frac{1}{3}$$

A positive excess ionic charge implies Lewis acidity whereas a negative charge implies Lewis basicity. For a neutral surface, the sum of the excess positive and negative charges must equal zero:

$$\sum_i e_i^{+} + \sum_j e_j^{-} = 0$$

For a polar surface, there is an overall excess charge. The energy costs required for creating polar surfaces are quite high because of the large energy penalty one must pay in separating the strong electrostatic interactions between the positively and negatively charged surfaces.

Whereas the (111) surface of MgO is polar and hence unstable, the (100) surface is neutral and thus fairly stable.

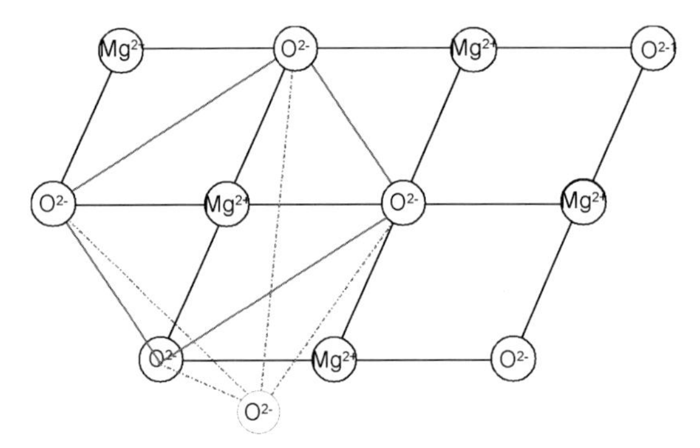

Figure 5.1. Coordination of Mg^{2+} on the MgO (100) surface.

Brønsted acidity or basicity develops when a molecule of water dissociates on such an oxide surface.

The question of whether H_2O dissociates or not on the (100) surface of MgO was debated for some time. First-principle DFT calculations were carried out on slab models for the MgO surface and showed that H_2O alone will not dissociate on the MgO (100) surface[3]. The activation of H_2O on MgO leading to surface hydroxylation requires coordinatively unsaturated surface sites that can arise as surface vacancies or step edge sites. In many cases, significant reconstruction can occur, ultimately leading to the formation of a surface that is close to that of $Mg(OH)_2$.

Here we will use the results of quantum-chemical studies of H_2O dissociation on the surfaces of alumina and titania, materials that are widely used as catalytic support surfaces and typically show limited reconstruction. The non-hydroxylated and hydroxylated (100) surfaces of γ-Al_2O_3 are shown in Fig. 5.2[4].

The bulk Al_2O_3 contains Al octahedra that share oxygen atoms. The Pauling ionic bond strength of an Al–O bond with respect to Al is

$$S_{\mathrm{Al}}^{+} = \frac{+3}{6} = \frac{1}{2}$$

Each oxygen atom has four alumina cation neighbors so that $S_{\mathrm{O}}^{-} = S_{\mathrm{Al}}^{+}$, which is consistent with a stable oxide.

The Al cation in the (100) surface (shown in Fig. 5.2) has a coordination number of five, thus

$$e_{\mathrm{Al}}^{+} = +\frac{1}{2}$$

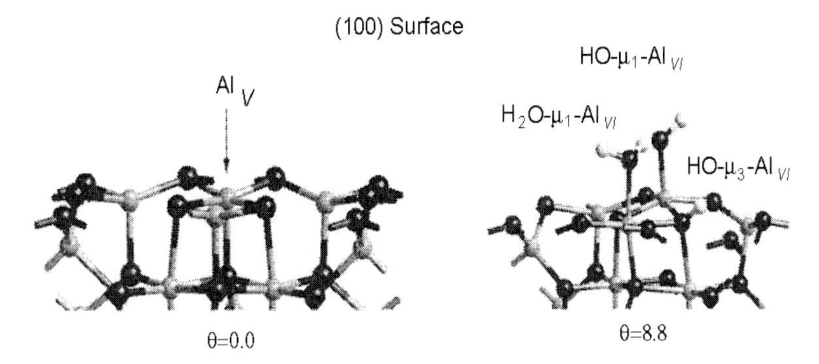

Figure 5.2. Relaxed configurations of γ-Al_2O_3 (100) surface for different hydroxyl coverages (θ in OH nm^{-2}. The most relevant surface sites are quoted. Al_n stands for aluminum atoms surrounded by n oxygen atoms, and HO–μ_m for OH groups linked to m aluminum atoms. Oxygen atoms are black, aluminum atoms are shown in gray whereas hydrogen atoms are white[4].

The surface oxygen atoms each have three Al neighbor atoms, one of which is below the surface, leading to an excess ion charge of:

$$e_{O}^{-} = -\frac{1}{2}$$

```
      (1)H
        O     H(2)
-----O — Al — O — Al — O-----
         |     |     |
         O    Al     O
```

Scheme 5.1 Schematic highlighting of the products from the dissociation of water on alumina.

The charge excess here makes the surface more reactive. Water dissociation can therefore proceed heterolytically as shown in Scheme 5.1 to form the two different hydroxyl intermediates labeled (1)and (2).

As illustrated in this scheme, hydroxylated surfaces can contain various different hydroxyl groups, each of which may present different vibrational frequencies (see Fig. 5.3). The high-frequency bond is usually assigned to hydroxyl groups coordinated to a single aluminum ion, and labeled type 1. Their chemical properties are usually Brønsted basic. The low frequency O–H stretch is usually assigned to the hydroxyl coordinated to several cations. As we will see, its chemical reactivity is Brønsted acidic.

Figure 5.2 shows the appearance of the two different hydroxyl intermediates, as well as the stabilization of adsorbed H_2O, which is often linked to hydroxyl (1) through the formation of a hydrogen bond.

The Brønsted acidity or basicity of the hydroxyl groups in Scheme 5.1 can be deduced from the Pauling ionic charge excesses:

$$\epsilon^{-}(O_1) = -2 + 1 + \frac{1}{2} = -\frac{1}{2}$$
$$\epsilon^{-}(O_2) = -2 + 1 + \frac{3}{2} = +\frac{1}{2}$$

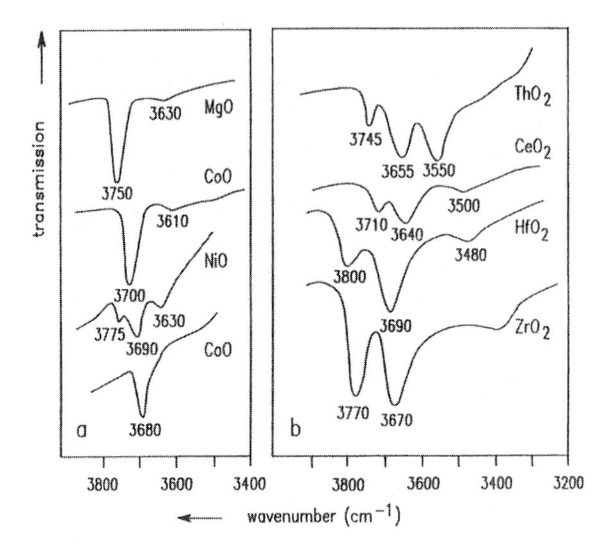

Figure 5.3. Vibrational frequencies of free hydroxyls on oxide surfaces with different crystal structures[5].

The increased number of Al cation centers around O(2) induces an effective repulsion on the proton, which helps to explain its acidic character.

This leads to an important general conclusion. When oxidic surfaces become exposed to water, basic hydroxyls, as well as Brønsted acidic protons, may be generated upon dissociation by water. The actual chemical reactivity will depend on the charge and radii of the cation to which they bind. Simple ion charge excess estimates are quite useful in deducing the chemical nature of the hydroxyl intermediates that can form.

For instance, for the silanol group, which is present on hydroxylated silica surfaces, and is thought to be catalytically active but has an excess charge of zero:

$$\epsilon^{-}(\mathrm{O}) = -2 + 1 + 1 = 0$$

The structure of the surface silanol is as follows:

```
   /H
  O
  |
  Si—O—
 / \
O   O
/     \
```

The zero value, which by the charge excess theory would suggest that these silanol groups are inactive, illustrates the approximate nature of the charge excess approach. The actual nature of the silanol can only be deduced by more accurate quantum mechanical calculations which show that the silanol group behaves as a weak Brønsted acid.

For a zeolitic proton, the charge excess on the oxygen is

$$\epsilon^{-}(\mathrm{O}) = -2 + 1 + 1 + \frac{3}{4} = +\frac{3}{4}$$

The structure around the zeolitic proton is:

H
O
Si
Al

The large excess in ionic charge indicates a destabilization of the proton, which therefore makes it much more Brønsted acidic.

Very often, metal oxide surfaces will reconstruct when exposed to water as was discussed for MgO. We will illustrate the importance of such a reconstruction by comparing the reactivity of the (001) surface of anatase (TiO_2) with that of the rutile (110) surface of TiO_2. Both are the surfaces of lowest energy for the respective phases. The difference between anatase and rutile relates to the stacking of the TiO_6 octahedra in both phases. In anatase, the octahedra stack tetrahedrally, whereas in rutile they stack parallel. Quantum-chemical results are available for both phases as well as for their corresponding surfaces. We refer here to the work of Arrouvel et al. [6a] on the (100) surface of TiO_2 anatase and of Lindan et al. [6b] on the (110) surface of rutile. The respective surfaces are shown in Fig. 5.4a and b.

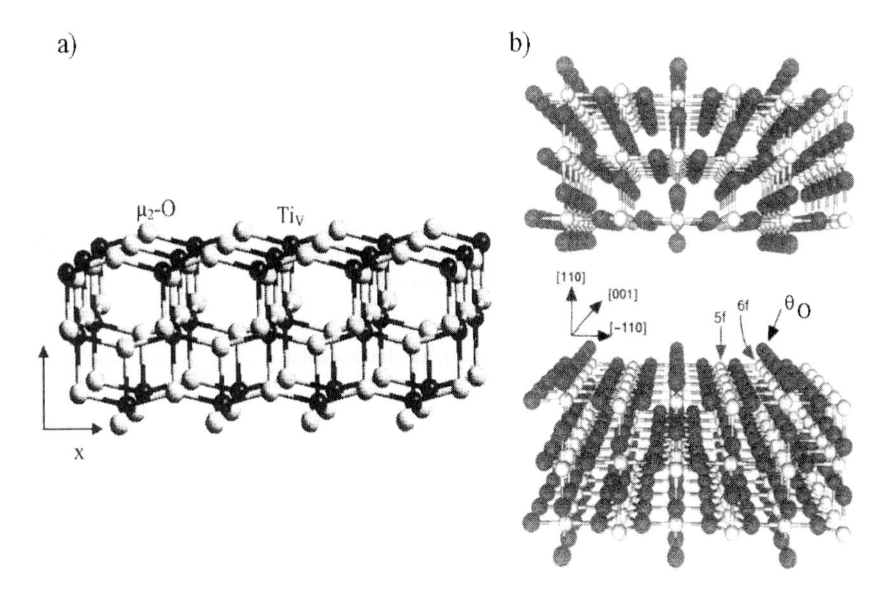

Figure 5.4. (a) Structure of the dehydrated TiO_2(001) surface after relaxation (θ_{001}) of anatase. Titanium atoms are black, oxygen atoms are gray. (b) Simulated system geometry of the TiO_2(110) rutile surface. Light and dark spheres indicate titanium and oxygen atoms, respectively. Perspective view showing the slab geometry used. The 36-ion cell is extended for display purposes[6a].

On the Ti-anatase (001) surface, Ti is five-coordinated and surrounded by two different types of surface oxygen atoms. Half of them have three Ti ions as neighbors as in the bulk (type I) whereas the other half have two Ti neighbors (type II).

The corresponding ion charge excesses are

$$\epsilon_{001}^{+}(\text{Ti}) = +\frac{2}{3}$$
$$\epsilon_{001}^{-}\Big(\text{O(I)}\Big) = 0$$
$$\epsilon_{001}^{-}\Big(\text{O(II)}\Big) = -\frac{2}{3}$$

Hence one expects H_2O to dissociate with the generation of an end-on OH bonded to Ti and a proton attached to O(II).

Figure 5.5 shows the resulting surface configurations for a completely relaxed system as a function of water coverage.

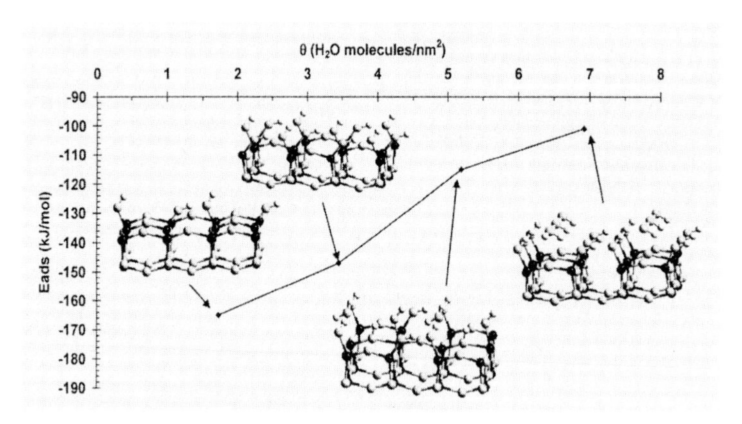

Figure 5.5. Adsorption energy of water on the (001) surfaces as a function of coverage (titanium = black, oxygen = gray and hydrogen = white. Insets give a ball-and-stick representation of the local structures are shown[6a].

Clearly, the following reaction events occur:

$$H_2O + \cdots\text{Ti}-\text{O}-\text{Ti} \longrightarrow \cdots\text{Ti}(\text{OH})-\text{O(H)}-\text{Ti}\cdots \quad \text{(a)}$$

$$\cdots\text{Ti}(\text{OH})-\text{O(H)}\cdots \longrightarrow \cdots\text{Ti}-\text{O(H)}\cdots\text{H}\cdots\text{O}-\text{Ti}\cdots \quad \text{(b)}$$

The Ti–O(II) bond opens up, thus allowing two surface hydroxyls to form, and generating hydrogen bonds.

The Brønsted acidity of hydroxylated anatase is due to the additional stabilization of the weakly acidic end-on OH groups that form due to hydrogen bonding. Such effects have been experimentally reported for analogous hydrogen-bonded silanol nests[7].

The (110) rutile surface has two different types of oxygen present. The four oxygen atoms that surround the coordinatively unsaturated Ti cation (5f, Fig. 5.4b) have no charge excess. The apical bridging oxygen atoms (θ_O, Fig. 5.4b), however, have an ion charge excess $\epsilon^-(O) = -\frac{2}{3}$.

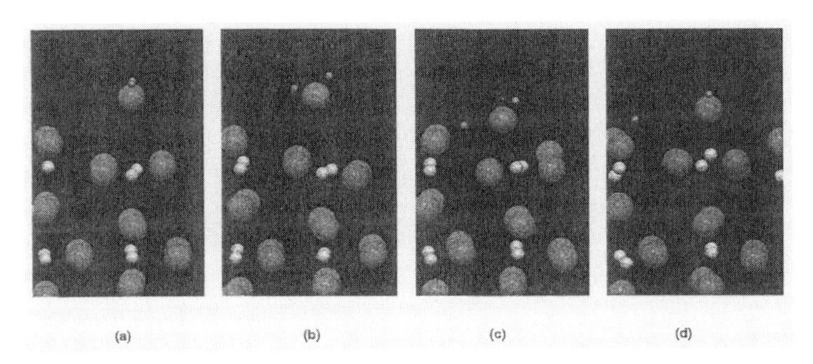

Figure 5.6. Snapshots of the ionic configuration taken from a dynamic simulation of water dissociation. (a) Initial configuration in which the water molecule lies in the (110) plane. The large gray, small white and small gray speres represent oxygen, titanium and hydrogen, respectively[6b].

In agreement with our prediction, H_2O dissociates, such that the proton attaches to an apical bridging oxygen atom, θ_O, and OH^- to the coordinatively unsaturated Ti atom. Figure 5.6 shows that this indeed happens, with little relaxation of the surface. Because of the large distances, there is no hydrogen bonding between the two hydroxyls. The larger Brønsted acidity of rutile has to be ascribed to the charge excess of proton-coordinated θ_O: $\epsilon^-(O, \theta_O) = +\frac{1}{3}$.

Pacchioni[8] summarized recently the results of quantum-chemical studies and experiments aimed at characterizing the surfaces of ionic solids such as MgO and related oxides. We present a number of the salient results from this study here. A schematic representation of the MgO surface which displays a surface terrace, step and edge sites and their respective coordination numbers is shown in Fig. 5.7.

Adsorption experiments with CO have conclusively shown that CO adsorbs weakly on the MgO terrace but more strongly on edges and corners sites. Using QM cluster calculations, Petterson et al.[10] predicted CO adsorption values of 8, 18 and 48 kJ/mol for the terrace, corner and edge sites, respectively. The adsorption of CO at terrace, edge, and kink sites was found to lead to an upward shift in the CO stretching frequency by +9, +27 and +50 cm^{-1}, respectively. These results are consistent with the generally accepted experimental data on this system of Wichtendahl et al.[11].

Interestingly, the adsorption energy of CO to Mg^{2+} located at an MgO corner site, attached through three neighbor oxygen atoms, is close to the adsorption energy of CO bound to an Mg^{2+} ion, ion-exchanged into a zeolite and charge compensated by a zeolitic four ring sof tetrahedra (see Chapter 4 pages 179, 180). The upwards shift of the CO frequency is also similar. The charges on the Mg ions located on the terrace and corner sites are very similar. The Madelung constants, however, are proportional to the electrostatic potentials and, hence, are quite different, as is shown in the caption to Fig. 5.7. The

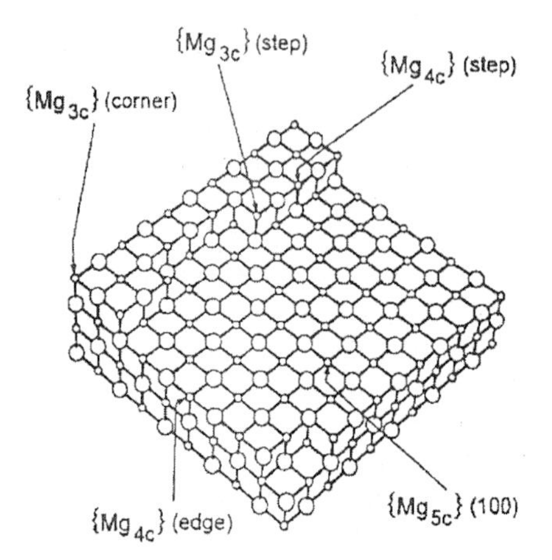

Figure 5.7. Schematic representation of an MgO surface; Mg^{2+} ions are represented as small spheres, the O^{2-} anions as large spheres. The subscript indicates the coordination number of each cation site. The Madelung constants computed for these sites are: 1.681 Mg_{5c} (terrace), 1.591 $_{4c}$ (edge), 1.566 Mg_{4c} (step), 1.344 Mg_{3c} (corner), 0.873 Mg_{3c} (step). The Madelung constant for a bulk Mg_{6c} is 1.747. From Sauer et al.[9].

Madelung constant M is the proportionality constant for the ionic energy of stabilization of cation–anion pairs in an ionic solid: $E_{ion} = -M\frac{q^2}{r}$, where q is the ion charge and r the nearest-neighbor cation–anion distance. The reduced polarization of CO at terrace sites over those at corner sites is the main contribution to the reduced Mg–CO bond energy. This reduction is due to the larger number of O^{2-} ions coordinated to Mg^{2+} at the terrace site. The interaction with the O^{2-} anions reduces the electrostatic field at CO, because their negative field counteracts the positive field from the Mg^{2+} cation. The decreased effective electrostatic local fields are reflected by the decreased Pauling charge excesses that we discussed above. The low reactivity of the MgO(100) for the dissociative adsorption of H_2 as well as H_2O, together with the observations of reactivity at step edges for these reactions, are consistent with these conclusions. The electrostatic basis to the reactivity of the MgO surface is further elucidated by a comparison with CaO[12].

Based on theoretical studies, it is concluded that the MgO(001) surface is unreactive towards CO_2 and SO_2. These surfaces, however, are known to contain defect sites such as step edges and kink sites. The defect sites can be quite reactive to CO_2 and SO_2, thus leading to the formation of surface carbonates and sulfites, respectively. The reactivities of CaO, SrO and BaO are significantly greater than that of MgO whereby both CO_2 and SO_2 can react at the (100) terraces of these surfaces. This increase in activity is the result of increasing basicity as one moves from MgO to CaO, SrO or BaO. This can be explained by the differences in electrostatic potential between MgO and CaO, SrO or BaO. Oxygen is more easily donated by oxides which have smaller Madelung potentials and hence have weaker cation–anion interactions. This reduction is due to the larger cation–anion distances for oxides with larger cation radii. For instance, the Madelung constant for the bulk oxygen anion in MgO is 23.9 eV. CaO has the same cubic structure as MgO but has a much larger lattice constant (2.399 Å in CaO versus 2.106 Å in MgO). The Madelung constant for CaO is 20.2 eV, which is 3.7 eV lower than that for MgO.

5.3 The Contribution of Covalency to the Ionic Surface Chemical Bond

In this section, we extend our treatment on ionic bonding to include covalent contributions and their relevance to oxidation catalysis. We provide a more detailed molecular orbital analysis of the properties of these oxides by relating the electronic structure of cations at the surface of the oxide with that found in corresponding organometallic cluster complexes. As such, we can use more classical hybridization schemes to understand their reactivity. Accurate calculations are available for the RuO_2 system that have been used as input to dynamic Monte Carlo simulations of the CO oxidation reaction to be discussed in the next subsection.

5.3.1 CO Oxidation by RuO_2

In order to begin to examine the effects of covalent bonding, we move from the more classical ionic oxides such as MgO, Al_2O_3 and TiO_2 to RuO_2, which provides more covalent chemical bonding contributions. RuO_2 is a natural extension for our discussions since both the rutile and the anatase surface structures have been extensively examined in the previous section. RuO_2 is known to be efficient in the oxidation of CO down to as low as room temperature[14]. CO oxidation is catalyzed by a range of other transition-metal catalysts such as Pt (see also Chapter 8). At low temperatures on metals such as Pt, however, the rate is greatly suppressed. CO is strongly bound to Pt and results in blocking of active sites, which subsequently prevents O_2 dissociation. This is not the case on Ru since Ru readily forms RuO_2, which turns out to be the catalytically active phase. The dissociative adsorption of oxygen and the adsorption of CO tend to compete on the RuO_2 surface.

We analyze, here the interaction of CO with the RuO_2 surface in detail, since CO is often used as a probe molecule to estimate the Lewis acidity (see Section 4.2.3.1) The chemisorption of CO on the Ru^{4+} ions of the rutile $RuO_2(110)$ surface [which has the same structure of rutile $TiO_2(110)$ (Fig. 5.4b)] was calculated to be 120 kJ/mol [13,14]. The oxygen atoms around the Ru ions are all coordinatively saturated since they are attached to three Ru neighbors. The binding energy of CO on the more stable $RuO_2(001)$ anatase type surface (Fig. 5.4a) is only 70 kJ/mol. The decreased interaction energy is due to the fact that two of the four surface oxygen atoms around Ru are now coordinatively unsaturated, thus leading to two stronger Ru–O bonds. Therefore, as follows from the application of bond order conservation, the Ru^{4+}–CO interaction energy on the $RuO_2(001)$surface is decreased compared with that on the $RuO_2(110)$ surface. The adsorption energies are substantially less than the predicted DFT values on the metallic Ru surface, where the adsorption energy is 180 kJ/mol atop Ru. This reduction in the CO adsorption energy is mainly due to the substantially reduced interaction of CO with the Ru metal s and p electrons on the oxide surface. In contrast the latter contribute significantly to the metal–oxygen bond strength. The adsorption energies of CO adsorbed on Ru^{4+} are relatively high compared with the interaction with a hard or soft cation as found in the zeolites (Section 4.2.3.1). This is due to the relatively high charge on Ru^{4+} and to the additional interactions with the d-electronic orbitals.

The electronic structure of RuO_2 can be deduced by inspection of Fig. 5.8a and b.

Figure 5.8a compares the LDOS on oxygen in the RuO_2 bulk with that of oxygen at the surface. The main contribution to the LDOS of oxygen is from the 2p atomic orbitals. In Fig. 5.8a one notes a gap in the LDOS at –2 eV below the Fermi level. The density of states below –2 eV is assigned to electrons localized mainly in bonding orbitals with oxygen. The density of states above –2 eV is due to electron density in antibonding

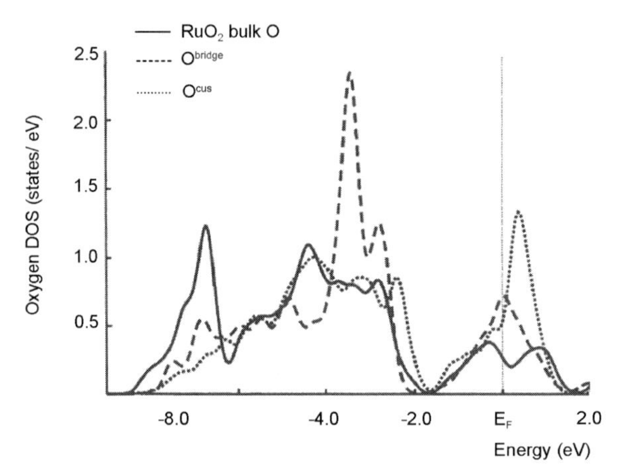

Figure 5.8a. Electronic structure of RuO_2, comparison of local density of states (LDOS) of oxygen in bulk and at the (110) surface. Adapted from Y.D. Kim et al.[15].

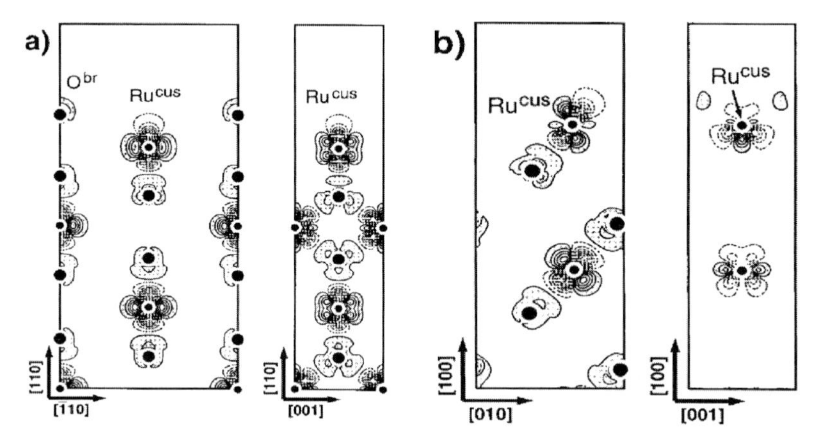

Figure 5.8b. Pseudo valence charge density contour plots of the (a) $RuO_2(110)$ surface in comparison with (b) the $RuO_2(001)$ surface cut through the cus-Ru atoms. These plots are defined as the difference between the total valence electron density and a linear superposition of radially symmetric atomic charge densities. Contours of constant charge density are separated by 0.15 eV/$Å^3$. Electron depletion and accumulation are marked by dashed and solid lines, respectively. In addition, regions of electron accumulation are shadowed[15].

orbitals mainly localized on Ru. One deduces from Fig. 5.8a a higher electron density on the oxygen atoms at the bridging sites (br) and a lower electron density on the oxygen on the coordinatively unsaturated (cus) sites (see Fig. 5.9a for notation). The smaller number of cation neighbors leads to less electron donation. The decreased band gap also indicates a lower electron density on surface Ru. The Ru atoms have less than half of their atomic density unoccupied, in agreement with their high positive charge. This is confirmed by the charge density difference plots of the RuO_2 (110) and (001) surfaces, respectively (Fig. 5.8b). One notes the decrease in charge densities on Ru and increased densities on oxygen.

The mechanism for CO oxidation over the RuO_2 (110) surface is fairly well understood

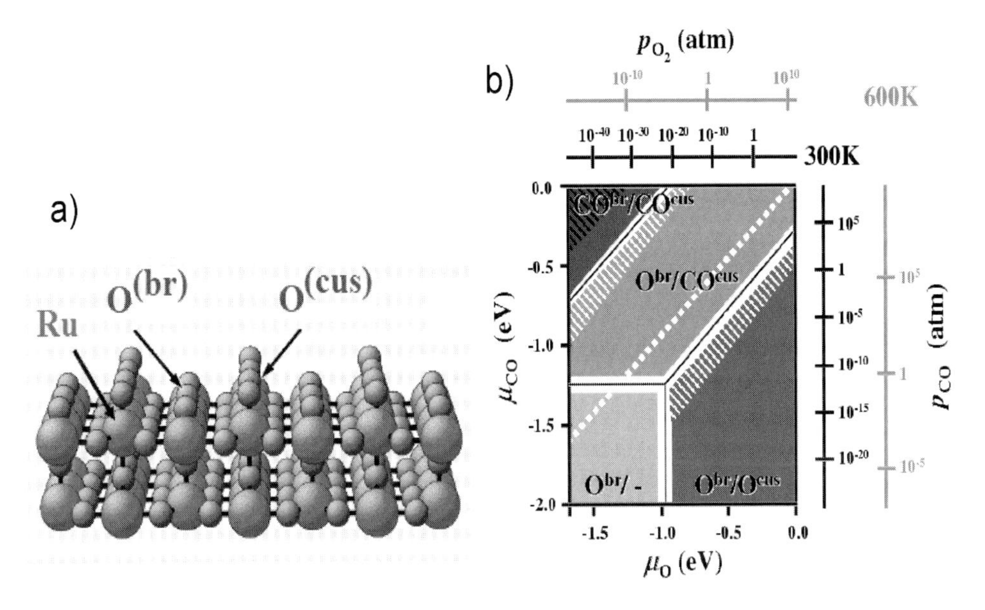

Figure 5.9. a) Perspective view of the $RuO_2(110)$ surface, illustrating the two prominent adsorption sites in the rectangular surface unit cell. These sites are labeled as the br(bridge) and the cus(coordinatively unsaturated) site, and both are occupied with oxygen atoms. b) Regions of the lowest–energy structures in (μ_0O, μ_{CO} space. The additional axes give the corresponding pressure scales at T=300 and 600 K. In the hatched region gas phase CO is transformed into graphite. Regions that are particularly strongly affected by kinetics are marked by white hatching. [16]

[16]. The surface thermodynamics and the kinetics have been studied by first-principle DFT calculations in combination with dynamic Monte Carlo simulations. Figure 5.9 shows the reactive O atoms and vacant adsorption sites on the $RuO_2(110)$ surface. Two Ru atoms are present at the surface. The Ru^{cus} atom has five oxygen atom neighbors and the other Ru atom is coordinated to six oxygen atoms. The four surface O atoms (O^{br} that coordinate to Ru^{cus} are all coordinatively saturated.

Table 5.1. DFT binding energies, E_b, for CO and O [with respect to $(1/2)O_2$] at br and cus sites (Fig. 5.9a), diffusion energy barriers, $\Delta E^b_{\mathrm{diff}}$, to neighboring br and cus sites, and reaction energy barriers, $\Delta E^b_{\mathrm{reac}}$, of neighboring species at br and cus sites. All values are in eV[16b]

	E_b	$\Delta E^b_{\mathrm{diff}}$		$\Delta E^b_{\mathrm{reac}}$	
		to br	to cus	with CO^{br}	CO^{cus}
CO^{br}	–1.6	0.6	1.6		...
CO^{cus}	–1.3	1.3	1.7		
O^{br}	–2.3	0.7	2.3	1.5	1.2
O^{cus}	–1.0	1.0	1.6	0.8	0.9

O^{cus} sites terminate the oxygen octahedron that surrounds the coordinatively saturated Ru atom. During reaction Obr may exchange position with an adsorbing CO molecule (CO^{br}) and, both CO and O can adsorb to the coordinatively unsaturated Ru^{cus} atom giving CO^{cus} and O^{cus}, respectively. The calculated binding energies, diffusion barriers and activation energies for different surface reaction steps on the $RuO_2(110)$ surface are given in Table 5.1. One recognizes the small interaction energy of O_{ad} as O^{cus} and the relative strong bonds of three-fold coordinated CO^{br} and O^{br}. Figure 5.9b illustrates the different surface phases that are formed in thermodynamic equilibrium with gas-phase CO and O_2. At high O_2 pressure and low CO pressure, the ideally terminated RuO_2 surface is stable (O^{br}/O^{cus}). When the CO pressure increases, CO initially adsorbs as CO^{cus} and at very high pressure begins to substitute not only for O^{cus} but also for O^{br}. The Dynamic Monte Carlo results show that maximum CO oxidation occurs at a surface that is to be described as a disordered phase. This phase occurs in the boundary region of the stable phases shown in Fig. 5.9b, in particular between the O^{br}/CO^{cus} and O^{br}/O^{cus} phases. At low CO partial pressure and relatively high O_2 pressure the O^{br}/O^{cus} phase is stable, on which CO only rarely adsorbs and the CO_2 rate is extremely low. When the CO pressure increases, at $T = 600$ K, $P_{CO} = 20$ atm and $P_{O_2} = 1$ atm, the average occupation numbers are

$N_{CO}^{br} = 0.11$, $N_{CO}^{cus} = 0.70$, $N_{O}^{br} = 0.89$ and $N_{O}^{cus} = 0.29$. The rate of CO_2 formation is high and dominated by the reaction between CO^{cus} adsorbed to coordinatively unsaturated Ru and O^{cus}, the coordinatively unsaturated apex O atom. At very high CO pressure the CO^{br}/CO^{cus} surface is formed, close to the state of the reduced Ru surface. The rate of CO_2 formation is now suppressed because the surface is poisoned with CO_{ad}.

5.3.2 Atomic Orbital Hybridization at Surfaces; Hydration Energies

The differences in chemisorption between transition metal and transition-metal oxide surfaces, as discussed above, are quite general and are very similar for the other oxides and for sulfides and other ionic solids. The electronic factors that control this behavior can be captured by relating the cations in the oxide to the metal atoms in well-defined coordination complexes.

By way of example, we probe the adsorption and dissociation of water. Water is also known to dissociate heterolytically on the RuO_2 surface. The 5s and 5p electron energies of Ru (a second-row transition metal atom, located in Group VIII, with higher ionization energies) are lower than the corresponding 4s and 4p electronic energies of Ti because of the higher effective nuclear charge of Ru. Hence the OH^- anion will be more strongly bonded to the unoccupied lone-pair orbital of Ru^{4+} than Ti^{4+}. The Ru-oxide surface will hence, be more Lewis acidic than that of Ti oxide. On RuO_2 the RuO bonds are stronger than in TiO_2 and, hence the Brønsted acidity of the hydroxylated RuO_2 (110) surface is also greater. The overall result will be that H_2O dissociates more easily on RuO_2 than TiO_2.

The chemical reactivity of transition-metal oxides or sulfides can be understood in terms of elementary quantum-chemical concepts by using the hybridization schemes introduced earlier for the metal–carbonyl clusters analyzed in Section 3.3.1. The following hybridization schemes are proposed for different geometric arrangements: six d^2sp^3-orbitals for octahedrally coordinated transition metals with three nonbonding d-orbitals; four-sp^3 orbitals with five nonbonding d-atomic orbitals and also four dsp^2 orbitals for planar four-coordinated metal atoms and two nonbonding d-atomic orbitals (see the Addendum in Chapter 3 for a background on hybridization).

This can be generalized to bonding in the oxides and sulfides, when each anion neighbor of a cation is considered to contribute two electrons to σ-type cation–anion covalent bond.

The electronic structure on the cation is very similar to that in the coordination com-

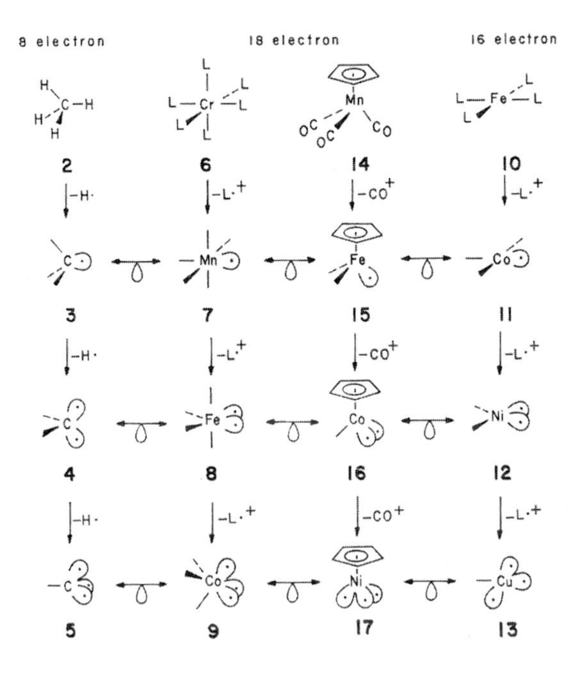

Figure 5.10. Isolable orbital schemes indicating similarity in reactivity based electron count and coordination[17].

plexes. Therefore, the electronic structure of the surface cations can be deduced by using the same hybridization schemes but counting now two σ electrons for each oxygen or sulfur atom. For example, in Co_2O_3, Co^{3+} is coordinated to six oxygen atoms. Within this model description, the six Co d^2sp^3 orbitals strongly overlap with six σ-type oxygen orbitals, which results in six low-energy bonding and six high-energy antibonding σ-type directed orbitals. The 12 oxygen electrons will occupy the six bonding σ-type hybridized bonding orbitals. The six electrons left will occupy the three nonbonding d-atomic orbitals continue. At the (111) surface of Co_2O_3, one oxygen–cobalt bond will be broken. Co^{3+} now creates a σ-type empty d^2sp^3-orbital that acts as a dangling bond along the surface normal. In Co^{2+}, this dangling bond would be occupied with one electron.

In Figure 5.10, orbital schemes are summarized for coordination complexes predicting the dangling bonds that arise when ligands are removed. A systematic analysis has been developed based on the total electron count and dependent on geometry [17]. There is a so-called isolable correspondence between different molecular systems with different coordinatively saturated structures having the classical 8, 18 or 16 valence electron count (Fig. 5.10, compounds 2, 6, 14 and 10). For example, column two in Fig. 5.10 shows the development of lone-pair orbitals according to the d^2sp^3 scheme. In the Cr complex six σ bonding orbitals with ligands occur. Each ligand contributes two electrons. The nonbonding d-atomic orbitals of Cr contain six electrons. On moving to Mn, a position lower in the column concerned, one electron is added to the system that in the MnL_6 complex would have to occupy an antibonding σ-type orbital. The complex prefers instead to remove a ligand with formation of MnL_5 and a lone–pair orbital occupied with one electron. Replacing Mn by Fe adds another electron. Double occupation of a lone-pair orbital is disfavored over losing another ligand, hence FeL_4 is stable, with two lone-pair orbitals. The isolable correspondence refers to the similarity of the lone-pair orbital arrangements one discovers when comparing complexes in the same row, deduced from different parent structures. The CH_2 fragment can be considered isolable with FeL_4 or NiL_2. Figure 5.10 helps not only to predict the reactivity of coordinatively unsaturated transition metal–ligand complexes, but also to clarify the reactivity of surface cations with related local topology and similar electron counts.

For instance, the electronic structures of TiO_2 and RuO_2 are analogous to the clusters in the CrL_6 row in Fig. 5.10, with the exception that unoccupied nonbonding d-states of the metal result in a formal charge of four on Ti. On Ru, four electrons occupy the three nonbonding d–atomic orbitals according to this hybridization scheme. The octahedral coordination with oxygen gives bonding and antibonding σ orbitals between Ti and O, of which the six bonding orbitals are occupied. At the anatase (001) surface a non-occupied d^2sp^3 orbital will appear as a dangling bond. When Ti^{4+} is reduced to Ti^{3+}, this surface dangling bond would become occupied with one electron. Replacing Ti by V would create an empty surface dangling bond for V^{5+}, a surface radical orbital for V^{4+} and an occupied V^{3+} dangling bond with two electrons. The last situation would be unstable and could initiate a reconstruction with reduction of oxygen coordination.

Figure 5.10[17] indicates that five-coordinated Ti has one empty dangling orbital perpendicular to the surface, two empty dangling orbitals for four-coordinated Ti at the anatase or rutile surface etc.

On the RuO_2 (110) surface the σ-type lone-pair orbital is nicely recognizable in Fig. 5.8b. When CO binds to the rutile RuO_2 (110) surface, the σ-type orbitals will have a bonding interaction with this empty dangling bond. The d-electrons in nonbonding orbitals will back-donate electrons into the CO $2\pi^*$ orbitals. Back-donation will be limited because of the high charge on Ru.

We will conclude this section by summarizing the interaction energy with cations, using as illustration the hydration enthalpy of M^{2+} ions of the first-row elements[18]. The electrostatic interaction increases with decreasing cation radius and increasing cation charge.

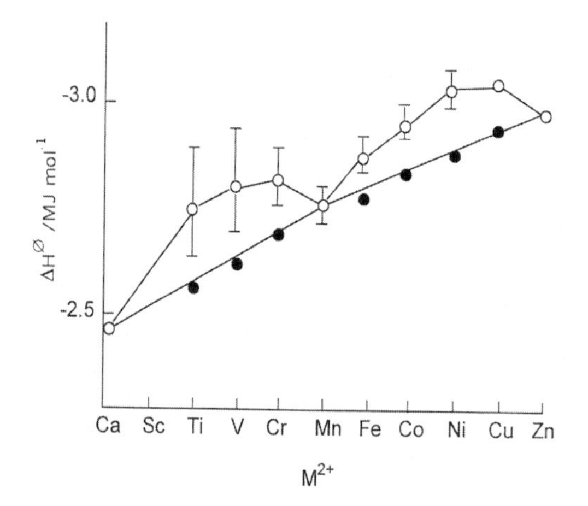

Figure 5.11. The hydration enthalpy of M^{2+} ions of the first row of the d elements. The straight lines show the trend when the ligand field stabilization energy has been subtracted from the observed values[18].

As can be seen in Fig. 5.11, the hydration enthalpy increases with decreasing cation size across a row in the periodic table. Hence an OH^- group attached to Ni^{2+} is expected to be more strongly bonded than one to Ca^{2+}. For similar surface topologies, the $Ni^{2+}(OH)^-$ group is predicted to be less basic than $Ca^{2+}(OH)^-$. In the (100) surface of the rock-salt structure, the ion charge excesses on the corresponding oxygen atoms are the same for nickel oxide and calcium oxide, $\epsilon = \frac{1}{3}$, indicative of a basic group. Ion charge excesses are only sensitive to surface topology, and cannot predict properties that are cation or anion dependent other than when the charge is changed.

In addition to the electrostatic effect, there is the contribution to the chemical bond which is the result of the distribution of electrons over the metal d-atomic orbitals, the ligand field effect[19]. The contribution of this effect is smallest when only five electrons are present and distributed over the five d-atomic orbitals. The each electron occupies one atomic orbital (the high-spin state) thus resulting in a spherical electron distribution. When the electron distribution is nonspherical, additional stabilizing interactions occur due to the redistribution of electrons over the d-atomic orbitals. This effect is a maximum for d-electron counts of three or eight.

Interestingly, the catalytic activity for different first-row metal oxides shows two maxima across the periodic table for catalytic reactions involving C–C and C–H activation. Dowden and Wells[20] showed that $3d^5$ (Mn^{2+}, Fe^{3+}), $3d^0$ (Ti^{4+}) and $3d^{10}$ (Zn^{2+}) cation-containing oxides show the lowest activity, whereas $3d^3$ (Cr^{3+}), $3d^6$ (Co^{3+}), $3d^7$ (Co^{2+}) and $3d^8$ (Ni^{2+}) cation containing oxides have the highest activity (Fig. 5.12).

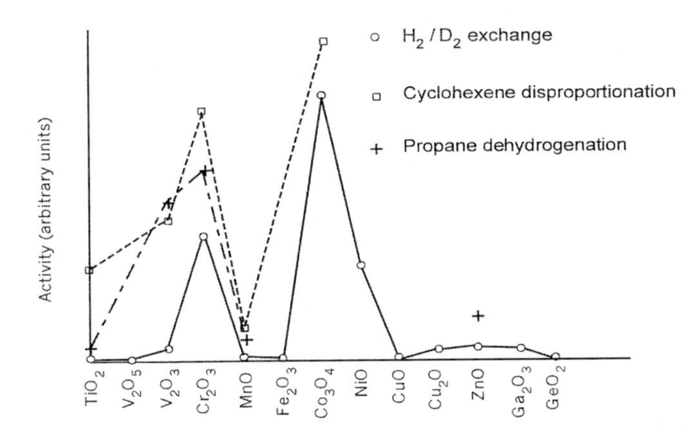

Figure 5.12. The catalytic activity of metal oxides along the first row of the periodic table. Adapted from D.A. Dowden[20].

In this section we have related the reactivity of oxides to cation properties such as charge, radius and ligand field stabilization. In Section 5.1 we have seen that differences in the reactivity of anionic-oxygen atoms as well as electrostatic effects are important also. In Section 5.6.8 we will emphasize reactivity differences in selective oxidation further in relation to the type of surface. We will again see that it is important to distinguish between coordinatively saturated surfaces and surfaces formed by metal–oxygen bond cleavage.

5.4 Medium Effects on Brønsted Acidity

In order to understand he physicochemical basis of Brønsted acidity, it is important to distinguish between homolytic and heterolytic dissociation and the corresponding energies related to each process. p

$$E_{\text{diss}}(\text{HOMO}) = E_{\text{HX}} - E_{\text{H}} - E_{\text{X}}$$
$$E_{\text{diss}}(\text{HETERO}) = E_{\text{diss}}(\text{HOMO}) + I.P._{\text{H}} + E.A._{\text{X}}$$

The homolytic dissociation energy is the energy cost to separate a molecular bond into neutral atoms. The heterolytic dissociation energy is the energy required to split a molecular bond into oppositely charged ions. This involves the energy necessary to dissociate the molecule homolytically into neutral fragments [E_{diss}(HOMO)] plus the energy required to remove an electron from the more electropositive fragment (ionization potential $I.P._{\text{H}}$] and place it in the more electronegative fragment (electron affinity [$E.A._{\text{X}}$]) as is shown above.

The acidity of a molecule strongly depends on the medium in which it is studied. Most studies are typically carried out in water. The concentration of the hydronium ions (H_3O^+) that form in water determines the solution pH. The following equilibrium is key.

$$\text{HX(aq)} + \text{H}_2\text{O(e)} \rightleftharpoons \text{H}_3\text{O}^+\text{(aq)} + \text{X}^-\text{(aq)}$$

This equilibrium is controlled by the dissociation energy of HX and the stability of the solvated H_3O^+ and X^- intermediates that form.

In the halides, the electron affinity does not vary much, hence the changes in the homolytic dissociation energy of HX determines the changes in the equilibrium constants. The same holds in comparing H_2S with H_2O. The results, however, are quite different if one compares the acidity across a row in the periodic table, such as H_2O compared with HF or H_2S compared with HCl.

Tables 5.2 and 5.3 present the energies for the homolytic dissociation of various HX species [E_{diss}(HOMO)]and the electron affinities [$E.A._{\text{X}}$])] for their corresponding anions (X^-), respectively.

Table 5.2. The homolytic dissociation energy for E_{diss} HX in kcal/mol

H_2O	119	HF	136
H_2S	90	HCl	103

Table 5.3. The electron affinity [$E.A._{\text{X}}$])] of the X^- anion in kcal/mol

OH	42.4	F	79.6
SH	50.4	Cl	83.2

The difference in electron affinity between OH and F or SH and Cl is much larger than the corresponding homolytic bond energies. HF in H_2O is more acidic than H_2O itself, because F^- is more stable than OH^-. The same holds for the difference in acidity of H_2S and HCl.

The stability of the hydronium ion determines the pH of water or that of other acids dissolved in water. This changes when we compare the acidity in different solvents. Neat acids such as HF and H_2SO_4 are much more acidic than when they are dissolved in water. This is because different protonated species are present. For HF dissolved in HF there is the equilibrium

$$2HF_{solv} \rightleftharpoons [HFH]^+ + F^-$$

The difference from HF dissolved in water is the relative stability of $[HFH]^+$ versus H_3O^+. The deprotonation energies for HFH^+ and H_3O^+ were calculated to be 652.7 and 957.2 kJ/mol, respectively. The differences in relative stability imply a much smaller equilibrium concentration of $[HFH]^+$ than $[H_3O]^+$ normalized on the same HF concentration. The much smaller deprotonation energy of HFH^+ compared with H_3O^+ implies a much higher reactivity of H^+ attached to HF than attached to H_2O.

Such medium effects imply a dramatic difference between acid catalysis carried out over acidic surfaces in a gas-phase medium and that carried out in a polar solution phase medium in the absence of a heterogeneous solid acid surface. On a solid surface, the activation of an adsorbate occurs from a neutral state since the carbenium ion states are typically unstable and serve as transition states. On the other hand, protonation in a solution can occur from an ionized state, in which reactant molecules have already been protonated. This explains the much higher reactivity for protonation reactions which occur in polar media.

An interesting comparison for acid dissociation carried out in neat and in aqueous media was presented by Kazansky[21] for the sulfuric acid system.

in water : $$HA \rightleftharpoons H_3O_{solvated}{}^+ + A_{solvated}{}^-$$

in neat sulfuric acid : $$(H_2SO_4)_2 \rightleftharpoons H_3SO_{4solvated}{}^+ + HSO_{4solvated}{}^-$$

The dissociation of sulfuric acid is exothermic in water ($\Delta E = -83.6$ kJ/mol), whereas in neat sulfuric acid the heat of autodissociation is slightly endothermic ($\Delta E = +18.8$ kJ/mol). Notwithstanding their low concentration, the catalytically active species in the solvent will be H_3O^+ or $H_3SO_4{}^+$. The high reactivity of $H_3SO_4{}^+$ arises from the much lower deprotonation energy for $H_3SO_4{}^+$ than H_3O^+.

The essential difference between proton transfer from a neutral complex and protonated state becomes clear in comparing the activation of isobutene by the neutral $(H_2SO_4)_2$ cluster shown in Fig. 5.13a and $H_3SO_4{}^+$ shown in Fig. 5.13b[17]. There is a high activation barrier for proton transfer in the neutral system (100 kJ/mol for the H_2SO_4 monomer as shown in Fig. 5.13a) as compared with the low activation barrier (14 kJ/mol as shown in Fig. 5.13b) and high exothermicity for protonation in the charged system.

The interaction energy of a carbenium ion with a neutral sulfuric acid molecule is only 64 kJ/mol, compared with a C–O bond dissociation energy of 760 kJ/mol for the neutral tert–butyl sulfuric acid. The QM results are corrected for solvation energy affects by using a continuum model (see also Chapter 6. page 290) that describes the dielectric response of the solvent medium on the energy of dissociation for the reaction:

$$[\text{t-butyl}H_2SO_4]_{solvated}{}^+ \longrightarrow [\text{t-butyl}]_s{}^+ + H_2SO_{4solvated}$$

The reaction now becomes exothermic at –57 kJ/mol and, hence, the protonated isobutene is a stable compound in solution. This result illustrates the importance of solvation on reactions in polar media with high dielectric constants.

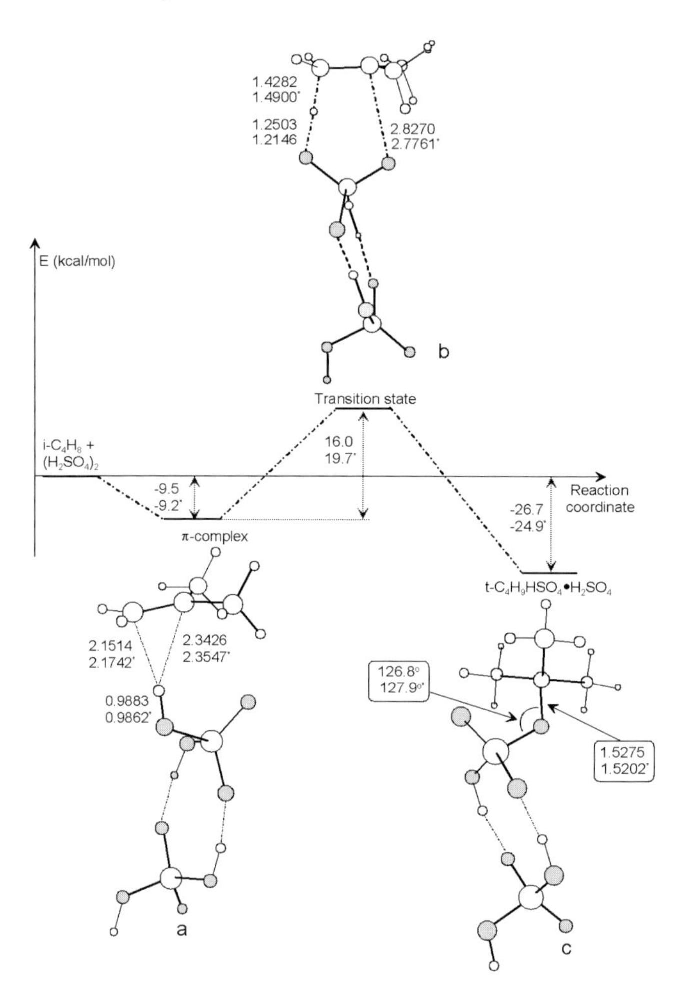

Figure 5.13a. Energy profile for isobutene interaction with dimeric sulfuric acid $(H_2SO_4)_2$. (a) π-Complex; (b) transition state. (c) t-$C_4H_2SO_4.H_2SO_4$. * The data on the reaction with monomeric sulfuric acid are given for comparison[22].

In contrast to the low activation barrier for the protonation of isobutene in sulfuric acid, the computed activation energies for isobutene protonation in a zeolite using comparable models are much higher, namely ~70 kJ/mol. In Chapter 4, the formation of protonated isobutene in mordenite is considered in more detail. It is concluded that protonation is an activated process with respect to the adsorbed π-complex and that the protonated isobutene forms an alkoxy complex. This chemistry is very similar to that discussed here for protonation of isobutene with $(H_2SO_4)_2$ (Fig. 5.13a).

Protonation in zeolites is also extensively discussed in Chapter 4. Zeolitic protons (see page 163) are part of the zeolite (SiO_2/Al_2O_3) microporous system. The zeolite is made up of a rather rigid medium with a low dielectric constant ($\varepsilon \approx 2$). Chemistry in zeolites is, therefore, closer to that which is carried out in a vacuum rather than in solution. Solvation effects remain limited to the relatively small adjustments of the lattice atom positions that occur in the direct environment of adsorbed cations or anions. Electrostatic charges are

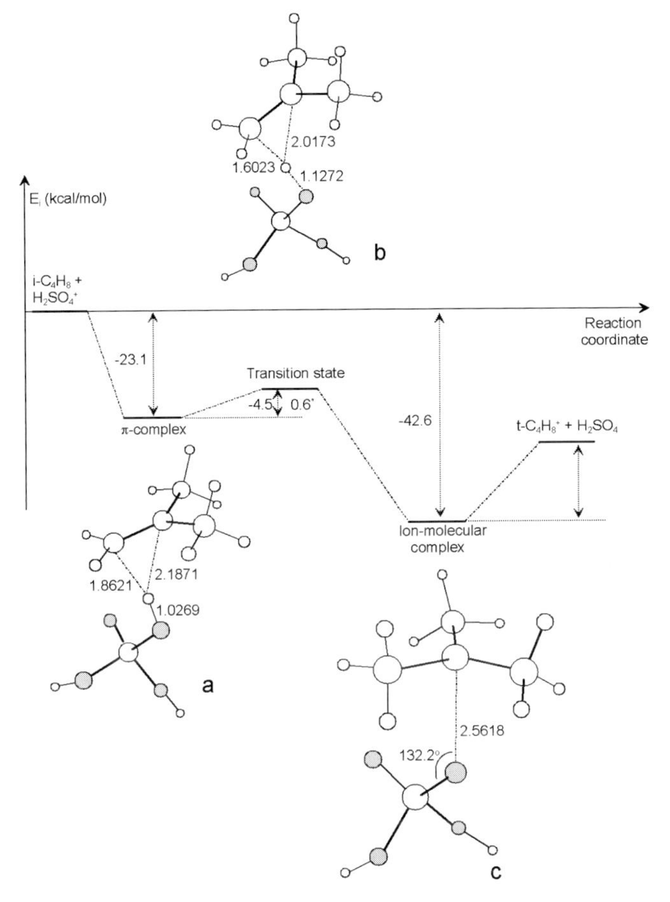

Figure 5.13b. Energy profile for isobutene interaction with protonated sulfuric acid ($H_3SO_4^+$). (a) π-Complex; (b) transition state; (c) Ion–molecular complex[22].

only screened by polarization of nearby lattice oxygen atoms. In a protonation reaction, the interaction between positively charged reactant cation state and the negative charge on the zeolite lattice is a critical in establishing the energy or the transition-state energy. In zeolite catalysis, the protonated intermediate is always a highly activated complex and usually part of the transition state.

For reactions in H_2O an analogous discussion applies as for reactions with the protonated sulfuric acid form. The reactive intermediate now, however, is H_3O^+. H_3O^+ has a much stronger OH bond than $H_3SO_4{}^+$, and hence its reactivity is much less. The species that is formed by reaction with alkene in an aqueous solution can best be compared with protonated alcohols forms typically known as alkyl oxonium ions[23]. Protonated alkene is hydrated in aqueous media. The formation of alkyloxonium ions instead of the carbenium ions in H_2O can be viewed as due to the basicity of water. The major difference between solid acids and acidic solutions arises because the hydrogen atoms in solid acids are part of strong covalent bonds and are not present as protons that are present in in the solution phase. In a solution there is a equilibrium between non-dissociated acid molecules and the ionized ions, solvated by the solvent. Their equilibrium defines the pH of the solution. In a solid acid there is no counterpart of the pH. Acidic chemistry evolves only upon contact with a reactant. The protonated intermediates often only exist as part of transition states. Acidity in solid acid catalysis has to be considered a kinetic phenomenon.

5.5 Acidity of Heteropolyacids

Heteropolyacids are unique molecular metal oxide clusters which contain acid, base and redox functionality that can be varied over a fairly broad range in order to tune their potential catalytic behavior. The polytungstate forms have Hammett acidity values of $H_o = -13.1$ which, ranks them as being more acidic than 100% H_2SO_4 ($H_o = 12$). They are, therefore, known as superacids. However, the use of Hammett indicators is based on color changes of the indicator molecule by protonation. In solution their is a equilibrium between protonated indicator molecules and non-dissociated acid molecules. On the solid acid the protonated indicator molecule is stabilized by the negative cgarge of the ionized surface. This is very different from solvation phenomena in solutions. Hence interpretation of Hammett indicator measurements on solid acids is not straightforward and immediate comparison with their use in liquid acids is not justified. We discussed an analogous situation in Chapter 4 for acid catalysis of zeolites. There we noted that adsorption effects obscure intrinsic acidity similarities or differences. The heteropolyacids are stable both in solution and in the solid state. They can also be readily supported on silica or other metal oxide supports. They have therefore been considered as possible alternatives to replace the highly corrosive and environmentally toxic HF and H_2SO_4 liquid acids presented in the previous section for low-temperature acid-catalyzed reactions such as isomerization and alkylation[24]. The most predominant form studied in the literature is the Keggin structure. The phosphotungstic form of the Keggin unit is shown in Fig. 5.14. The Keggin structure is comprised of primary, secondary and tertiary features. The primary Keggin unit (KU) is made up of a central metal oxide tetrahedron comprised of a central atom such as P or Si which is directly bound to four oxygen atoms. The charge imbalance between the central atom and its oxygen ligands leads a negatively charged anionic core.

Their central anionic tetrahedron core is surrounded by twelve outer metal oxide octahedra that form an outer metal oxide shell that acts to delocalize the charge on the central cluster. These outer metal–oxide octahedra are comprised of a metal atom that sits in the center, known as the addenda atom, surrounded by six oxygen atoms. The negative charge in this system is then delocalized over the outer metal oxide shell. The resulting cage structure contains three uniquely different oxygen atoms. The first are the terminally bound tungstenyl (W=O) oxygen atoms. The other two refer to bridging oxygen atoms from either the corner- or edge-sharing octahedra that make up the outer shell. The heteropolyanion form is subsequently converted to the acid form by the addition of protons at external oxygen atoms of the cage. The protons can reside at either the corner- or edge-sharing oxygen bridges or the terminal oxygen atoms. The calculated energy difference indicates that there is typically a small site preference for the bridge sites but that the energy difference between each of these sites is not very large (< 15 kJ/mol). The charge of the anion can essentially be considered to be delocalized over the entire outer shell of the Keggin structure. This delocalization gives rise to a lower proton affinity, which would imply a greater acidity for these materials.

In the solid state, the primary Keggin structures pack into secondary and tertiary forms. In the secondary structure, six water molecules of hydration come together to aid in assembling primary Keggin structures into a hexahydrate BCC form. The waters of hydration can form around an internal proton which exists in the form of the $H_5O_2{}^+$ Zundel ion in the center of the hexahydrate cage. These "inner" protons may be considered inaccessible to reagents that cannot penetrate into the secondary structure.

Thermogravimetric analysis indicates that the waters of hydration are removed at temperatures above around 500 K. This can lead to a dehydrated secondary form. Ab initio calculations indicate[25] that the loss of secondary water molecules will begin to isolate protons on the bridging positions between two Keggin units (Fig. 5.16), which for the most part will be inaccessible for reaction.

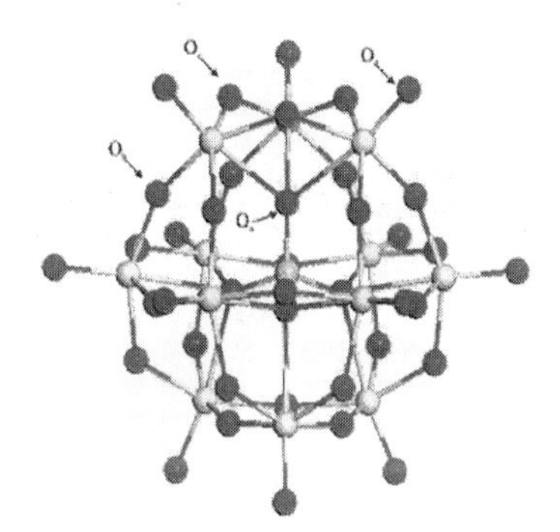

Figure 5.14. The Keggin structure of the $PW_{12}O_{40}{}^{3-}$, identifying the three types of oxygen atoms in the structures[25]. The hexahydrate heteropolyacid contains typically six water molecules per Keggin unit. The water molecules essentially help to assemble and pack the primary structure into an organized secondary hierarchy. A computed representation of the secondary $H_3PW_{12}O_{40}.6H_2O$ hexahydrate form is given in Fig. 5.15.

The use of HPAs to catalyze specific reactions heterogeneously requires anchoring these molecular cages to a particular support. Silica currently appears to be the favorite used in the experimental literature. The supported Keggin structure can readily be modeled by examining just the interactions between the reactants and the primary Keggin structure. DFT calculations were carried out to establish the proton affinities for at various sites on the Keggin structure and for different compositions for both the addenda and the central atoms. The results for the adsorption of the proton on the primary $PW_{12}O_{40}{}^{3-}$ Keggin structure are depicted in Fig. 5.14[26] and summarized in Table 5.4[26].

The results demonstrate that the proton attached to the Keggin unit with a charge of 3^- is, as should be expected, the strongest interaction resulting in a proton affinity of 1591 kJ/mol. The presence of additional protons on the metal oxide cage ultimately compensates for the negative charge and lowers the proton affinity. The proton affinity of the third proton is lowest at 1077 kJ/mol. This is significantly lower than those for zeolites with low aluminum concentration. The affinity of the HPA for the second proton (1349 kJ/mol) is actually closer to the affinity found for zeolitic protons. The strength of these bonds is thus indicative of strong Brønsted acidity for the HPAs.

From Table 5.4, one can see that the whereas the proton prefers the bridging oxygen atoms, the difference between the bridge and atop sites is fairly small, which would suggest that the proton may be quite mobile, having the ability to diffuse between different sites.

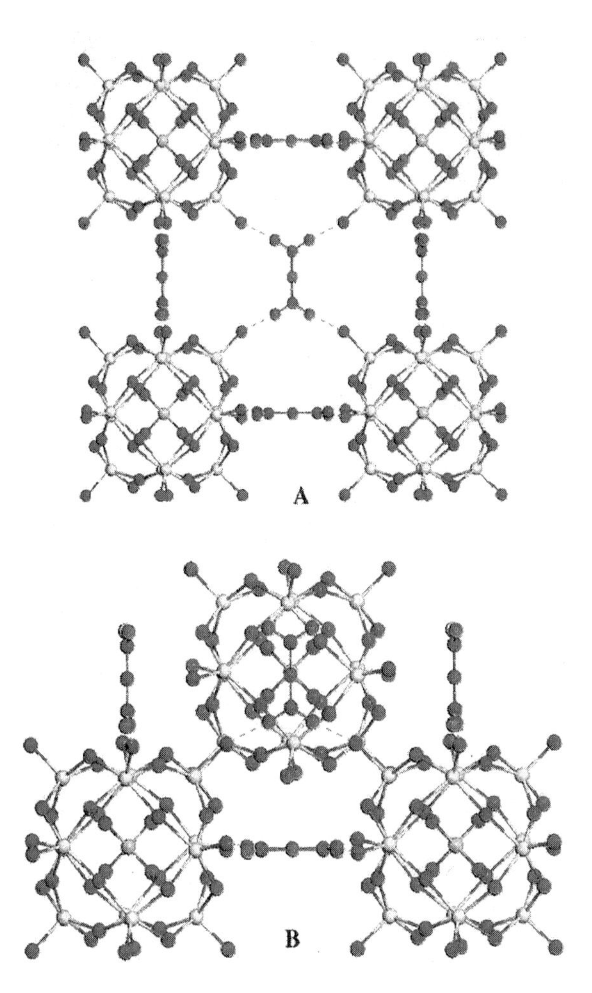

Figure 5.15. The optimized BCC structure of the $H_3PW_{12}O_{40}.6H_2O$ hexahydrate form. In this structure, all O_d atoms are bound to an $H_5O_2{}^+$ species; however, many of these species have been omitted for simplification. (A) The view of one face of the BCC structure. (B) The relationship between KUs on an edge of the cube and the nearest-neighbor body-centered Keggin unit[25].

The high Brønsted acidity found for heteropolyacids is further confirmed by ammonia adsorption studies on isolated anhydrous HPW structures[27]. The adsorption of ammonia on Mo- and W-based heterpolyacids were compared. The calculations of ammonia bound to the trimer HPA complex reveals chemisorption energies of –103 and –141 kJ/mol for the HPMo and HPW structures, respectively, comparable to values found for zeolites with a low concentration of alumina. The results compare very favorably with the values from microcalorimetric NH_3 adsorption experiments. Ammonia adsorbs in a bidentate configuration, thus leading to a proton transfer and the formation of the ammonium ion upon adsorption. Ammonium formation tends to occur when ammonia adsorbs at the higher coordination sites. The energy cost required for charge separation must be compensated for by the stabilizing electrostatic interactions [27]. The weaker acidity for

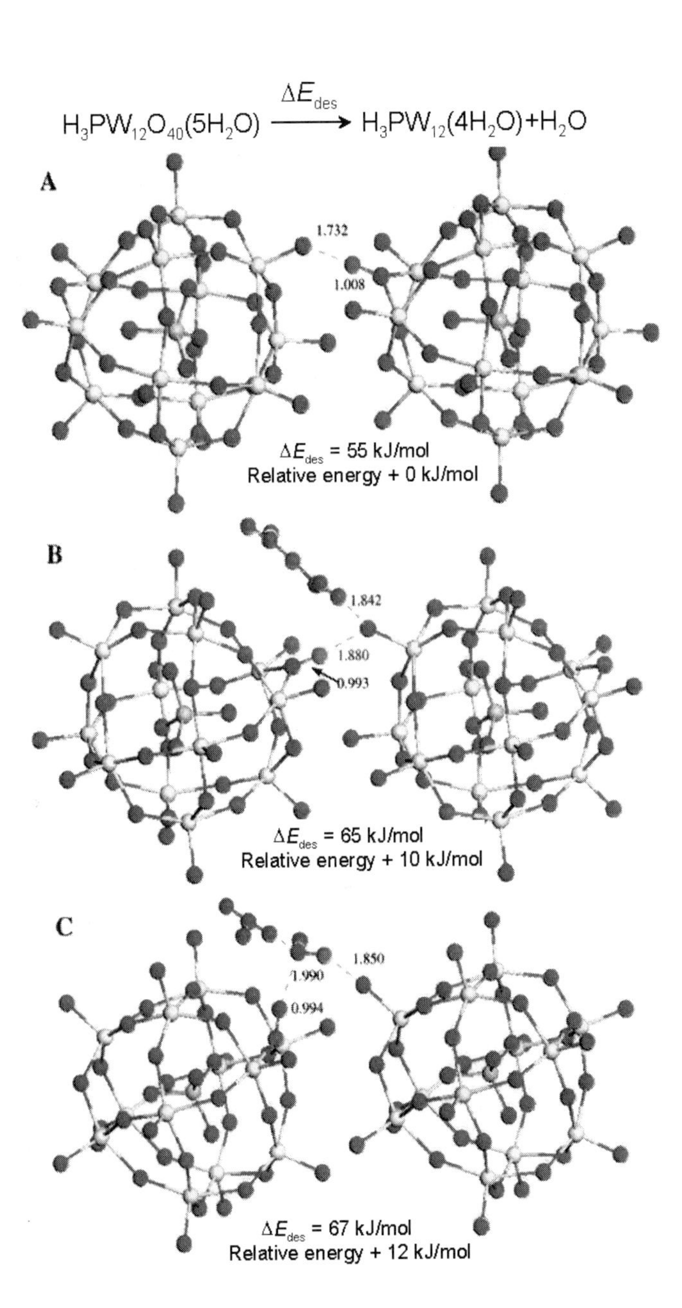

Figure 5.16. Structures resulting from the removal of two of the six water molecules of hydration from $H_3PW_{12}O_{40}.6H_2O$. The remaining anhydrous H^+ atom locations are at (A) bridging an O_d atom and an O_c atom of the nearest-neighbor body-centered KU, and (B) bridging an O_b atom and an oxygen atom of a water molecule of hydration bound to the nearest-neighbor body-centered KU. Relative energies are given with respect to the optimal configuration (A). All distances are given in å. There are two remaining $H_5O_2{}^+$ species per KU and only those in close proximity to the anhydrous proton are shown[25].

the HPMo compared with that of the HPW is consistent with the higher ionization potentials for sixth row transition metals compared with third row transition metals and relates to partial screening of nuclear charge by the atomic f-electrons.

The acidity of the $PW_{12}O_{40}{}^{3-}$ is such that it can stabilize the adsorption of an alkene. Janik et al.[28] have shown that the barriers for the protonation of ethylene, propene, 2-butene and isobutene are 69, 50.5, 51 and 14 kJ/mol, respectively. The results indicate that the formation of the carbenium ion from the π adsorbed state is favored on the Keggin structure over the similar corresponding state in the chabazite (see Chapter 4). For example, the barrier for the protonation of propene in chabazite was calculated to be +56 kJ/mol in comparison with the value of 50.5 kJ/mol calculated on the phosphotungstic (HPW) Keggin structure. The barriers, however, for the reaction of a surface alkoxy to the carbenium ion state are all higher on phosphotungstic acid than on chabazite.

The barriers for formation of ethylene, propene, 2-butene and isobutene from the alkoxy HPW states were calculated to be 114, 93.7, 87.8 and 64 kJ/mol, respectively, as compared with the value of 83 kJ/mol for propene over chabazite. Cleavage of the alkoxide bond initiates consecutive reaction steps necessary for many hydrocarbon conversion reactions.

The catalytic differences between protonic zeolites and the HPW structures are likely due to the greater stability of the alkoxide reactant state on the HPW structure than that on chabazite.

Table 5.4. Calculated proton affinity values for progressive addition of three protons to each of the oxygen types of $PW_{12}O_{40}{}^{3-}$[26]

	Proton affinity (kJ/mol)		
Site	1st proton	2nd proton[a]	3rd proton[b]
O_b	1579	1349	1077
O_c	1591	1340	1079
O_d	1564	1321	1071

[a] The calculations for finding the position of the second proton were performed with the 1st proton on its preferred O_c site.

[b] The 3rd proton calculations were performed with both the 1st and 2nd on their preferred O_c and O_b sites.

5.6 Oxidation Catalysis

5.6.1 Introduction

Metal oxides are quite diverse and posses a wide range of different properties which make them active materials for the conversion and selective oxidation of a wide range of different reagents. They can be used as acid, base, reducible, non-reducible and bifunctional catalytsts to carry out specific reactions. The mechanisms that govern the reactions over these materials, however, can be quite different. Here we will predominantly discuss hydrocarbon oxidation. The selective oxidation of hydrocarbons tends to proceed via a sequence of elementary steps which include: the activation of a C–H bond, oxygen insertion into activated hydrocarbon intermediate, subsequent C–H activation steps, desorption of products, and the regeneration of the active site[29]. The initial C–H activation step is thought to be the rate-determining step for a range of different oxidation processes.

Subsequent oxidation and C–H activation steps, however, tend to control the selectivity and the slate of final products that form. C–H activation can proceed through either homolytic or heterolytic processes depending upon the catalyst that is used and the nature of the active site. Homolytic C–H activation processes usually lead to the production of free radical intermediates whereas heterolytic activation leads to the formation of charged complexes. The heterolytic activation of a C–H bond can occur through the formation of a carbanion intermediate which is bound to a surface metal cation and a proton which binds to a nearby surface oxygen atom, thus forming a surface hydroxyl intermediate. Alternatively, the heterolytic activation of the C–H bond can lead to the formation of a metal hydride (M–H) and a surface alkoxide intermediate.

The overall selectivity for oxidation depends not only on the composition of the catalytic material, but also on the type of active oxygen species that are present at the surface. Oxygen on the surface can exist in various different forms ranging from adsorbed O_2 to fully oxidized atomic oxygen (O^{2-}) as is shown in the Eq. (5.1) below [29–32].

$$O_2^* < O_2^- < O^- < O^{2-} \tag{5.1}$$

The first three of these oxygens ($O_2{}^*$, $O_2{}^-$, O^-) are typically electrophilic. They tend to activate C–H bonds and lead to total oxidation of the reactants. O^{2-}, however, is more nucleophilic and can insert into the carbon–carbon bonds of the reactant molecules, thus favoring more selective oxidation paths.

The activation path along which one proceeds is strongly dependent on the nature of the active catalyst and the support that is used. As mentioned already, selective oxidation can occur over both reducible and non-reducible oxides. Oxidation over reducible oxides undergoes direct O_2 activation at a metal center, thus leading to changes in the oxidation state of the metal. The metal is subsequently reduced upon reaction with the hydrocarbon. The active catalyst then cycles between reduced and oxidized states. Kulkarni and Wachs[33], for example, have shown that redox activity over different metal oxides can vary by over six orders of magnitude. Catalysis over non-reducible oxides, however, does not involve a change in the oxidation state of the metal. Instead, it requires the presence of promoters, defect sites, or sites on the oxide that do not require changes in the redox character of the cations[34] to help activate hydrocarbon reagents. Regardless of which path is ultimately taken, there are a number of properties for the different oxide materials that can ultimately be tuned to influence the actual catalytic behavior. The factors that typically control the properties of these catalytic systems, ultimately dictating the specific chemistry that occurs, depend upon the structure, composition, and electronic properties of the metal oxide surface. We describe here the most direct effects on catalysis which arise from changes in the structural or electronic features of the active site directly and the effects of changing the overall environment of the system. This includes:

1) *Degree of coordination of the active site.*
   In many instances the metal ions can be thought of as individual or isolated molecular centers. As such, a number of the fundamental rules established from organometallic chemistry apply[35]. The relative degree of coordination, or lack thereof, is important in controlling the metal–adsorbate bond strength. This can govern the relative propensity for either C–H activation or subsequent oxygen insertion. Sites which are more coordinatively unsaturated will, in general, bind adsorbates more strongly and thus aid in the activation of adsorbate bonds. Analogously, sites which are more coordinatively saturated will bind adsorbates more weakly and tend to favor bond-making processes.

2) *Band gap of the oxide.*
   The band gap of the oxide is important in that it is a measure of redox capabilities of the actual oxide. It describes the electronic propertiesy of catalytic oxide material that "align" the lowest energy states in their conduction band and the highest energy states of the valence band of the oxide with the highest occupied molecular orbital of the adsorbate and the highest energy states of the valence band and the lowest unoccupied molecular orbital for O_2, respectively. This provides for the best overlap that would permit charge transfer and reactivity[32]. The band gap also relates to the reducibility of the oxide, which in some cases can correlate with both the activity and the selectivity[35]. The band gap of oxide can, in principle, be tuned by changing the domain size of the active surface ensembles or by changing the properties of the metal oxide support.

3) *Oxidation state of the metal ion.*
   As in organometallic chemistry, the oxidation state of the metal ion can have a profound effect on its ultimate reactivity. The oxidation state strongly influences the redox and acid–base properties of the metal ion center[35]. Since most elementary reactions intimately involve the metal, changes in its oxidation state will strongly control its ability to carry out specific reaction steps.

4) *Acid–base properties available at different adsorption sites.*
   The Lewis and Brønsted acid and base properties dictate whether a molecule will adsorb at a particular site along with the specific reaction chemistry that may occur at that site. Many probe molecules have been used to try to help identify the acid and basic characteristics for particular reactions. The best probe molecule is one which specifically tests for the specific reaction of interest. The strength best adsorption probe for acid and base site strength will clearly depend on the degree of acidity or basicity needed for the specific reaction. By using isopropanol as a probe molecule to characterize both acid and base sites, Kulkarni and Wachs[33] showed that the turnover frequency for isopropanol oxidation changed by over eight orders of magnitude with changes in the metal oxide. The surface acidity is therefore a highly tunable property[36] that may be used to control reactivity. Macht et al.[[37] used 2,6-di-tert-butylpyridine as a selective probe of the Brønsted acid sites that form in situ via the dehydration of butanol over supported WO_x complexes. Their results demonstrate strong support effects on the overall rate. The intrinsic turnover frequency per acid site for butanol dehydration, however, remained relatively constant. The acid site density was found to be strongly tied to the nature of the support. They showed a general correlation between the Sanderson electronegativity of the cation in the support and the measured Brønsted acid regioselectivities.

5) *Chemical bonding in oxides: covalent vs. ionic*
   The nature of the bonds that form within the oxide control its oxidation state and sets its ability to participate in carrying out specific reactions. In addition, the bond types tend to control the mechanisms by which reactions proceed. These effects can either be direct or occur more subtly. Some oxides actually take on both covalent as well as ionic characteristics, as was discussed in Section 5.3[34].

6) *Defect sites.*
   Various defects are known to form within the surface of an oxide including the presence of anion vacancies (F-centers), cation vacancies, interstitials, trapped electrons, or holes.[35] These defects change the local electronic structure and can significantly

impact surface reactivity[29,32,38–40]. The surface chemistry for a range of different systems may actually occur at defect sites since they expose coordinatively unsaturated centers.

In Chapter 2, we indicated that some of the most viable routes for the selective oxidation of propane to acrylonitrile today appear to come from substituted vanandium antimonates (VSbO). There have been various studies carried out to understand the influence of cation substitution. Andersson et al. [41] demonstrated increased catalytic performance for acrylonitrile production when Ti was substituted into VSbO oxides. They suggested that the increased performance was due to the prevention of V_2O_5 formation and the stabilization of site-isolated vanadium. Xiong et al.[42] followed up on these ideas in an elegant study in which they combined UV–Raman spectroscopy, X-ray diffraction and periodic DFT calculations to elucidate the influence of Ti on vanadium antimony oxide and its redox behavior. The combination of UV spectroscopy and DFT frequency calculations was used in this work to identify the bulk rutile structure, the characteristic modes of V_2O_5 and the formation of unique bands at 880 and 1016 cm^{-1}. Theory together with experiment was used to identify these bands as stretching bands for two-coordinate oxygen species which reside at SbOSb bridges (880 cm^{-1}) and SbOV bridges (1016 cm^{-1})that sit near a cation vacancy. Oxygen, in the rutile structure, is predominantly surrounded by three cations and is thus considered three-coordinate. The twocoordinate oxygen sites are the result of oxygen adsorption near the cation vacancies. Exposure of the surface to ammonia shows the complete removal of these bands, which suggests that ammonia is predominantly activated at these sites to produce water. These bands reappear, however, when the sample is exposed to air, which indicates that they are readily reoxidized. The incorporation of Ti into the vanadium antimonate structure was found to enhance the number of sites as measured by the increased intensity of the UV band and by the ease of formation. This suggests that Ti appears to improve redox properties. The authors speculate that ammonia reacts with the bridging oxygen species at the SbOSb sites, thus removing water and forming SbNHSb sites, which take part in the ammoxidation of the hydrocarbon. They speculate that these Sb sites are active in the formation of acrylonitrile. The substitutional addition of Ti was also found to help stabilize V^{3+} and V^{4+} cations and hence prevent the breakdown of vanandium antimonate into V_2O_5 surface structures.

Other factors which are also important include the following:

7) *Surface morphology and structure.*
   Metal oxides can demonstrate remarkably different behavior for reactions carried out over different surfaces. The phase, morphology and the exposed surface facets can all be important. For many reaction systems there are well-prescribed phases and crystal faces which appear to dominate activity and/or selectivity. For example, the active phase for butane oxidation to maleic anhydride is thought to be $(VO)_2P_2O_7$. The (100) surface for this system appears to be the most active. The nature of both the oxygen and the vanadium cations in VPO can be quite different depending upon which surface is exposed[29,40]. In addition, the surface structure itself can be quite different depending on the nature of the support. The surface structure can dramatically change as the reaction proceeds and conditions change.

8) Surface termination.

Depending upon the reaction conditions, the active surface can take on various different forms. Under highly oxidizing conditions, most of the sites may be covered by oxygen. More reducing conditions, on the other hand, will likely expose coordinatively unsaturated sites that take on more metallic-like properties. Intermediate conditions can give rise to interesting interfaces between oxide and metal-like configurations[44]. Most of the theoretical, and surface science studies, however, have been carried out over model surface structures which may potentially oversimplify the true nature of the surface. Fundamental studies have looked at both neutral- and polar-terminated surfaces and demonstrate that these two surfaces can lead to rather different kinetics[44,45].

9) Site isolation and domain size.

Under the conditions of interest, a number of mixed metal oxides may exist as site-isolated metal oxide complexes or as two-dimensional chains that wet the surface of the support. The nature of the bridging metal–oxygen–support bonds can ultimately begin to control the surface chemistry. The optimal active domain size strongly depends upon the reaction of interest. Wachs and co-workers[46,47] and Corma and Garcia[36] have shown that isolated MO_x sites are selective and, in some cases, more active in carrying out different selective oxidation reactions. Macht et al. [37] have demonstrated that the optimal supported-WO_x domains for alcohol dehydration are two-dimensional domains of size 9–10 W/nm^2.

10) Influence of water.

Water can play an important role in changing the surface composition, thus leading to an increase in the number of surface hydroxyl intermediates as well as Brønsted acid sites. This can enhance the kinetics for various different reactions[33,48–50]. In other instances, water can act to block sites and impede surface kinetics.

11) Metal oxide support and ligand effects.

The metal–oxygen bonds as well as the properties of the oxygen can change significantly if they are attached to a second metal such as in the M–O–S or M–O–M' systems where S refers to the support and M' refers to the second metal. Wachs and co-workers, for example, have shown that there are significant changes in activity with changes in the metal oxide support[33,48–50]. As such, the support can be thought of as a potential ligand which can be tuned to tailor the activity and selectivity of the molecular oxide surface layer. Wang and Wachs demonstrate a 12-fold increase in the methanol decomposition rate over V_2O_5 supported on silica, alumina, titania, and ceria[50]. The origin of the increase in activity is thought to be due to the ligand effect whereby the changes can be correlated with the electronegativity of the cation that corresponds to the ligand. The more electropositive cations such as Ti and Ce have substantially higher turnover frequencies than the more electronegative Al and Si cations.

In the following sections, we elaborate in more detail on how the structural and electronic properties of the oxide control its catalytic behavior by following a few example systems. We describe the applications to oxidation chemistry over reducible and non-reducible oxides, and acid chemistry via specific example catalytic systems.

5.6.2 Lessons Learned from Surface Science

Before discussing the link between the properties and performance of supported metal oxide particles, we first review how some of the factors discussed above control the chemistry on well-defined metal oxide surfaces. Surface science has provided a wealth of information which has been used to advance our understanding of the elementary processes that occur on transition metal surfaces and the factors that control them. This was discussed in detail in Chapters 2 and 3. Our understanding of metal oxides, however, is much less mature. This is due to the complexity of the structure and properties of metal oxides and the difficulty of carrying out well-defined UHV experiments that are not masked by the complexity of the system. More recent efforts have shown that many of these difficulties can be overcome. This has helped to establish much of the current interest in this area. While the fundamental surface science on metal oxides is a less mature subject, it this has become a topic of much greater interest over the past decade. The interested reader is pointed to elegant reviews on the fundamental properties of single-crystal metal oxide surfaces by Heinrich and Cox[51], Freund[43], and Barteau[35].

Here we follow the elegant analysis presented by Barteau [35] as it discusses some of the critical factors that control the activity and selectivity of metal oxides. In addition, it illustrates how the surface structure and properties can be controlled even under reducing UHV conditions in order to impart specific reactivity.

The surface structure and properties can be modulated in order to establish specific surface ensembles active for carrying out a range of different oxygenate and hydrocarbon coupling reactions. However, by specifically controlling

1) the coordination number of the surface cations
2) the oxidation states of the cations
3) the redox properties of a well-defined oxide surface

Barteau[35,52] was able to demonstrate for well-defined TiO_2 and ZnO surfaces the activity and selectivity for C2 oxygenate (carboxylates and aldehydes) and hydrocarbon (alkynes) coupling reactions over these model metal oxide surfaces under UHV conditions that is typically only seen in organometallic systems in solution.

They found that three different C–C bond formation paths could occur over different TiO_2 surfaces depending upon the nature of the surface, its properties and specific properties of these surfaces as well as the reaction conditions. The reaction paths identified included carboxylate ketonization, base-catalyzed aldol condenstation and reductive carbonyl coupling. Each reaction was found to be sensitive to the specific electronic and structural properties of the cation sites available at the surface of the oxide as well as the nature of the sites local environment. Carboxylate coupling or ketonization was shown to occur most favorably over the TiO_2 (001)-[114] faceted surface which contains a distribution of 4-, 5-, and 6-coordinate Ti sites. The high number of the coordinatively unsaturated Ti sites were thought to be responsible for the higher selectivity of the bimolecular coupling of acetate groups over their unimolecular decomposition paths. This is a stable surface that is formed at higher temperatures. Cation sites which contain pairs of vacancies were thought to be necessary in the active ensemble in order to accommodate the electron pairs that result from the coadsorption of the two ligands to the same center.

Aldol condensation, on the other hand, is favorably carried out over basic oxygen sites on the TiO_2 surface and does not appear to require coordinatively unsaturated Ti sites. The reaction is thought to proceed initially by the proton abstraction at the carbonyl carbon from the aldehyde reagent molecule by a basic oxygen site on the surface and the subsequent formation of an adsorbed enolate intermediate. This is followed by a nucleophilic attack on the adsorbed enolate by a second aldehyde reactant molecule. The coupled intermediate that forms then undergoes a dehydration in order to from the $\alpha\beta$-unsaturated aldehyde product. This occurs much more like that of an "outer-sphere" mechanism typically found in homogeneous solution-phase catalysis (discussed in more detail in Chapter 6) whereby there is a reaction between the adsorbed enolate ion and the gas phase reagent molecule (ion–molecule reaction). The reaction does not require the initial formation of a bimolecular surface ensemble. The carboxylate ketonization reaction discussed early, however, requires the explicit reaction between two surface intermediates (ion–ion reaction). This reaction they can therefore be considered to follow more of an "inner-sphere" mechanism.

The aldol condensation reaction demonstrates conversions on the order of 60% and selectivities to coupling products of 96% when carried out over the TiO_2 (001)-[114] surface. The yield is increased even further on going to the [001] faceted TiO_2 (001)-[001] surface and thus approaches those found over TiO_2 powders (run under similar conditions). The patent literature shows yields that are only slightly higher with 80% conversion and selectivities of 90%.

The third and final coupling path involves the reductive coupling of two carbonyl groups which requires the ability to facilitate a four-electron reduction process. Redox now is much more important. This reaction therefore requires the presence of reduced lower valent cations. This can be established, however, by having ensembles of lower valent cations. The mechanism is thought to involve the reductive coupling of two aldehyde reagents to form a diolate, also known as the pinacolate intermediate, which is bound to the surface through its two oxygen atoms. This intermediate subsequently dissociates to form an alkene leaving behind two oxygen atoms at the surface. XPS studies clearly reveal the presence of Ti^{+}, Ti^{+2} and Ti^{+3} oxidation states and their importance in the reaction chemistry.

5.6.3 Redox Considerations

Haber and Witko[32] developed a simple picture of redox processes based on frontier orbital concepts that were presented in Chapter 3 in order to begin to assess metal-oxide bonding and the ability of the oxide to carry out redox chemistry. Many oxidation reactions are carried out via a Mars–van Krevelen-like mechanism, which involves the activation of an organic molecule (RH) and the subsequent insertion of either a surface oxygen atom or lattice oxygen into the hydrocarbon surface intermediate. Oxygen at the surface is then replenished by the activation of O_2. This process requires that cations in oxide be able to readily change oxidation states. This can be described quite simply by the following redox couple:

$$\mathrm{RH} + \mathrm{O}^{2-} \longrightarrow \mathrm{R{-}O}^{-} + \mathrm{H}^{+} + 2\mathrm{e}^{-} \tag{5.2a}$$

$$\frac{1}{2}\,\mathrm{O}_2 + 2\mathrm{e}^{-} \longrightarrow \mathrm{O}^{2-} \tag{5.2b}$$

In order for the redox cycle depicted in Eqs. (5.2a) and (5.2b) above to proceed spontaneously, two conditions must be satisfied. First, the highest molecular orbital of the organic molecule must be above the Fermi level and aligned with electronic states at the

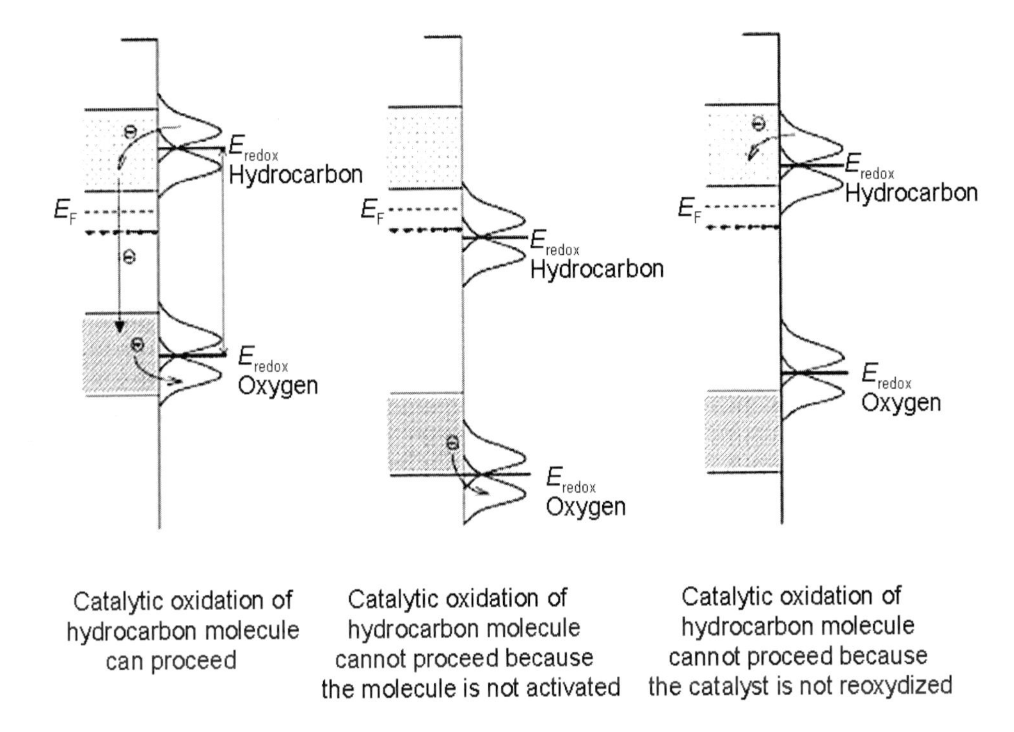

Figure 5.17. Schematic comparisons of the different electronic interactions between a gas-phase hydrocarbon molecule, oxygen and an oxide surface. A necessary, but not sufficient, requirement for a redox reaction to occur is that the hydrocarbon can donate electron density into the bottom of the conduction band which lies above the Fermi level whereas electron density can be transferred from the top of the valence band into splitting O_2 at the surface. This requires the appropriate alignment of the highest occupied orbitals from the hydrocarbon with the bottom of the conduction band and the lowest unoccupied orbitals of O_2 with the top of the valence band as in the schematic on the left. The inappropriate alignment of states is shown in the scheme at the center and in the scheme on the right hand side, both of which would prevent reaction from occurring[32].

bottom of the unoccupied conduction band so that electron density can be transferred from the organic reactant to the surface states. The second condition is that the unoccupied molecular orbitals of the O_2 molecule must sit below the Fermi level in line with occupied states in the valence band. The appropriate alignment is shown in the schematic on the left hand side of Fig. 5.17. The alignment of the states in the other two schematics is such that electron transfer will not proceed. The potentials of the conduction band can be changed by the creation of:

(a) an interface with a second metal oxide surface,
(b) addition of different valence ions,
(c) formation of defect sites, or an applied potential.

These ideas present a simple picture of how one can influence the reactivity in redox-based systems. While the picture is somewhat oversimplified, it offers a useful tool for conceptually understanding the reactivity of redox systems.

5.6.4 Bifunctional Systems

The active sites involved in the activation of various alkanes require some degree of bifunctionalty whereby the best materials appear to couple redox properties together with acid or base properties[29,34,36]. Alkane activation steps, in general, are comprised of a complex set of elementary reaction processes including C–H bond activation or hydrogen abstraction from the substrate, oxygen insertion into hydrocarbon and, in addition, oxygen migration processes. Many of these reactions also require electron transfer. The oxidation of butane to form maleic anhydride, for example, requires an overall transfer of 18 electrons. The surface chemistry, however, is still described quite locally. This suggests that the atomic ensemble that makes up the active site is rather large, or that the local environment about the active site plays an important role and the communication between the active site and its environment is rapid. In addition to redox properties, the operative surface structure can also contain cations which serve as Lewis acid sites along with O^{2-} and OH^- sites which serve as base sites. The chemistry that results from the oxygen atoms near the active site is difficult to resolve since the oxygens can take on electrophilic as well as nucleophilic characteristics depending upon the environment in which they sit. The acid characteristics of the metal cations and base characteristics of the oxygen influence both the adsorption and the activation of the hydrocarbon. In addition, the acid–base characteristics also change the product desorption rates (see also chapter 2). Lastly, acid or base sites can be formed in situ as the reaction proceeds. This was shown by Macht et al.[37] for butanol dehydration over supported-HPW structures which were reduced in the activation of the alcohol thus forming Brønsted acid sites in situ.

5.6.5 Butane Oxidation to Maleic Anhydride

Butane oxidation is the most widely discussed alkane oxidation system. This is predominantly because, to date, it is the only operative commercial process. Butane oxidation to maleic anhydride is thought to occur selectively over crystalline vanadium pyrophosphate surfaces. The mechanism by which the reaction proceeds, however, has been very difficult to elucidate owing to the complexity of VPO structure and its changes during processing. The performance of the catalyst is known to be quite sensitive to the specific synthesis procedure, the precursor phase that is formed and the activation and conditioning of the precursor. These procedures ultimately control the active phase, the specific surface structure, the surface composition and the oxidation state of the metal, all of which will influence dictate the measured catalytic activity[39,53–59]. A wide range of different V^{5+} phases which include α_I-, α_{II}-, β-, γ-, δ-$VOPO_4(2H_2O)$, V^{4+} phases which include $VOHPO_4(4H_2O)$, $VOHPO_4(0.5H_2O)$, $VO(H_2PO_4)_2$, $(VO)_2P_2O_7$, $VO(PO_3)_2$ can all form as the result of synthesis, activation or conditioning. These structures are polymorphic and can display a range of different surface facets that may be active for oxidation[55]. The active surface structure, active site and oxidation state have been found to be quite complex and are still highly debated. While the (100) surface of $(VO)_2P_2O_7$ was thought to be the reactive form, time-resolved x-ray absorption spectroscopy[54], near-edge x-ray absorption spectroscopy (NEXAFS)[56], laser Raman spectroscopy[58], ^{32}P magic angle spinning NMR[58], and other techniques have made it clear that the surface is dynamic and changes with changes in reaction composition. Coulston et al.[53,54] clearly demonstrated that under dynamic conditions the rate of maleic anyhydride formation closely follows the rate of decay of the V^{5+} signal. In situ Raman spectroscopy was also used to show that $VOPO_4$ (V^{5+}) was present under working conditions. The authors proposed that the V^{5+} centers are responsible for the initial abstraction of hydrogen from butane. The presence of radical type V^{4+} sites were suggested to be responsible for by-product formation. Subsequent studies by Birkeland et al.[58] on the steady-state analysis of butane oxidation over VPO supported on SiO_2 indicated that the working surface may contain more than one phase. The selectivity under steady-state conditions increases with increase in the P/V ratio from 1 to 2. The corresponding activity, however, decreases. They suggested that the higher oxidation state or the presence of more oxygen atoms available at the reaction site enhances the rate of overoxidation and leads to the formation of CO_2. Their results suggest that although the V^{5+} sites are active, they are unselective. The V^{4+} sites are necessary for the selective reaction to form maleic anhydride.

In addition to the V^{5+}–V^{4+} redox couple, the V^{4+}–V^{3+} couple has also been suggested to be important in the chemistry but is still actively debated. There is also evidence for the presence of V^{3+} and a suggestion that it is also active in the catalysis[61].

It is fairly well established that enrichments of the surface layer with phosphorus tends to improve the selectivity of butane to maleic anhydride as it avoids overoxidation of the intermediates that form. Increases in the P/V ratio, however, decrease the reducibility of the catalyst and thus its ability to subsequently reoxidize. The optimal balance of activity and selectivity therefore requires an optimal P/V ratio. Centi et al.[60] suggest that this should be in the range 1.0–1.1.

Although the nature of the active site is still actively debated, there are some general characteristics that tend to be agreed upon. The active surface requires at least 7 lattice oxygen atoms to carry out the hydrogen abstraction, 3 surface oxygen atoms for insertion and 14 electron transfers[40,53–55,58,59]. As such, the active ensemble may be comprised of four vanadium sites arranged in dimer pairs which are separated by pyrophosphate or phosphate ligands. The unit cell for bulk-vanadyl pyrophosphate, $(VO)_2P_2O_7$, structure is comprised of 104 atoms, 16 of which are vanadium and 16 of which are P[67].

The dynamics of the surface structure and its adaptation to changes in the gas-phase composition have considerably obscured our ability to understand the influence of oxidation states, P/V ratio and the development of structure–property relationships. The surface structure adapts to applied process conditions. The surface structure is metastable and changes with the temperature, partial pressure of oxygen and partial pressure of butane. Gai and Kourtakis used in situ environmental transmission electron microscopy to examine the surface structure of vanadium pyrophosphate under actual operating conditions and showed direct evidence for the formation of glide shear defects along with the formation of anion vacancies and the loss of oxygen[38,39]. These vacancies appear to form in the basal plane between the phosphate tetrahedra and the vanadium oxide octahedra. Despite the formation of defects, the X-ray structure appear to remain the same. These defects have been suggested to be responsible for the activation of butane[39].

The presence of water also appears to be crucial. Calculations by Neurock et al.[62] show that these glide shear planes may be the result of hydrolysis of V–OV bonds. The results show that the V_2O_2 dimer structures can dissociate along V–O bonds and form tetrahedral vanadyl fragments by way of OH formation. These surface intermediates begin to take on isolated tetrahedral vanadyl structures with the presence of defect sites. The sites may be responsible for butane activation. Water can also structurally or compositionally alter the state of the working catalyst surface. In addition, water can also aid in driving some of the more strongly adsorbed intermediates and products from the surface in order to prevent surface deactivation.

The mechanism by which butane is oxidized is still intensely debated (see discussions in references 1,8,10,11,21–24,26–30 and 63) It is clear, however, that the oxidative capabilities of the surface, Lewis acid vanadium sites and Brønsted acid sites on the surface pyrophosphate groups play a role in the overall activity and selectivity. The interplay between redox capabilities and the acid properties is important for establishing the kinetics. The presence of promoter ions such as Nb, Si, Ti, Zr and Cr have been shown to increase the local Lewis acidity and thus aid in increasing selectivity for butane activation[67,64].

Theory has been used predominantly to probe the nature of the sites on vanadium clusters and model vanadium oxide surfaces. Cluster[65–67] and periodic DFT calculations [68,69] have been carried out in order to understand the electronic and structural properties of the exposed (100) surface of $(VO)_2P_2O_7$. Both cluster and slab calculations reveal that surface vanadium sites can act as both local acid and base sites, thus enhancing the selective activation of n-butane as well as the adsorption of 1-butene. Vanadium accepts electron density from methylene carbon atoms and, thus aids in the subsequent activation of other C–H bonds. Calculations reveal that that the terminal P=O bonds lie close to the Fermi level and thus present the most nucleophilic oxygen species present at the surface for both the stoichiometric as well as phosphate-terminated surfaces. These sites may be involved in the nucleophilic activation of subsequent C–H bonds necessary in the selective oxidative conversion of butane into maleic anhydride. Full relaxation of the surface, however, tends to lead to a contraction of the terminal P=O bonds and a lengthening of the P–OV bonds. This pushes the P–OV states, initially centered on the oxygen atoms, higher in energy and thus increases their tendency to be involved in nucleophilic attack[65–69].

DFT cluster calculations indicate that the adsorption of water at higher temperatures leads to the formation of defect sites which may be the sites involved in activating the alkane[62] as discussed earlier. Water can adsorb and heterolytically activate at the V–O bridge sites, thus forming surface hydroxyl groups. This can make the vanadium sites more accessible, change the oxidation state of vanadium and/or lead to the formation of defect sites on the vanadium which would serve as adsorption sites for the alkane and ultimately aid in its activation.

5.6.6 Methanol Oxidation

Alcohols can react via a range of different paths and, in some systems, can be used as probes to distinguish acid, base and redox catalysis, as was discussed earlier. In a more industrial sense, these paths can also be manipulated by controlling catalyst features so as to enhance the selectivity to a particular path and formation of a particular product in high yield. Liu and Iglesia[70] examined the oxidation of methanol over $H_5PV_2Mo_{10}O_{40}$ (HPVMo) Keggin structures supported on ZrO_2, TiO_2, SiO_2 and Al_2O_3. Methanol was found to react catalytically via condensation, dehydrogenation and bimolecular dehydration pathways to form dimethoxymethane (DMM), methyl formate (MF) and dimethyl ether (DME), respectively, as major reaction products. The overall selectivities for which of these paths predominates was strongly influenced by the acid, base and redox properties of the Keggin structure and by the properties of the support. They found relatively high selectivities for the formation of DMM for reactions carried out over the HPVMo Keggin supported on SiO_2. The reactivity is controlled by bifunctional acid–base properties and redox centers.

Significant changes in the activity and the selectivity in this system can be made by varying the nature of the support. The use of amphoteric oxide supports such as ZrO_2 and TiO_2 catalyzed the selective formation of methyl formate, whereas the use of non-reducible weakly acidic oxides such as SiO_2 led to the formation of DMM. While high selectivity to DME were reported on Al_2O_3-supported HPVMo structures, the Al_2O_3 support also promoted the decomposition of the HPVMo structure into MoO_x and VO_x oligomers. These changes in activity and selectivity were ascribed to changes in the acid site density and reducibility of the Keggin via changes in the support. The accessibility of the protons for the supported HPVMo structures demonstrated the following order based on the oxide support.

$$SiO_2 > ZrO_2, TiO_2 >> Al_2O_3$$

The redox properties for these systems were also found to be quite sensitive the nature of the interaction between the HPVMo Keggin structure and the oxide support.

5.6.7 Isobutyric Acid Oxidative Dehydrogenation

Isobutyric acid is dehydrogenated over iron hydroxyphosphates to form methylacrylic acid. The reaction proceeds via a Mars–van Krevelen mechanism. The active phase is thought to be $Fe_2{}^{3+}Fe^{2+}(P_2O_7)_2$, which is comprised of face-sharing FeO_6 octahedral trimers. These octahedral iron oxide trimers are made up of Fe^{3+} and Fe^{2+} cation centers in the iron pyrophosphate which are thought to be the active sites. Water plays an important role in establishing hydroxylated surface sites which are necessary in carrying out the reaction. Water is thought to be necessary in converting the pyrophosphate into the hydroxylated $Fe_2{}^{3+}Fe^{2+}(PO_3OH)_4$ surface[29,71–75]. The impregnation of Pd metal into this catalyst helps to facilitate the reactivity by reducing $FePO_4$ present under the reaction conditions to the more active $Fe_2P_2O_7$ surface phase[36,76]

5.6.8 Oxidative Dehydrogenation of Propane

The oxidative dehydrogenation of propane is currently a very active area of research. Here we present just a few of the salient features as they relate to bifunctional redox and acid–base catalysis. More specifically, we discuss some of the efforts on VMgO as well as on vanadium-supported metal oxides. The oxidative dehydrogenation of propane to propylene readily occurs over the orthovanadate $Mg_3V_2O_8$, pyrovanadate $Mg_2V_2O_7$ and metavanadate MgV_2O_6 phases of VMgO. There appears to be some discrepancy about which of these phases is active in the conversion of C_3 and C_4 alkanes. Reactions with isopropanol and acetone were used to probe the relative acidity and basicity, respectively. The vanadium cations provide Lewis acidity whereas the oxygen anion sites are basic sites. The results indicate that these systems are fairly basic in character and that it is this basicity along with the specific atomic arrangement that aids the redox process in catalyzing oxidative dehydrogenation[29,36,77–79] Supported vanadium phosphates have also demonstrated reasonable performance for the catalytic oxydehydrogenation of propane. The presence of a base such as water or ammonia in this system tends to decrease propane conversion but significantly increases the selectivity to propene[29,36].

Supported vanadium oxides are also active in the activation of propane and critically depend upon the vanadium dispersion and the surface acidity, which appears to be controlled by its interaction with the support. The activation energy for C–H bond activation is thought to increase in the following order[80a]:

$$VO_x/ZrO_2 < MoO_x/ZrO_2 < WO_x/ZrO_2$$

Wachs[80a] used methanol as a probe of propane oxidative dehydrogenation as it closely mimics the activity. C–H bond activation is thought to be rate determining both for the activation of propane and for methanol. Methanol was therefore used to probe of the nature of redox, acid and base properties of the catalyst. The results indicate that the activity is quite sensitive to the metal oxide and the oxidation state. The redox activity of supported Nb, Te, V and Mo oxides and mixed metal oxides shows the following trend for decreasing reactivity:

$$V^{5+} > Mo^{6+} \gg Nb^{5+}, Te^{4+}$$

V^{5+}, Mo^{6+} and Te^{4+} demonstrate redox behavior whereas Nb^{5+} demonstrates more Lewis acid character. The results suggest that the $Mo_{1.0}V_{3.0}Te_{0.16}Nb_{0.12}O_x$ catalysts used in the conversion of propane contain both redox sites and acid sites which must be in close proximity in order to aid catalytic conversion (see also Section 2.3.5). The Nb^{5+} and Te^{4+} are thought to act as ligands that promote the activity at the V^{5+} and Mo^{6+} sites.

5.6.9 Chemical Reactivity of Reducible Oxides.

A key question in selective oxidation catalysis is whether the reactive surface-oxygen species is used to activate C–H bonds or is used, instead, to aid in the addition of oxygen to the reactive surface intermediates to form oxygenated products. We analyze the results for studies carried out over both MoO_3 or V_2O_5 surfaces.

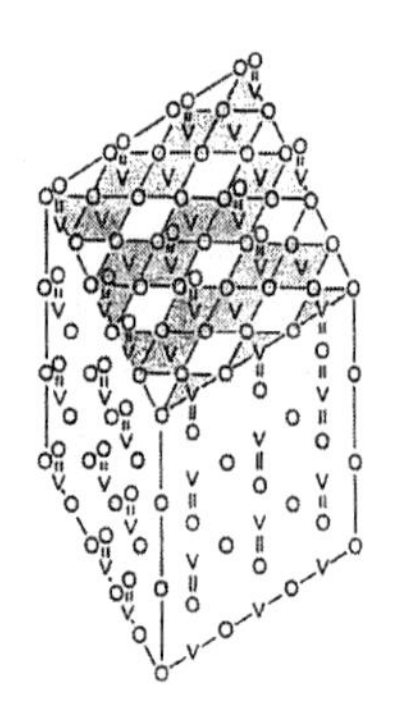

Figure 5.18. Different crystal faces of V_2O_5[80b].

The topmost layer in Fig. 5.18 is made up of the neutral coordinatively saturated cations in the (001) surface for the layered V_2O_5 bulk material. In contrast, the (110) surface is formed by the cleavage of covalent V–O bonds at the surface, thus resulting in coordinatively unsaturated V and O atoms at its side faces as shown in Fig. 5.18. The less reactive (001) surface is thought to be responsible for the selective oxidation of hydrocarbons such as toluene, whereas the reactive (110) surface predominantly results in total oxidation. This surface is also easily hydroxylated and, hence, contains both Brønsted acid and base sites.

Two different types of oxygen can be distinguished on the (001) surface. The first refers to the terminal vanadyl oxygen sites whereas the latter refers bridging oxygen atoms that connect either two or three vanadium centers. Quantum-chemical calculations indicate that there is a higher electron density on the O atoms which have the greatest number of cation neighbors. The electron density on the oxygen atom is thus increased as we increase the number of electron-donating vanadium atom neighbors. The charge density on the bridging oxygen atoms is therefore found to be significantly higher than that on the terminal vanadyl oxygen atoms.

Ab initio quantum mechanical results [81] show that hydrogen addition to the terminal oxygen of the V=O bond is preferred over hydrogen addition to the oxygen in the V–O–V bridge. This agrees with ion charge excess-based estimates. As we discussed in the previous section, the ion charge excess is larger on the single coordinated hydroxyl than towards a bridging-oxygen atom. The proton attached to the bridging-oxygen atom would, of course, be more acidic and, hence, less strongly bound.

Interestingly, the bridging-oxygen atom rather than the terminal (end-on) oxygen atom is thought to be more favorable for its insertion into C–H bonds. Oxygen insertion is controlled by the more weakly bonded oxygen atom. The vanadyl or molybdenyl oxygen bonds appear to be stabilized by strong donation of electrons from the 2p oxygen orbitals into empty cation d-valence orbitals. On the MoO_x surface, the Mo≡O bond order is close to three, as in the CO molecule[82].

The Mo≡O bond is so strong that even hydrogen abstraction from a hydrocarbon does not tend to occur readily. The V=O bond, however, is significantly weaker, and can thus stabilize the proton to form an acidic hydroxyl intermediate. The energy for the removal of the oxygen atoms that bridge between the cationic vanadium or molybdenum centers is compensated for by a reordering of the Mo- or V-containing (half) octahedra. Sharing of apices is replaced by sharing of edges. Sheared phases of the oxides can then be formed. For a general introduction to the mechanisms that govern heterogeneous catalyzed selective oxidation processes, we refer the interested reader to reviews by Grasselli [83a] and Vedrine[83b].

5.6.10 Selective Catalytic Reduction of NO with NH_3

As a final example in selective oxidation, we summarize the current understanding of the mechanism involved in the removal of NO with NH_3 over vanadium oxide[84]. The reaction is important for the selective catalytic reduction (SCR) processes which target reducing vehicular emissions of combustion systems rich in air.

The terminal V=O groups of vanadium oxide appear to be important in carrying out this chemistry since they are the energetically favored sites, and in addition, are in positions that are accessible for protons to transfer to in the formation of Brønsted acid sites. Ammonia can readily adsorb on these sites to form an ammonium intermediate, which subsequently reacts with coadsorbed NO. Figure 5.19 shows the formation of $NH_4{}^+$ on a $V_4O_{16}H$ cluster used to model the V_2O_5 surface. DFT-cluster calculations indicate that the energy required to protonate NH_3 here is 110 kJ/mol. The positively charged $NH_4{}^+$ intermediate is stabilized by the negative charges which are distributed over the two coordinating V=O groups.

The reaction proceeds by the coadsorption of NO and $NH_4{}^+$, which subsequently react to the form the adsorbed $[NH_3\text{–}NHO]^+$ intermediate. The N–H bond in $NH_4{}^+$ is broken as it reacts with NO. The proton which transfers from the $NH_4{}^+$ to the V_2O_5 surface during this initial activation step is subsequently transferred back to the NO molecule from the V_2O_5 surface as shown in Fig. 5.19. Ultimately this proton transfers back again to a terminal V=O group together with a second proton, which attaches itself to neighboring V=O group. This produces gas-phase NH_2NO, which can be further converted by consecutive reactions over vanadium oxide. The surface reaction energy schemes for these

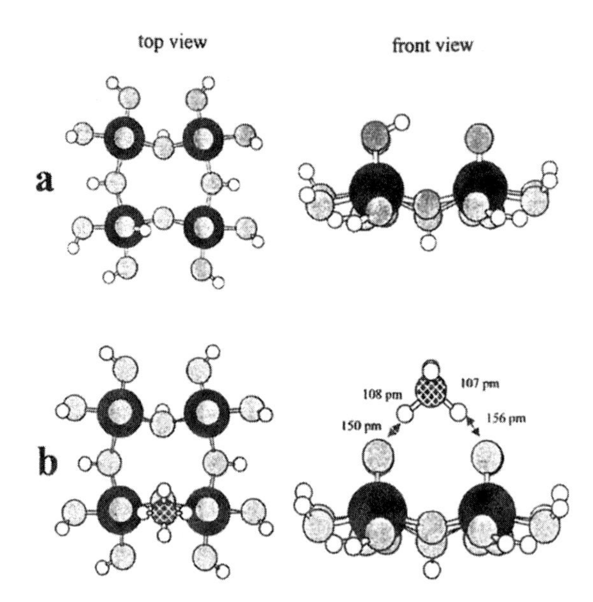

Figure 5.19a. Top and front view of (a) the V_4 cluster used in this study and (b) the V_4 cluster with adsorbed NH_4. The following shading scheme is used in this figure: H, white; O, gray; N, gray with cross hatching; V, black[84].

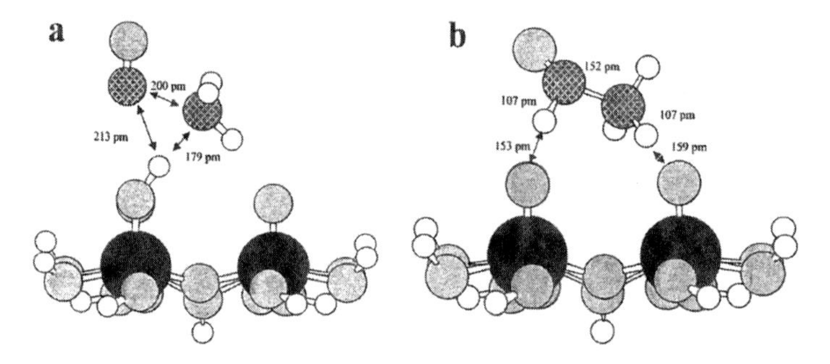

Figure 5.19b. (a) Approximate transition state in the reaction of NO with NH_4 and (b) the NH_3NHO species formed in this reaction. The specific atoms are labeled by their shading in the caption for Fig. 5.14[84].

paths are given in Fig. 5.21. The uniqueness of the NO reduction reaction with NH_3 is that it can occur in the presence of excess O_2. The thermal de-NO_x process, without catalyst, also proceeds via an intermediate formation of NH_2. While the $ONNH_2$ formation reaction is relatively facile at higher temperatures, the corresponding reaction of N_2 with O_2 is ten orders of magnitude slower.

As observed from Fig. 5.21, the NH cleavage reaction is the rate-limiting step that occurs on the surface in the formation of NH_2NO. This reaction proceeds after a succession of isomerization steps, which are assisted by the presence of acidic VOH groups. The overall calculated activation barrier for this reaction is close to that of the NH_2NO formation reaction. The preference of the catalytic reaction over the gas-phase reaction

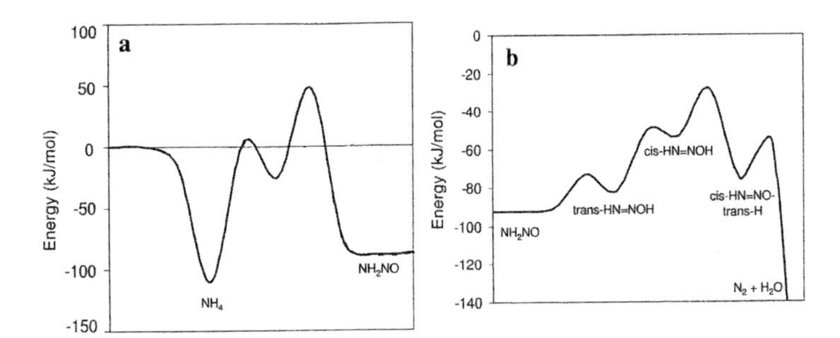

Figure 5.20. Potential energy profiles of (a) reaction of NH_3 with NO to form NH_2NO over the V_4 cluster and (b) decomposition of NH_2NO over the V_4 cluster[84].

is likely due to the low reaction barrier of the second elementary reaction sequence in which, the barrier for internal hydrogen transfer was substantially lowered.

The energetics predicted may depend on the cluster size used to model this system or on the actual particle size. For oxidative dehydrogenation of alkanes catalyzed by vanadia, it has been shown that the activity per vanadium atom increases with increasing size of the vanadium particle. This is due to reduced electron transfer of oxygen to vanadium on the smaller particle and, hence, to a the lower reducibility of the smaller nano-sized particles[85]. The lower coordination number of vanadium implies that an increase in charge is less easily accommodated on the smaller particle.

5.6.11 Oxidation by Non-Reducible Oxides

In addition to the wide range of metal oxide catalysts that can carry out oxidation via redox catalysis, there are a host of other materials that can carry out oxidation over non-reducible metal oxides. The oxidation mechanisms over non-reducible metal oxides are quite different and typically involve the production of free radical intermediates. The mechanisms tend to contain both heterogeneous and homogeneous activation and functionality. The oxide is used to activate a free radical process that can then proceed in the gas phase or at the surface. Li-substituted MgO and the rare earth metal oxides are two classes of materials that are considered non-reducible oxidation catalysts. Here we will specifically focus on the activation of alkanes over non-reducible metal oxides.

A wealth of papers have been published on the activity of Li-substituted MgO for the activation of alkanes[87–95]. The defect sites which form as the result of charge imbalance upon doping different metal ions into the host MgO framework lead to the formation of O^- sites or sites for O^{2-} which can readily abstract hydrogen. These sites are stabilized by the electronic structure of the host oxide and changes in the host due to metal atom doping. While much is known experimentally about methane activation over different non-reducible oxides[87–95], there have been relatively fewer theoretical studies. The activation of methane over Li/MgO appears to be controlled by the presence of oxygen vacancies that form upon substitution of Li into MgO matrix. Leveles et al.[87], for example, demonstrated a strong correlation between oxygen removal and propane activation. Theoretical studies by the groups of Catlow[96], Gillan[97], Truong[98] and others[99–101] demonstrate that the O^- site that forms as part of the $[Li^+–O^-]$ pair is the active oxygen species in breaking the C–H bond of methane. The Li dopant leads to a local hole state which gives rise to the formation of this active O^- site. There is considerable structural relaxation at this site in order to stabilize it.

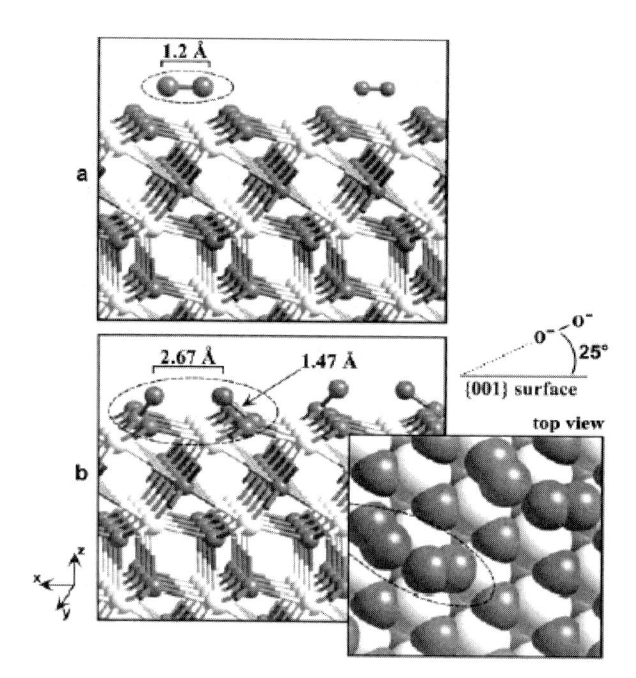

Figure 5.21. The formation of surface peroxide ($O_2{}^-$) sites on La_2O_3 which are thought to be active for methane activation.

The reactivity of rare earthmetal oxides show many features similar to those of Li-substituted MgO[108–112]. The activity for the oxidation of ethane increases experimentally in the following order over the lanthanides[108]:

$$La_2O_3 > Sm_2O_3 \gg Pr_6O_{11} \sim CeO_2$$

The presence of redox functionality in these materials tends to reduce the selectivity to the alkene product.

Theoretical results by Palmer et al. discussed earlier in Chapter 2, indicate that the active surface oxygen species are $O_2{}^-$ intermediates that form as the result of O_2 adsorption at anion vacancies or by the activation of O_2(g) over the oxide surface[86,103]. The formation of the $O_2{}^-$ surface complex is shown in Fig. 5.21. The formation of anion defects is not likely since a considerable amount of energy is necessary in order to active the La–O bond in forming the vacancy. The theoretical results indicate that surface peroxides are more likely responsible[86,103]. This is consistent with various experimental results such as those reported by Sinev et al.[90] and Otsuka et al.[104,105], who showed that simple peroxides such as Na_2O_2, BaO and SrO_2 are able to activate methane, and Lundsford who identified the $O_2{}^-$ intermediates by in situ EPR and Raman measurements[106,107]. Ab initio calculations suggest that the surface oxygen site that is formed by the direct Sr^{2+}/La^{3+} exchange mimics the O^- site, which is very active for methane activation[86]. The doping effects may be similar to those found in the Li-substituted Mg discussed above.

5.7 Heterogeneous Sulfide Catalysts

5.7.1 Introduction

The classical catalysts used in hydrotreating heavy oils and residua in petroleum processing consist of MoS_2 or WS_2 promoted with Ni or Co ions. These catalysts have been developed for hydrodesulfurization and hydrodenitrogenation of heavy oils or coal liquids which involves the hydrogenation of the unsaturated aromatic rings coupled with the hydrogenolysis of the C–S and C-N bonds of refractive nitrogen- and sulfur-containing molecules. Sulfur and nitrogen are subsequently removed as H_2S and NH_3, respectively. Even if reducible oxides are used as the initial catalyst, they will ultimately go on to form the more stable sulfide phases under reaction conditions. These phases are active for hydrodesulfurization.

The catalytically active materials are typically present as relatively small nanoparticles distributed over a porous Al_2O_3 support. The shape and size of the nanoparticles that form are strongly dependent on gas-phase conditions. The size, shape and morphology of the particle dictate the surfaces that are exposed. The nature of the exposed surface along with the particle size can critically impact catalytic performance. The particle shape and morphology also depend on the presence of promoting cations. The effect of promoters on the reactivity at the sulfide interface is only just beginning to be understood. Recent ab initio calculations together with well-defined model experiments are providing key insights into the active surface structure and sites under different conditions. This has allowed for a more detailed understanding on the role of surface structure and composition on catalytic behavior. This will be discussed later in this chapter, where we will try to answer the question: "What makes the promoter system used so unique?".

Finally, it is important to describe the effects of H_2S on the reactivity of different surface sulfide phases. The influence of H_2S can be four-fold. It can react with an oxide or transition metal thus converting it to the sulfide:

$$M_xO_y + yH_2S \longrightarrow M_xS_y + yH_2O$$

or

$$M_x + yH_2S \longrightarrow M_xS_y + yH_2$$

On non-reducible oxides, H_2S can exchange with surface hydroxyl groups.

$$\text{M–O–M(OH)–OM} + H_2S \longrightarrow \text{M–OM(SH)–OM} + H_2O$$

The S–H bond is weaker than the O–H bond, which should increase the acidity of the sulfide. H_2S can also create acidic sulfhydryl groups via dissociative adsorption:

$$\text{M–}\overset{\square}{\text{S}}\text{–M–S–M} + H_2S \longrightarrow \text{M}\overset{\text{H}}{\overset{|}{\text{S}}}\text{M–}\overset{\text{H}}{\overset{|}{\text{S}}}\text{–M}$$

and finally it can adsorb as molecular H_2S on Lewis acidic surface sites.

$$\text{M–S–M–S–M} + H_2S \longrightarrow \text{M–S–}\overset{H_2S}{\overset{\vdots}{\text{M}}}\text{–S–M}$$

We will encounter these three different influences on the reactivity that result from the presence of H_2S.

5.7.2 The Sulfide Surface

The most widely studied sulfide surface is that of MoS_2. A short review of the most salient features of the MoS_2 surface is presented here. An important question that illustrates many aspects of the sulfide surface is that of the shape and morphology of the particles that form when the particles are dispersed on an inert support. Figure 5.22 shows that there are predominantly two different types of reactive surfaces present.

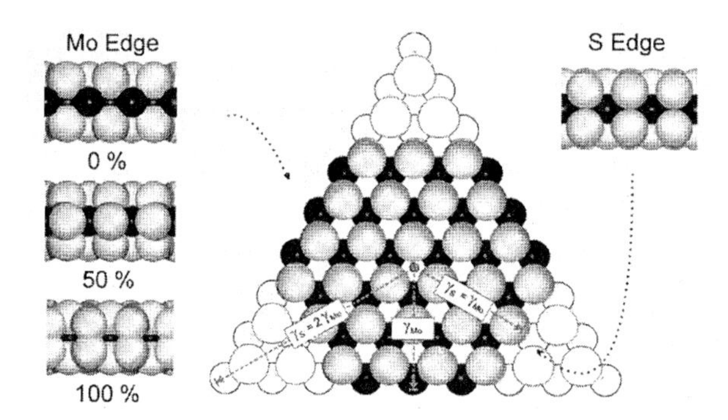

Figure 5.22. Atomic ball model (top view) showing a hypothetical, bulk-truncated MoS_2 hexagon exposing the two types of low-index edges, the S edges and Mo edges. The Mo atoms (dark) at the Mo edge are coordinated to only four S atoms (light). To the left, the stripped Mo edge is shown in a side view together with two more stable configurations with S adsorbed in positions predicted from theory; the 50% covered (monomer) and a 100% covered Mo edge (dimer). To the right is shown a side view of the S edge with a full coordination of six sulfurs per Mo atom. Plotted on the MoS_2 hexagons are vectors with length corresponding to the edge free energies γ_{Mo} for the $(10\bar{1}0)$ Mo edge and γ_S edge. The envelope of tangent lines drawn at the end of each such vector constructs a hexagon if γ_S equals γ_{Mo}. If $\gamma_S > 2 \times \gamma_{Mo}$ the result is a triangle (outlined shape) terminated exclusively by the Mo edge, or vice versa for the S edge. Intermediate values result in clusters with a hexagonal symmetry[113].

MoS_2 consists of layers in which Mo cations, with formal ionic charge +4, are sandwiched between layers of sulfur atoms. Each Mo atom is prismatically surrounded by six sulfur atoms, each of which has three Mo atoms as neighbors. The interaction between the MoS_2 layers is controlled by the weak van der Waals interactions and, hence, the MoS_2 particles are slabs terminated by the coordinatively saturated sulfur atoms (see the central part of Fig. 5.22). We examine here the (100) surface.

The sulfided (100) surface is unreactive. Two different reactive surfaces are formed if the MoS_2 is cut perpendicular to the surface plane. The first surface from such a cut is terminated with Mo atoms (Mo edge), whereas the second surface is S terminated (S edge). These are polar surfaces, one contains excess positive charge (Mo edge), the other excess negative charge (S edge). At the Mo edge, the Pauling ion charge excess of Mo is $e^+ = +\frac{4}{3}$ with no negative charge excess on S. At the S edge, the S ion however has a charge excess $e^- = -\frac{2}{3}$.

In Section 5.2, we discussed the fact that such polar surfaces are unstable and tend to reconstruct. The MoS_2 edges become stabilized in vacuum because charge transfer between Mo and S is reduced. The Mo atoms at the Mo edge reduce to a state close to Mo^{3+}. The sulfur atoms at the S edge reduce to a state close to S^-, a state as we will see which is highly reactive for hydrogen. Exposed to a mixture of H_2S and H_2, this surface becomes covered with SH^- groups.

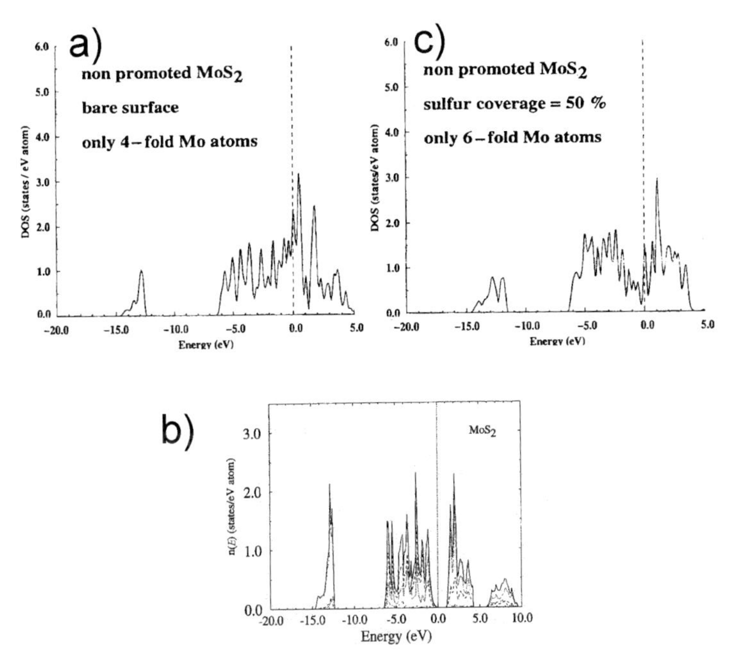

Figure 5.23. Local densities of states projected on the Mo sites (a) Mo edge 0% S, (b) bulk MoS_2[24], the dark lines here refer to Mo DOS, (c) Mo edge 50% S[117b].

This reduction of MoS_2 at the Mo edge follows immediately from a comparison of the electron density of states of bulk MoS_2 (Fig. 5.23b) and that on the atoms at the edge (Fig. 5.23a). In the density of states (DOS) plot for bulk MoS_2, we note a gap at the Fermi level (0.0 eV). The DOS below the Fermi level consists of electrons mainly localized on sulfur and the DOS above the Fermi level is located on Mo. The gap at the Fermi level implies that the system is an insulator with no conductivity. This is consistent with a description in which Mo has a formal charge of +4 and sulfur a formal charge of −2 Figure 5.23a shows a very different situation around the Fermi level; the highest occupied orbitals are now located at the MoS_2 edge. The gap is no longer present, consistent with a conductive material implying that Mo is reduced. This reduction in charge is consistent with the reduction in the number of S neighbors from six to four, implying less electron donation from Mo atoms to sulfur atoms.

A similar conclusion is drawn when one considers the system to be completely ionic (a poor description) and computes the Madelung potentials at the Mo cation in bulk MoS_2 compared with that on the surface. The decrease from six negatively charged neighbors to four implies a decrease in the relative stability of positive charge on the Mo atoms and, hence less of a driving force for electron donation from Mo to sulfur.

This reduction of charge is a general effect that occurs on many reducible oxides or sulfides.

As illustrated in the center part of Fig. 5.22, the shape of the particle is triangular, terminated with solely Mo or S edges, and is hexagonal when the surface energies of the two edges are similar. The morphology of the particle is determined by Wulff[116] construction according to the law

$$\frac{\gamma(\text{kke})}{d(\text{kke})} = c$$

The distance d of a surface with respect to the center of a crystal is proportional to its surface energy. Hence surfaces with large surface energies are only present at the crystal surface to a small extent. The planes with low surface energies tend to dominate.

Hence, when one of the surface energies changes (as expected in the presence of an H_2S/H_2 atmosphere), the particle shape will begin to deviate. At the reduced Mo edge, H_2S will react and S atoms will attach as shown in Fig. 5.22 (left) (Mo edge at 50% S coverage).

$$\text{Mo}\square + \text{H}_2\text{S} \longrightarrow \text{MoS} + \text{H}_2$$

The S atoms adsorb between two edge Mo atoms. The reaction energy for H_2S decomposition at the Mo edge remains exothermic up to the point where the S coverage reaches 50%. Each Mo center is once again six-fold coordinated to Mo and has a formal 4+ charge [see Fig. 5.22(left)]. The electron energy gap shown in Fig. 5.13b reappears at the Fermi level.

An extensive set of calculations have been carried out to produce the phase diagrams shown in Fig. 5.24 for the Mo and S edges as a function of H_2S and H_2 pressure, expressed as relative chemical potentials[24].

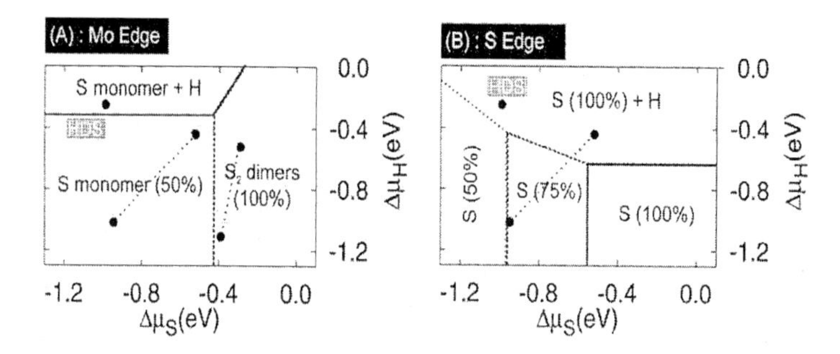

Figure 5.24. Phase diagrams for (A) the Mo edge and (B) the S edge with regions indicating the stable configurations for realistic values of $\Delta\mu_S$ and $\Delta\mu_H$. The chemical potentials have been normalized with respect to the energy of bulk sulfur and half the energy of a single, isolated H_2 molecule. Typical HDS conditions are indicated in the figures[113].

The chemical potential of hydrogen is defined with respect to P_{H_2}. $\Delta\mu_H < -0.6$ eV implies that no hydrogen is present. The chemical potential of H_2S is determined with respect to the chemical potential of S in MoS_2. $\Delta\mu_S = 0$ implies a high H_2S partial pressure.

We note from the phase diagram that in the presence of hydrogen, the S edge remains 100% saturated over a wide hydrogen pressure interval, but is covered with SH groups. The Mo edge (50%) is more difficult to hydrogenate. At a maximum, half of the S atoms added to this edge then contain SH groups. Because the surface energies of these edges are different, the shape of the particle differs from the ideal hexagon, see Fig. 5.25.

The conclusion from these studies is that under practical sulfide catalysis conditions, both the Mo edge and the S edge contain coordinatively saturated Mo atoms. The bright rim of white spots in the STM picture (Fig. 5.25) near the sulfide particle edge is indicative

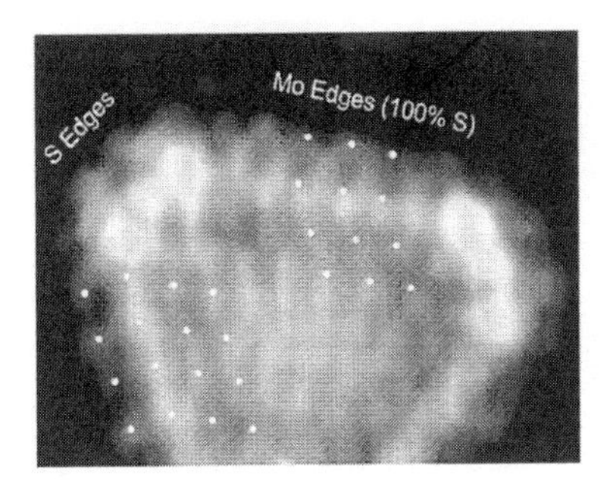

Figure 5.25. An atom-resolved STM image of a typical single-layer MoS_2 hexagon exposed to H_2S gas at 573 K. The white dots indicate the registry at the longer (Mo) edges. The registry of edge protrusions and the bright rim reveals that the Mo edges have become resulfided to the fully saturated S_2 dimer configuration[113].

of a high electron density at the Fermi level on the Mo atoms at the edge. The Mo atoms are partially reduced. Lauritsen et al.[118] have shown that the electronic state at these Mo-edge atoms is metallic, forming a one-dimensional chain. In contrast to the interaction with bulk Mo atoms in the basal plane, thiophene interacts weakly with rim Mo atoms, notwithstanding these full coordinative saturation with S atoms. Below 200 K thiophene adsorbs in a flat five-coordinate η_5 adsorption geometry through its π-electrons parallel to the plane of Mo atoms. When hydrogen is coadsorbed to the edge S-atoms, exposure of thiophene at 500 K leads to partial hydrogenation of thiophene and cleavage of a C–S bond. Theoretical calculations are in agreement with STM data and show that *cis*-but-2-enethiolate is formed. The hydrogen molecule will not dissociate under these conditions at the rim sites. In the experiment, hydrogen atoms had to be produced by dissociating hydrogen on a glowing tungsten filament.

5.7.3 Promoted Sulfide Catalysts

On an atomic level, the mechanistic basis to the promoting action of divalent reducible cations such as Co^{2+} or Ni^{2+} can best be illustrated by discussing in detail the H_2/D_2 exchange reaction on promoted and non-promoted sulfide systems[118]. Reactivity aspects of the terminated MoS_2 surface have been investigated by several research groups[117–119]. We will limit our discussion to the Co^{2+}/MoS_2 system.

The Mo edge (50%) and the Co/Mo edge in which part of the center Mo atoms have been substituted by Co are shown in Fig. 5.26. Both edges are in equilibrium with a mixture of H_2S and H_2 of the same partial pressure. Whereas the non-promoted Mo edge (50%) contains only coordinatively saturated Mo, one notes that under the same condition the Co ions have a maximum coordination of five S atoms. This implies that a Co ion and a neighboring Mo cation remain coordinatively unsaturated. This can be readily understood once one realizes that Co^{2+} substitution of Mo^{4+} is not a charge-neutral substitution.

A charge-neutral substitution occurs when an $[Mo^{4+}S^{2-}]^{2+}$ unit substitutes for Co^{2+} at the Mo edge, then Co^{2+} becomes directly accessible to S. Its dominant surface state has a coordination number of 5, in line with the expectation based on the charge neutrality argument. The activation of H_2 on a coordinatively unsaturated edge of Mo or Co is energetically very different. Figure 5.26 illustrates adsorption modes for dissociatively adsorbed H_2. Adsorption is heterolytic. One hydrogen atom adsorbs on the cation and the other on S. The reactivity of the S atom depends strongly on whether it is adsorbed

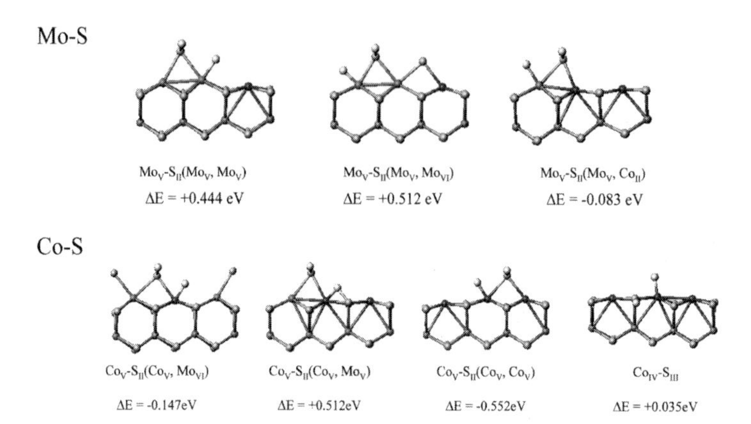

Figure 5.26. Hydrogen adsorption on metal–sulfur pairs. Adapted from A. Travert et al.[119].

between two surface atoms, between two Co atoms or between an Mo and a Co atom. Adsorption is exothermic to Co and a S atom between two Co ions, or to a Co atom and an S atom between Mo and Co. Adsorption is endothermic to an Mo and an S atom between two Mo atoms. The calculation of activation barriers for the dissociative adsorption of H_2 gives the following result:

$$E_{\text{act}}^{\text{H}_2}\left(\text{Co/S}\right) = 34\,\text{kJ/mol} \qquad E_{\text{act}}^{\text{recomb}}\left(\text{Co/S}\right) = 100\text{kJ/mol}$$

$$E_{\text{act}}^{\text{H}_2}\left(\text{Mo/S}\right) = 80\text{kJ/mol} \qquad E_{\text{act}}^{\text{recomb}}\left(\text{Mo/S}\right) = 20\text{kJ/mol}$$

The activation energy for H diffusion is 30 kJ/mol. One notes that the activation energy for H_2 dissociation on Mo is higher than the activation energy for recombination. On the Co site, this situation is reversed. On Co the rate of hydrogen atom recombination is rate limiting. Since the activation energy for diffusion is low, these facts are consistent with the experimental observation that on MoS_2 the H_2/D_2 exchange reaction rate is first order in H_2 or D_2 pressure, whereas on Co/MoS_2 it is only half order[120].

The increased reactivity of the CoS surface site stems mainly from the difference in reactivity of an S atom coordinated between two Mo atoms as MoSMo compared with S between two Co atoms or Co and Mo. This is due to the weaker CoS bond energy which increases the SH bond of S bonded to Co instead of Mo.

Under conditions where desulfurization occurs at finite H_2S/H_2 ratio, the surface of MoS_2 will be covered with S and not contain any vacancies, whereas on the promoted catalyst the presence of vacancies to adsorb a hydrocarbon will be present.

A comparison of the kinetics of a model desulfurization reaction:

$$\text{thiophene} + 2\text{H}_2 \longrightarrow \text{H}_2\text{S} + \text{butadiene}$$

catalyzed by different metals is shown in Fig. 5.27[121].

Kinetic data are given in Table 5.5. The catalysts are sulfide metal-containing particles dispersed on a carbon support. The rates are normalized on metal content. As can be

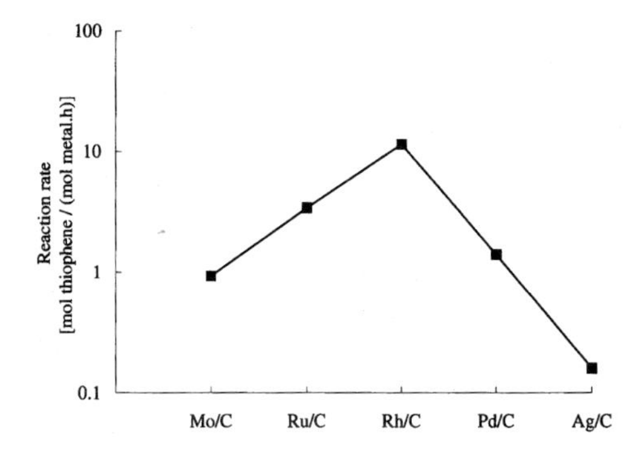

Figure 5.27. Thiophene HDS activity for the different carbon-supported transition-metal sulfides under standard conditions (3.33 kPa thiophene, 1 kPa H_2S, and $T = 573$ K). Adapted from E.J.M. Hensen[121].

deduced from the orders for this reaction shown in Table 5.5, the coverage with reactive S increases in the sequence Mo, Ru, Rh, Pd and Ag. A high reaction order in H_2S implies a low coverage with reactive sulfur and a low reaction order in H_2S implies a high coverage with reactive sulfur. Thiophene and H_2S compete for surface vacancies and, hence, between one another. In order to activate thiophene it has to adsorb with its sulfur atom on a metal surface atom. A high reaction order in thiophene implies that it does not easily compete with H_2S adsorption or decomposition, and a low reaction order implies easy competition. One notes for the systems of maximum reactivity (the RhS system has a similar reactivity to the Co/MoS_2 system) that the order of thiophene is a minimum, implying that for that system competition with H_2S is optimum. This is consistent with earlier discussed equilibrium surface states of MoS_2 and Co/MoS_2.

Table 5.5. Reaction orders of thiophene (n_T), H_2S (n_S) and H_2 (n_H) under different conditions[a][121]

Catalyst	T = 573 K					T = 623 K		
	n_T^b	n_T^c	n_S	n_H^b	n_H^c	n_T	n_S	n_H^b
Mo/C	0.40	0.50	–0.32	0.54	0.57	0.65	–0.34	0.74
Ru/C	0.28	0.39	–0.25	0.56	0.53	0.57	–0.27	0.93
Rh/C	0.21	0.31	–0.83	0.71	0.93	0.53	–0.59	1.03
Pd/C	0.50	0.65	–1.04	0.77	0.99	0.77	–0.97	1.42
CoMo/C	0.10	0.12	–0.46	0.61	0.78	0.28	–0.30	0.92

[a] 95% confidence interval for n_T ±0.05, n_s ±0.07, n_H ±0.02.
[b] Inlet H_2S partial pressure: 0 kPa.
[c] Inlet H_2S partial pressure: 1 kPa.

Whereas, the Mo (edge) does not contain vacancies, the Co-promoted system does and can therefore be covered to a high extent with reaction intermediates under the same reaction conditions.

5.8 Summary

Quantum chemistry has reached the state where the reactivity of well-defined metal oxide and metal sulfide surfaces can be probed with reliable accuracy and chemical relevance. This becomes particularly clear in the first section of this chapter, where we analyzed the surface structures of acidic, covalent and ionic materials and their interaction with probe molecules. First-principle quantum mechanical results can readily be analyzed and understood in terms of semi-classical models. Semi-classical models such as the Pauling charge excess model provide an easy way in which to examine electrostatic effects on surface acidity or basicity. The metal surface cations are analogous to those found in organometallic clusters. They can therefore be analyzed using essentially the same time-honored methods and theoretical frameworks that have been developed for organometallic systems with little modification. Isolable molecular orbital analyses were used herein to demonstrate how changes in the surface hybridization of atomic orbitals influence the general catalytic activity.

The semi-empirical models nicely demonstrate the differences between various different oxide surfaces. Systems such as MgO in which differences in the energies of adsorption sites are dominated by electrostatic potential differences were explicitly distinguished from more covalent oxide systems such as alumina or silica surfaces. In addition, the models help to understand why surface reconstruction is often important when Lewis acid and base centers are converted into their corresponding Brønsted base and acid sites.

As was discussed more generally in Chapter 2, the influence of the reaction environment can significantly influence the intrinsic catalytic kinetics for both metal oxides and metal sulfides. Medium effects were shown in this chapter to be extremely important in comparing gas-phase acidity with the acidity in solvents, and in addition, solid acidity with acidity in solution.

One of the greatest advances that theory has made over the past decade has been its ability to examine the sensitivity of the state of the working surface to changes in reaction conditions and surface structure. This has required the ability to integrate ab initio-derived thermodynamic and kinetic results into phase equilibrium as well as atomistic kinetic simulations. The oxide and sulfide surfaces are sufficiently stable that useful studies can be carried out. CO oxidation catalyzed by RuO_2 demonstrated that the maximum turnover rate occurs under conditions where the surface is in a disordered state at the boundaries of two phases, one of which is completely covered with oxygen adatoms and a second which is partially covered with CO.

In a similar way, theory has been used not only to establish the nature of active sites on the surface but, in addition, to map out a full phase diagram for MoS_2 as a function of the H_2/H_2S ratio. The results provided the ability to predict the state of the surface at specific conditions and to establish the region of phase space. This can ultimately be used in order to aid in the design of optimal operating conditions.

Mixed oxides and sulfides are very often used as catalysts to carry out a range of different reactions. On the sulfides it has become clear that the use of cations with different valencies creates surface vacancies not present on the non-promoted systems under the reaction conditions.

References

1. H. Freund, *Angew. Chem. Int. Ed. Engl.* 36, 452 (1997)
2. L. Pauling, *The Nature of the Chemical Bond*, Oxford University Press, Oxford (1950);
   R.A. van Santen, *Theoretical Heterogeneous Catalysis*, World Scientific, Singapore (1991)
3. R. Refsin, R.A. Wogelius, D.G. Fraser, M.C. Payne, M.H. Lee, V. Milman, *Phys. Rev. B*, 52, 10823 (1995);
   W. Langel, M. Parinello, *J. Chem. Phys.* 103, 3240 (1995)
4. M. Digne, P. Sautet, P. Raybaud, P. Euzen, H. Toulhoat, *J. Catal.* 211, 1 (2002)
5. A.A. Tsychanenko, V.N. Filimonov, *J. Mol. Struct.* 19, 579 (1972)
6. (a) C. Arrouvel, M. Digne, M. Breysse, H. Toulhoat, P. Raybaud, *J. Catal.* 222, 152 (2004);
   (b) P.J.D. Lindan, N.M. Harrison, J.M. Holender, M.J. Gillan, *Chem. Phys. Lett.* 261, 246 (1991)
7. T.W. Dijkstra, R. Duchateau, R.A. van Santen, A. Meetsma, G.P.A. Gap, *J. Am. Chem. Soc.* 124, 9856 (2002)
8. G.Pacchioni, *Surf. Rev. Lett.* 7, 277 (2000)
9. J. Sauer, P. Ugliengo, E. Garrone, V.R. Saunders, *Chem. Rev.* 94, 2096 (1994)
10. L.G.M. Petterson, M. Nyberg, J.L. Pascual, M.A. Nygren, in *Chemisorption and Reactivity on Supported Clusters and Thin Films*, R.M. Lambert and G. Pacchioni (eds.), NATO ASI Series E, Vol. 331 p. 425, Kluwer, Dordrecht (1997)
11. R. Wichtendahl, M. Rodriguez–Rodrigo, U. Härtel, H. Kuhlenbeck, H.J. Freund, *Surf. Sci.* 423, 90 (1999)
12. G. Pacchioni, J.M. Ricart, F. Illas, *J. Am. Chem. Soc.* 116, 10152 (1994);
    G. Pacchioni, A. Clotet, J.M. Ricart, *Surf. Sci.* 315, 337 (1994)
13. A.P. Seitsonen, Y.D. Kim M. Knapp, S. Wendt, H. Over, *Phys. Rev. B*, 65, 35413 (2001)
14. L. Zang, H. Kisch, *Angew. Chem. Int. Ed.* 112, 4075 (2000)
15. Y.D. Kim, S. Schwegmann, A.P. Seitsonen, H. Over, *J. Phys. Chem. B*, 105, 2205 (2001)
16. a) K. Reuter, M. Scheffer, *Phys. Rev. Lett.* 90, 46103 (2003);
    b) K. Reuter, D. Frenkel, M. Scheffer, *Phys. Rev. Lett.* 93, 116105 (2004)
17. T.A. Albright, J.K. Burdett, M.H. Whangbo, *Orbital Interactions in Chemistry*, Wiley, New York, p. 404 (1985)
18. D.F. Shriver, P.W. Atkins, C.H. Langford, *Inorganic Chemistry*, Oxford, Oxford University Press, p. 248 (1994)
19. C.J. Ballhausen, *Introduction to Ligand Field Theory*, McGraw-Hill, New York (1962)
20. D.A. Dowden, D. Wells, *Actes 2. Congress Int. Catalysis*, Paris, p. 1499 (1961);
    D.A. Dowden, *Endeavor*, 24, 69 (1965)
21. V.B. Kazansky, *Top. Catal.*, 11–12, 55 (2000)
22. D.A. Zhurko, M.V. Frash, V.B. Kazansky, *Catal. Lett.* 55, 7 (1998)
23. V.B. Kazansky, *Catal. Rev.* 43, 199 (2001)
24. T. Okuhara, N. Mizuno, *Adv. Catal.* 41. 113 (1996)
25. M.J. Janik, R.J. Davis, M. Neurock, *J. Phys. Chem. B*, 108, 12292 (2004)

26. M.J. Janik, K.A. Campbell, B.B. Bardin, R.J. Davis, M. Neurock, *Appl. Catal. A*, 256, 51 (2003)
27. B.B. Bardin, R.J. Davis, M. Neurock, *J. Phys. Chem. B*, 104, 3556 (2000)
28. M. Janik, R.J. Davis, M. Neurock, *J. Catal.* to be submitted
29. J.C. Vedrine, J.M.M. Millet, J.C. Volta, *Catal. Today,* 32, 115 (1996)
30. M. Che, *Adv. Catal.* 31, 78 (1982)
31. M. Che, A. Tench, *Adv. Catal.* 32, 1 (1983)
32. J. Haber, M.J. Witko, *J. Catal.* 216, 416 (2003)
33. D. Kulkarni, I. Wachs, *Appl. Catal.* 6162, 1 (2002)
34. M. Banares, *Catal. Today,* 51, 319 (1999)
35. M. Barteau, *Chem. Rev.* 96, 1413 (1996)
36. A. Corma, H. Garcia, *Chem. Rev.* 102, 3837 (2002)
37. J. Macht, C.D. Baertsch, M. MayLozano, S. L. Soled, Y. Wang, E. Iglesia, *J. Catal.* 227, 479 (2004)
38. P.L. Gai, *Top. Catal.* 8, 97 (1999)
39. P.L. Gai, K. Kourtakis, *Science*, 267, 661 (1995)
40. K. Kourtakis, P.L. Gai, *J. Mol. Catal. A Chem.* 220, 93 (2004)
41. A. Andersson, S. Hansen, A. Wickman, *Top. Catal.* 15, 103 (2001)
42. G. Xiong, V. Sullivan, P.C. Stair, G.W. Zajac, S.S. Trail, J.A. Kaduk, J.T. Golab, J. Brazdil, *J. Catal.* 230, 317 (2005)
43. H. Freund, *Faraday Discus.* 114, 1 (2000)
44. K. Reuter, C. Stampfl, M. Scheffler, in *Ab Initio Atomistic Thermodynamics and Statistical Mechanics of Surface Properties and Functions,* S. Yip, (ed.) (2004) (to be published)
45. H. Hu, I.E. Wachs, *J. Phys. Chem.* 99, 10911 (1995)
46. Y. Chen, I.E. Wachs, *J. Catal.* 217, 468 (2003)
47. I.E. Wachs, Y. Chen, J.M. Jehng, L.E. Briand, T. Tanaka, *Catal. Today* 78, 13 (2003)
48. D.E. Fein, I.E. Wachs, *J. Catal.* 210, 241 (2002)
49. X. Wang, I.E. Wachs, *Catal. Today,* 96, 211 (2004)
50. L.J. Burchan, M. Badlani, I.E. Wachs, *J. Catal.* 203, 104 (2001)
51. V. E. Henrich, P.A. Cox, *The Surface Science of Metal Oxides,* Cambridge University Press, Cambridge (1994)
52. M. Barteau, *Adv. Catal.* 45, 262 (2000)
53. G.W. Coulston, E.A. Thompson, N. Herron, *J. Catal.* 163, 122 (1996)
54. G.W. Coulston, S.R. Bare, H. Kung, K. Birkeland, G.K. Bethke, R. Harlow, N. Herron, P.L. Lee, *Science* 275, 191 (1997)
55. J.C. Volta, *C.R. Acad. Sci. Paris, Ser. IIC, Chimie* 3, 717 (2000)
56. M. Havecker, A. Knop-Gericke, H. Bluhm, E. Kleimenov, R.W. Mayer, M. Fait, R. Schlogl, *Appl. Surf. Sci.* 230, 272 (2004)
57. J.K. Bartley, J.A. Lopez-Sanchez, G.J. Hutchings, *Catal. Today,* 81, 197 (2003)
58. K.E. Birkeland, S.M. Babitz, G.K. Bethke, H.H. Kung, G.W. Coulston, S.R. Bare, *J. Phys. Chem. B*, 101, 6895 (1997)
59. V.V. Guliants, S.A. Holmes, *J. Mol. Catal. A: Chem.* 175, 227 (2001)
60. G. Centi, F. Cavani, F. Trifiro *Selective Oxidation by Heterogeneous Catalysis,* Kluwer, New York (2001)
61. G. Centi, F. Trifiro, J.R. Ebner, V.M. Franchetti, *Chem. Rev.* 88, 55 (1988)
62. M. Neurock, G. Coulston, D.A. Dixon, unpublished work (1993

63. E. Bordes, *Acad. Sci. Ser. IIc: Chim.* 3, 725 (2000)
64. G. Koyano, T. Saito, M. Misono, *J. Mol. Catal. A*, 155, 31 (2000)
65. A. Haras, H.A. Duarte, D.R. Salahub, M. Witko, *Surf. Sci.* 513, 367 (2002)
66. D.J. Thompson, M.O. Fanning, B.K. Hodnett, *J. Mol. Catal. A: Chem.* 198, 125 (2003)
67. M. Witko, R. Tokarz, J. Haber, K. Hermann, *J. Mol. Catal. A: Chem.* 166, 59 (2001)
68. D.J. Thompson, I.M. Ciobica, B.K. Hodnett, R.A. van Santen, M.O. Fanning, *Surf. Sci.* 547, 438 (2003)
69. D.J. Thompson, I.M. Ciobica, B.K. Hodnett, R.A. van Santen, M.O. Fanning, *Catal. Today,* 91–92, 177 (2004)
70. H. Liu, E. Iglesia, *J. Catal.* 223, 161 (2004)
71. J.M.M. Millet, J.C. Vedrine, G. Hecquet, *Stud. Surf. Sci. Catal.* 55, 883 (1990)
72. J.M.M. Millet, J.C. Vedrine, *Appl. Catal.* 76, 209 (1991)
73. J.M.M. Millet, D. Rouzies, J.C. Vedrine, *Appl. Catal. A: General* 124, 205 (1995)
74. C. Virely, M. Forissier, J.M.M. Millet, J.C Vedrine, *J. Mol. Catal.* 71, 199 (1992)
75. M.M. Gadgil, S.K. Kulshreshtha, *J. Solid State Chem.* 111, 357 (1994)
76. M.A. Chaar, D. Patel, M.C. Kung, H.H. Kung, *J. Catal.* 105, 483 (1987)
77. M.A. Chaar, D. Patel, H.H. Kung, *J. Catal.* 109, 463 (1988)
78. D.S. Sam, V. Soenen, J.C. Volta, *J. Catal.* 123, 417 (1990)
79. K. Chen, A.T. Bell, E. Iglesia, *J. Phys. Chem.* 104, 1292 (2000)
80. (a) I.E. Wachs, *J. Phys. Chem. B*, 109, 2275 (2005);
    (b) J. Haber, in, R.A. Sheldon, R.A. van Santen (eds.), *Catalytic Oxidation*, World Scientific, Singapore, p.40 (1995)
81. N.Y. Topsøe, M. Anstrom, J.A. Dumesic, *Catal. Lett.* 76, 11 (2001);
    K. Hermann, M. Witko, R. Druzinic, R. Tokarz, *Appl. Phys. A*, 72, 429 (2001)
82. J.N. Allison, W.A. Goddard, in *Solid State Chemistry and Catalysis*, R.R. Grasselli, J. Z. Brazdil (eds.), ACS Symposium Series, No. p. 279, 23. American Chemical Society, Washington, DC (1985)
83. a) R.K. Grasselli, *Top. Catal.* 21, 79 (2002);
    b) J.C. Vedrine, *Top. Catal.* 21, 97 (2002)
84. M. Anstrom, N.Y. Topsøe, J.A. Dumesic, *J. Catal.* 213, 115 (2003)
85. K. Chen, A.T. Bell, E. Iglesia, *J. Catal.* 209, 35 (2002)
86. M.S. Palmer, M. Neurock, M.M. Olken, *J. Am. Chem. Soc.* 124, 8452 (2002)
87. L. Leveles, K. Seshan, J.A. Lercher, L. Lefferts, *J. Catal.* 218, 307 (2003)
88. Y. Amenomiya, V. Birss, M. Goledzinowski, J. Galuszka,A. Sanger, *Catal. Rev. Sci. Eng.* 32, 163 (1990)
89. J.H. Lunsford, *Angew. Chem. Int. Ed. Engl.* 34, 970 (1995)
90. M.Y. Sinev, V.N. Korchak, O.V. Krylov, *Kinet. Katal.* 27, 1274 (1986)
91. M.Y. Sinev, *Catal. Today,* 13, 561 (1992)
92. M.Y. Sinev, *Catal. Today,* 24, 389 (1995)
93. M.C. Wu, C.M. Truong, K. Coulter, D.W. Goodman, *J. Catal.* 140, 344 (1993)
94. M.C. Wu, C.M. Truong, K. Coulter, D.W. Goodman, *J. Vac. Sci. Technol.* 11, 2174 (1993)
95. D.J. Driscoll, W. Martir, J.X. Wang, J.H. Lunsford, *J. Am. Chem. Soc.* 107, 58 (1985)
96. L. Ackermann, J.D. Gale, C.R.A. Catlow, *J. Phys. Chem. B*, 101, 10028 (1997)
97. L.K. Dash, M.J. Gillan, *Surf. Sci.* 549, 217 (2004)

98. M.A. Johnson, E.V. Stefanovich, T.N. Truong, *J. Phys. Chem. B*, 101, 3196 (1997)
99. R. Orlando, F. Cora, R. Millini, G. Perego, R. Dovesi, *J. Chem. Phys.* 105, 351 (1999)
100. K.J. Borve, L.G. Pettersson, *J. Phys. Chem.* 95, 3214 (1991)
101. M.L. Anchell, K. Morokuma, A.C. Hess, *J. Chem. Phys.* 99, 6004 (1993)
102. J.M. DeBoy,R.R. Hicks, *J. Chem. Soc. Chem. Commun.* 982 (1988)
103. M.S. Palmer, M. Neurock, M.M. Olken, *J. Phys. Chem. B*, 106, 6543 (2002)
104. K. Otsuka, A.A. Said, K. Jinno, T. Komatsu, *Chem. Lett.* 77 (1987)
105. K. Otsuka, Y. Murakami, Y. Wada, A.A. Said, A. Morikawa, *J. Catal.* 121, 122 (1990)
106. C. Lin, K.D. Campbell, J.X. Wang, J.H. Lunsford, *J. Chem. Soc. Chem. Commun.* 90, 534 (1986)
107. J.X. Wang, J.H. Lunsford, *J. Phys. Chem. B*, 90, 3890 (1986)
108. M Banares, *Catal. Today*, 51, 319 (1988)
109. E. M. Kennedy, N.W. Cant, *Appl. Catal.* 75, 321 (1991)
110. V.T. Amorebieta, A.J. Colussi, *J. Am. Chem. Soc.* 118, 10236 (1996)
111. A. Shamsi, *Ind. Eng. Chem. Res.* 32, 1877 (1993)
112. R. Burch, G.D. Squire, S.C. Tsang, *Appl. Catal.* 43, 105 (1988)
113. J.V. Lauritsen, M.V. Bollinger, E. Lægsgaard, K.W. Jacobsen, J.K. Nørskov, B.S. Clausen, H. Topsøe, F. Besenbacher, *J. Catal.* 221, 510 (2004)
114. J. Raybaud, J. Hafner, G. Kresse, H. Toulhoat, *J. Phys. Condens Matter,* 9, 11107 (1997)
115. J. Raybaud, J. Hafner, G. Kresse, , S. Kasztelan,H. Toulhoat, *J. Catal.* 190, 128 (2000)
116. G. Wulff, *Z. Kristallographic*, 34, 449 (1901);
W.K. Burton, N. Cabrera, F.C. Frank, *Philos. Trans. Roy. Soc. A*, 243, 299 (1951)
117. (a) L.S. Byskov, J.K. Nørskov, B.S. Clausen, H. Topsøe, *J. Catal.* 187, 109 (1999); (b) P. Raybaud, J. Hafner, G. Kresse, S. Kasztelan, H. Toulhoat, *J. Catal.* 189, 129 (2000)
118. J.V. Lauritsen, M. Nyberg, J.K. Nørskov, B.S. Clausen, H. Topsøe, E. Lægsgaard, F. Besenbacher, *J. Catal.* 224, 94 (2004).
119. A. Travert, H. Nakamura, R.A. van Santen, S. Cristol, J.-F Paul, E. Payen, *J. Am. Chem. Soc.* 124, 7084 (2002)
120. E.J.M. Hensen, G.M.H.J. Lardinois, V.H.J. de Beer, J.A.R. van Veen, R.A. van Santen, *J. Catal.* 187, 95 (1999)
121. E.J.M. Hensen, H.J.A, Brans, G.M.H.J. Lardinois, V.H.J. de Beer, J.A.R. van Veen, R.A. van Santen, *J. Catal.* 192, 98 (2000)
122. H. Knözinger, S. Huber, *Faraday Trans.* 94, 2047 (1998)

CHAPTER 6
Mechanisms for Aqueous Phase Heterogeneous Catalysis and Electrocatalysis; A Comparison with Heterogeneous Catalytic Reactions

6.1 General Introduction

A wide range of different heterogeneous catalytic reactions which are carried out in liquid or aqueous media show marked changes in activity and/or selectivity over the same reactions carried out in the vapor phase. The rate can be reduced owing to reactant availability at the catalyst surface as a result of external mass transfer or solubility limitations. These are classical, well-characterized extrinsic kinetic limitations. The solution phase can also act to alter the intrinsic chemical kinetics associated with the elementary adsorption, surface reaction, surface diffusion and desorption processes. It is well established in physical organic chemistry that the solution phase can influence a reaction by stabilizing or destabilizing its transition and products states over those of the reactant state. For example, both aqueous and polar solvents tend to stabilize reactions that have transition states which are more polar than their corresponding reactant state. The aqueous or polar medium stabilizes a charge transfer process and the corresponding transition state that forms. Similarly, the presence of an aqueous phase or polar medium tends to enhance heterolytic bond activation reactions that result in electron transfer and/or proton transfer processes. These reactions are at the heart of both electrocatalysis and enzyme catalysis. In addition to their influence on charge stabilization or destabilization, the aqueous medium or protic solvent can also directly participate in the catalytic action via proton and electron transfer processes.

A fundamental understanding of the atomic and electronic processes that govern reactions in solution has eluded the heterogeneous catalysis community to a large extent owing to the complexity of the aqueous medium and its interaction with heterogeneous substrates. Previous experimental efforts aimed at characterizing the interface between the metal and an aqueous solution, for example, have been met with very limited success owing to the difficulty in resolving molecular scale information. The critical spectroscopic bands that might be used to follow specific reaction modes are typically masked by the spectral features that arise from the bulk solution.

The past decade, however, has witnessed the development of a host of spectroscopic methods that can begin to elucidate molecular structure at the aqueous/metal interface. Surface Enhanced Raman Spectroscopy (SERS), Surface Enhanced Infrared Spectroscopy (SEIRS), and Sum-Frequency-Generation (SFG) can begin to separate out the surface-adsorbate characteristics that arise fromfrom those from the bulk solution and, thus provide the vibrational properties of molecules adsorbed on a metal substrate to be followed along the course of a catalytic reaction. In addition, the exponential growth of computational resources along with novel algorithm development has made it possible to begin to simulate the surface structure and the elementary processes that occur at the metal/solution interface.

In this chapter we extend our treatment of mechanisms for metal-catalyzed reactions in the vapor phase to heterogeneous catalytic reactions carried out in aqueous media and electrocatalytic reactions. More specifically, we discuss what is known about the water/metal interface, its reactivity, and the influence of the aqueous phase on elementary surface processes including adsorption, reaction, diffusion and desorption and solution-phase kinetic processes. We advance these ideas into the discussion of the mechanisms

Molecular Heterogeneous Catalysis. Rutger Anthony van Santen and Matthew Neurock

ISBN: 3-527-29662-X

that govern four different example reactions, namely the synthesis of vinyl acetate, the low-temperature oxidation of ammonia, NO reduction, and CO oxidation. We will draw ties between organometallic coordination complexes and their reactivity, gas-phase heterogeneous catalysis, heterogeneous catalysis in solution and electrocatalysis.

6.2 The Chemistry of Water on Transition Metal Surfaces

The structure and chemistry at the water/metal interface is critical in dictating the properties and ultimately controlling the catalytic performance of aqueous phase heterogeneous, electro-, homogeneous, and bio-catalytic systems. In addition, they have great relevance to understanding corrosion as well as a host of other materials issues. A number of outstanding reviews exist which describe the structure and reactivity of water over metal and metal oxide substrates[1,2]. Rather than repeat these analyses, we try to summarize some of the important factors and describe the systematic changes that occur in the structure, adsorption and reactivity of water on metal substrates as we move from the adsorption of water in the vapor phase to the solution phase and then on to the influence of applied potentials.

6.2.1 Reactions in Solution

The presence of solution can dramatically alter the resulting chemistry at the solution-metal interface. This is clearly present even in the neat liquid phase dissociation processes alone. For example, the dissociation of acetic acid proceeds in the vapor phase at higher temperatures via a homolytic process that leads to the formation of $CH_3CO_2\bullet$ and $H\bullet$ free radical intermediates. This homolytic activation of the O–H bond in acetic acid costs 440 kJ/mol[3a]. The heterolytic activation of acetic acid to form CH_3CO_2- and H^+ intermediates, however, is significantly more endothermic, costing +1532 kJ/mol.

This reaction is simply the reverse of the proton affinity for acetate. The dramatic increase in the endothermicity for the heterolytic reaction is due to the fact that the charged products which form are unstable alone in the vapor phase. The neighboring water molecules begin to hydrogen bond with acetic acid and stabilize the charged complexes. The results in Fig. 6.1 indicate that there is a precipitous drop in the endothermicity of the dissociation reaction that occurs by adding just 1, 2, or 3 water molecules. The water molecules effectively stabilize both the proton and the acetate anion (CH_3COO^-) that form. As we get to 12 water molecules or more, the dissociation actually becomes thermoneutral. The stabilization of the charged products is much stronger than the energy required for the heterolytic activation, hence the dissociation energy becomes thermoneutral. The dissociation energy changes from +1532 kJ/mol in the vapor phase to 0 kJ/mol in solution. This agrees quite well with experimental results, which indicate that the proton affinity is 1540 kJ/mol whereas the dissociation of acetic acid in solution is 1.6 kJ/mol.

6.2.2 The Adsorption of Water on Metal Surfaces

The adsorption of water on most metal surfaces is typically rather weak and controlled by a balance between the strength of the metal–water bond and the water–water[1,2,4,5] interactions. Molecular water adsorbs on metal and metal oxide substrates through the donation and back-donation of electrons between the frontier molecular orbitals of water and the states of the metal near the Fermi level.

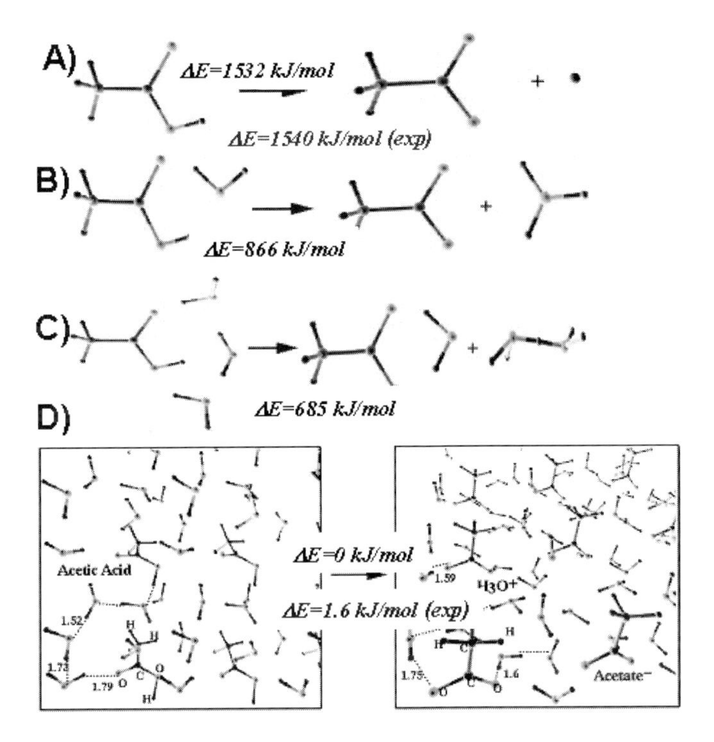

Figure 6.1. The effect of water on the energies for the heterolytic dissociation of acetic acid. (a) Gas phase, (b) reaction with one water molecule, (c) reaction with three water molecules, (d) reaction in bulk water.

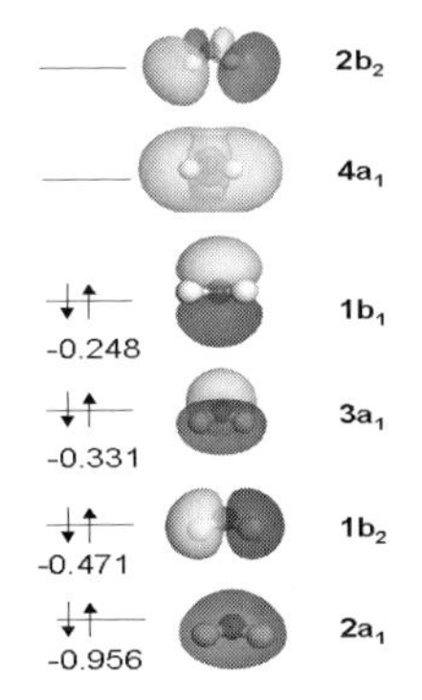

Figure 6.2. Molecular orbitals and their corresponding energies for water.

On metal oxides weak hydrogen bond interactions can also evolve. The molecular orbital energy diagram for the gas-phase water molecule alone is shown in Fig. 6.2.

The frontier orbitals are the $1b_1$, $3a_1$ and the $1b_2$ states. We recognize these orbitals as two p_x- and p_y-type oxygen atomic orbitals interacting with symmetric and antisymmetric hydrogen atomic orbital combinations ($1b_2$ and $3a_1$, respectively). The $1b_1$ molecular orbital is the non-bonded $2p_z$ atomic orbital on oxygen. All three of these states are typically seen in valence band photoemission spectroscopy. Which of these states controls adsorption is difficult to discern from the shifts in photoemission spectroscopy alone [2].

Michaelides and co-workers[4,5] have shown theoretically that for monolayers and bilayers of water on Ru, the predominant overlap occurs between the $2p_x$ type $3a_1$ orbital of water with a d_{z^2} state on Ru. In addition there is overlap between the lone–pair $1b_1$ orbital on water and the Ru d_{z2} state. The overlap and mixing of the metal surface state with the $1b_1$ orbital is significantly stronger. An electronic analysis shows that there is a charge depletion from the $1b_1$ and Ru d_{z^2} states with a charge accumulation on the lower lying d_{xz} and d_{yz} states. There is also a small charge increase between the O and the Ru[4].

The adsorption of water at low coverage on most metal surfaces is fairly weak, whereby water prefers to sit atop a metal atom so as to avoid Pauli repulsive interactions between the lone pair of electrons on water and the filled states of the metal. The bond lengths between the metal and the oxygen are within the range 2.1–2.3 Å[3b,5]. Valence bond theory indicates that water should tilt significantly with respect to the surface. The degree of tilt with respect to the surface normal varies with the metal and the nature of the surface. At low coverages, water can adsorb as isolated species; form 2D or 3D clusters; assemble into well-ordered surface structures commensurate with the registry of metal lattice spacing; or dissociate into OH and H. The structures that form depend upon the balance between the metal–water bond strength and water–water bond strength for the specific system of interest in addition the system conditions, i.e. temperature and pressure. For systems where the metal-water interactions are weaker than the water–water interactions, water prefers to form clusters. For systems where the metal–water interactions are stronger than the water–water interactions, water will tend to form commensurate surface structures. For systems where the metal-water interactions are much stronger than the water–water interactions, water can begin to dissociate[2].

The adsorption of water onto the metal in the vapor phase is typically quite weak with adsorption energies on the order of 30–50 kJ/mol, with some notable exceptions[3b,5]. DFT slab calculations predict the adsorption of a single water molecule to be 50 kJ/mol on Ag(111), 30 kJ/mol on Pd(111), 30 kJ/mol on Pt(111), 37 kJ/mol on Rh(111), and 100 kJ/mol on Cu(110). Noteworthy is the increased hydrophilicity of the Group IB noble metals compared with that of the group VIII metals.

As the coverage is increased, water can form monolayers, bilayers or 3D water[2] clusters. In addition, water can form either ice-like crystalline surface structures or amorphous liquid water. This, once again, is highly dependent upon the balance between metal–water and water–water bond strengths and also the system conditions. Vassilev et al.[6], for example, found that as the coverage of water was increased from 1/3 to 2/3 ML, the effective adsorption energy per water molecule increased from 37 to 56 kJ/mol. At 2/3 ML coverage, the water layer is actually a bilayer which takes on considerable hydrogen-bonding interactions between neighboring water molecules. When the binding energy of the entire bilayer is calculated, it is found that the interaction energy with the metal surface is essentially vanishing. As mentioned above, a very similar conclusion follows from the work of Desai and Neurock[3b].

The adsorption of water on most close-packed metal surfaces leads to the formation of the well-known bilayer structure. In this system, water prefers to adsorb on the metal substrate in a hexagonal ring structure which matches the registry of the metal substrate, as shown in Fig. 6.3. Three water molecules within the ring are bound to the substrate through their oxygen ends. The two hydrogen atoms are directed towards the oxygen end of two neighboring water molecules, thus forming two hydrogen bonds to the solution phase. Each of the three water molecules bound to metal surface within the hexagonal ring brings in one more water molecule due to hydrogen bonding, thus resulting in six water molecules that make up the hexagonal ring structure. These secondary waters are further away from the surface where they form hydrogen bonds with those water molecules that are directly bound to the substrate.

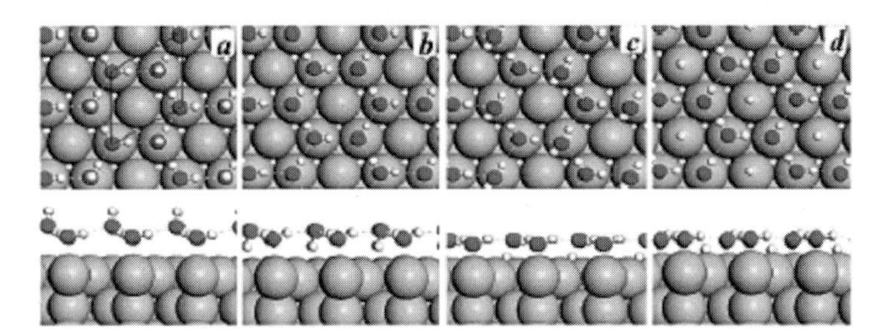

Figure 6.3. The general bilayer structure of water adsorbed on a close-packed transition-metal surface. (a) H-up structure, (b) H-down structure, (c) transition state for bilayer dissociation, and (d) partially dissociated bilyaer structure[4].

This leads to the formation of two distinct layers of water molecules[2,5]. The first layer is directly bound to the surface via metal oxygen bonds whereas the second layer is indirectly bound to the substrate via the formation of hydrogen bonds. Two different conformations of the bilayer are possible. In the first conformation, one of the hydrogen atoms from the water molecules in the second layer is pointed up (H-up) into the vacuum. In the second, the hydrogen atom from the water molecule in the second layer is directed down toward the metal substrate (H-down).

Water which is bound to most transition-metal surfaces tends to form this bilayer structure on close-packed surfaces. Experimental results for water on Ru(0001), however, were found to be characteristically different than those on other transition-metal surfaces. Held and Menzel[7] identified the oxygen atoms from water molecules in two different layers and speculated on the formation of a "bilayer" structure for water adsorbed on Ru(0001) in a ($\sqrt{3}$ x $\sqrt{3}$)R30° state separated by only 0.1 Å[7]. They speculated that this was the result of the formation of a compressed water bilayer. Feibelman[8] later showed via theoretical calculations that the only water structures that fit the short 0.1 Åseparation distance between oxygen atoms of the bilayers are those that involve partially dissociated water along with neighboring water molecules. Feibelman[8] and Michaelides et al.[4] showed that the barrier for water dissociation was reduced from 0.8 eV to 0.5 eV of moving from isolated water molecules to water molecules contained within the bilayer structure. The partially dissociated states also had the lowest energies. Water can also dissociate over various other metal surfaces, especially at higher temperatures.

The dissociation of an isolated water molecule occurs via the insertion of metal atom into one of its O–H bonds. As the O–H bond is initially activated, the energy level of the unoccupied σ^* antibonding O–H orbital lowers, thus allowing for electron transfer from the metal into this state. This subsequently facilitates the activation of water. Dissociation of an isolated water molecules adsorbed from the gas phase at the metal/vapor interface leads to the formation of a surface hydroxyl as well as a surface hydride. Desai and Neurock[3b] showed that the activation of water over Pt(111) would be highly endothermic (+90 kJ/mol) and require overcoming a barrier of +140 kJ/mol. The reaction path is shown in Fig. 6.4 (see also Chapter 3. pages 133, 134).

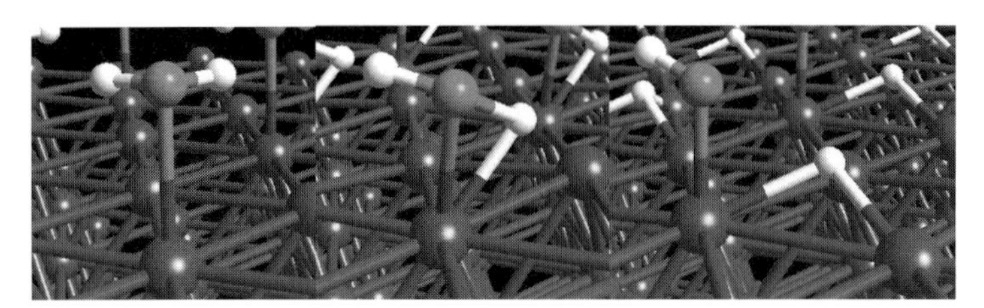

Figure 6.4. The homolytic activation of water over Pt(111). On the left: water adsorbed reactant state. In the center the transition state for O–H activation. On the right the surface hydroxyl and hydride products[3b].

Under UHV conditions, water would preferentially desorb rather than react. The barrier for the activation of water over Ru, however, is considerably lower at +90 kJ/mol. The reaction is now slightly exothermic. The barrier, however, is still too high to overcome under UHV conditions. The barrier however is reduced significantly as the density of water increases from the monomer structure to the bilayer and then on to multilayer adsorption, as will be discussed.

The dissociation of water was examined over a range of different close-packed transition-metal surface structures in order to establish periodic trends. The results shown in Fig. 6.5 indicate that it becomes easier to activate water over metals that lie to the left of the periodic table which have more vacancies in the d-band.

Fe	Co	Ni -45.1	Cu 26.8
Ru -39.7	Rh 2.5	Pd 70.1	Ag 89.7
Os	Ir 71.7	Pt 110.2	Au 152.5

$H_2O^* \rightarrow H^* + OH^*$

Figure 6.5. Periodic trends in the reaction energies for the homolytic surface dissociation of water at the vapor/metal interface. Energies are in kJ/mol.

In addition, water is also more readily activated over metals that lie higher up in the periodic table. The relative ordering of the metals, examined in Fig. 6.5, that most readily activate water is:

$$\mathrm{Ni} > \mathrm{Ru} > \mathrm{Rh} > \mathrm{Cu}$$

The activation of water at the aqueous metal interface can be characteristically different from that at the vapor metal interface owing to changes in the dielectric constant of the medium, increased hydrogen bonding network and potential changes to the metal substrate. Density functional theoretical results for multilayers of water adsorbed on Pt(111) indicate that the water molecules near the surface take on a bilayer-like structure whereas water molecules in solution either form an ordered ice-like structure if they are optimized at 0 K, or form a more amorphous random structure as the result of ab initio MD simulations at higher temperatures. The homolytic activation of water to form adsorbed hydrogen and a surface hydroxyl intermediate is stabilized owing to the presence of aqueous water but the change is only 10 kJ/mol. This is likely due to the fact that the transition state for the homolytic activation is not very polar. A second, more likely, path would involve the heterolytic activation of water to form a proton (H^+) which is stabilized by the aqueous medium and a negative hydroxide ion on the metal substrate. The extra electron can readily transfer to the metal. The proton migrates away from the surface via proton transfer through the network of water molecules at the metal surface. Water directly participates in the mechanism by providing a conduit for proton transfer. This lowers the barrier by over 40 kJ/mol. The reactant, transition and product states for this reaction over Pt(111) are very similar to those over the $Pt_{66.6\%}Ru_{33.3\%}$ surface which, is shown in Fig. 6.6[3b].

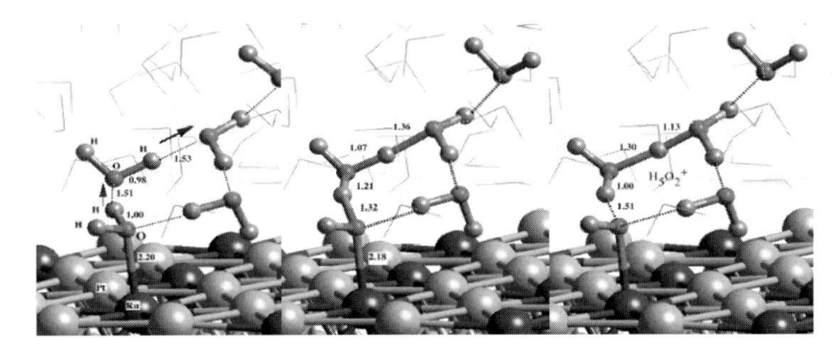

Figure 6.6. The heterolytic activation of water over $Pt_{66.6\%}Ru_{33.3\%}$ alloy in the presence of aqueous water. (a) Water adsorbed reactant state, (b) the transition state for O–H activation, and (c) the surface hydroxyl and aqueous hydronium ion product states.[3b]

The addition of Ru to the Pt surface can influence both the vapor phase and the aqueous phase reaction energies.[3b] The substitution of just 1 out of every 3 Pt atoms in Pt surface layer leads to a significantly lower activation barrier. The results in the vapor phase indicate that the barrier falls by nearly 45 kJ/mol to 108 kJ/mol from the Pt(111) surface. This is due to the fact that water preferentially adsorbs and activates at the Ru sites. The activation barrier for water over the $Pt_{66.6\%}Ru_{33.3\%}$ surface is 106 kJ/mol, which is only 11 kJ/mol higher than that on the Ru(0001) surface. This is due to the fact that the reaction is local, involving a single metal atom insertion. The surrounding Pt atoms have a weak electronic effect on the chemistry at the Ru site. The activation of water over the $Pt_{66.6\%}Ru_{33.3\%}$ alloy in an aqueous medium, however, is quite different. The reaction proceeds heterolytically, thus forming a solvated proton, a surface hydroxyl intermediate and an electron that is delocalized over the metal. The transition state for the heterolytic reaction is much more polar and therefore can be stabilized by the presence on an aqueous medium. The barrier for this reaction over the alloy surface in solution is only 26 kJ/mol. This is considerably lower than the energy for the vapor phase reactions and those reactions carried out over Pt and Ru substrates in the aqueous phase. There appears to be strongly enhanced synergy between Pt, Ru and the aqueous medium. The reactant, transition, and product states for this reaction are shown in Fig. 6.7.

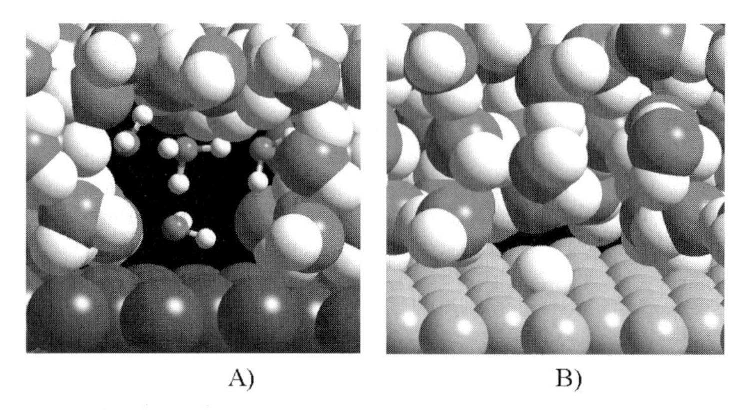

Figure 6.7. The adsorption of atomic hydrogen on (A) Pt(111) in aqueous solution leads to proton formation whereas the adsorbed atomic hydrogen on (B) Ru(0001) remains as adsorbed hydrogen in the presence of an aqueous solution. [3b]

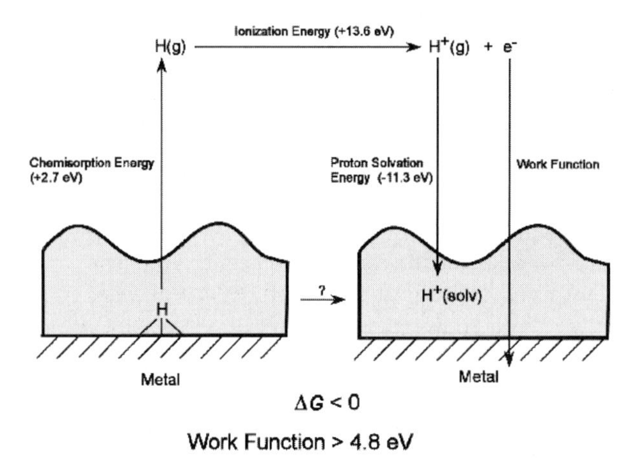

Figure 6.8. The overall Born–Haber thermodynamic cycle for the free energy required to create hydronium ions from adsorbed hydrogen in the presence of water[3b,10].

A key step in the aqueous phase activation of water involves the generation of protons. Ab initio calculations showed that atomic hydrogen adsorbed on Pt(111) would transfer its electron into the metal and then migrate into solution as a proton. Wagner and Moylan[9] and later Kizhakevariam and Stuve[10] showed experimentally that adsorbed hydrogen in the presence of solution on Pt(111) forms hydronium ions. Similarly, Desai and Neurock[3b] used theory to show that adsorbed hydrogen on Pt(111) can readily transfer its electron to the metal and thus generate protons. The net results, however, would ulrimately require that proton and the electron recombine. The results from this reaction are shown in Fig. 6.7A. These same calculations performed on the Ru(0001) surface indicate that the hydrogen remains bound to Ru as a hydride (Fig. 6.7B). These differences can be explained by constructing a simple Born–Haber cycle for the process such as that shown in Fig. 6.8:

$$H(a) + H_2O(aq) + M \longrightarrow H_3O^+(aq) + M + 1 \tag{6.1}$$

The desorption of hydrogen and its dissolution into the aqueous phase require the following steps: (1) desorption of H• into gas phase, (2) the ionization of H^+, (3) the solvation of the proton into water, and 4) the capture of the resulting electron by the metal (i.e. electron affinity). The only steps which change on moving to different metals are the desorption of hydrogen and the work function of the metal. Steps 2 and 3 are identical regardless of the metal. While the desorption of hydrogen changes with the metal substrate used, the change is small compared with the changes in the work function of the metal. Kizhakevariam and Stuve[10] estimated that the free energy for proton transfer would be greater than or zero for metals with work functions greater than 4.88 eV. Theoretical results from Desai and Neurock[3b] showed that Pt and Pd in the presence of adsorbed water had work functions that were greater than 4.88 eV. The overall energy for proton formation was calculated to be –57 and –14 kJ/mol exothermic over Pt(111) and Pd(111) substrates, respectively. The work function for Ru(0001) was calculated to be below 4.88 eV and thus the reaction to form the hydronium ion was found to be endothermic by +40 kJ/mol. This suggests that while Ru may activate water it will have a stronger tendency to hold onto the hydrogen atoms as a hydride.

Interestingly, the $Pt_{66.6\%}Ru_{33.3\%}$ surface alloy has a work function that is 4.92 eV, which would suggest that this metal could result in the formation of protons. The calculations confirm that hydrogen atoms desorb as protons on this surface. The key here is that the addition of Pt to the Ru lattice helps to aid in the heterolytic activation of water since Pt increases the work function of the metal, thus enhancing its ability to accept electrons. The Pt, Ru and aqueous solution form a unique active nanoscale environment whereby Ru is necessary to adsorb and activate water, Pt increases the work function of the metal which enhances electron transfer, and finally the aqueous solution phase promotes proton transfer into solution.

It is clear that the activation of water at the aqueous phase/metal interface is quite different than the activation of water at the vapor phase/metal interface owing to charge stabilization and electron transfer. The ideas presented on PtRu can be extended to the activation of water in the aqueous phase over various other metals. The results in Fig. 6.9 were calculated by using a Born–Haber cycle similar to that shown for hydrogen dissolution described earlier[11]. In this analysis, the ionization energy of H•, the solvation energy for H^+, the work function of the metal and the change in hydrogen-bonding contributions are used to correct the energies for the vapor-phase activation of water to account for the effects of solution. These thermodynamic estimates of the overall reaction energies for the dissociation of water at the aqueous/metal interface are shown in Fig. 6.9 for various close-packed transition-metal surfaces. The best metals are those which demonstrate favorable energies for the vapor-phase activation of water and, in addition, have work functions that are high enough to promote the heterolytic activation of water. The best metals show the following trend with respect to their overall favorability to activate water:

$$\mathrm{Ni} > \mathrm{Rh} > \mathrm{Ru} > \mathrm{Cu}$$

The higher work function for Rh over Ru here changes the relative ordering of the metals in terms of favorability from that found for vapor-phase activation. Actual ab initio calculations which explicitly examine the reactants and products at the metal/solution interface yield overall energies for the dissociation of water on Cu and Pt in an aqueous medium of 30 and 90 kJ/mol, respectively. These values are consistent with the estimates

Fe	Co	Ni -6.5	Cu 42.5
Ru 25.7	Rh 15.7	Pd 82.5	Ag 158.0
Os	Ir	Pt 93.9	Au 106.5

$H_2O^* \rightarrow H^+(aq) + OH^* + e$

Figure 6.9. Periodic trends in the reaction energy for the protolysis of water to form surface-adsorbed hydroxyl intermediates and hydronium ions in solution.

reported in Fig. 6.9.

In addition to its influence on stabilizing charged states, water can also directly participate in elementary physicochemical processes such as reaction and diffusion. We have already described the fact that water can be directly involved in a chemical reaction by providing a conduit for proton transfer. Similarly water can aid in the diffusion of charged surface intermediates. For example, Vassilev et al.[6] found that adsorbed OH intermediates demonstrated enhanced mobility on the Rh(111) surface when additional water molecules were coadsorbed. The increased diffusion was the result of fast proton transfer through neighboring water molecules and the adsorbed OH. Protons transfer in the opposite direction, thus giving the impression OH migration in the forward direction (see Fig. 6.10). This is a Grotthus-like mechanism which has been used to describe hydroxyl mobility in H_2O[12]. Desai and Neurock[3b] found similar results for the activation of water in an aqueous solution adsorbed on a PtRu surface. Their ab initio molecular dynamics results show that the OH hopping is quite rapid, occurring in only a few picoseconds. The barrier for proton transfer was found to be very low.

6.2.3 Influence of Potential

The results for the dissociation of water at the metal/solution interface show the well-known double-layer structure that is at the heart of most electrochemical systems. While the negative charge is delocalized, it still acts to polarize the surface. The proton which forms exists as either a hydronium (H_3O^+) or a Zundel ($H_5O_2{}^+$) ion both of which are about one solvation shell removed from the surface. This is known as the inner-layer Helmholtz layer. The chemistry that occurs at the interface polarizes the surface, which ultimately leads to a potential across the interface. In an actual system, the electrolyte plays an important role in establishing the potential as well as in potentially altering the structure and chemistry that occur at the interface.

Modeling electrochemical systems from first principles presents a considerable challenge. Quantum mechanical simulations are typically carried out within the canonical ensemble formalism where the number of electrons remains constant. The free energy is calculated with a constant temperature, volume and number of electrons $F(T, V, Ne)$. Electrochemical systems, on the other-hand, are typically performed at a constant chemical potential in the grand canonical ensemble where $\mu(T, V, Ne)$ is a constant. Throughout this book we have presented examples where the number of electrons is preserved upon chemical reaction. In order to model an electrochemical system, we would have to model the structure and chemistry that occur at the anode simultaneously with that which occurs at the cathode, the rate of electron transfer through the circuit that connects the two, along with the rate of ion transport across the cell and local changes in electrolyte composition. This is not currently possible.

Various approximate approaches, however, have been taken in order to simulate electrochemistry at an electrode surface. Some of the first quantum mechanical models were developed by Anderson[13], who used non-charge self-consistent atom superposition and electron delocalization (ASED) molecular orbital theory to probe changes in the polarization and hybridization and their influence on surface bonding. Bagus and Pacchioni[14], Illas and Mele[15], Lambert[16], Curulla and Clotet[17], Head-Gordon and Tully[18] and more recently Wasileski et al. [19] demonstrated both qualitatively and quantitatively the influence of electrode potential on the intramolecular bond stretching known as the Stark effect.

There have been very few ab initio efforts, however, aimed at examining the influence of an applied potential on electrochemical reactions. Anderson and co-workers have developed models to determine the reversible potentials for reactions occurring at the outer-sphere as well as within the double layer, and the activation barriers for electron transfer reactions.[20],

Anderson and co-workers calculated reversible potentials for outer-sphere processes by using the following equation for the Gibbs free energy:

$$\Delta G^o = -nFU^o \tag{6.2}$$

where n is the number of electrons that transfer and F the Faraday constant. They derived the following semiempirical expression in order to estimate the internal energy:

$$U^0 = (E_r eV^{-1} + c)V \tag{6.3}$$

which linearly relates the reaction energy E_r to the internal energy. The reaction energy can be calculated using ab initio methods quite easily. The term c, in Eq. (6.3) refers to an empirical parameter which can capture some of the features of the local environment and, hence allow for such a linear trend. The value of c for reactions involving O_2 reduction in acid is 0.49 for MP2-type calculations and also B3LYP using 6–31G**. The constant will change based on the system which is being analyzed. This approach is typically limited to outer-sphere reactions run in acidic media.

The second method extends these ideas and is able to calculate reversible potentials for reactions that occur within the double layer. The model can be used to simulate both oxidation and reduction within the double layer. The influence of counterions from the electrolyte on the reactions can also be included. This is accomplished by using point charges and a Madelung sum in order to calculate the longer range electrostatic interactions and the field that arises from these ions and their influence on the reaction center.

The reversible potential models are limited to overall reaction energies. The activation barriers for electron transfer at electrode surfaces were modeled by establishing a reaction center and following the radiationless electron transfer into or out of an open system. The reaction center refers to the specific structural system where oxidation or reduction occur. The system is open, thus allowing electrons to transfer in and out. Electron transfer in the system occurs without an activation barrier when the electron affinities (reduction) or ionization potentials (oxidation) directly match the thermodynamic work function of the electrode ($U = 0$). For the standard hydrogen electrode this can be written as:

$$EA = eU + 4.6\,\text{eV} \quad \text{(for reduction)} \tag{6.4}$$

$$IP = eU + 4.6\,\text{eV} \quad \text{(for oxidation)} \tag{6.5}$$

In order to calculate the activation barrier for a specific potential, one has to establish the structure and corresponding energy of the reaction complex for constant electron affinity (EA) or ionization potential (IP) surfaces. The minimum energy point on the corresponding surface is then the activation barrier.

The method was first used to treat outer Helmholtz plane reactions such as

$$\text{Pt–OH}_2 \ldots \text{OH}_2(\text{OH}_2)_2 \longleftrightarrow \text{Pt–OH}_2 \ldots \text{H}^+\text{–OH}(\text{OH}_2)_2 + \text{e}^-(U) \tag{6.6}$$

The activation energies can be calculated as the internal energy required for electron transfer to occur and are therefore activation internal energies rather than the activation free energies described by Marcus[21].

The approach was used to study the electrochemical dissociation of H_2O at an anode at electrode potentials of 0.6 V (NHE) on Pt. The reaction sequence used is

$$\text{H}_2\text{O} \longrightarrow \text{OH}_{\text{ads}} + \text{H}^+_{\text{solv}} + \text{e}^-(U) \tag{6.7}$$

where U is the electrode potential.

The solvation structure of the proton was taken into account by modeling it as a solvated hydronium ion $H_3O^+(H_2O)_2$. Ab initio SCF-HF MP2 theory was used along with small Pt atom clusters to show that the dissociative adsorption of H_2O requires the direct assistance of additional H_2O molecules:

$$\text{Pt–HOH} \ldots \text{OH}_2(\text{OH}_2)_2 \longrightarrow \text{PtOH} \ldots \text{H}^+ \cdot \text{OH}_2(\text{OH}_2) + \text{e}^-(U) \tag{6.8}$$

The approach was later extended to include explicit solvent molecules and Madelung potentials in order to begin to model the influence of solution and electrolyte. The approach begins to capture features of the elementary electrochemistry but does not include the effects due to electrode potential on bond polarization or more complex reaction environments.

Halley and Mazzolo[22] developed a first-principles-based direct dynamics method to examine the water/copper metal interface. Previous models on the electrochemical metal/water interface published in the literature could not straightforwardly describe the asymmetry of the capacitance measured experimentally in the double layer. In approach taken by Halley and Mazollo, the electrons in the metal are modeled quantum mechanically using a jellium-type free electron model where only the s-electrons in copper are treated. Pseudopotentials are used to describe the electron interactions with water. The water solution phase is decoupled from the electronic structure and treated by molecular dynamics simulations with explicit water molecules using classical force fields. Gouy–Chapman theory is used to treat ionic screening. The electronic structure at the interface between the metal and the water is carefully matched by performing electronic structure calculations on the metal substrate after each time step in the water MD simulation. The approach was used to examine the influence of applied potential on the structure of the metal-water interface. The system size examined was very large, including a solution layer comprised of 245 water molecules sandwiched between two metal layer slabs, each of which contains 36 metal atoms per layer. The simulations explicitly account for electronic structure contributions and on the influence of the double layer observed interfacial structure. The results show that more strongly bound water can lead to metastable charging at the interface. Water molecules that sit at atop sites tend to be very sensitive to potential and may help to explain previous X-ray structural data for water adsorbed on Cu and the asymmetry in the capacitance. The results also suggest that the macroscopic field is effectively screened to a significant degree near the interface.

Nørskov et al.[67] recently developed a simple approach to examine electrocatalytic reduction of oxygen over Pt. More specifically, they examined the reaction sequence proposed in Eqs. (6.9–6.11) as a model of the chemistry at the cathode.

$$\frac{1}{2}O_2(g) + * \longrightarrow O* \tag{6.9}$$

$$O * + H^+ + e \longrightarrow HO* \tag{6.10}$$

$$HO * + H^+ + e \longrightarrow H_2O + * \tag{6.11}$$

In their approach, they set a reference potential equal to that of the standard hydrogen electrode:

$$\frac{1}{2}H_2(g) \longrightarrow H^+ + e^-$$

. This can then be used to relate electrocatalytic reactions proposed in Eqs. (6.9–6.11) to the simple surface reaction energies defined in Eqs. (6.12–6.13).

$$H_2O + * \longrightarrow HO * + 1/2H_2(g) \tag{6.12}$$

$$H_2O + * \longrightarrow O * + H_2(g) \tag{6.13}$$

by separating out the electron transfer terms. The free energies for the electrocatalytic reactions (6.10) and (6.11) can be equated with the reverse reaction energies of Eqs. (6.12) and (6.13) at the electrode potential $U = 0$ by taking their relative energies with respect to the standard hydrogen electrode

$$H^+ + e \longrightarrow \frac{1}{2}H_2(g)$$

The energies for the individual surface intermediates are simply shifted by the electrode potential, $-eU$. They explicitly examine the influence of surface water molecules and the effects of coverage in their calculations of the intermediate binding energies. The solution pH is accounted for by calculating the value of the free energy with respect to the system pH by the following classical expression:

$$G(\text{pH}) = -kT \ln[H^+] \tag{6.14}$$

The gas-phase surface energy calculations were then used to calculate the influence of electrode potential on the O_2 reduction scheme presented above. They then used this methodology to examine a range of different transition metal surfaces in order to map out periodic trends and establish metals that would be most effective. In addition to the mechanism proposed in Eqs. (6.9–6.11), they also examined a mechanism in which the O_2 bond remains intact before the addition of hydrogen.

More recently, Lozovio et al.[23] developed a first-principles periodic DFT supercell approach for simulating the metal vapor phase interface at constant chemical potential. In this approach, charge is either added to or removed from the metal in order to fix an applied electrochemical potential. Charge neutrality within the supercell is maintained by adding an opposite background charge into the vacuum region. The background charge is added to a reference plane which lies parallel to the surface at some distance into the vacuum region. For simplicity, this can be chosen to be the center of the vacuum region. The electrostatic potential along this plane is then defined as zero. Placing the charge sheet at a fixed distance into the background has two important effects. It acts as a reference electrode and it terminates the field at some finite distance from the surface. In addition, it defines the energy zero by which to reference different systems. The chemical potential is then defined as the energy to move an electron from the metal to this fixed reference electrode or take an electron from the reference electrode and place it within the metal.

The approach requires that both the energy and the potential be corrected for carrying out such a process. The correction to the DFT-calculated energy is

$$E_{es} = E_{es}^{\text{DFT}} + \frac{q}{A_0}\left\langle V^{\text{DFT}} \right\rangle_{\Omega_0} + \frac{\pi q^2}{A_0^2}\left(\wedge - \frac{L_z}{3} \right) \tag{6.15}$$

where E^{DFT} and V^{DFT} refer to the uncorrected electrostatic energy and potential, respectively, that are the direct result from the periodic DFT calculations, $\langle V^{\text{DFT}} \rangle_{\Omega_0}$ is the average electrostatic potential which can be calculated by integrating the potential derived from the electronic structure calculations over space, q is the net charge that is added or removed from the system, A_0 is the cross-sectional area of the unit cell in the $z = 0$ plane, L_z is the total length of the cell in the z direction, L is the distance from the surface ($z = 0$) to the background reference electrode, Ω_0 refers to the volume of the cell and $\wedge$ is the distance from the surface into the double layer. To test the system for Pt, the Helmholtz free energy was plotted with respect to the chemical potential. The results which are shown in Fig. 6.10 indicate a parabolic behavior whereby the minimum free energy lies close to $\mu = 0$, as should be expected. The approach has more recently been used to examine the stability and possibility of reconstruction of Pt(110) and Au(110) surfaces under different electrochemical conditions [23].

Filhol and Neurock[24] established a similar constant charge (canonical) approach in order to simulate electrochemistry at aqueous/metal interfaces. A potential across the interface is induced by tuning the charge in the metal. This change in charge of the metal is compensated for by the addition of an equal, but opposite, change in charge, distributed homogeneously over the background of the cell. The homogeneous background used in these simulations is similar to that defined by others for other solid-state systems[23,25]. The addition (or removal) of charge at the surface of the metal subsequently polarizes the homogeneous background charge density in the solution layer and thus orients the water molecules at the water/metal interface. Heterolytic reactions subsequently go on to form the well-known double-layer structure at the interface.

The approach subsequently uses a double-reference system in order to determine the potential. By comparing the energies for different structures at a given potential, one

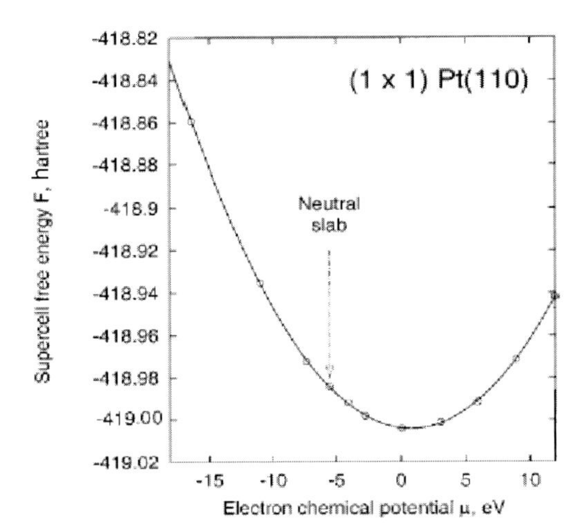

Figure 6.10. Ab initio calculations of the Helmholtz free energy with respect to the chemical potential [23].

is able to determine the free energy difference between these given states at a constant potential, thus allowing for constant chemical potential calculations. The approach has been used to examine various electrochemical reactions at the metal/solution interface.

In the Filhol–Neurock approach, the total energy (E_{DFT}) of the charged slab with n_e electrons and a background charge of n_{bg} is described by the following expression:

$$E_{\mathrm{DFT}}(n_e, n_{bg}) = E_{\mathrm{slab}}(n_e) + E_{\mathrm{slab}-bg}(n_e, n_{bg}) + E_{bg}(n_{bg}) \tag{6.16}$$

where E_{slab} is the energy of the slab without a background, E_{bg} is the energy of the background without a slab and $E_{\mathrm{slab}-bg}$ the interaction energy between the slab and the background.

The charge which is delocalized over the surface of the slab and the compensating background charge interact with one another and therefore must be corrected for in order to determine the true total energy. The values for n_e and n_{bg} are not independent as $mn_e = -n_{bg} = q$. The general relationship between the chemical potential μ and the total energy (E) therefore does not hold true for the total energy of the unit cell:

$$\mu = \left[\frac{\partial E(n_e, n_{bg})}{\partial n_e}\right]_{n_{bg}} \neq \left[\frac{\partial E_{\mathrm{DFT}}(q, -q)}{\partial_q}\right] \tag{6.17}$$

The description of the total DFT energy, E_{DFT}, must therefore be corrected for the interaction between the electrons in the slab and the background charge as was done in Eq. (6.16).

The slab and background are ultimately decoupled and the total electron energy can be defined as

$$E_{\mathrm{elec}} = E_{\mathrm{DFT}} + \int_0^q \left\langle \overline{V_{\mathrm{tot}}}(Q) \right\rangle \mathrm{d}Q \tag{6.18}$$

where

$$\overline{V}(q) = \frac{1}{\sum_{unit\ cell}} \int\int\int V(\vec{r})\mathrm{d}\,\vec{r} \tag{6.19}$$

This satisfies the relationship given in Eq. (6.11). The total free energy of the system (E_{Free}), which includes contributions for the excess electrons (q) at the Fermi potential ϕ_{vac} and is equivalent to

$$E_{\mathrm{Free}} = E_{\mathrm{DFT}} + \int_0^q \left\langle \overline{V_{\mathrm{tot}}}(Q) \right\rangle \mathrm{d}Q - q\phi_{\mathrm{vac}} \tag{6.20}$$

The potential-dependent energies presented here refer to the E_{Free} values.

The chemical potential, ϕ_{vac}, referred to as the vacuum level is obtained directly from the calculations by using the Janak[26] theorem. The corresponding potential, U, which is referred to the standars hydrogen electrode, is extrapolated from the chemical potential according to the experimental relationship [27]:

$$U = -4.85 - \frac{\phi}{e} \tag{6.21}$$

in order to allow for the direct comparison with the experiments. The value of U for the standard hydrogen electrode was calculated by Taylor et al.[32] to be 4.51 at 0 K and 4.67 at 300 K, which are in very good agreement with experimental values.

To simulate constant potential systems, one maps out the free energies between different states of the system over a range of different potentials. The free energy between states can then be calculated at any fixed potential. This approach has been used to examine the structure and reactivity of the aqueous water/metal interface under electrochemical conditions. Some of the results are summarized in the next section.

6.2.4 Electrochemical Activation of Water

We start our discussion with the adsorption and activation of water over Pd(111) and then advance to other metals. Filhol and Neurock[24] used the approach described above to examine the potential–dependent behavior of water over Pd(111). In the absence of an electrochemical field or potential, water adsorbs on Pd(111) with its oxygen end directed down towards the surface and its molecular plane tilted away from the surface normal vector by approximately 60°, as shown in Fig. 6.11.

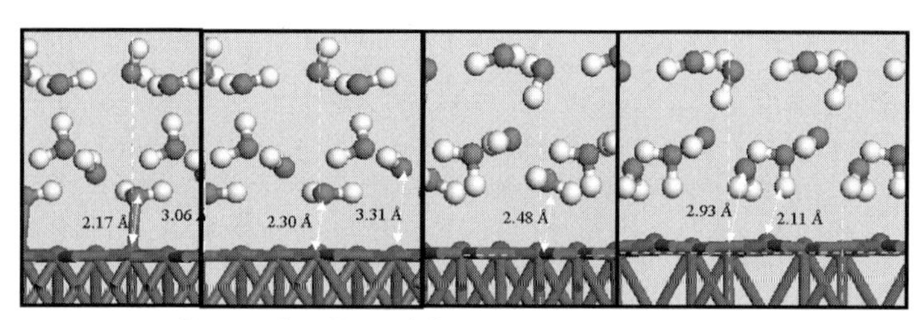

Figure 6.12. The structure of the adsorbed water in an aqueous medium as a function of electrode potential. The snapshots follow the well-known water flip-flop mechanism as one moves from a potential of zero charge to more cathodic potentials. The dark atoms refer to oxygen and the light atoms refer to hydrogen[24].

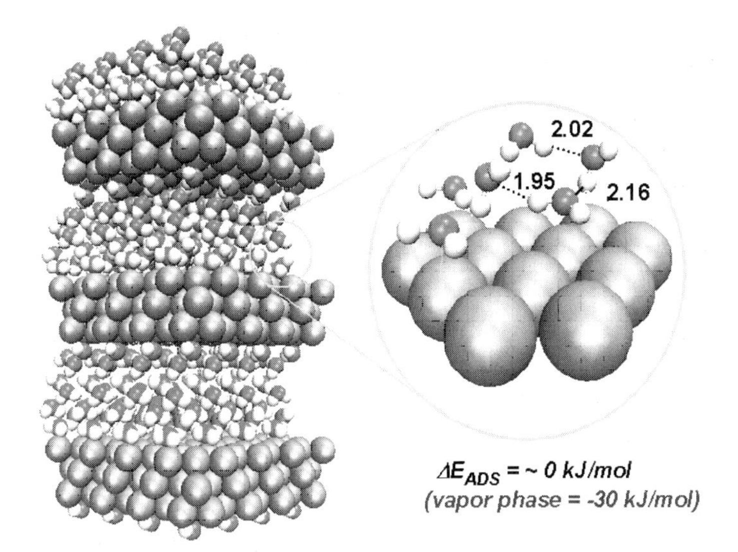

Figure 6.11. The adsorption of liquid water on Pd(111). The binding energy for water on Pd(111) in the vapor phase is 30 kJ/mol. The binding energy for liquid water is –2.5 kJ/mol[3b]. The dark atoms refer to oxygen and the light atoms refer to hydrogen.

This is consistent with the results described earlier for the single bilayer structure of water on different transition-metal surfaces. As the potential is decreased to more negative values, water flips over, whereby its hydrogen atoms are now directed toward the surface. This is seen in the sequence of structures for decreasing potential in Fig. 6.12. This change in the water structure is well known in electrochemistry as the "water flip-flop" mechanism [28]. Further decreases in the potential result in the stretching of the O–H bond. The O–H bond ultimately breaks to form an OH^- intermediate along with a surface hydride. The hydroxyl ion readily migrates into solution. A further decrease in the potential drives the OH^- species further away from the metal surface. This is the result of field-induced electromigration (diffusion). The phase transition for the activation of water to form the surface hydride appears to occur at 550 mV. This is consistent with electroadsorption of hydrogen experiments over Pd(111) which indicate that the reaction

$$(\mathrm{Pd{-}H_2O}) \longrightarrow (\mathrm{Pd{-}OH}) + \mathrm{H^+} + \mathrm{e^-} \tag{6.22}$$

proceeds at 0.4 V with respect to the hydrogen electrode[29].

At potentials which are slightly greater than the potential of zero charge, water is adsorbed with its oxygen end pointed down towards the surface. Upon further increase in the potential, water reacts to form a surface hydroxide intermediate along with a proton, which migrates into the aqueous phase to form either a hydronium or a Zundel ion, as showns in Fig. 6.13. The phase change for the conversion of water to hydroxide on Pd(111) appears to occur at 1100 mV (pH = 7). The experimental results for this system indicate that water reacts to form adsorbed hydroxide and protons at 0.8 V. The experimental results, however, were obtained in an acidic medium. Correcting for pH could shift the calculated potential closer to the actual experimental potential of 800 mV[22,28]. The composition under electrochemical conditions, however, is ill-defined, thus making it difficult to compare theory and experiment directly.

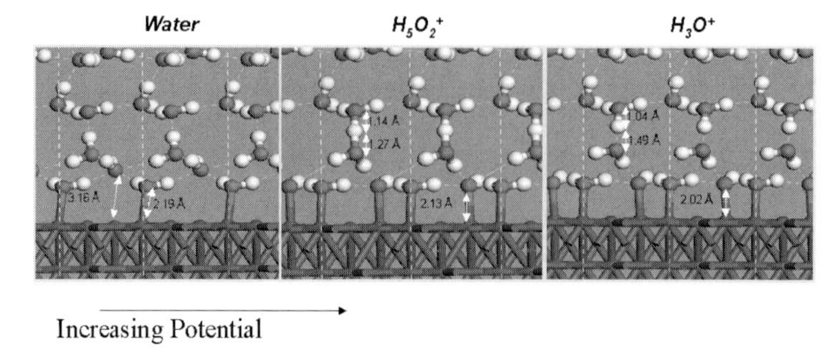

Figure 6.13. The anodic activation of water on Pd(111) to form adsorbed hydroxyl intermediates and hydronium ions. The dark atoms refer to oxygen and the light atoms refer to hydrogen[21].

The phase diagram for the activation of water to form surface hydride and surface hydroxide phases is shown in Fig. 6.14. The comparison of different structures at a given potential can be compared in order to determine the free energy differences between these phases at a constant chemical potential.

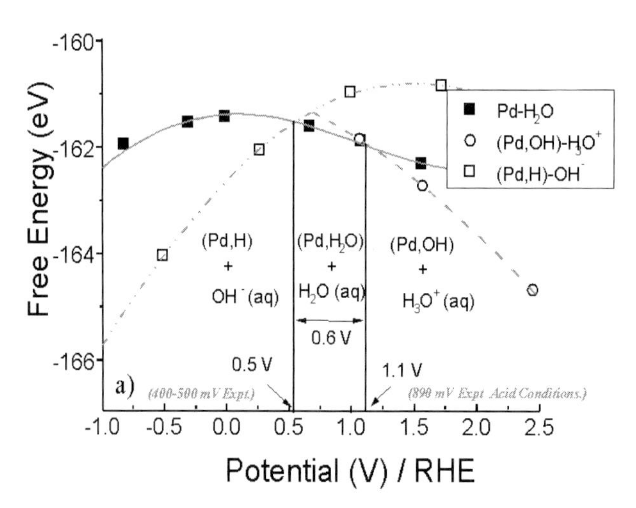

Figure 6.14. Ab initio-calculated electrochemical phase diagram for the activation of water over Pd(111) to form either the surface hydride phase along with aqueous OH^- ions or a surface hydroxide phase along with hydronium ion[21].

Similar calculations were carried out for water activation over Cu(111), Ni(111) and Pt(111) surfaces[31]. The calculations for water on Cu(111) go into much more detail and show that the surface structure changes continuously as a function of applied potential [31,32]. The results show that various additional surface phases can form that were not explored over Pd(111). The snapshots outlined in Fig. 6.15 are the result of increasing the charge in the system from what would be negative potentials to positive potentials. In order to establish explicitly whether or not one of these phases exists would require mapping out a detailed phase diagram for all of the phases, which is currently in progress. Even without the entire phase diagram, the results presented in Fig. 6.15 are rather en-

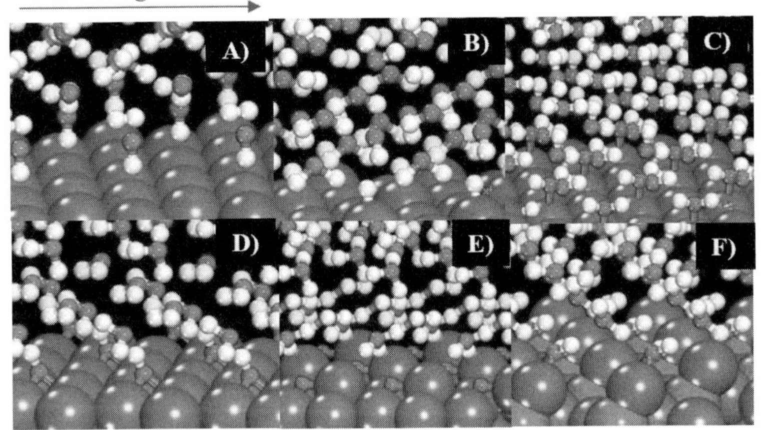

Figure 6.15. The structural changes of water and its reactivity on Cu(111) as a function of applied potential.[32].

lightening as they show the full range of different structures that can form. In moving from neutral potentials to more negative potentials, water flips over as was already discussed for water on Pd(111). At more negative potentials, water dissociates to form a surface hydride along with hydroxyl intermediates that move into solution. This occurs at −0.95 V for basic conditions. At more negative potentials, subsequent water molecules can dissociate to begin to form OHOH chains in solution along with the evolution H_2.

As the potential is increased from the potential of zero charge, water becomes more oxidized and moves from the favored atop site to a bridge site. Further increases in the potential lead to the activation of an OH bond and the formation of a surface hydroxide which binds in a two-fold hollow site and proton which migrates into the solution to form a hydronium ion. On Cu(111), the potential for this to occur was calculated to be −0.5 V for basic conditions. At even higher potentials, the OH intermediate moves from the two-fold bridge site to the 3-fold fcc site. Further increases in the potential lead to the formation of a surface oxide layer and ultimately dissolution of the metal. At even higher potentials, place exchange between adsorbed atomic oxygen and surface metal atom can proceed, thus allowing for the dissolution of Cu. While the specific potential at which this occurs has yet to be determined, the results suggest that it is possible to begin to examine the reactivity of the aqueous/metal interface and the stability of such interfaces to changes in the potential.

The examples that follow below outline how we can begin to extend the general mechanistic information gleaned from gas-phase surface reactions to more complex liquid phase systems and electrochemical systems.

6.3 The Synthesis of Vinyl Acetate via the Acetoxylation of Ethylene

Vinyl acetate is synthesized via the selective oxidation of ethylene and acetic acid over supported Pd and PdAu catalysts via the reaction

$$CH_2{=}CH_2 + CH_3COOH + \frac{1}{2}O_2 \rightleftharpoons CH_2{=}CHO(O)CCH_3 + H_2O \qquad (6.23)$$

The reaction is quite exothermic (–178 kJ/mol) and carried out in a multitubular fixed bed reactor at temperatures that range from 140 to 190 °C, pressures between 5 and 12 atm and ratios of acetic acid/ethylene/CO_2/O_2 of 10–20/50/10–30/8. [33]. Potassium acetate is typically added as a promoter to enhance reaction selectivity. The selectivity to vinyl acetate is typically greater than 96%. Despite its industrial relevance, there have been very few fundamental studies on the mechanism by which this reaction proceeds. There is a long-standing debate as to whether the chemistry is carried out on the metal surface or within a supported liquid phase that can form by the condensation of acetic acid on the support[34]. It is well established in the homogeneous catalysis literature that organometallic palladium acetate clusters catalyze the formation of vinyl acetate when run in the presence of glacial acetic acid[35,36]. Similar palladium acetate clusters can also form in the supported liquid phase of heterogeneous catalytic systems and may be responsible for the activity seen over the supported catalysts. We compare the results from gas-phase experimental and theoretical studies carried out over reduced Pd metal with those carried out in the liquid phase over homogeneous palladium ion complexes.

We start by first analyzing the mechanism for this reaction as it may occur over the metal surface and then describe mechanistic aspects of how it proceeds over in the homogeneous solution phase on organometallic Pd clusters.

The vapor-phase path which is carried out over supported Pd metal particles is thought to occur via one of two different mechanisms. In the first, acetic acid adsorbs and readily dissociates to form surface acetate intermediates as was proposed by Nakamura and Yasui[37]. Ethylene also adsorbs and can react to form a surface vinyl intermediate ($CH_2{=}CH^*$). The vinyl intermediate subsequently reacts with surface acetate directly, forming the vinyl acetate product. Molecular oxygen is activated to form atomic oxygen, which acts as a thermodynamic sink to pick up any of the surface hydrogen that forms from the activation of acetic acid or ethylene. The rates for the general steps for the Nakamura and Yasui mechanism are outlined in Fig. 6.16 with and without direct involvement of adsorbed oxygen. Oxygen can play also a role kinetically in assisting the C–H activation of ethylene and/or the O–H activation of acetic acid, thus resulting in the formation of water as a primary product which must desorb from the surface.

Nakamura/Mosieev	*Nakamura Kinetic O_2*	*Samanos*
$AcOH + * \rightarrow AcOH^*$	$AcOH + * \rightarrow AcOH^*$	$AcOH + * \rightarrow AcOH^*$
$C_2H_4(g) + * \rightarrow C_2H_4^*$	$C_2H_4(g) + * \rightarrow C_2H_4^*$	$1/2\,[\,O_2(g) + 2^* \rightarrow 2O^*\,]$
$1/2\,[\,O_2(g) + 2^* \rightarrow 2O^*\,]$	$O_2(g) + 2^* \rightarrow 2O^*$	$AcOH^* + O^* \rightarrow AcO^* + OH^*$
$AcOH^* + * \rightarrow AcO^* + H^*$	$AcOH^* + O^* \rightarrow AcO^* + OH^*$	$C_2H_4(g) + * \rightarrow C_2H_4^*$
$C_2H_4^* + * \rightarrow C_2H_3^* + H^*$	$C_2H_4^* + O^* \rightarrow C_2H_3^* + OH^*$	$C_2H_4^* + AcO^* \rightarrow C_2H_4OAc^* + *$
$C_2H_3^* + AcO^* \rightarrow VAM^* + *$	$C_2H_3^* + AcO^* \rightarrow VAM^* + *$	$C_2H_4OAc^* + * \rightarrow VAM^* + H^*$
$VAM^* \rightarrow VAM + *$	$VAM^* \rightarrow VAM + *$	$VAM^* \rightarrow VAM + *$
$O^* + H^* \rightarrow OH^* + *$	$2OH^* \rightarrow H_2O^* + O^*$	$OH^* + H^* \rightarrow H_2O^* + *$
$OH^* + H^* \rightarrow H_2O^* + *$	$H_2O^* \rightarrow H_2O(g) + *$	$H_2O^* \rightarrow H_2O(g) + *$
$H_2O^* \rightarrow H_2O(g) + *$		

Figure 6.16. The different surface reaction paths proposed for the synthesis of vinyl acetate.

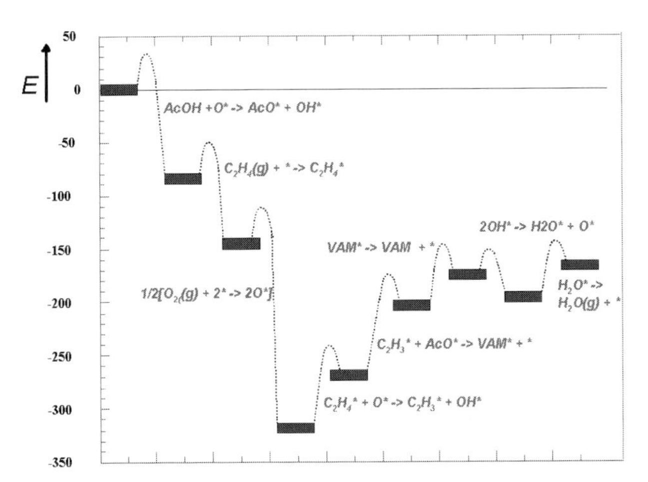

Figure 6.17. Reaction energy diagram for surface oxygen atom-assisted formation of vinyl acetate catalyzed by Pd(111)[39]. Energy is given in kJ/mol.

The primary surface reaction steps which include the activation of ethylene to vinyl and the subsequent coupling of vinyl with acetate surface intermediates have been cited as potential rate-determining steps in the Nakamura and Yasui route. Moiseev and Vargaftik carried out experiments over giant palladium clusters comprised of 561 atoms and arrived at a similar set of pathways[38]. They suggested, however, that the rate-controlling step for this process involves the shift of ethylene from the π-bound mode to a di-σ-bound mode.

Periodic density functional theoretical calculations were performed by Neurock and Kragten[39] to examine the overall reaction energies and selected activation barriers for the elementary steps for this particular mechanism in both the presence and absence of oxygen. The overall reaction energies for the elementary steps for the oxygen-assisted path are shown in Fig. 6.17. The two potential rate-limiting steps appear to be the reaction of ethylene to vinyl and the coupling of vinyl and acetate to form VAM. The transition states for the proposed limiting steps and their corresponding activation barriers were calculated using ab initio DFT calculations. The results indicate that both the C–H bond activation of ethylene and the coupling of vinyl and acetate have rather high barriers at +120 and +110 kJ/mol, respectively. Atomic oxygen readily dissociates to form atomic oxygen, which is clearly shown to be the stable thermodynamic sink, thus allowing for the formation of water.

The second general mechanism proposed for the vapor VAM synthesis involves the direct reaction between ethylene and adsorbed acetate as proposed by Samanos and Bountry[40]. Both acetate and oxygen are found to adsorb very strongly to the Pd surface, thus forming a partially oxidized surface. Ethylene can adsorb in either a π- or a di-σ-configuration or react directly from the vapor phase with adsorbed acetate in a process similar to that suggested for the homogeneous path over $Pd_3(OAc)_6$ (see Section 6.3.1). The surface reaction of ethylene with acetate involves the insertion of ethylene into the Pd–OAc bond to form the acetoxyethyl $[CH_3CO(O)CH_2CH_2^*]$ intermediate (labeled C_2H_4OAc in Fig. 6.16). The $CH_3CO(O)CH_2CH_2^*$ intermediate can subsequently undergo a β-hydride elimination, thus leading to the formation of vinyl acetate, $CH_3CO(O)CH{=}CH_2$. The hydrogen that forms can react with surface oxygen and desorb as water. This mechanism is very similar to that which was proposed by Zaidi[41]. Zaidi, however, suggested that the reaction actually proceeds in the solution phase directly at the reactive Pd–acetate complex that forms in the liquid solution layer. The potential for supported liquid-phase catalysis is described in more detail below.

The primary difference between the Nakamura ans Yasui and Samanos and Bountry mechanisms specifically involves whether ethylene stays intact before reacting with acetate or dissociates to form the vinyl intermediate. The overall energies calculated from DFT indicate that ethylene insertion into acetate and the β-C–H bond activation of the surface ethyl acetate could be rate limiting reaction steps. The overall energy for the insertion of ethylene into acetate was calculated to be 60 kJ/mol whereas the overall energy for β-C–H activation is exothermic at 30 kJ/mol[39]. The barriers for these steps, however, have not been determined.

6.3.1 Homogeneous Catalyzed Vinyl Acetate Synthesis

The acetoxylation of ethylene to form the vinyl acetate monomer (VAM) can be catalyzed by homogeneous catalysts comprised of PdCl/CuCl salts and carried out in glacial acetic acid[38]. The reaction is catalyzed by the Pd^{2+} species which are reduced by the adsorption of ethylene and subsequently reoxidized by oxygen and copper chloride. The speculated mechanism is as follows

$$\begin{aligned}
C_2H_4 + PdX_2 + HX &\longrightarrow [C_2H_4PdX_3]^- + H^+ \\
[C_2H_4PdX_3]^- + AcOH &\longrightarrow [C_2H_4PdX_2(OAc)]^- + HX \\
[C_2H_4PdX_2(OAc)]^- &\longrightarrow [C_2H_4PdX(OAc)] + X^- \\
[C_2H_4PdX(OAc)] &\longrightarrow CH_2CH\text{–}OAc + Pd + HX
\end{aligned}$$

where X is a chloroacetate species[34,35]. Yields of 90% (based on ethylene) have been reported for the processes developed by Hoechst and Bayer[33].

The reaction can also be carried out by starting out directly with palladium acetate complexes. Alkali metal salts are typically used as promoters to help increase both the activity and the selectivity. In the presence of an alkali metal promoter such as Na, palladium acetate can exist as either a trimer (Pd_3OAc_6), a dimer ($Na_2Pd_2OAc_6$) or a monomer (Na_2PdOAc_4). These three different cluster forms are in equilibrium and can be interchanged by changing the concentration of alkali metal. The rate appears to be controlled by the β-hydrogen transfer from the bound β-acetoxyethyl intermediate to form VAM. The apparent activation energy for VAM formation was reported to be 70 kJ/mol[42].

In the presence of water, Wacker chemistry tends to predominate, whereby ethylene reacts with water rather than acetic acid and forms acetaldehyde as the primary product. Henry and Pandey[44] showed that the addition of alkali metal acetates can help to shift the product spectrum in order to form vinyl acetate in higher yields. The reaction is thought to involve the nucleophillic attack of ethylene by acetate to form a C_2H_4–OAc–Pd complex which subsequently undergoes β-C–H activation to form VAM. Acetic acid or HCl can desorb from the complex to form $Pd^{(0)}$, which is reoxidized back to copper chloride to regenerate Pd^{2+}.

Zaidi[41] and Samanos[40] suggested that a similar path, which is shown below, may give rise to catalysis in the supported liquid layer for heterogeneous Pd catalysts:

$$\mathrm{Pd} + \frac{1}{2}\mathrm{O_2} + 2\mathrm{AcOH} \rightleftharpoons \mathrm{Pd(OAc)_2} + \mathrm{H_2O} \tag{6.24}$$

$$\mathrm{Pd(OAc)_2} + \mathrm{AcO^-} \rightleftharpoons \mathrm{Pd(OAc)_3}^- \tag{6.25}$$

$$\mathrm{Pd(OAc)_3}^- + \mathrm{C_2H_4} \rightleftharpoons \mathrm{AcOCH{=}CH_2} + \mathrm{AcOH} + \mathrm{AcO^-} + \mathrm{Pd^0} \tag{6.26}$$

Pd is first oxidized to form a homogeneous Pd^{2+} cluster. Ethylene can then react via an inner- or outer-sphere mechanism to form the acetoxyethyl intermediate (see next section). The acetoxyethyl intermediate then undergoes a β-hydride elimination to form VAM. DFT calculations were performed in order to understand the potential presence of the liquid layer on a Pd surface and the energetics for the proposed homogeneous pathways.

6.3.2 Elementary Reaction Steps of Vinyl Acetate in the Liquid Phase

In order to understand the influence of a liquid layer, Desai et al.[3a]. carried out periodic DFT calculations on the dissociative adsorption of acetic acid in both the presence and absence of multilayers of water over Pd. Water was used as a simple model to mimic the solution phase. The calculations were carried out by placing enough water molecules throughout the unit cell in order to minimize the energy and approach the bulk density of liquid water. Various reaction channels for acetic acid were subsequently explored. The results indicate that the dissociative adsorption of acetic acid over Pd(111) is 28 kJ/mol endothermic in the gas phase. This reaction proceeds via a homolytic process, resulting in products that are radical-like. The dissociation of acetic acid in the presence of an aqueous medium, however, proceeds via a heterolytic process, resulting in products that take on ionic characteristics. The dissociation energy of acetic acid in aqueous water over Pd(111) to produce adsorbed acetate and a hydronium ion in the aqueous phase is +37 kJ/mol. The products form a double layer at the metal–solution interface, as shown in Fig. 6.18. The acetate anions are adsorbed at the surface whereas the protons that form are removed from the surface by one solvation shell. Although this structure is thermodynamically stable at the interface, it is not the lowest energy structure that can form. The acetate anion would also prefer to be located in solution. The water molecules in solution are more effective in solvating the anion than the metal surface is. The energy difference between adsorbed acetate and free acetate anions in solution is exothermic at –57 kJ/mol, which indicates that the acetate anions prefer to reside in solution. The ability to access this state, however, requires energy to surmount the activation barrier associated with desorption and solvent reorganization. At higher temperatures, the proton from the solution phase recombines with the surface acetate species to form acetic acid. Water subsequently displaces acetic acid in a concerted effort. Acetic acid ultimately re-dissociates in solution to form an acetate anion and a proton. This process requires the energy necessary to overcome the barrier associated with solvent reorganization. Similarly, Campbell[43] found experimentally that the presence of an ethanol solution phase gave rise to a significant increase in the activation barrier for the adsorption and desorption of different alkanethiols on to the Au substrate. There is no measurable barrier for alkanethiol to adsorb on to Au under UHV conditions. In solution, the alkanethiol must first displace

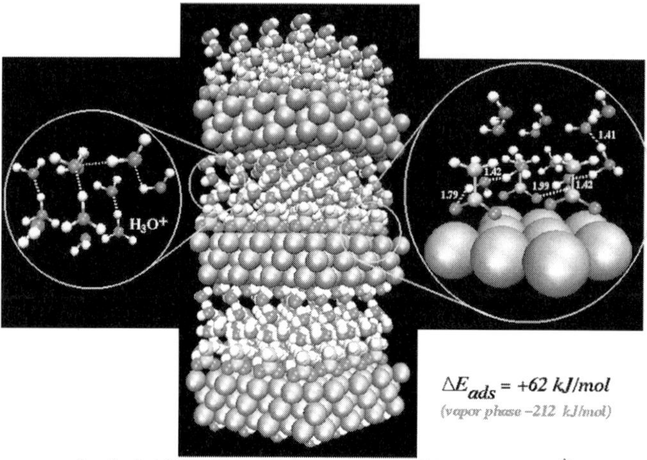

Figure 6.18. The adsorption of liquid water on Pd(111). The Pd–water binding energy on Pd(111) in the vapor phase is 30 kJ/mol. The Pd–water binding energy for liquid water is reduced to –2.5 kJ/mol. The overall binding energy, however, increases owing to the formation of hydrogen bonds with the aqueous phase[3a].

the ethanol before it can adsorb. This solution reorganization energy ultimately leads to the measured barrier.

The aqueous solution layer that forms at the metal interface can ultimately provide a medium for the dissolution of Pd ions or oxidized Pd clusters into the supported liquid layer where they can then act as homogeneous catalysts. As was discussed earlier, the acetoxylation of ethylene can be carried out over various Pd_xOAc_y clusters where alkali metal acetates are typically used as promoters. DFT calculations were carried out on both the $Pd_2(OAc)_2$ and $Pd_3(OAc)_6$ clusters in order to examine the paths that control the solution-phase chemistry. The $Pd_3(OAc)_6$ cluster is the most stable structure but is known experimentally to react to form the $Pd_2(OAc)_2$ dimer and monomer complexes in the presence of alkali metal acetates. The reaction proceeds by the dissociative adsorption of acetic acid to form acetate ligands. Ethylene subsequently inserts into a Pd–acetate bond. The cation is then reduced by the reaction to form the neutral Pd^0. The reaction is analogous to the Wacker reaction in which ethylene is oxidized over Pd^{2+} to form acetaldehyde. Pd^0 is subsequently reoxidized by oxygen to form Pd^{2+}[35,36,44].

The reductive elimination of vinyl acetate can then proceed through either an inner- or an outer-sphere reaction channel. The inner-sphere mechanism involves the reaction between two ligands that are both already coordinated to the metal center. The outer-sphere mechanism involves the reaction between ethylene which is coordinated to the metal center and an acetate anion which resides in solution. A sketch of both the inner- and outer-sphere reactions is given in Fig. 6.19.

Kragten et al.[36] carried out DFT calculations to determine the reaction energies and the activation barriers for a sequence of elementary steps that make up both the inner- and outer-sphere mechanisms. The effects of solution are included via the explicit introduction of one or two acetic acid molecules along with an overall reaction field. The solute is modeled by a cluster in which the charge is balanced by the coordination of protons to

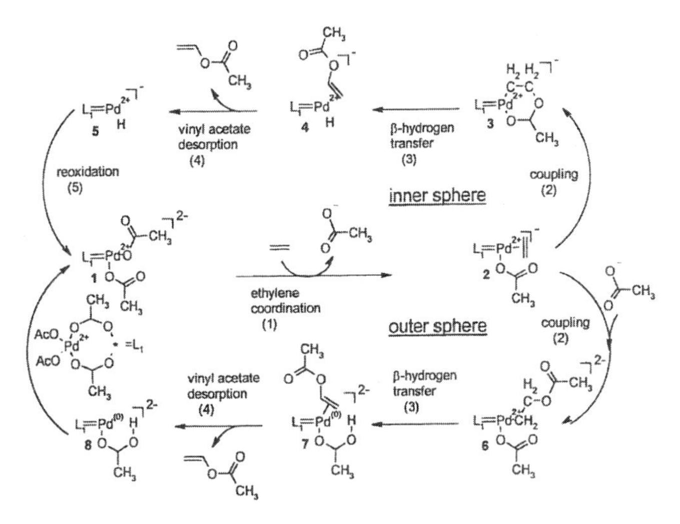

Figure 6.19. Proposed catalytic cycles for the inner- and outer-sphere Wacker-like mechanisms for the acetoxylation of ethylene to vinyl acetate[36].

the cluster. The field is modeled via an overall continuum comprised of a specific dielectric constant.

The overall potential energy profiles were calculated for both the inner- and outer-sphere mechanisms for the homogeneous acetoxylation of ethylene to vinyl acetate. The results, which include the effect of solvation, are shown in Fig. 6.20.

Both mechanisms initially proceed by the direct coordination of ethylene to a Pd^{2+} cation vacancy site. In the inner-sphere path, the coordinated ethylene species reacts with an adjacent acetate ligand. This results in the formation of the acetoxy ethyl intermediate coordinated to the Pd^{2+} center via the formation of a six-membeed ring structure. The acetoxy ethyl intermediate subsequently undergoes β-CH bond activation over the palladium ion and forms vinyl acetate along with a hydride ligand (Pd–H). The calculated barrier for C–H activation is 68 kJ/mol. Vinyl acetate subsequently desorbs, thus regenerating a free site. The resaturation of this vacancy with a ligand costs 85 kJ/mol. The desorption of VAM therefore appears to compete with β-C-H transfer as the rate-limiting process.

In the outer-sphere mechanism, the coordinated ethylene reacts with an acetate anion in the solution phase to form the acetoxy ethyl intermediate at the Pd center. Acetoxy ethyl subsequently undergoes a β-C–H transfer step which involves the coordinative transfer of a proton from ethyl acetate to an adjacent acetate anion bound to the Pd^{2+} center. This step liberates vinyl acetate along with acetic acid.

Figure 6.20 shows quite clearly that the outer-sphere mechanism is much less endothermic than the inner-sphere mechanism and hence the more likely path. β-Hydrogen transfer appears to be the limiting step for both mechanisms. Although the presence of solution helps to stabilize the transition state for this step in the outer-sphere process, the reaction is still fairly endothermic. In the outer-sphere mechanism, hydrogen does not have to transfer to a vacant site but can undergo a ligand-to-ligand transfer. This coupling reaction is clearly much more favored in the outer-sphere path. The reaction is negative order in acetate. This suggests that an acetate vacancy is necessary to carry out the reaction.

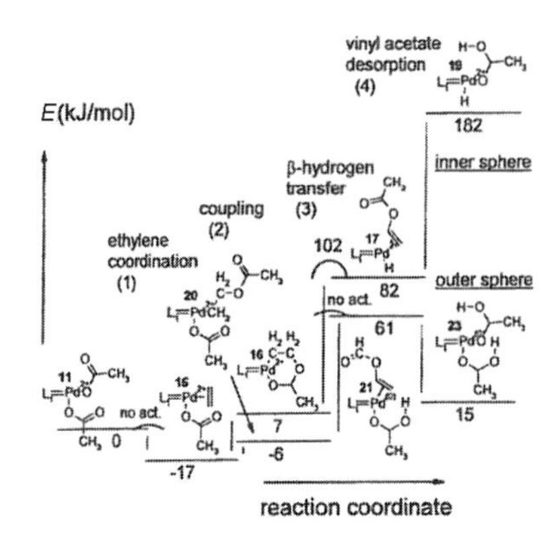

Figure 6.20. Energy diagram of the inner- and outer-sphere mechanisms, including solvent effects. Activation barriers are indicated by the small arcs; a barrier as low as the reaction energy is designated by "no act." The structure number refers to models shown in Fig. 6.19[36].

The results are consistent with the literature, which indicates that VAM synthesis proceeds over homogeneous palladium acetate complexes via either an inner- or an outer-sphere mechanism[35,44]. The reaction process is quite similar to that for Wacker chemistry, which is thought to proceed via an outer-sphere process. The rate-limiting step for both the inner- and outer-sphere mechanisms appears to be the β-hydrogen transfer. This is consistent with experimental studies for vinyl acetate synthesis over homogeneous Pd acetate and with other theoretical studies carried out on oxidized Pd clusters. The activation barrier for the β-hydrogen transfer was calculated by DFT to be 67 kJ/mol, which is in good agreement with the experimental value of 70 kJ/mol found by Tamura and Yasui[44] for VAM synthesis over homogeneous Pd acetate. These comparisons, should be made very carefully, however, since the results are highly dependent on the actual reaction conditions.

Solvent effects were found to be quite important in stabilizing charged complexes, especially for the outer-sphere mechanism. In general, outer-sphere reactions are found to have lower activation barriers than those for inner-sphere reactions. This is due to the enhanced charge stabilization due to the presence of solution. Solvent effects play a small role in altering the energy landscape. The influence of solvation has been included and has been estimated using the reaction field theory expressions. Solvation corrections may vary between -1 and -12 kJ/mol.

In addition to introducing the concepts of inner- and outer-sphere mechanisms, we have also described proton transfer between ligands without direct interference with the Pd^{2+} cation. This coordinative proton transfer helps to lower the activation barrier for β-CH transfer in the outer-sphere mechanism, thus providing a lower energy path than that available for the inner-sphere reaction route. Vinyl acetate desorption then becomes easier in the outer-sphere route with a desorption energy of only 45 kJ/mol. The outer-sphere mechanism is therefore preferred. The rate-controlling step in the outer-sphere mechanism appears to be that for β-CH transfer. The activation energy for the ethylene–acetate coupling reaction is significantly lower than that found in the inner-sphere. This is because the presence of solution stabilizes the anion, thus opening up a low-energy nucleophillic attack by the anion.

6.3.3 VAM Synthesis: Homogeneous or Heterogeneous?

It is clear that there is now evidence that VAM synthesis can occur via both homogeneous and heterogeneous pathways. Which of these mechanisms predominates is still actively debated. Elegant reviews comparing the two mechanisms were given by Kragten[45] and by Reilly and Lerou[34]. Mosieev and Vargaftik[38] indicated that the active sites are the Pd^0 species and that these sites lead selectively to vinyl acetate with essentially no production of acetaldehyde. Nakamura and Yasui[37] and Debellefontine and Besombes–Vailhe[46] et al., on the other hand, suggest that Pd^{1+} species in the form of Pd–OAc are the active species that catalyze the reaction. There is evidence that a supported liquid layer can form under reaction conditions[40] and that VAM synthesis predominantly occurs over the Pd^{2+} species that form in this layer. Augustine and Blitz[47] showed that $Pd(OAc)_2$ can form over Pd crystallites. Crathorne et al.[48] suggested that a liquid layer which is 3 ML of acetate/acetic acid thick can form and that the presence of alkali metal acetates increases the absorption of AcOH. KOAc is speculated to be essential for VAM formation. KOAc is thought to immobilize acetic acid in the form of a KOAc melt, which can solubilize homogenous Pd^{2+}-acetate complexes that then carry out the chemistry. The presence of potassium and acetic acid help to limit the production of acetaldehyde. In addition, the liquid layer that forms can block the metal surface to ethylene exposure and thus suppresses ethylene combustion. Others, however, have indicated that the Pd^{2+} species that form in the liquid layer are not active and can actually lead to agglomeration and sintering since they are more mobile.

Recent in situ spectroscopic studies on the reactions of ethylene and acetic acid over homogeneous complexes and reactions carried out over single-crystal Pd(111) surfaces tend to suggest that the dominant path is over the metal surface. In situ ultraviolet–visible, Raman, and infrared experiments carried out on the homogeneous Pd acetate clusters by Kragten et al.[36b] show that although vinyl acetate is formed, the primary product is actually acetaldehyde, which occurs via Wacker chemistry. Water, which is the product from VAM synthesis, readily reacts with the homogeneous complexes to form acetaldehyde, thus significantly decreasing the overall selectivity. The thought is that although these homogeneous complexes may form, they are unlikely to catalyze the primary route to VAM.

Recently, Tysoe and Stacchiola[49a], showed that VAM can be readily formed under UHV conditions by the addition of ethylene over an acetate-covered Pd(111) surface. The acetate intermediates were formed by the dissociative adsorption of acetic acid at low temperature on the surface. The subsequent addition of ethylene leads to the titration of acetate from the surface. The primary product identified by mass spectrometry was vinyl acetate. The reaction was carried out over a narrow temperature range indicating an activation barrier of about 65 kJ/mol. More recently, the same group[49b] used in situ IR spectroscopy to show that reactions with labeled ethylene resulted in a shift of the IR band at 1788 cm^{-1}, which is indicative of VAM, to 1718 cm^{-1}. DFT calculations indicate that the band at 1788 cm^{-1} is from adsorbed VAM whereas the band at 1718 cm^{-1} is much more likely the adsorbed acetoxyethyl intermediate. This suggests that the Samanos mechanism, which proceeds via the direct coupling of ethylene with acetate over the Pd surface, is the dominant path rather than that which proceeds through the formation of surface vinyl intermediates and their coupling to form VAM.

The in situ Raman and UHV studies indicate that it is more likely that the metal surface, and not solution-phase homogeneous complexes, catalyzes the acetoxylation of ethylene to form vinyl acetate.

6.4 Low-Temperature Ammonia Oxidation

In a second example, we examine reactions that relate to the Ostwald oxidation process, where NH_3 is converted to NO with high selectivity. The reaction is typically run at high temperatures of around 1100 K over Pt/Rh alloy catalysts. The NO that forms is subsequently converted into nitric acid via a series of consecutive reaction steps. At lower temperatures, ammonia reacts to form N_2 and N_2O instead. The low-temperature conversion of ammonia to N_2 would be much more desirable in that it would lower energy costs and, in addition, replace NO, an atmospheric pollutant, with N_2, which is environmentally benign. We will describe here the low-temperature catalytic conversion of ammonia to form N_2. For a review of high-temperature oxidation, in which coupling with gas-phase radical chemistry plays an important role, we refer to Ref. [50].

Here, we compare three different catalytic systems used to carry out the catalytic oxidation of ammonia: gas-phase oxidation over transition-metal heterogeneous catalysts, gas-phase oxidation within zeolites and electrocatalytic oxidation of ammonia.

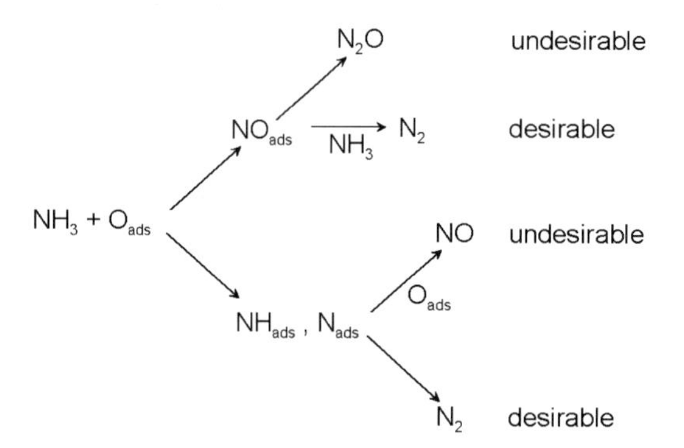

Figure 6.21. Low-temperature oxidation of NH_3 using transition-metal catalysts.

In this first section, we focus on heterogeneous transition-metal catalysis. Two different reaction paths can be distinguished in the activation of NH_3 (see Fig. 6.21). The first distinguishes the direct path through the dissociation of ammonia and subsequent recombination of N_{ads} atoms. The second reaction path involves the reaction of NH_3, NO and O to form N_2 and H_2O (see also Chapter 5, Section 5.6.10).

Ammonia adsorbed on the transition-metal surface can react to form different NH_x species. Both the presence of adsorbed oxygen and hydroxyl adsorbed on the metal surface can help promote this reaction (see Chapter 3, Section 3.8).

Adsorbed oxygen readily acts to promote the activation of NH_3 to NH_2. Surface hydroxyl intermediates, however, were found to be even more reactive than adsorbed oxygen and lower the barriers of all NH_x decomposition reactions. At high temperatures all of the NH_x intermediates dissociate to form nitrogen atoms, which can then recombine and desorb as N_2 or react with adsorbed O and desorb as NO. In Table 6.1 we present a comparison of DFT-calculated energies found in the literature for some of the elementary reaction steps over different Pt surfaces. The results indicate that there is a very high sensitivity of the activation energies to surface structure. The reactivity of the (100) surface is dramatically increased over that of the (111) surface. As was discussed in Chapter 3, this increase in the rate is due to the removal of metal atom sharing between the fragments that form in the transition state.

Table 6.1. Computed activation energies in kJ/mol for elementary surface reactions relevant to ammonia oxidation.

	Surface			
Reaction	[111]	[211]	[533]	[100]
$NO + N \rightarrow N_2O$	178[a];200[b]		43[h]	142*
$NO + O \rightarrow NO_2$	152[a]			
$NO + NO \rightarrow N_2O + O$	160[b]			115*
$NO \rightarrow N + O$	210[b];235[e], 246**	65[f]		107[f] 110[i]
$N + O \rightarrow NO$	220**	73[e]		21[f]
$N + N \rightarrow N_2$	255**	73[g]	104[h]	10[i]
$O_2 \rightarrow 2O$	86[c]			
$O + H \rightarrow OH$	87[d]			
$OH + H \rightarrow H_2O$	22[d]			

* M. Neurock, S.A. Wasileski, D. Mei, *Chem. Eng. Sci.* 59, 4703 (2004)
** W. Offermans, R.A. van Santen, to be published.
[a] R. Burch, S. T. Daniells, P. Hu, *J. Chem. Phys.* 117, 2902 (2002)
[b] A. Bogicevic, K. C. Hass, *Surf. Sci.* 506, L237 (2002)
[c] A. Eichler, F. Mittendorfer, J. Hafner, *Phys. Rev. B* 62, 4744 (2000)
[d] A. Michaelides, P. Hu, *J. Am. Chem. Soc.* 122, 9866 (2000)
[e] M. Neurock, *From First Principles To Catalytic Performance: Tracking Molecular Transformations*, Presentation at Schloss Ringberg Conference, Tegernsee 11 september 2003
[f] Q. Ge, M. Neurock, *J. Am. Chem. Soc.* 126, 1551 (2004)
[g] E.H.G. Backus, A. Eichler, M.L. Grecea, A.W. Kleyn, M. Bonn, *J. Chem. Phys.* 121, 7946 (2004)
[h] H. Wang, R.G. Tobin, C.L. DiMaggio, G.B. Fisher, D.K. Lambert, *J. Chem. Phys.* 107, 9569 (1997)
[i] A. Eichler, J. Hafner, *Chem. Phys. Lett.* 343, 383 (2001)

The activation of ammonia with oxygen to produce NO is endothermic, whereas the activation of ammonia to form N_2 in the gas phase is exothermic. The selectivity towards NO will, therefore, increase as the temperature is increased. However, the activation barrier to form adsorbed NO from adsorbed N and O is comparable to that for the recombination of two nitrogen adatoms to form N_2. The only data that can be directly compared are those for the (111) surface, where there is a large enough dataset and the methodology applied is similar. These reactions preferentially occur at step edges. NO strongly adsorbs to the surface, so that at low temperatures NO likely does not appear as a product. Only N_2 and N_2O are formed as products at low temperature. As is seen from Table 6.1, N_2O formation readily occurs at low temperature due to the recombination of adsorbed NO. N_2O is weakly adsorbed and therefore desorbs once it is formed.

As was just mentioned, in addition to its reaction with ammonia, NO can react with itself or with a surface nitrogen adatom to form N_2O:

$$NO_{ads} + NO_{ads} \longrightarrow N_2O + O_{ads} \tag{6.27}$$
$$NO_{ads} + N_{ads} \longrightarrow N_2O \tag{6.28}$$

Theoretical studies on Cu[51] and Pt[52] have shown that the associative mechanism is the preferred reaction (see Table 6.1 also) As indicated in the upper portion of Fig. 6.21, NO can in principle also react with NH_3 to form N_2. In ammonia oxidation catalysis, this is known as the Fogel mechanism. The reaction of NO and NH_3 in the presence of oxygen is a well-known reaction used commercially at higher temperatures to remove NO from exhaust emission streams by its injection with ammonia as a sacrificial reductant. The catalysts that are typically used to carry out selective catalytic reduction (SCR) are based on MoO_3, V_2O_5, WO_3 and Cr_2O_3 (see Chapter 5, Section 5.6.10). Although these catalysts are less reactive than Pt, they offer much higher selectivities at higher temperatures.

Reaction	Activation energy (kJ/mol)	Overall energy (kJ/mol)
$NH_3(g) \longrightarrow NH_2(g) + H(g)$	+498	+498
$NH_3^* \longrightarrow NH_2^* + H^*$	+344	+176
$NH_3^* + O^* \longrightarrow NH_2^* + OH^*$	+132	+86
$NH_3^* + O^* \longrightarrow NH^* + H_2O(g)$	>204	+92
$NH_3^* + O_2^* \longrightarrow NH_2^* + OOH^*$	+67	−84
$NH_3^* + O_2^* \longrightarrow NH^* + O^* + H_2O(g)$	+134	−184

Figure 6.22. Ammonia oxidation activation energies and overall reaction energies on Cu(111)[53].

Figure 6.22 highlights the reaction energies and the activation energies for the oxidation of ammonia over the Cu(111) surface. Note the exothermicity of the reactions that proceed directly via reaction with coadsorbed molecular oxygen.The experimental selectivity for the ammonia conversion reaction as a function of temperature[54] over a Cu catalyst is shown in Figure 6.23.

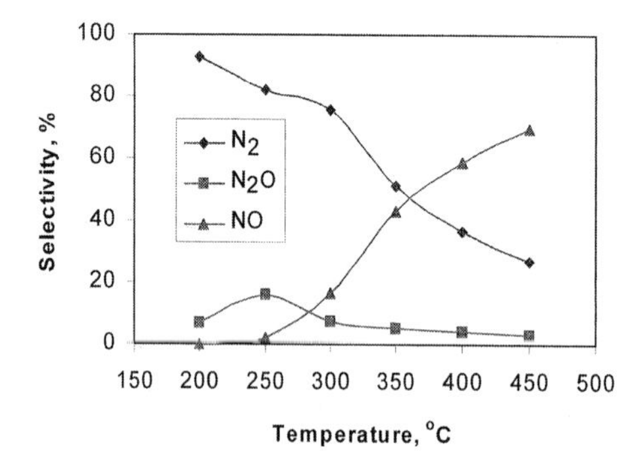

Figure 6.23. Ammonia oxidation on Cu/Al_2O_3. Reaction conditions: NH_3 = 1000 ppm, O_2 = 10%, GHSV = 90,000[54].

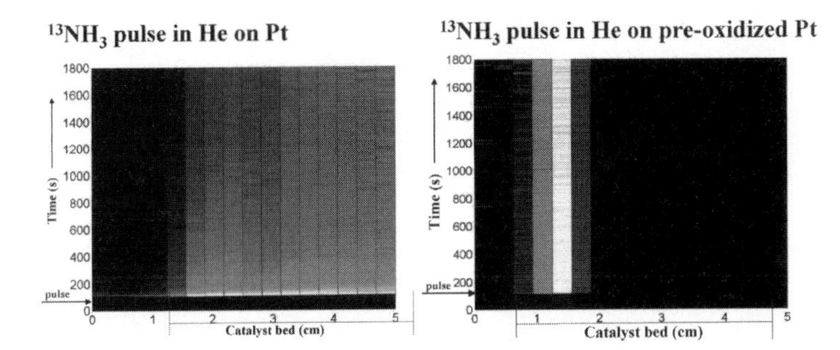

Figure 6.24. Adsorption of ammonia on Pt and pre-oxidized Pt sponge at 323 K. Pre-oxidized Pt: treated with 3% of O_2 in a helium flow at 473 K for 1 hour[54b].

At low temperatures, the only products that form are N_2O and N_2. In situ spectroscopic studies of working Cu and Ag catalysts show that apart from adsorbed oxygen, there is a high surface coverage of nitrite and nitrate species[55]. Hence, on these metals at low temperature, N_2 and N_2O production is likely the result of consecutive reactions of NO_x, the most abundant reaction intermediate (MARI), with NH_3. N_2 is formed by the reaction of nitrite with NH_3, whereas N_2O can also form via reaction of nitrate with ammonia (see also Section 6.4.1).

The low-temperature activation of NH_3 over a Pt sponge[54,57] illustrates the complexity of this reaction on platinum. On Pt under steady-state conditions, the surface is predominantly covered with NH_x and OH_x species[56]. This is quite different from the surface state on Cu and Ag. As shown in Fig. 6.24, only a Pt catalyst precovered with oxygen will activate NH_3, and at low temperature the products remain adsorbed as NH_x intermediates. The results in Fig. 6.24 on the left show little ammonia reaction from a pulse of NH_3 in He on Pt. In the absence of oxygen, NH_3 does not activate, therefore no NH_x product species are formed. The $^{13}NH_3$ pulse rapidly moves through the catalyst bed. The reactivity of an ammonia–oxygen mixture on the reduced Pt at 320 K is initially quite high but rapidly declines after only a few minutes (see Fig. 6.25). The initial product, N_2, subsequently becomes replaced by N_2O. The production of both species rapidly decreases with time[54b]. These features are consistent with recombination of N adatoms to form N_2 and N and O adatoms to form NO through low activation energy reaction channels. These reactions tend to be favored at the step edge sites. Nitrogen will desorb, but NO does not desorb at low temperatures. Adsorbed NO will diffuse away from the step edges to the terraces and will only desorb once N_2O has been formed. The initial period of high activity at low temperature becomes inhibited at longer times. The terraces become covered with a high concentration of non-reactive NH_x species. Radiochemical experiments at low temperature show that 20% of reacted NH_3 remains adsorbed on the catalyst (Fig. 6.25) as an irreversible adsorbed N species.

Interestingly, single crystal experiments on Pt executed at low pressures have never observed the presence of N_2O as has been seen in experiments carried out at higher pressures on Pt sponges. This is indicative of very different surface conditions in the vacuum experiment to those in the atmospheric pressure experiments presented above. These differences are very likely the result of the different surface compositions and coverages that form in reactions carried out under vacuum conditions and those at atmospheric pressure.

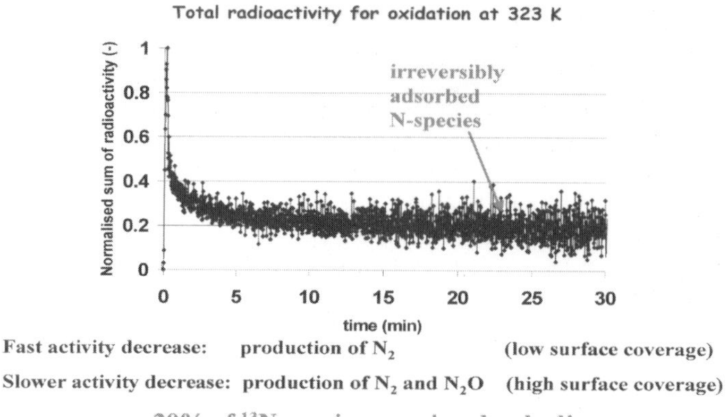

Figure 6.25. Radioactivity from adsorbed $^{13}NH_x$ species , oxidation at 323 K after an initial $^{13}NH_3$ pulse in an oxygen flow. 20% of ^{13}NH species remain adsorbed[54b].

Ab initio kinetic Monte Carlo simulations carried out by Neurock and co-workers[52] on Pt(100) surface, for example, show that under UHV conditions NO does not react to form N_2O but as the pressure is increased to atmospheric conditions N_2O becomes one of the major reaction products along with N_2. In these simulations NO dissociates when a surface has a low coverage with NO. N_2O is formed when the coverage with NO is such that the dissociation of NO becomes suppressed.

Single-crystal UHV experiments, however, demonstrated the unique finding that N_2 can be formed at temperatures as low as 210 K[57] if the reactions are carried out on Pt(100). The Pt(100) surface was found to be active, in that it contains uniquely active 4-fold sites that allow for the dissociation of NO at significantly lower barriers than other surfaces. Similarly, the barrier for the recombination of N_{ads} adatoms to form N_2 is also rather low. The unusually low activation energies reported here are due to the elimination of sites which poison as the result of their reactants and/or products due to the elimination of metal atom sharing in the transition state. As was discussed in previous chapters, NO activation over Pt(100) and also at step edges is due to the removal of lateral repulsive interactions that arise from metal atom sharing.

Additional experimental evidence on the reactivity of different surface intermediates initially formed on the Pt sponge has also been obtained via radiochemical temperature-programmed reaction experiments. These experiments probe the state of Pt and pre–oxidized Pt surfaces and their composition after exposure to a pulse of ammonia. A dramatically different reaction pattern is found as a function of temperature when in a subsequent experiment such a pretreated surface is exposed to a flow of H_2, NO or NH_3 [54] (see Fig. 6.26). The results from temperature-programmed desorption experiments show the desorption of a small amount of N_2 after surface treatment with NH_3 at 373 K and a small amount of NO desorption at 540 K (Fig. 6.26a). In contrast, experiments performed at fixed temperatures lead to the formation of much larger amounts of N_2O and N_2, which form immediately after H_2 addition (Fig. 6.26b). This result is consistent with the proposal that in the presence of H_2, the non-reactive surface oxygen (with respect to adsorbed NH_x) is converted to surface hydroxyl groups which demonstrate a much higher reactivity for activating the NH bonds than atomic oxygen does. The formation of N_2 and NO occurs from stepped or reactive surfaces as the (100) surfaces or step edges. The

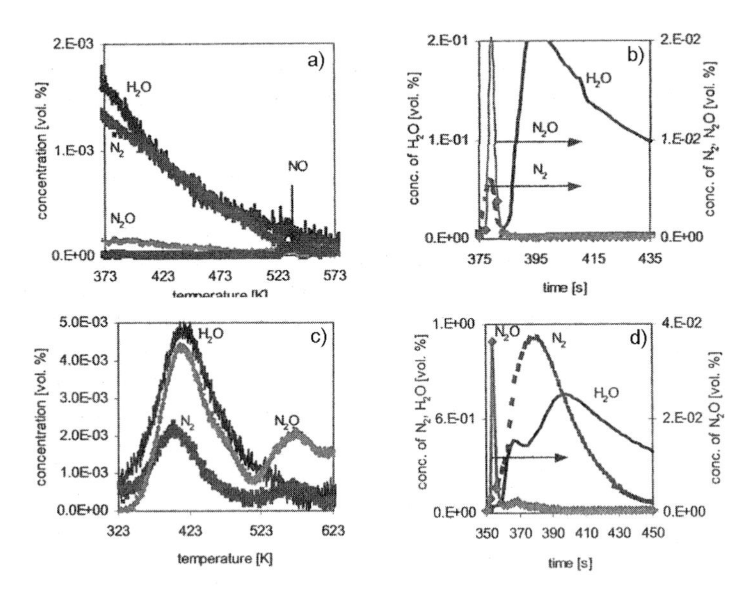

Figure 6.26. (a) A [^{13}N]NH_3 pulse was adsorbed on the pre-oxidized platinum sponge kept under an He flow at 373 K followed after 170 s by temperature-programmed desorption. (b) A [^{13}N]NH_3 pulse was adsorbed on a pre-oxidized platinum sponge kept under an He flow followed after 380 s by a removal of N species with hydrogen: mass spectrum shown from the moment of hydrogen addition. (c) Formation of N_2, N_2O and H_2O measured by online mass spectrometry in temperature-programmed NO reaction after oxidation with ammonia at 323 K. (d) A [^{13}N]NH_3 pulse was adsorbed on the pre-oxidized platinum sponge kept under an He flow followed after 350 s by a removal of N species with ammonia. Mass spectrum shown from the moment of NH_3 addition.

reaction of NO responsible for low temperature N_2O formation is

$$2NO_{ads} \longrightarrow N_2O + O_{ads} \tag{6.29}$$

Figure 6.26c shows the result of NO treatment of the Pt surface after exposure to ammonia and oxygen at low temperature. The products appear in two temperature regimes. The first product peaks appear at about 400 K and are the result of N_2, N_2O and H_2O formation. A second set of peaks appear at 570 K, which correspond to the formation of N_2O and N_2. Reaction (6.29) is consistent with low-temperature N_2O formation. The high-temperature N_2O peak is consistent with a more difficult reaction sequence:

$$NO_{ads} + N_{ads} \longrightarrow N_2O \tag{6.30}$$

NO tends to desorb at the same higher temperature hence, surface vacancies become available for NO dissociation. Finally, Fig. 6.26d shows that the addition of NH_3 initially produces a small amount of N_2O and after some time delay also produces N_2. The experiment represented by Fig. 6.26d is also consistent with a second reaction channel for N_2 formation apart from the direct recombination of adsorbed nitrogen atoms. The N_2 produced in this experiment is most likely due to the SCR reaction:

$$NO_{ads} + NH_{2ads} \longrightarrow N_2 + H_2O$$

This helps to explain the coproduction of H_2O. The difference in selectivity on Pt for the ammonia oxidation reaction, which produces N_2 at low temperature and NO at high temperature is due to the very different surface composition at these conditions. At low temperature, where N_2 recombination is slow, the surface is covered mainly by NH_x species. Above the temperature of N_2 desorption, oxygen is the dominating surface-adsorbate species. Once NH_3 decomposes then the probabilities for NO formation are high.

6.4.1 Ammonia Oxidation with Pt^{2+} Ion-Exchanged Zeolite Catalysts; Catalysis Through Coordination Chemistry

Ammonia oxidation can also be carried out over Pt-exchanged zeolite catalysts. In these systems, it is desirable to produce highly dispersed Pt along the exterior or inside the micropores of the acidic zeolite. This is accomplished by exchange of cations located inside the zeolite channels by positively charged Pt^{2+} complexes and the consecutive reduction of these complexes. The $Pt^{2+}(NH_3)_4$ complex is typically the preferred ion of choice for carrying out the exchange with zeolite channel cations. If one wishes to prepare a highly dispersed Pt catalyst on an alumina or magnesium oxide surface, it is worth remembering that these surfaces are predominantly covered with basic OH^- groups. The preferred complex for ion exchange is then ${PtCl_4}^{2-}$. The ${Pt(NH_3)_4}^{2+}$ ion-exchanged zeolite can be converted into an excellent low temperature oxidation catalyst for the production of N_2 from NH_3. The results for conversion of ammonia in the presence and absence of H_2O are shown in Fig. 6.27.[56].

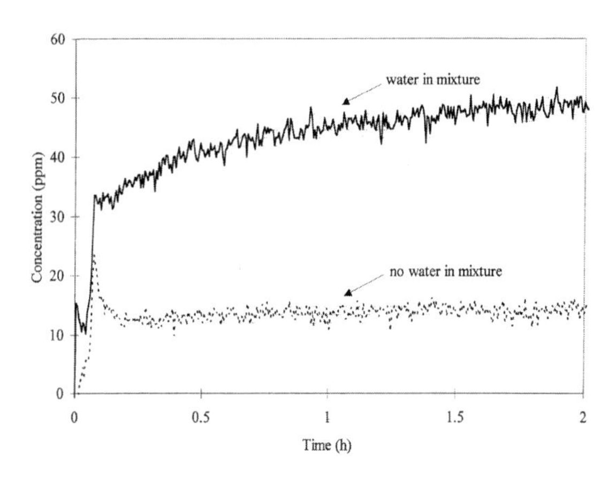

Figure 6.27. The concentration of nitrogen in the oxidation of ammonia over $Pt(NH_3)_4$ containing ZSM-5 at 473 K. Water 0%, dotted line; 1.7%, full line. The N_2O production (~5 ppm) is unaffected by addition of water[58].

In contrast to the noble metal catalysts discussed in the previous section, the presence of water in the bifunctional Pt zeolite does not suppress the activity of the reaction but actually enhances the rate of ammonia conversion and also the reaction selectivity to N_2. The two primary reaction products are, once again, N_2O and N_2. Water enhances the selectivity for the formation of N_2.

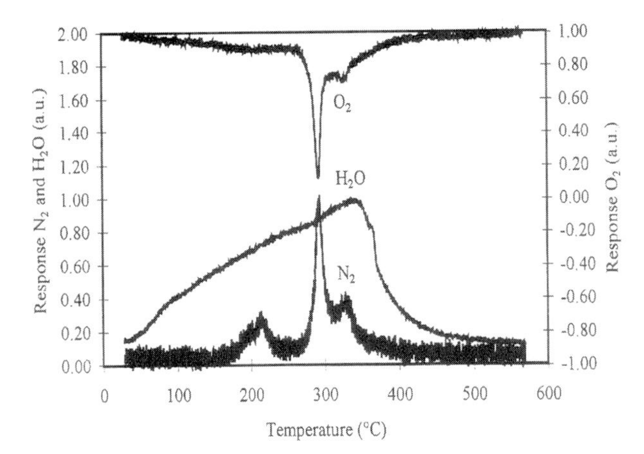

Figure 6.28. Temperature-programmed mass spectrometry of the decomposition of $Pt^{2+}(NH_3)_4$ ion exchanged in zeolite HZSM-5[58b]. The corresponding reaction steps are indicated. Three peaks can be distinguished in the TPD spectrum. Oxygen is consumed in only two of the N_2 formation peaks. The mechanism that explains the occurrence of these three peaks is consistent with proposals made earlier on the homogeneous oxidation of Ru–amine complexes in basic solution[59] and the reaction of NO with Ru or Os complexes[60].

Temperature-programmed desorption data reveal three different temperatures at which the $Pt(NH_3)_4{}^{2+}$ complex reacts with O_2 (Fig. 6.28)[58].

The zeolite lattice carries a negative charge, which compensates for the positive charge of the ion-exchanged Pt complex. The zeolite lattice oxygen atoms therefore act as basic sites. At low temperature the key initial reaction step is

$$Pt(NH_3)_4{}^{2+} + O_2 \longrightarrow [Pt(NH_3)_3NO]^+ + H^+ + H_2O \qquad (6.31)$$

In the Pt–amine complex, the NO ligand formally has a charge of −1. The Pt(d^8)-containing $Pt^{2+}(NH_3)_4$ complex was determined to be planar. Therefore, the initial reaction intermediate of O_2 and $Pt^{2+}(NH_3)_4$ probably involves a complex similar to that showns in Fig. 6.29.

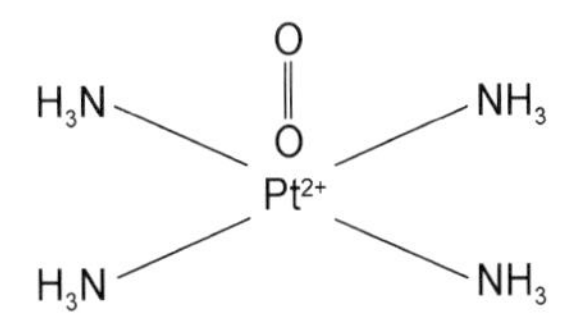

Figure 6.29. Schematic illustration of a potential intermediate that forms upon the adsorption of O_2 on $Pt^{2+}(NH_3)_4$ (schematic).

where O_2 is stabilized by the back-donation of electron density from filled molecular orbitals into d_{xz}, d_{yz} states. The reaction of NH_3 with O_2 ultimately leads to the formation of NO. This is due to the fact that a single N atom is not stabilized on a single Pt^{2+} center and must therefore react with other potential intermediates. After the NO formation step, five different reactions can occur:

Reaction step 1:

$$[Pt(NH_3)_3(NO)]^+ \longrightarrow [Pt(NH_3)_2] + N_2 + H^+ + H_2O \tag{6.32}$$

This involves the reductive elimination of N_2, which is initiated by the activation of NH_3 with NO to form NH_2. It is the analogue of the Fogel reaction discussed in the previous section.

Reaction step 2:

$$[Pt(NH_3)_3(NO)]^+ + H_2O \longrightarrow [Pt(NH_3)(NO_2)]^- + 2H^+ \tag{6.33}$$

Here, Pt^{2+} acts as an oxidant and liberates atomic O from N_2O. When followed by step 4, this intermediate will also lead to H_2 production but now promoted by H_2O.

Reaction step 3:

$$[Pt(NH_3)_3(NO)]^+ + O_2 \longrightarrow [Pt(NH_3)_3(NO_3)]^+ \tag{6.34}$$

In this reaction, $NO_3{}^-$ formed by oxygen consumption is responsible for N_2O formation (see reaction step 5).

Reaction step 4:

$$[Pt(NH_3)_3(NO_2)]^- + H^+ \longrightarrow Pt(NH_3)_2 + N_2 + 2H_2O \tag{6.35}$$

In this reaction, a proton is consumed to produce N_2.

Reaction step 5:

$$[Pt(NH_3)_3(NO_3)]^+ + H^+ + 2NH_3 \longrightarrow Pt^{2+}(NH_3)_4 + N_2O \tag{6.36}$$

Here, $Pt^{2+}(NH_3)_4$ is regenerated by a reaction between the zeolitic protons of $Pt(NH_3)_2$ and NH_3:

$$Pt(NH_3)_2 + 2NH_3 + 2H^+ \longrightarrow Pt^{2+}(NH_3)_4 + H_2 \tag{6.37}$$

The same reaction regenerates the $Pt^{2+}(NH_3)_4$ complex after reactions (6.32) and (6.35). In the presence of oxygen, hydrogen will be oxidized to form H_2O, which drives this reaction thermodynamically. In the overall reaction scheme, N_2 and N_2O formation compete in the absence of water through reactions (6.32) and (6.36). In the presence of water, N_2 formation is promoted because of reactions (6.33) and (6.35). There is ample evidence that the reduction of the zeolitic protons by reduced metal atoms such as Pt is actually an easy reaction step[61]. In the temperature-programmed desorption spectra (Fig. 6.28), N_2O formation occurs predominantly in the peak in which excess oxygen is consumed. The other two peaks follow N_2 formation. In the absence of water ,the low-temperature peak that corresponds to the reaction sequence initiated by reaction step (6.33) is suppressed.

6.4.2 Electrocatalytic NH_3 Oxidation

Two important factors that act to control the electrocatalytic oxidation of ammonia at room temperature are the NH_3 conversion activity (current) and the build-up of surface species[61]. Figure 6.30 follows these two important features experimentally as a function of potential. The results clearly show that as the rate is increased, the surface becomes increasingly covered with surface intermediates. This occurs up to a potential of 0.6 V versus RHE, where the activity goes through a maximum. The nitrogen adatom coverage at this point approach, about 0.5 ML. Poisonous surface intermediates begin to form at potentials greater than 0.6 V and thus lower the activity.

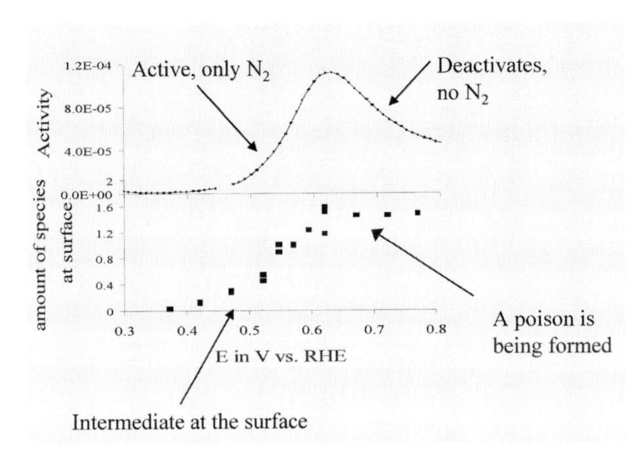

Figure 6.30. The electrochemistry of NH_3 oxidation: current against potential coverage[62].

A primary and important conclusion from this experiment is that NH_3 is converted to N_2 at potentials lower than where surface oxygen species are generated due to the anodic oxidation of H_2O. This implies that the initial step of NH_3 activation is

$$NH_3 \longrightarrow NH_{2ads} + H_{solvated}{}^{+} + e \tag{6.38}$$

The activation of NH_3 by the reaction with a co-adsorbed oxygen atom or hydroxyl group that occurs over the metal in the gas-phase reaction is now replaced by the acceptance of protons by H_2O molecules from solution (Fig. 6.31).

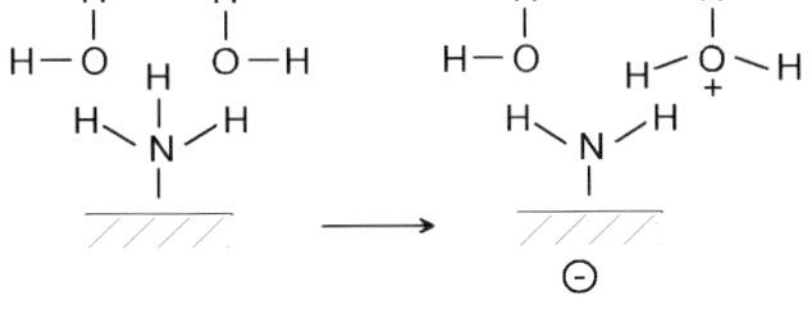

Figure 6.31. Protolysis reaction of ammonia.

The protons are stabilized here by solvation in solution. The key initial elementary step is heterolytic activation of NH_3 to give surface $NH_2{}^-$ and an aqueous H_3O^+, a reaction that reminds us of the initial activation of NH_3 adsorbed on Pt^{2+} by the basic zeolite lattice (6.32) discussed previously. For the electrochemical decomposition of H_2O, theoretical results showing the analogous reaction were discussed in Section 6.1. From the slope of Tafel plots (see Addendum to this chapter), it is deduced that the rate-limiting step of the reaction is the oxidation of NH_2 to a surface intermediate that, in a consecutive step, is converted to N_2:

$$NH_{2ads} \rightarrow N_2H_x + (4-x)H^+ + (4-x)e \text{ (rate-determining step)} \tag{6.39}$$

$$N_2H_x \rightarrow N_2 + xH^+ + xe + \text{ free site} \tag{6.40}$$

This reaction towards N_2 is of course assisted by the H_2O phase that can accept the protons that are released in the N_2 formation step. The rate of N_2 formation is 0.01 N_2 mol/s/Pt_{atom}. Adsorbed nitrogen atoms are found to lead to catalyst poisoning. They are formed in the following reaction steps:

$$NH_{2ads} \longrightarrow NH_{ads} + H^+ + e \tag{6.41}$$

$$NH_{ads} \longrightarrow N_{ads} + H^+ + e \tag{6.42}$$

The recombination of nitrogen atoms does not occur under the conditions of the electrochemical experiment. At higher potentials other reactions tend to take place. Oxygen develops at the electrode and ammonia is oxidized to N_2O, NO, $NO_2{}^-$ and $NO_3{}^-$. The selection of path (6.400 versus (6.420 correlates with the heat of adsorption of N_{ads}. The governing mechanism over Pd, Rh, Ir and Ru metal surfaces is similar to that which is seen on Pt. The steps outlined in (6.40) do not appear to occur over Pd, Rh or Ru. Reaction (6.42) is much faster than reaction (6.40) and therefore tends to dominate over these metals. These catalysts appear to be selective catalysts for the oxidation of NH_3 to the oxidized products.

Other metals such as gold, silver or copper are not active for either reaction (6.40) or (6.42). Instead, the metal tends to undergo oxidative dissolution. It is of interest to compare the potentials at which the nitrogen adatoms poison the surface versus their heat of adsorption.

The computed sequence of decreasing N_{ads} binding energies is as follows:

$$\text{Ru} > \text{Ir} > \text{Rh} > \text{Pd} > \text{Pt} > \text{Cu} > \text{Au} > \text{Ag}$$

The electrochemical experiments indicate the following sequence measured according to the potential at which the metal poisons:

$$\text{Ru} > \text{Rh} > \text{Pd} > \text{Ir} > \text{Pt} >> \text{Cu, Au, Ag}$$

The two trends are quite similar, with Ir being the exception.

Here we have shown that the presence of water and electrochemical potential can significantly alter both the activity and the selectivity of the NH_x reaction paths. Care must be taken in comparing the results obtained between the vapor phase and those under electrochemical conditions in the electrochemical experiments. Second, explicit evidence has been found that NH_x species are involved in the N_2 formation reaction.

6.5 Electrochemical NO Reduction

As was noted in Chapter 3, at the vapor metal surface, interface the dissociative adsorption of NO is strongly site dependent. The activation barriers for NO dissociation are significantly reduced at step edges. Such a dependence does not appear to play a role in electrochemical reduction experiments on NO. Product selectivity is sensitive to the presence or absence of NO in solution. On polycrystalline Pt, the reduction of adsorbed NO in the presence of NO in the solution yields N_2O as the only product at potentials between 0.4 and 0.8 V vs RHE[63].

The mechanism for this reaction is not of the Langmuir–Hinshelwood form, but rather involves the reaction between a surface-bonded NO molecule with an NO molecule in the solution phase along with a simultaneous electron transfer. The Tafel slope in acidic solution is $(120\ \mathrm{mV})^{-1}$, which implies that the first electron-transfer step is rate-determining (see Addendum to this chapter). The reaction is first order in H^+ and shows no apparent isotope effect. The following mechanism has been proposed:

$$\mathrm{NO_{aq}} \longrightarrow \mathrm{NO_{ads}} \qquad \text{(fast)} \tag{6.43}$$

$$\mathrm{NO_{ads}} + \mathrm{NO_{aq}} + \mathrm{H^+} + \mathrm{e} \longrightarrow \mathrm{HN_2O} \qquad \text{(rds)} \tag{6.44}$$

$$\mathrm{HN_2O_2} + \mathrm{H^+} + \mathrm{e} \longrightarrow \mathrm{N_2O_{aq}} + \mathrm{H_2O} \qquad \text{(fast)} \tag{6.45}$$

The rate-determining step (rds) consists of a pre-equilibrium involving the protonation followed by a rate-determining electron transfer step.

Stripping voltammetry of adsorbed NO in acidic solutions exhibits three peaks at 0.15, 0.23 and 0.7 V vs RHE. The only product formed is ${NH_4}^+$. The third peak vanished when the ${SO_4}^{2-}$ in solution is replaced by ${ClO_4}^-$. This suggests that two types of surface-bound NO exist.

The voltammetric results on stepped and non-stepped surfaces appear to be very similar[64]. The Tafel slopes of the first two peaks in the adsorbate solution is ca $(40\ \mathrm{mV})^{-1}$ which is typical for the electrochemical equilibrium followed by a rate-determining potential-independent chemical step. The following reaction scheme is proposed:

$$\mathrm{NO_{ads}} + \mathrm{H^+} + \mathrm{e} \longrightarrow \mathrm{HNO_{ads}} \qquad \text{(equilibrium)} \tag{6.46}$$

$$\mathrm{HNO_{ads}} + \mathrm{H^+} + \mathrm{e} \longrightarrow \mathrm{H_2HNO_{ads}} \qquad \text{(rds)} \tag{6.47}$$

$$\mathrm{H_2NO_{ads}} + 4\mathrm{H^+} + 3\mathrm{e} \longrightarrow \mathrm{NH_4}^+ + \mathrm{H_2O} \qquad \text{(fast)} \tag{6.48}$$

HNO is suggested as the hydrogenated intermediate rather than NOH because gas-phase HNO is nearly 100 kJ/mol more stable than gas-phase NOH. H_2NO is chosen over HNOH as the second hydrogenated intermediate, since calculations show that gas phase H_2NO is over 500 kJ/mol more stable than HNOH. No direct activation by the surface is involved.

All metals show a high selectivity to N_2O at higher potentials and a high selectivity to NH_3 at low potentials. N_2 is predominantly formed at intermediate potentials. The formation of N_2 produced at potentials between the formation of N_2O and NH_3 most likely takes place by the reduction of previously formed N_2O. The activation of NO in the electrochemical system is very different from that which takes place over the metal in the gas phase, in that there may not be a beneficial effect of the presence of steps on the kinetics.

The important differences between the electrochemical and gas-phase systems are the result of:

(1) direct contributions from the adsorbed liquid phase.
(2) direct activation of the molecule by the aqueous phase protons, leading to the bond cleavage and formation of water without the direct activation of the NO bond by contact with the surface atoms.

6.6 Electrocatalytic Oxidation of CO

The electrochemical reduction of NO discussed in the previous section was found to be independent of surface steps. By way of comparison, we discuss in this final section the electrochemical oxidation of CO which, in contrast, is strongly affected by the presence of surface steps.

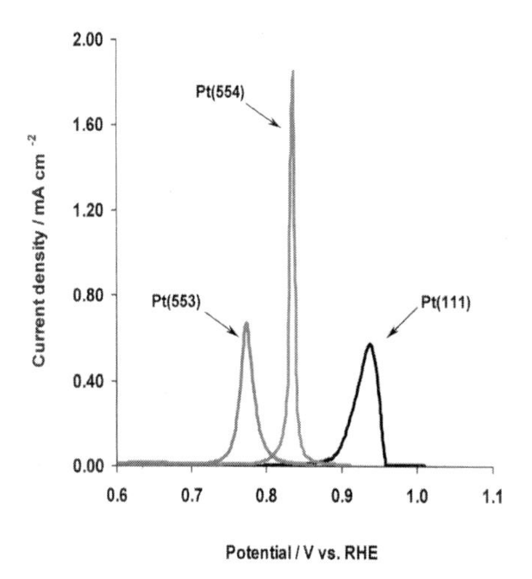

Figure 6.32. The stripping profiles from a saturated CO adlayer over different Pt single-crystal surfaces, sweep rate 50 mV/s[65]. CO adlayer oxidation is structure sensitive.

Figure 6.32 compares the voltage at which CO oxidation starts[65] over the Pt(111), Pt(553) and Pt(554) surfaces which contain planar terraces and steps which are 10 or 5 atoms wide, respectively. The steps are oriented in the (110) direction. The overall reaction scheme is

$$H_2O_{ads} + * \longrightarrow OH_{ads} + H^+ + e \tag{6.49}$$

$$CO_{ads} + OH_{ads} \longrightarrow COOH_{ads} \tag{6.50}$$

$$COOH_{ads} \longrightarrow CO_2 + H^+ + e \tag{6.51}$$

The potential dependence of the apparent rate constant is found to be structure insensitive and gives a Tafel slope of $(80\ \mathrm{mV})^{-1}$, which is consistent with step (6.50) being rate limiting (see Addendum to this chapter).

The reaction preferably takes place at the steps, whereas the terraces supply CO through fast surface diffusion. Infrared spectra acquired during he oxidation of the CO adlayer indicate that CO adsorbed on the (111) terraces is more reactive than that adsorbed on the steps. The infrared frequencies for CO adsorbed on terraces initially declines during course of reaction. CO adsorbed on steps reacts later.

The apparent reaction rates fall in the range 10^{-2}–10^{2} s^{-1}. The oxidative activation of water to form OH groups appears to occur at the step sites. The OH intermediates can subsequently go on to oxidize CO. The intrinsic catalytic reactivity of the steps is shown to be independent of the step density. In the oxidation of a saturated CO adlayer, two processes are distinguished: reaction initiation, which is a zero-order process in which CO desorbs to generate reaction centers, and an oxidation process that is of the Langmuir–Hinshelwood form with competitive adsorption of CO and OH. CO diffusion is thought to be fast $d_{CO} \sim 10^{-11}$ cm^2/s) on the terraces towards the step where OH is generated.

Step (6.49) preferably occurs at the step edges and is fast. This implies that OH_{ads} builds up concentration at the step edges. The CO binds to strong on step edges to react. The more weakly bonded, mobile CO adsorbed to terrace, therefore, reacts first with adsorbed OH.

6.7 Summary

In previous chapters, we aluded to the importance of solvent effects on different catalytic reactions. More specifically in Section 5.4 we discussed the molecular basis for the differences that arise between gas-phase and liquid-phase acidity. In this chapter, we provide a further analysis of solvent effects on surface-catalyzed reactions focusing on the role of protonic solvents such as water on elementary physicochemical steps. An important part of this chapter deals with electrocatalysis. One of the key concepts that arises in both chemical and electrochemical processes carried out in solution is that water (or the solvent) itself can often behave as a co-reactant. For instance, in the gas phase the dissociation of protic reagents such as water or ammonia over a transition-metal surface occurs homolytically, leading to the generation of surface intermediates such as hydroxyl (OH^*) or amine ($NH_2{}^*$) fragments along with a surface hydride (H^*). In the presence of solution, however, the dissociation can occur heterolytically, thus forming charged products such as OH^- and H^+ or $NH_2{}^-$ and H^+. The anion fragments tend to transfer their electron to the metal surface whereas the protons that form can be stabilized as hydronium or Zundel ions in solution. The work function of the metal plays an important role in electron capture thus dictating whether or not heterolytic dissociation can occur at least in the absence of an applied potential. The activation of bonds can now proceed without the intermediate adsorption of both the molecular fragments on the surface as we have seen for electrochemical NO reduction. In this system, there is little influence of step sites on the reaction kinetics. NO dissociation instead proceeds first through the protonation of NO by hydronium ions. This weakens the NO bond appreciably and enhances its activation.

Intrinsic to reactions in polar media is the stabilization of the charge separation that forms in the transition state by the large dielectric constant of the medium. The activation barriers for reactions that proceed via the formation of polar transition states are therefore significantly lowered owing to the charge stabilization of the transition state from the polar medium. The charged product intermediates are also stabilized by the solution phase. Low-temperature ammonia oxidation over Pt, for example, occurs heterogeneously in the gas-phase as well as electrocatalytically. In the gas phase, the reaction requires co-adsorbed oxygen atoms to be present on the surface in order to help activate the NH bond in ammonia. NH_2 and NH species are activated by OH_{ads}. On the electrode, the role of coadsorbed oxygen is taken over by the aqueous phase which mediates N–H activation. In addition, the gas-phase oxidation of ammonia leads to the production of N_2 and NO as the result of the recombination of surface nitrogen species. At low temperatures NO does not desorb. It reacts to give N_2O, which readily desorbs. The electrocatalytic oxidation of ammonia, on the other hand, evolves from partially hydrogenated hydrophilic NH_x intermediates. Nitrogen atom formation appears to poison the electrochemical reaction. The reactivities of reducible cations such as Pt^{2+} ion-exchanged into zeolites have some similarities to the solution phase chemistry with the important difference being that the dielectric constant of the zeolite is significantly lower than that of a polar solvent.

The ammonia oxidation reaction, which is catalyzed by a Pt^{2+} complex, proceeds via the rapid formation of a $Pt(NH_3)_3NO^{2+}$ complex. This complex is generated by the oxidation of $Pt(NH_3)_4^{2+}$ in which an H^+ is accepted by a basic lattice oxygen atom in the zeolite and H_2O is generated. The zeolite lattice oxygen atom takes over the role of the basic ligands in organometallic chemistry or the role of the proton-accepting water molecules.

In gas-phase metal catalysis, water tends to poison the reaction by competitive adsorption. Instead, in the coordination complex, in the zeolite cavity, water can promote the reaction by oxidizing NO to NO_2 with further generation of zeolitic protons. In a subsequent reaction NO_2 reacts with $NH_4{}^+$ and water.

The coordination complex chemistry in zeolites provides a very useful conceptual bridge to coordination-chemistry controlled catalysis in the liquid phase. This is discussed in this chapter for the oxidation of ethylene to produce vinyl acetate from acetic acid and ethylene. The catalytic system appears to consist of dimeric or trimeric Pd complexes. The elementary reaction steps can take place in the direct contact with the metal centers, the so-called inner-sphere mechanism. The reaction can also proceed through an outer-sphere mechanism in which proton transfer between reactants and acetate plays an essential role.

Addendum: The Tafel Slope and Reaction Mechanism in Electrocatalysis

The Tafel slope can be used to determine the rate-determining steps in electrochemical systems[65]. We describe Tafel slope plots and illustrate their use here. We start by analyzing the mechanisms for hydrogen evolution reactions by plotting the exchange current density as a function of overpotential. Two different mechanisms, labeled here as A and B, can be considered:

A.
$$H^+ + e + M \longrightarrow MH \qquad (1)$$
$$2MH \longrightarrow 2M + H_2 \qquad (2)$$

B.
$$H^+ + e + M \longrightarrow MH \qquad (1)$$
$$MH + H^+ + e \longrightarrow M + H_2 \qquad (3)$$

When the electron-transfer step (1) is rate limiting, one cannot distinguish between the two mechanisms. The rate of reaction (1) can be written as

$$\vec{V}_1 = \vec{k}_1 \, c_{H^+}(1-\theta)$$

where $\vec{k}_1$ is the rate constant of reaction (1) in the forward direction which is potential dependent. This reaction involves the transfer of an electron. When reaction (1) is the slow step, by implication the hydrogen coverage θ_H will be low and $\vec{V}_1$ can then be approximated as

$$\vec{V}_2 = \vec{k}_1 \, c_{H^+}$$

The current for H_2 evolution is given by:

$$-I = F \, \vec{k}_1 \, c_{H^+}$$

where F is the Faraday constant.

One can then assume a Brønsted–Evans–Polanyi relationship to relate the dependence of rate constant $\vec{k}_1$ to the overpotential E:

$$\vec{k}_1 = \vec{k}_{11} \, \exp\left(- \frac{\alpha F}{RT} \cdot E \right)$$

The Tafel slope is defined as

$$-\frac{\delta(\log -I)}{\delta E} = \alpha \frac{F}{2.3RT} = \alpha \cdot 60\,\mathrm{mV}^{-1} = 120\,\mathrm{mV}^{-1}$$

The value of α is typically assumed to be equal to 0.5.
If step 2, in mechanism A is rate determining:

$$-I = 2F \, \vec{k}_2 \cdot \theta^2$$

the recombination rate constant $\vec{k}_2$ is calculated to be independent of potential, but at equilibrium θ is

$$\frac{\theta_H}{1-\theta_H} = \frac{\vec{k}_1}{\overleftarrow{k}_2} c_{H^+} = K_1 \exp\left(\frac{-F}{RT} \cdot E \right) \cdot c_{H^+}$$

and the Tafel slope in this case becomes

$$-\frac{\delta(\log -I)}{\delta E} = 2\alpha \frac{F}{2.3RT} = 2 \cdot (60\,\mathrm{mV})^{-1} = (30\,\mathrm{mV})^{-1}$$

This Tafel slope is found when the electron transfer step is at equilibrium, but the chemical step is rate determining and second order in the surface concentration.

If mechanism B controls the reaction path and step (3) is considered rate determining, the rate can be written as

$$V_3 = \vec{k}_3 \, c_{H^+} \cdot \theta$$

According to this mechanism, one electron transfer step is equilibrated and the other is rate determining. At low overpotential, one deduces a Tafel slope of $(40\,\mathrm{mV})^{-1}$.

In principle two additional cases can be considered:

$$\mathrm{A + e \rightleftharpoons B}$$
$$\mathrm{B \longrightarrow C}$$

The electrochemical step at equilibrium, the chemical step first order in surface concentration, gives a Tafel slope of $(60\ \mathrm{mV})^{-1}$. Of the three successive electron transfer steps, two are at equilibrium with Tafel slope $(20\ \mathrm{mV})^{-1}$:

$$\mathrm{A + e \rightleftharpoons B}$$
$$\mathrm{B + e \rightleftharpoons C}$$
$$\mathrm{C + e \longrightarrow D}$$

References

1. P.A. Thiel, T.E. Madey, *Surf. Sci. Rep.* 7, 211 (1987)
2. M.A. Henderson, *Surf. Sci. Rep.* 46, 1 (2002)
3. (a) S.K. Desai, P. Venkataraman, M. Neurock, *J. Phys. Chem. B*, 105, 9171 (2001);
   (b) S.K. Desai, M. Neurock, unpublished results;
   S.K. Desai, M. Neurock, *Phys. Rev. B*, 68, 075420 (2003)
4. A. Michaelides, A. Alavi, D.A. King *J. Am. Chem. Soc.* 125, 2746 (2003)
5. A. Michaelides, V.A. Ranea, P.L. de Andres, D.A. King *Phys. Rev. Lett.* 90, 216102 (2003)
6. P. Vassilev, M.T.M. Koper, R.A. van Santen, *Chem. Phys. Lett.* 359, 337 (2002)
7. G. Held, D. Menzel, *Surf. Sci.* 316, 92 (1994);
   G. Held, D. Menzel, *Phys. Rev. Lett.* 74, 4221 (1994);
   G. Held, D. Menzel, *Surf. Sci.* 327, 301 (1995)
8. P.J. Feibelman, *Science*, 295, 99 (2002)
9. F.T. Wagner, T.E. Moylan, *Surf. Sci.* 206, 187 (1988)
10. N. Kizhakevariam, E.M. Stuve, *Surf. Sci.* 275, 223 (1992)
11. M. Neurock, S. Wasileski, unpublished results (2004)
12. C.V. Grotthus, *Ann. Chim. Paris*, 58, 54 (1815)
13. A.B. Anderson, *J. Chem. Phys.* 62, 1187 (1975)
14. P.S. Bagus, G. Pacchioni, *Electrochim. Acta*, 36, 1669 (1991)
15. F. Illas, F. Mele, *Electrochim. Acta*, 44, 1213 (1998)
16. D.K. Lambert, *Electrochim. Acta*, 41, 623 (1996)
17. D. Curulla, A. Clotet, *Electrochim. Acta*, 45, 639 (1999)
18. M. Head-Gordon, J.C. Tully, *Chem. Phys.* 175, 37 (1993)
19. S.A. Wasileski, M.T.M. Koper, *J. Phys. Chem. B*, 105, 3518 (2001)
20. (a) A.B. Anderson, D.B. Kang, *J. Phys. Chem. A*, 102, 5993 (1998);
   (b) A.B. Anderson, T.V. Albu *Electrochem. Commun.* 1, 203 (1999);
   (c) A.B. Anderson, T.V. Albu, *J. Am. Chem Soc.* 121, 11855 (1999);
   (d) A.B. Anderson, *J. Electrochem. Soc.* 147, 4229 (2000);
   (e) A.B. Anderson, *Electrochim. Acta*, 47, 3759 (2002);

(f) A.B. Anderson, N.M. Neshev, *Electrochim. Acta*, 47, 2999 (2002);
(g) A.B. Anderson, R.A. Sidik, *J. Phys. Chem. B*, 107, 4618 (2003);
(h) A.B. Anderson, *Electrochim. Acta*, 48, 3743 (2003)
21. (a) R.A. Marcus, *J. Chem. Phys.*, 43, 679 (1965);
(b)R.A. Marcus, *J. Electroanal. Chem.*, 438, 251 (1997)
22. J.W. Halley, A. Mazzolo, *J. Electroanal. Chem.* 450, 273 (1998)
23. A.Y. Lozovio, A. Alavi, J. Kohanoff, R.M. Lynden-Bell *J. Chem. Phys.* 115, 1661 (2001)
24. J.S. Filhol, M. Neurock, in preparation
25. G.Makov, M. Payne, *Phys. Rev. B*, 51, 4014 (1995)
26. J.F. Janak, *Phys. Rev. B*, 18, 7165 (1978)
27. E.R. Kötz, H. Neff, *J. Electroanal. Chem.* 215, 331 (1986)
28. J.O.M. Bockris, A.K.N. Reddy, *Modern Electrochemistry,) Vol. 2A*, Kluwer, New York (1998)
29. A. Czerwinski, I. Kiersztyn, *J. Electroanal. Chem.* 491, 128 (2000);
A. Czerwinski, I. Kiersztyn, *J. Electroanal. Chem.* 471, 190 (1999)
30. J.S. Filhol, M. Neurock, *Angew. Chem. Int. Ed.* submitted (2005)
31. C. Taylor, S. Wasileski,J. Fanjoy, J.S. Filhol, M. Neurock *Phys. Rev. B*, submitted (2005)
32. C. Taylor, R. Kelly, M. Neurok, to be submitted (2005)
33. W. Schwerdtel, *Hydrocarbon Proc.* 47, 187 (1968)
34. C. Reilly, J.J. Lerou, *Catal. Today* 41, 433 (1998)
35. P. Henry, *Catalysis by Metal Complexes: II. Palladium Catalyzed Oxidation of Hydrocarbons*, Reidel (ed.), Boston (1980)
36. (a) D.A. Kragten, R.A. van Santen, *Inorg. Chem.* 38, 331 (1999);
(b) D.A. Kragten, R.A. van Santen, M. Neurock, J.J. Lerou, *J. Phys. Chem. A*, 103, 2756 (1999)
37. S. Nakamura, T. Yasui, *J. Catal.* 17, 366 (1970);
S. Nakamura, T. Yasui, *J. Catal.* 23, 315 (1971)
38. I.I. Moiseev, M.N. Vargaftik, *Izr. Akad. Nauk. SSSR, Engl. Transl.* 133, 377 (1960);
I.I. Moiseev, M.N. Vargaftik, in *Perspectives in Catalysis*, J.M. Thomas, K.I. Zamaraev (eds.), Blackwell Scientific, Oxford, p. 91 (1992)
39. M. Neurock, D.A. Kragten, *J. Catal.* submitted (2004)
40. B. Samanos, P. Bountry, *J. Catal.* 23, 19 (1971)
41. S.A.H. Zaidi, *Appl. Catal.* 38, 353 (1988)
42. M. Tamura, T. Yasui, *Kagaku Zasshi*, 72, 561 (1969)
43. L.S.J. Campbell, *J. Phys. Chem. B*, 104, 11168 (2000)
44. P.M. Henry, R.N. Pandey, *Homogeneous Catalysis–II. Advances in Chemistry Series* Chicago (1973)
45. D.A. Kragten, *Thesis*, Eindhoven University of Technology (1999)
46. H. Debellefontaine, J. Besombes-Vailhe, *J. Chim. Phys.* 75, 801 (1978)
47. S.M. Augustine, J.P. Blitz, *J. Catal.* 142, 312 (1993)
48. E.A. Crathorne, D. MacGowan, S.R. Morris, A.P. Rawlinson *J. Catal.* 149, 254 (1994)
49. (a) W.T. Tysoe, D. Stacchiola, *J. Am. Chem. Soc.* 126, 15384 (2004);
(b) D. Stacchiola, F. Calaza, L. Burkholder, A. Schwabacher, M. Neurock, W. T. Tysoe, *Angew. Chem. Int. Ed.* 44, 2 (2005)
50. D.A. Hickman, L.D. Schmidt, *Ind. Eng. Chem. Res.* 30, 50 (1991)

51. A. Bogavic, K.C. Hass, *Surf. Sci.* 546, L237 (2002)
52. (a) L. Kieken, M. Neurock, D. Mei, *J. Phys. Chem. B*, 109, 2234 (2004);
    b) D. Mei, Q. Ge, M. Neurock, L. Kieken, and J. Lerou, *Mol. Phys.* 102, 361 (2004)
53. M. Neurock, R.A. van Santen, *J. Am. Chem. Soc.* 116, 4427 (1994)
54. (a) D.P. Sobczyk, A.M. de Jong, E.J.M. Hensen, R.A. van Santen, *J. Catal.* 219, 156 (2003)
    (b) D.P. Sobczyk, A.M. de Jong, E.J.M. Hensen, R.A. van Santen, *Top. Catal.* 23, 109 (2003)
55. (a) L. Gang, J. van Grondelle, R.A. van Santen, *J. Catal.* 186, 100 (1999);
    (b) L. Gang, J. van Grondelle, R.A. van Santen, *J. Catal.* 199, 107 (2001)
56. A.C.M. v.d. Broek, J. van Grondelle, R.A. van Santen, *J. Catal.* 185, 297 (1999)
57. M. Kim, S.J. Pratt, D.A. King, *J. Am. Chem. Soc.* 122, 2409 (2000)
58. A.C.M. van den Broek, J. van Grondelle, R.A. van Santen, *Catal. Lett.* 55, 79 (1998);
    (b) A.C.M. van den Broek, J. van Grondelle, R.A. van Santen, *J. Catal.* 167(Iss 2), 147 (1997)
59. S.D. Pell, J.N. Armor, *J. Am. Chem. Soc.* 97, 5012 (1975);
    Z. Assefa, D.M. Stanbur, *J. Am. Chem. Soc.* 119, 521 (1997)
60. S.D. Pell, J.N. Armor, *J. Am. Chem. Soc.* 94, 686 (1972);
    S.D. Pell, J.N. Armor, *J. Am. Chem. Soc.* 95, 7625 (1973);
    J.D. Buhr, H. Taube, *Inorg. Chem.* 19, 2425 (1980)
61. W.M.H. Sachtler, Z. Zhang, *Adv. Catal.* 39, 139 (1993)
62. A.C.A. de Vooys, M.T.M. Koper, R.A. van Santen, J.A.R. van Veen,
    J. Electroanal Chem. 506, 127 (2001)
63. A.C.A. de Vooys, M.T.M. Koper, R.A. van Santen, J.A.R. van Veen, *J. Catal.* 202, 387 (2001)
64. G.L. Beltramo, M.T.M. Koper, *Langmuir* 19, 8907 (2003)
65. N.P. Lebedeva, M.T.M. Koper, J.M. Feliu, R.A. van Santen, *J. Phys. Chem. B*, 106, 12938 (2002);
    N.P. Lebedeva, A. Rodes, J.M. Feliu, M.T.M. Koper, R.A. van Santen,
    J. Phys. Chem. B, 106, 9863 (2002)
66. D. Pletcher, F.C. Walsh, *Industrial Electrochemistry*, Blackie, Glasgow (1993)
67. J.K. Nørskov, J. Rossmeisal, A. Logadottir, L. Lindqvist, J. Kitchin, T. Bligaard, H. Jonsson,*J. Phys. Chem. B*, 108, 17886 (2004)

CHAPTER 7
Mechanisms in Biocatalysis; Relationship with Chemocatalysis

7.1 General Introduction

In this chapter we summarize key molecular concepts in biocatalysis and compare them with the molecular understanding of reaction mechanisms in heterogeneous catalysis that was developed in the previous chapters. The first four sections of this chapter emphasize enzyme catalysis. Biomimetic approaches are described in later sections.

More specifically we analyze the similarities and differences between bio- and chemo-catalytic systems in the sections on oxidation and reduction catalysis. We highlight the important mechanistic concepts and energetic requirements that were described in the previous chapters such as pre-transition-state orientation and solvation (the dielectric constant of an enzyme is as low as that of a zeolite) and the match of the shape and size of the transition state with the enzyme cavity. In this context, it will be important to compare enzyme action concepts such as "lock and key" and "induced fit" with some of the related ideas discussed in Chapter 2. We refer back to concepts in transition-metal surface catalysis in order to establish ideas on possible mechanisms.

The active site of the enzyme is located in a cleft of the peptide framework. The "site" may be considered to be equivalent to that for a homogeneous organometallic catalyst in that it involves a single center with a complex ligand sphere. The results can also be related back to those that occur at metal atom sites on the surface of a metal, metal oxide or metal sulfide. The classical relationship between the conservation of active sites and the turnover of a catalytic cycle suggests that enzyme kinetics should be analogous to those in homogeneous and heterogeneous catalysis.

The Michaelis–Menten rate expression for a monomolecular enzyme-catalyzed reaction is very similar to the Langmuir kinetic expression that we discussed previously for heterogeneous catalyzed systems. The Michaelis–Menten relationship is readily deduced for a simple model where the enzyme molecule (E) equilibrates with substrate molecule (S) to form the enzyme substrate complex (ES). The enzyme–substrate complex then reacts to the product molecule (P) and regenerates the active site of the enzyme (E) in what is considered to be the rate-limiting step of the proposed scheme:

$$\mathrm{E} + \mathrm{S} \underset{k_1}{\overset{k_2}{\leftrightarrows}} \mathrm{ES} \underset{k_3}{\rightarrow} \mathrm{E} + \mathrm{P}$$

Langmuir chemo kinetics should be equivelent to Michaelis–Menten kinetics since the total number of sites (the number of enzyme molecules) or the number of reaction sites on a surface is a constant. In addition, the often physically unrealistic assumption is made that $k_1, k_2 and k_3$ are independent of concentration.

The Michaelis–Menten rate expression as written in biochemistry is

$$V = V_{\max}\frac{[\mathrm{S}]}{[\mathrm{S}] + K_{\mathrm{M}}} \tag{1a}$$

$$= [\mathrm{E_T}].k_3\frac{{K_{\mathrm{M}}}^{-1}[\mathrm{S}]}{1 + {K_{\mathrm{M}}}^{-1}[\mathrm{S}]} \tag{1b}$$

where $K_{\mathrm{M}} = \frac{k_2+k_3}{k_1}$ and V is the overall rate of conversion and [S] is the substrate (reactant) concentration. One can rewrite V in terms of an expression which is analogous

Molecular Heterogeneous Catalysis. Rutger Anthony van Santen and Matthew Neurock

ISBN: 3-527-29662-X

to the site normalized rate expression used conventionally in chemocatalysis by dividing V by the total number of enzyme sites $[E_T]$:

$$r_{M.M} = \frac{V}{[E_T]} = k_r \theta_{M.M} \tag{1c}$$

where k_r is the elementary rate constant of the rate-limiting reaction step is to be identified with k_3. The inverse of the Michaelis–Menten constant ${K_M}^{-1}$ is the equivalent of the Henry coefficient or adsorption constant. $\theta_{M.M}$ is the fraction of enzyme molecules occupied by substrate:

$$\theta_{M.M} = \frac{[ES]}{ET}$$

When $\theta_{M.M} \cong 1$, the turnover number (or turnover frequency) of the reaction is determined by k_r. The turnover frequency determined by k_3 for most enzymes varies between 1 and 10^4 sec^{-1}. The highest value is found for the carbonic anhydrase enzyme, which will be discussed in Section 7.4: $k_3 = 6 \text{ x } 10^5 sec^{-1}$. It is interesting to compare this value for k_r with that measured for a fast heterogeneous reaction such as hexene isomerization over protonic zeolites, which has a corresponding k_r of only 10^{-2} sec^{-1}. The rates found in zeolite catalysis thus tend to be much lower than those in enzyme catalysis. This is due to the much higher activation energies required for the protonation reaction in a zeolite as compared with analogous reaction(s) carried out by the enzyme. This can be understood by referring to Chapter 4, page 203. We illustrated for chymotrypsin how well-positioned acid and base functional groups in the enzyme cavity were responsible for its high rate of reaction. There is an optimum match between several peptide reactive groups and the geometric structures of the transition state.

A unique feature of the enzymes is their hydrophobic interior and hydrophilic exterior character. Their hydrophilic external surface makes them water-soluble. Their hydrophobic interior and unique internal electrostatic properties permits reactivity not possible in an aqueous medium alone. It also provides for selectivity advantages since the adsorption of apolar molecules tends to be preferred over that of highly polar molecules. For example, the subsequent reactions that arise in oxidation catalysis from the readsorption of the polar product molecules formed via the oxidation of an apolar reactant are suppressed in the hydrophobic cavity of the enzyme. This significantly enhances the enzyme's overall selectivity.

Biocatalytic systems are part of biochemical systems that uniquely couple mass and energy transport. Enzymes are part of complex biochemical process systems. For instance, electron transport and proton transport drive catalytic systems electrochemically so that chemical potentials can be established for reactions to occur under mild conditions. Reduction processes, for example, can occur directly by electrochemically generating reducing agents such as NADH (nicotinamide adenine nucleotide, reduced form) or NADPH (nicotinamide adenine dinucleotide phosphate, reduced form). While it is outside the scope of this book to discuss these reactions in detail[1], some of the biocatalytic aspects will be discussed in later sections. A specific system that combines multiple reaction centers with rotational motion is the catalytic system adenosine triphosphate synthase, which we will introduce in Section 7.3. In this system, the chemical potential gradient that develops over the membrane acts control to proton transport. It can accept or generate the free energy of the chemical reaction.

Biochemical systems are also often reaction cycles where the key catalytic intermediate is generated, consumed, and subsequently regenerated as the part of the complex reaction cycle. Enzymes catalyze the formation of the intermediates that participate in the overall reaction cycle. These are intricately catalyzed reaction systems where the intermediates that form subsequently act as catalysts. The citric acid cycle, which was discussed in Section 2.1 (see also Fig. 2.5), is one such cycle. Two other complex reaction systems comprised of many catalytic steps that are of key metabolic importance are the Calvin and glycolytic cycles.

In the Calvin cycle, CO_2 is converted into hexose or fructose. 3-Phosphoglycerate is the key catalytic molecule that is consumed and generated in this cycle. In the citric acid cycle, citrate is the intermediate species in the sequence of oxidation steps that lead to CO_2.

In the glycolytic cycle, energy is regenerated by converting glucose to pyruvate. In the overall reaction, energy is produced and stored by the conversion of two molecules of ADP (adenosine diphosphate) to two molecules of ATP. Energy production and mass conversion are coupled. The autocatalyic system then involves self-organizing features together with kinetics that display oscillatory behavior. We will discuss this in more detail in Chapter 8. Autocatalytic reactions, which are part of the above-mentioned biochemical cycles, are essential to produce such self-organizing reaction systems.

A unique autocatalytic reaction mechanism can occur in biochemical systems due to enzyme modification as a consequence of pH changes during the catalytic cycle. Enzyme activity can be accelerated or inhibited by such pH changes, and in addition, by the enzyme interaction with specific molecules. An example of such an enzyme is ATP-synthase discussed in Section 7.3. The allosteric enzymes are a second class of enzymes that display unique reaction mechanisms in that they do not obey classical Michaelis–Menten kinetics. Allosteric enzymes are enzymes that contain several binding sites. Binding at one site affects binding at another site, for instance by conformational changes in the protein. In chemocatalysis this would lead to non-Langmuirian adsorption behavior. In enzymes this can lead to increased as well as decreased binding of reactant molecules. When molecules produced in a reaction provide for positive feedback through activation of such allosteric enzymes, the system is autocatalytic.

7.2 The Mechanism of Enzyme Action; the Induced Fit Model

Enzymes are comprised of proteins which have definitive primary structures that can organize into secondary, tertiary and even quaternary structures. The enzyme consists of a peptide framework that contains cavities in which the catalytically reactive centers are incorporated. The enzyme protein is typically about 20–40 nm in size. The catalytically reactive centers can be part of the peptide that builds the enzyme framework such as carboxylic acid groups, or consist of cationic centers such as the porphyrins or non-reducible cations as Zn^{2+} or Mg^{2+} directly attached to the protein cavity. As mentioned earlier, the external part of the enzyme is hydrophilic and the internal microcavity is hydrophobic, with the possible exception of the often polar reaction center. In enzymatic reactions, the match between shape and size of reactant or product and catalytic center cavity is important.

Upon reaction with a substrate molecule, the shape of the enzyme is altered. The flexibility of the peptide framework implies that changes in overall protein shape (induced, for instance, by allosteric effects) may affect the shape of the reaction center. This will

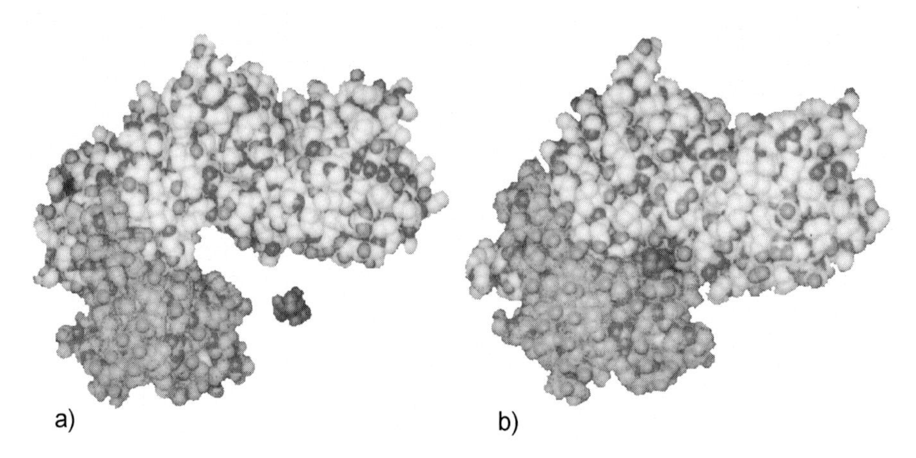

Figure 7.1. Model of the hexakinase enzyme: (a) the free cavity in hexakinase, (b) adsorbed glucose in the cavity[2c].

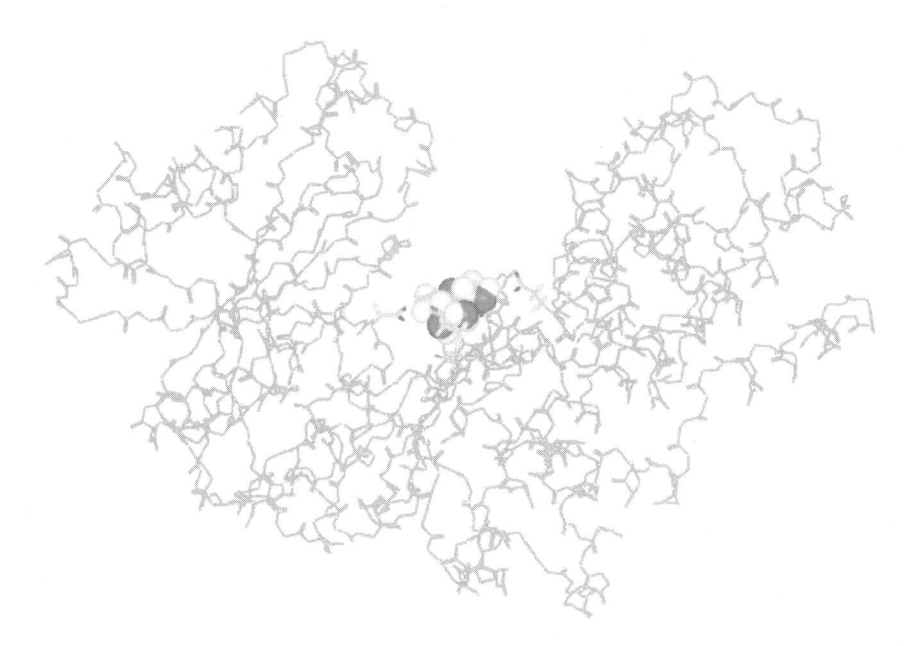

Figure 7.1c. Model of the hexakinase enzyme: adsorption of glucose[2a].

influence both its intrinsic electronic structure and its reaction environment, both of which will affect its catalytic performance. This will be described in the next section.

We will give a detailed description of recent insights obtained especially for the widely studied hexakinase enzyme, which catalyzes the phosphorylation of glucose. The phosphorylation of glucose by hexakinase has been modeled in detail by simulating the reaction events that can occur in the enzyme cleft[2]. The free hexakinase enzyme molecule is shown in Fig. 7.1a. The glucose molecule has not yet been attached. Its shape is significantly altered upon the adsorption of glucose, as seen in Fig. 7.1b. The reaction does not appear to follow the lock and key model[3], which was originally proposed to explain the stereoselectivity in enzymes. The lock and key model would suggest that glucose would exactly match the open cleft in the hexakinase enzyme. Instead, the enzyme cavity shape and hence overall three-dimensional structure adapt to accommodate optimally the glucose substrate molecule. This is consistent with the "induced fit model" proposed by Koshland[4]. In this model, small induced changes within the reaction complex lower the activation energy of reaction. The cavity shape adapts leading to an optimum fit between the reactive groups in the adjusted cavity and the reactant. Such changes cannot occur in the rigid inorganic frameworks of heterogeneous catalysts such as zeolites discussed earlier.

Upon the adsorption of the glucose molecule into the enzyme cavity, the cavity closes as observed in Fig. 7.1b. The cavity continues to change as the reaction proceeds, which helps to drive the reaction over the potential energy surface to the product state. Desorption of the product molecules requires reopening of the cavity in order to release them. This process can be aided by the coadsorption of an additional reactant molecule at a second peptide binding site (allosteric effect). This reduces the interaction between product molecules and cavity, assists desorption and decreases the tendency of the enzyme to become deactivated by product poisoning.

A unique feature of the enzyme is multi-point bonding of reaction intermediates to the enzyme cavity and the participation of several protein substituents often belonging to different amino acids in the activation of a substrate (Fig. 7.1c).

The optimum induced fit between the enzyme cavity and the reacting substrate enables the enzyme for a particular reaction to discriminate readily between reactants that differ in size. This is illustrated in Fig. 7.2 for the conversion of glucose and closely related molecules by the hexakinase enzyme.

A good example of this is found in the studies by Rose[6] on the conversion of malate to fumarate conversion. He showed for the enzyme fumarase that malate combines with a conformation of the enzyme that accommodates malate more easily than fumarate. As the chemical transition proceeds the cavity changes shape and provides for a better accommodation of the fumarate. For net catalysis to proceed, the fumarase must finally make a transition to the malate-preferring confirmation. In this step fumarate desorbs. The minimum scheme for catalysis by an enzyme is than as follows:

$$\mathrm{E}^0 + \mathrm{S} \longleftrightarrow \mathrm{E_sS} \longleftrightarrow \mathrm{E_{P.T}} \longleftrightarrow \mathrm{E_pP} \longleftrightarrow \mathrm{E_T^1P}$$

where E^0 is an enzyme form that has a better fit to substrate S than product P. E^0 changes its conformation to $\mathrm{E_s}$ when S adsorbs. When reaction proceeds, a pre transition (PT) intermediate is formed and enzyme conformation adapts. This state continues towards the formation of $\mathrm{E_p}$ which stabilizes the formation of the product just enough for the product P to form but not so far that it will not desorb. After reaction, the state of the enzyme $\mathrm{E_T^1}$ has to return to its original state, E^0.

Optimization of reaction complex topology implies the positioning of charged enzyme groups such that there is an optimum interaction with (activated) reactants. Significant protonation (in protolytic enzymes) or electron transfer (in redox catalysis) can already occur in the adsorbed state. This is due to the unique electrostatic properties of the enzyme, which is shielded from the solvent by its hydrophobic peptide framework.

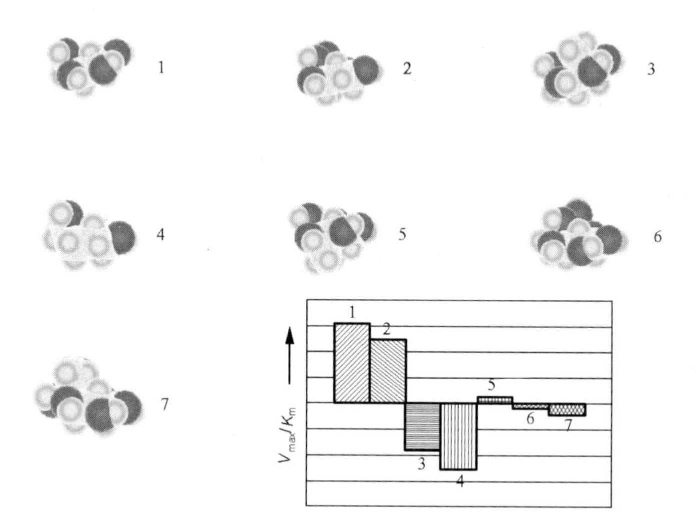

Figure 7.2. The rate of conversion of glucose related molecules by hexakinase. Dependence on molecule shape[5]. The molecules converted are labeled with a number.

The formation of the optimum pre-transition-state complex requires some energy since the enzyme restructures and the state of the substrate molecules becomes slightly deactivated compared to its most stable adsorption state. The activation energy with respect to the pretransition state may be only a few kJ/mol. For a hydrolysis reaction, the enzyme typically lowers the activation barriers by 30–50 kJ/mol compared with the acid- or base-catalyzed reaction in solution. This overall lowering of the apparent activation energy is the result of the optimized electrostatic interaction and hydrogen bonding that occurs in the initial reaction complex. The energy is only partially off-set by the cost in energy of restructuring of the enzyme peptide framework. The interaction between substrate and the atoms of the catalytic reactive center tends to weaken bonds within the catalytic center itself, thus initiating its relaxation and restructuring. As we have discussed previously for chemo catalysis, reaction energies not only depend on the changes in bond energies that occur within the substrate or between substrate and active site in the catalyst, but also within the catalyst itself.

The concept of a pretransition-state stabilization has also been extensively discussed in the chapter on zeolite catalysis (Chapter 4). In addition, the stereochemical selectivity found in chemocatalytic systems is often due to stabilization of reagents in a particular conformation before actual bond activation occurs.

Enzyme catalysts typically have low activation energies. The synchronized action of multi-atom displacements to optimize the interaction energy with the pretransition-state complex implies a decrease in the entropic state of the complex. This has been analyzed by Liktenstein[7], who formulated the principle of "optimum motion", in contrast to the principle of "minimum motion".

According to the latter, the less that nuclei have to change their position in the transition of an elementary reaction step, the lower is the reorganization energy required. This lowers both the overall energy and the activation energy for the reaction. The concerted motion of many atoms will stabilize the pretransition-state complex, thus leading to a small activation energy as the result. According to the principle of optimum motion, the number of nuclei whose configuration is changed in an elementary reaction step must be sufficiently large to lower the activation energy with respect to the adsorbed state. At the same time, this number has to be sufficiently small so that the synchronization probability along the reaction coordinate is not too low.

Liktenstein proposed a condition that has to be satisfied so that k_{syn} is faster than the reaction rate of the direct reaction k_{dir}:

$$\frac{k_{syn}}{k_{dir}} = \alpha_{syn}\exp\left(\frac{E_{dir} - E_{syn}}{kT}\right) > 1 \tag{7.2}$$

where α_{syn} is the synchronization factor, which is the ratio of the preexponential factors of the synchronous and direct processes

$$\alpha_{syn} = \frac{n}{2^{n-1}}\left(\frac{nkT}{\pi E_{syn}}\right)^{n-1} \qquad \varphi_{cr} > \varphi_0;\, E_{syn} > nkT \tag{7.2a}$$

$$\alpha_{syn} = \frac{n}{2^{n-1}} \qquad \varphi_{cr} < \varphi_0;\, E_{syn} > nkT \tag{7.2b}$$

where φ is a parameter that denotes the displacement of the nuclei, φ_{cr} is some critical value of this displacement and n is the number of vibrational degrees of freedom of the nuclei participating in the concerted transition.

For activation energies typical for enzymatic reactions, cooperation of each new nucleus in the transition state can lead to a ten-fold decrease in the reaction rate. Liktenstein used the Eq. (7.2)to analyze different mechanistic schemes proposed for an enzymatic reaction. One concludes that reaction paths involving a smaller number of synchronizing atoms, but proceeding through covalently bonded intermediates, as we discussed for chymotrypsin (Section 4.4, p. 203), are usually preferred.

When a reactant molecule adsorbs on a particular site, entropy is lost compared with the reactant state in solvent or gas phase. This was described earlier in the chapter on zeolites. Within the rigid lock and key model, this entropy loss would be maximum, thus reducing the free energy gained upon adsorption. This is an additional reason why an optimum fit between reactant and enzyme cavity is not preferred. When the fit between the reactant and the cavity is not optimum, the reactant will maintain some mobility in the adsorbed state, hence the entropy loss is less. The basic mechanistic principles for enzyme catalysis discussed so far include the induced fit of the enzyme cavity as a response to substrate shape and size, pretransition-state stabilization of activated molecules and the principle of optimum motion. A reaction that proceeds through intermediates via transient covalent bonds is preferred.

Enzyme catalysis, as discussed so far, occurs at a single site of the protein. Enzymes that modify proteins or macromolecules have to distinguish between similar substrate sites within larger dissimilar molecules. Discrimination between different substrates bearing the

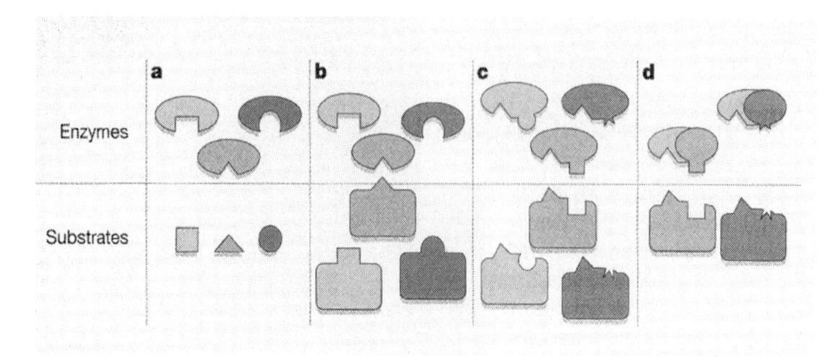

Figure 7.3. How enzymes select their substrates. (a, b), In general, enzymes recognize their targets through complementary structural features between the substrate and the enzyme's active site (indicated here by the shape of the "pocket"). Small substrates (a) and relatively small modification sites on proteins (b) can be recognized by this mechanism. (c) Some enzymes make additional, specific contacts with the substrate that enable them to distinguish between proteins that have identical or related sites of modification. (d) Cyclin-dependent protein kinases (CDKs) relegate that function to the exchangeable cyclin subunit, enabling a single CDK catalytic subunit to exist in numerous forms with different specificities. Adapted from Wittenberg, *Nature* 434, 35 (2005).

same motif occurs through additional specific interactions between sites other than the reaction center of the enzyme and substrate. This is illustrated in Fig. 7.3.

Such cooperativity has, for instance, been shown for cell cycle-regulating cyclin-dependent kinases (CDK) or baker's yeast. These CDKs consist of the combination of a subunit cyclin and a catalytic subunit. The presence of a particular hydrophobic structural motif on cyclin in combination with the catalytic subunit imports unique reactivity with respect to the phosphorylation of particular substrates with different size.

7.3 ATP-Synthase Mechanism; a Rotating Carousel with Multiple Catalytic Sites

The enzyme that catalyzes the hydrolysis or synthesis of adenosine triphosphate (ATP) demonstrates a unique feature in which a rotational motion of the protein that contains at least three reaction centers is coupled with different stages of the catalytic reaction occurring at each of these centers. The enzyme protein is attached to a proton translocating membrane. The energy changes due to ATP formation or hydrolysis is coupled with proton transport through the membrane, which is important to the biosystem's energy housekeeping. The rotary motion of the ATP synthase enzyme couples the chemical potential changes of the chemical reaction with chemical potential changes over the membrane. We will follow closely two important review papers[8,9] that help to elucidate the mechanism of catalytic action of this catalyst. A model representation of the F_1 enzyme protein attached to the membrane F_0 is given in Fig. 7.4.

F_1 is a globular aggregate attached through subunit γ to F_0, the membrane protein complex. γ is the shaft around which three $\alpha\beta$ subunits rotate. Proton transfer occurs through c and a channels. The three catalytic sites are located at α/β interfaces. Sequential opening and closing of the catalytic sites drives rotation of the γ subunit. Conformational signals travel from one catalytic site to another. ATP synthesized at a high-affinity catalytic site will desorb using energy produced by proton transfer through F_0. The F_1 unit then rotates by 120°. The original three-site model proposed by Cross[9] illustrating the sequence of reaction steps that occur is shown in Fig. 7.5.

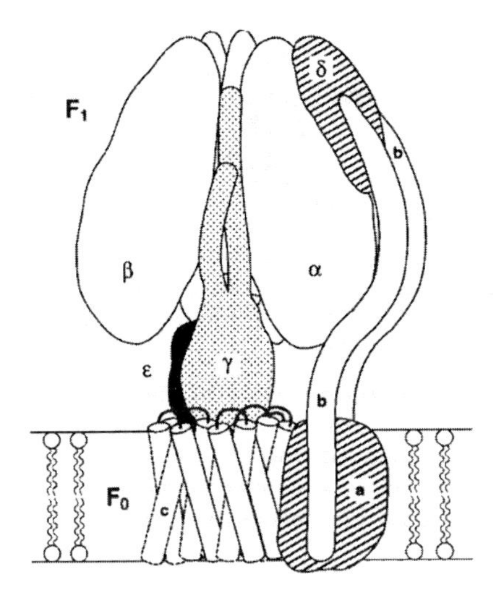

Figure 7.4. A model of the *E. coli* F_0F_1 ATP-synthase, after Ogilvie et al.[10].

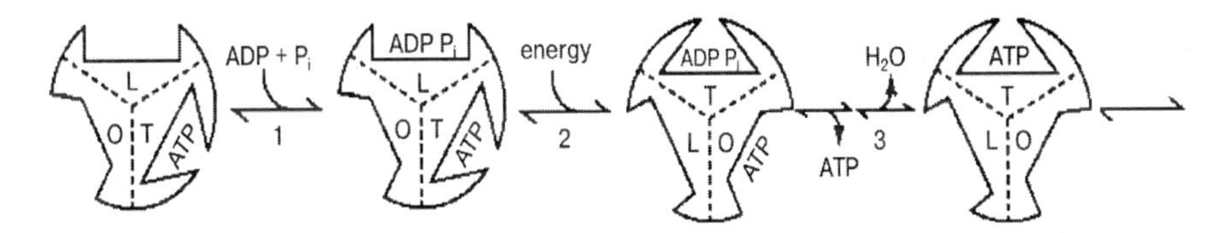

Figure 7.5. A three-site model for the binding change mechanism. Adapted from R.L. Cross[9].

Tight(T), loose(L) and open(O) phases of three catalytic centers are proposed. ATP will only be released from F_1 once ATP is formed at a different site by adsorption of ADP and phosphate. The conformational changes that occur in the enzyme cavities when the reaction proceeds are communicated through the subunits to the other sites. This allotropic communication implies synchronization of the different phases of the reaction at the different reaction centers.

The coupling of proton transfer through a membrane to the generation of rotary motion is known from other biochemical studies. For instance, it is also responsible for the flagellar rotation of swimming bacteria (see ref. [1]). A rotational velocity of 100 revolutions per second, consistent with the time scale of catalytic reactions, is typically reached. Each revolution is driven by a flow of 1000 protons across the membrane[11]. Rotary force is generated by the interplay of receptors on the rotating ring and the half-closed proton channels on the membrane.

The H^+ translocates through the membrane in three steps. First it adsorbs into a half channel of the membrane, then it connects with a proton acceptor site on the rotating F_1 ring. In a rotation, the proton is back-donated to the empty other half-channel of the membrane, that transfers the proton to the other side of the membrane. Rotational motion will only occur when one half-channel is unoccupied by a proton and the other half is empty.

7.4 Carbonic Anhydrase

As a first comparison of a chemocatalytic reaction with an analogous enzyme-catalyzed reaction, we discuss the hydrolysis of CO_2 by H_2O to give $HCO_3{}^-$ by the enzyme carbonic anhydrase. The reaction steps involved in the enzyme catalyzed mechanism will be compared with the chemocatalytic steps involved in the hydrolysis of acetonitrile by a Zn^{2+} containing zeolite as discussed on page 186 in Chapter 4. Similarly to the zeolite, the interior of the enzyme is hydrophobic except for the region close to the Zn^{2+} center. Its structure is shown in Fig. 7.6.

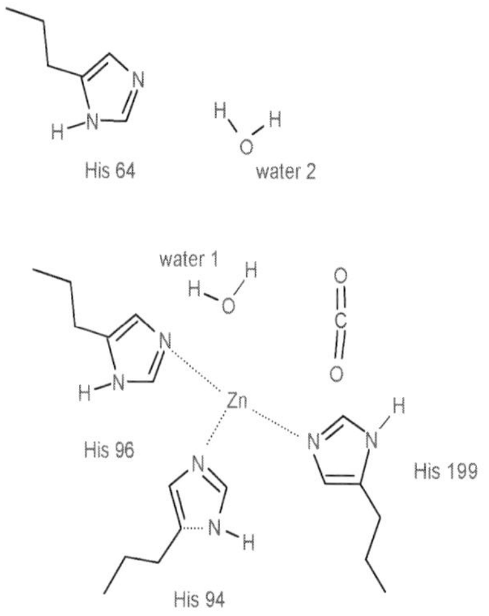

Figure 7.6. The catalytic center of carbonic anhydrase[12a].

The Zn^{2+} ion is attached to three basic nitrogen atoms of imidazole groups that are part of histidine peptides of the enzyme protien framework[12]. The positive charge of Zn^{2+} is compensated for by negatively charged glutamate residues in the second coordination shell of Zn^{2+}. A proton-transfer reaction between the negatively charged glutamate groups and imidazole occurs upon proton attachment that is the result of water dissociation. Two or three H_2O molecules are involved in the reaction[13]. One molecule dissociates on Zn^{2+} to give $ZnOH^+$ and a proton that attaches to imidazole which is stabilized upon the transfer of another proton from imidazole to neighboring glutamate. Similarly as in the zeolite (see Fig. 4.23), the second water molecule facilitates the initial proton-transfer reaction. The proton is transferred through the second water molecule to a basic proton-accepting site.

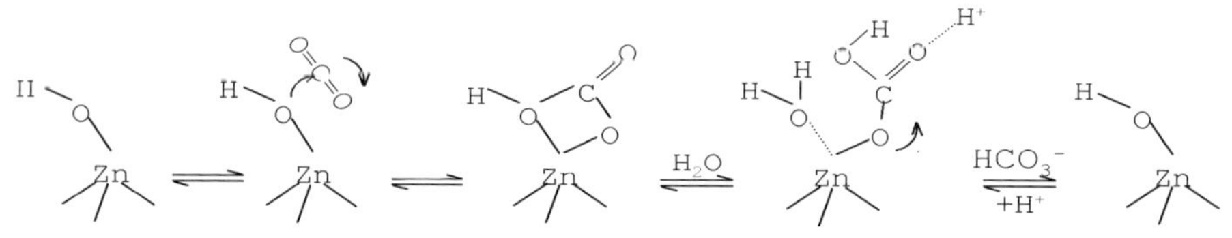

Figure 7.7. The hydrolysis reaction of CO_2 to bicarbonate[12b].

The OH^- bonded to Zn^{2+} reacts in a consecutive step with CO_2 to form a bicarbonate ion via the sequence shown in Fig. 7.7. The bicabonate is released when another water molecule adsorbs and dissociate to form new OH^- sites. The proton initially attached to the His64 peptide is then released through the water network to charge compensate the bicarbonate molecule. This closes the catalytic cycle. The reactive center is regenerated in the last adsorption-induced desorption step.

The uniqueness of the enzyme relates to the hydrophobic environment of Zn^{2+} and the protection of this site against deactivating reactions, for instance, leaching of Zn^{2+} by dissolution of Zn^{2+} into the water phase. The leaching of Zn^{2+} is a problem that tends to occur with the zeolite catalyst. In enzyme catalysis, bicarbonate is formed at neutral pH, which is very different from the basic conditions used in the non-catalytic system. In addition, the barrier for the dissociative adsorption of H_2O to liberate H_2CO_3 is lowered in a unique way, by the stabilization of the protons that are formed through a sequence of hydrogen bonds they create with the surrounding basic imidazole groups. Such a unique optimized environment is absent in the case of the zeolite.

The similarities between the nitrile hydrolysis and CO_2 hydrolysis system relate to the heterolytic H_2O-assisted splitting of H_2O and the importance of the adsorption-induced desorption of the reactant molecule. The primary difference between the enzyme and the zeolite relates to the specific interactions with the imidizole groups in the histidine framework, which can simultaniously interact with the reaction center to aid bond cleavage and bond formation reactions.

7.5 Biomimicking of Enzyme Catalysis

The development of chemocatalytic materials that mimic transformations carried out in enzyme catalysis has been a major goal of molecular imprinting approaches[14] and more generally biomimetic methods. The copolymerization of decomposable organic complexes into an inorganic framework should lead to the formation of cavities that have a prearranged shape that can be produced via the decomposition of a templating complex. The cavity then contains activating groups that are arranged such that a particular transition-state structure is stabilized.

In Chapter 4, we briefly discussed the hydrolysis of a peptide bond (Fig. 4.23a) by chymotrypsin, through the intermediate formation of a tetrahedral transition state. Figure 7.8 illustrates an approach to mimic such a reaction center for the hydrolysis of an ester using a polymer matrix. The reaction center is created using a templating phosphate-containing molecule that becomes part of the framework mimicking the transition state of the molecule to be hydrolyzed. The catalyst cavity is subsequently created by removal of the phosphate-containing molecular fragment before catalysis. In Chapter 8, self-organizing biocatalytic systems exploiting the immuno system will be discussed in which the reaction centers are generated by triggering the system using molecules which have shapes similar to those of the transition-state complexes for the reactions to be catalyzed. In addition directed synthesis approaches can also be used.

D'Souza and Bender[15] proposed the synthesis of a hydrolyzing cavity that mimicks chymotrypsin by attachment of selected substituents to a dextrine cavity (Fig. 7.9).

The two examples shown in Figs. 7.9 and 7.10 illustrate attempts at synthesizing a biomimetic flexible environment with well-controlled electrostatics strategically placed in a small cavity to provide concerted activation of probe molecules.

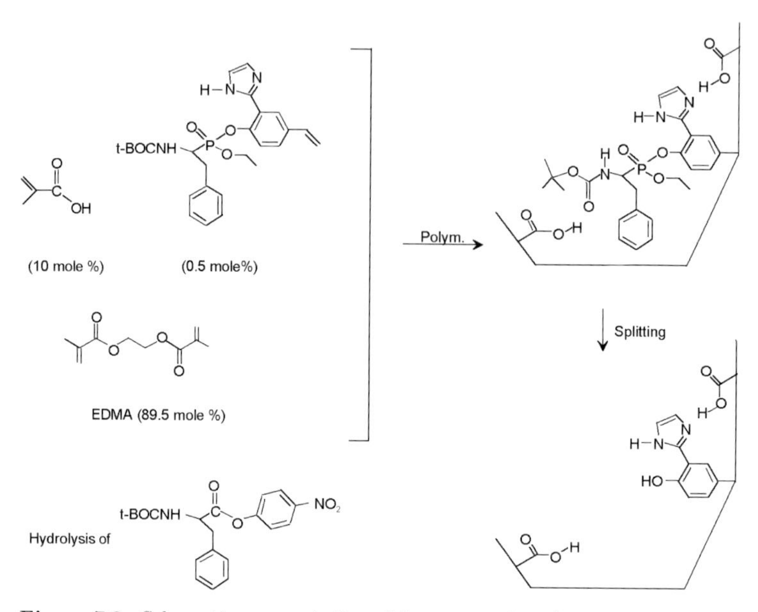

Figure 7.8. Schematic representation of the preparation of a catalyst by a transition-tate analogue using labile covalent binding and noncovalent binding[14].

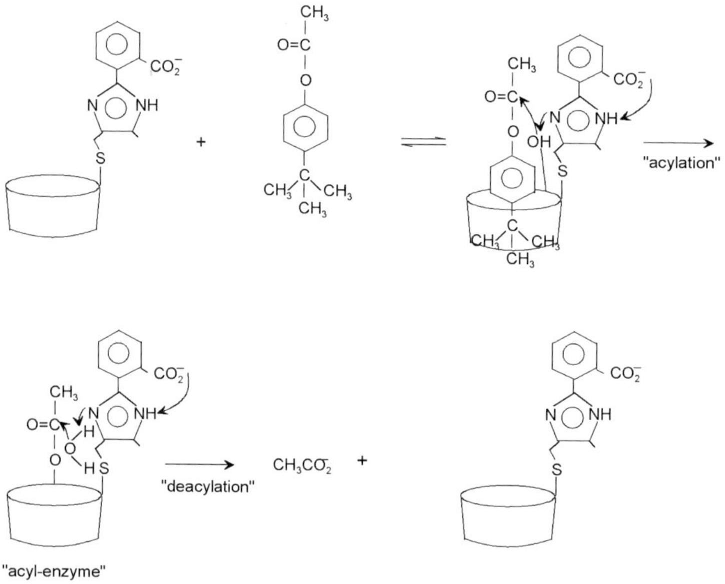

Figure 7.9. The mechanism of action of artificial chymotrypsin[15].

The principles of supramolecular catalysis relate to enzyme catalysis, because weak chemical bonds involving well defined hydrogen bonds are essential. Lehn[16] formulated two main steps required for supramolecular reactivity and catalysis:

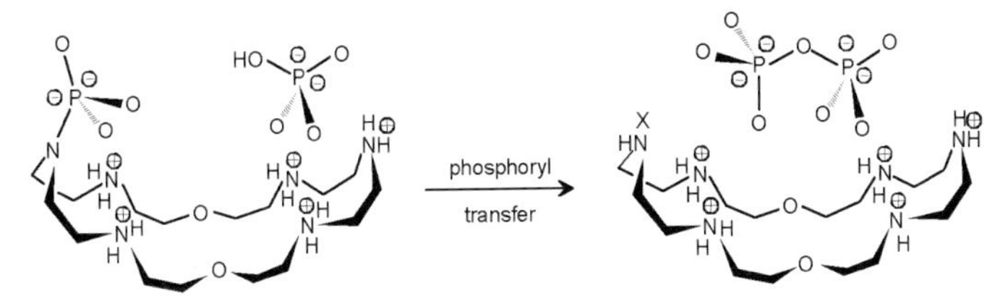

Figure 7.10. Cocatalysis: pyrophosphate synthesis by phosphoryl transfer[17].

- Binding which selects the substrate. Selectivity requires molecular recognition along with the optimization of the pretransition-state configuration.
- Transformation of the bound species into products. The supramolecular system has to transform reactants selectively to products that are accommodated in the supramolecule structure, but also readily desorb.

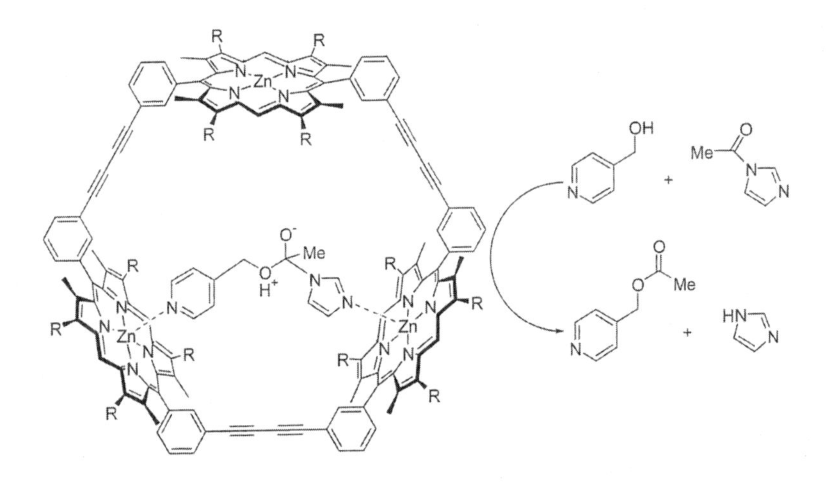

Figure 7.11. Tetrahedral intermediate of the acyl-transfer reaction bound inside the cavity of a porphyrin trimer. Adapted from V.F. Slagt [Thesis, Amsterdam (2004)].

The D'Souza and Bender system shows many of these features. Systems that induce bond formation require the presence of several binding and reactive groups. The catalytic molecule should act as a co-receptor bringing together reactant, substrate and intermediate complex. As an example, Hisselni and Lehn[17] demonstrated pyrophosphate formation from the intermediate phosphoramidate formed by phosphorylation of the macrocycle by ATP. In a second reaction step, the phosphoramidate reacts with a phosphate group to form pyrophosphate (see Fig. 7.10).

Enzyme-mimicking systems that contain metal cations have also been designed. A very elegant supramolecular assembly was designed by Sanders et al.[18] (see Fig. 7.11). They constructed trimeric porphyrin structures where Zn^{2+} porphyrin moieties function as templates for the organization of substrates into a conformationally optimal configuration that undergoes an efficient acyl-transfer reaction or that lead to Diels–Alder products.

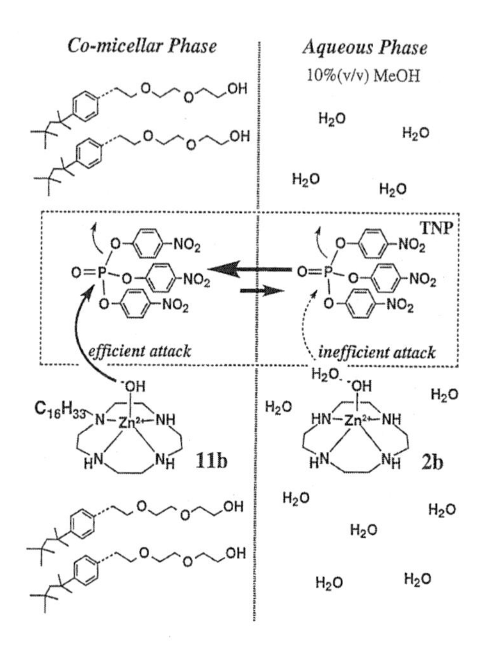

Figure 7.12. Proposed reaction mechanism for TNP hydrolysis catalyzed by lipophilic and hydrophilic zinc(II)–cyclen complexes[19].

It is important to remember that enzymes contain internally a hydrophobic cavity, but their exterior is hydrophilic, which makes the enzyme water soluble. This phase-property difference has been mimicked by Kimura and Koike[19] to develop an artificial organophosphorus hydrolase. Their system is illustrated in Fig. 7.12.

The system is designed to hydrolyze tris(4-nitrophenyl)phosphate (TNP), an insecticide. Both hydrophobic and hydrophilic Zn^{2+}–cyclen complexes were synthesized and tested for activity. The lipophilic Zn^{2+}–cyclen complex forms a co-micellar phase with triton, which creates a hydrophobic environment near the catalyst reactive center. The hydrophobic center allows for the efficient attack at the reaction center, thus leading to substantially enhanced catalytic reactivity. The hydrophilic catalyst, on the other hand, prevents an efficient attack at the reaction center and hence shows no reactivity. The nearly 100-fold difference in rate of the co-micellar catalyst system as compared with the aqueous system stems mainly from the solvability of TNP in the micellar phase and, to a partial extent, to the higher reactivity of Zn^{2+} due to the exclusion of H_2O.

Additional reviews on biomimetic catalytic systems are available[20, 21]. Shilov[20] reviews transition-metal complex systems that have related activities to biocatalytic systems. The review by de Vos et al.[21] compares the reactivities of zeolite and layered hydroxide-based enzyme-mimicking systems.

7.6 Bio-Electrocatalytic and Chemocatalytic Reduction Reactions

7.6.1 Oxidation Catalysis

When the two oxygen atoms of O_2 are incorporated into a product molecule, there must be a change in spin state. The reaction between triplet molecular oxygen and a singlet state reactant to produce singlet product molecules is spin forbidden. One way to overcome this spin forbidden event is if the reaction proceeds via free radicals. A transition-metal catalyst helps to overcome this spin forbidden event by an exchange of electrons between catalyst and oxygen molecules. Paramagnetic metal centers make the desired reaction towards singlet product molecules possible via a set of spin-conserving reaction steps. The spin state of the metal center is altered after reaction. A second major challenge in oxidation catalysis is to cleave the oxygen molecular bond. One has to distinguish between reactions in which one of the two oxygen atoms of the molecule is incorporated into the product molecule from reactions where both oxygen atoms are incorporated. In biocatalysis, enzymes that catalyze the former are called monooxygenases and those that catalyze the latter reaction are called dioxygenases. The monooxygenases use a co-reactant to remove the other oxygen atom so that the catalyst can be regenerated.

The heme-containing cytochrome P450 catalysts that hydroxylate CH bonds and are active as epoxidation catalysts are known to be monooxygenases. They utilize molecular oxygen as a source of two electrons to catalyze oxygen insertion into a CH bond. The overall electrocatalytic cycle for the oxidation reaction is shown in Scheme 7.1[22]

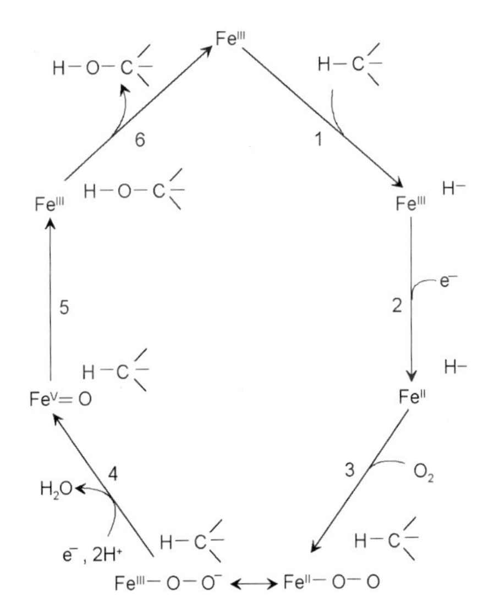

Scheme 7.1

Six steps occur in the cytochrome P450 reaction cycle. These steps appear to be universal for cytochromes P450 and are

1 Binding of substrate at the active site of low-spin hexacoordinate iron(III) form of the enzyme. This converts the low-spin Fe(III) into a high-spin pentacoordinate iron(III) enzyme. The substrate in Scheme 7.1 is represented simply by the C–H that is to be hydroxylated.
2 Addition of an electron to form the iron(II)-containing substrate form of the enzyme.
3 Addition of molecular oxygen to reduced the complex. Resonance forms exist for ferrous-dioxygen and ferric-superoxide with the latter favored.
4 Addition of a second electron. The addition of two protons to the O–O bond results in its cleavage and the subsequent elimination of water.

5 A radical-type hydrogen atom abstraction/oxygen rebound reaction occurs, the net effect of which is hydroxylation and re-formation of the iron(III) form of the enzyme.
6 Product dissociates from the active site.

Biochemical oxidation is part of a catalytic system in which electron and proton transport steps are coupled together in the generation of the specific oxygen atom necessary to carry out the specific selective oxidation reaction. Reactions that require hydrogen peroxide will similarly incorporate only one of the oxygen atoms into the desired product; the coproduct is H_2O. The decomposition of H_2O_2 occurs through a tetrahedral transition state stabilized by intermittent hydrogen bond formation and proceeds with proton and electron transfer. This is illustrated in Fig. 7.13. Note also the changes in the hydrogen bonds between the imidazolinium attached to Fe and the near basic groups.

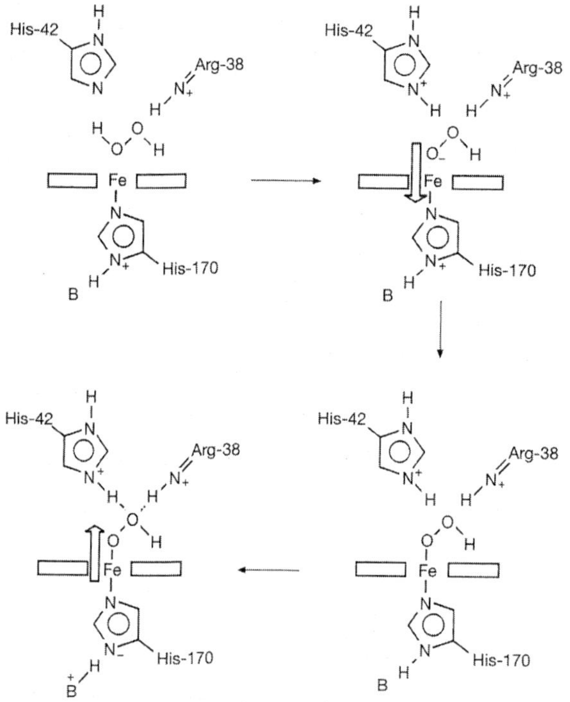

Figure 7.13. The decomposition of hydrogen peroxide by cytochrome P450 catalyst.

Dioxygenases form the second class of biochemical oxidation systems. They often oxidize hydrocarbons selectively to carboxylic acids. These catalysts proceed through low valence states of the catalytic metal center and through radical-type elementary steps. The metal centers donate electrons to the oxygen molecule, so as to assist oxygen bond cleavage.

Various heterogeneous analogues of dioxygenases exist such as the $(M_{1-x}{}^{2+}Al_xPO_4)$ zeolite redox systems discussed in Chapter 4 and the $(BiO_x)_2.(MoO_3)_y$ solid-state catalyst used to oxidize propylene to acrolein (see also Section 5.6.1). In this latter example, the two oxygen atoms generated by the dissociation of molecular oxygen at an Mo center, diffuse through the solid oxide material and then react such that one oxygen atom is used to produce H_2O, the hydrogen source being propylene, and the other inserts to give acrolein.

The Wacker reaction, introduced in Chapter 2, page 25, is another example of a chemocatalytic system that acts as a dioxygenase. In the homogeneous Wacker reaction catalyzed by the Pd^0/Pd^{2+}; Cu^+/Cu^{2+} couple the overall reaction is

$$H_2C{=}CH_2 + \frac{1}{2}O_2 \longrightarrow CH_3CHO$$

Mechanistically the reaction proceeds through the key reaction step

$$C_2H_4 + PdCl_4{}^{2-} + H_2O \longrightarrow CH_3CHO + Pd^0 + 2H^+ + 4Cl^-$$

The Pd^{2+} reaction center generates an oxygen atom through the activation of water to form an OH intermediate that subsequently inserts into ethylene. Molecular oxygen is then used to regenerate the Pd^{2+} through a redox cycle with Cu^+/Cu^{2+}. In the overall reaction the two atoms of O_2 become "incorporated" into acetaldehyde.

The biochemical oxomolybdooxidase enzymes operates similarly to the Pd^{2+} system with the important difference that reoxidation now proceeds electrochemically. As in the Wacker system, the selective oxygen atom is generated by the oxidation of water. A possible catalytic cycle for the sulfite oxidase system[24] is shown in Fig. 7.14.

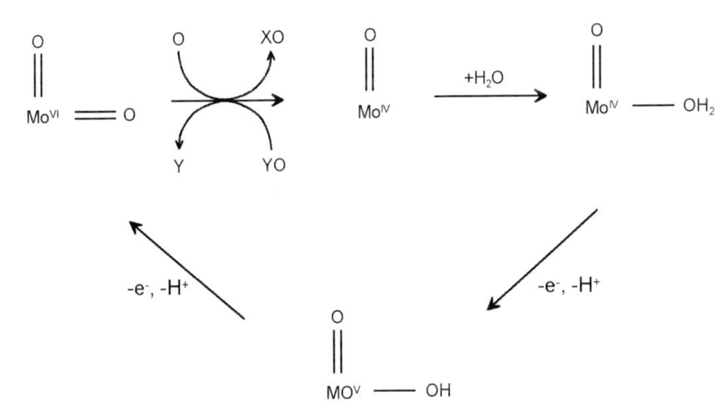

Figure 7.14. Possible catalytic cycle for sulfite (X) oxidase and dimethyl sulfoxide (DMSO) (Y) reductase[24].

The overall reaction is

$$R + H_2O \longrightarrow RO + 2H^+ + 2e^-$$

The coupling of oxidation catalysis in the biosystems with electrochemical systems enables such reactions to proceed under mild conditions. The electrochemical potential is adjusted to overcome the high activation barrier for generation of an oxygen atom from H_2O.

Propylene epoxidation over metallic Cu and ethylene epoxidation over Ag are two unique heterogeneous chemocatalytic systems which have no biochemical analogue. The epoxidation of ethylene over the Ag catalyst, which is briefly discussed here, was initially thought to proceed via the formation of an Ag superoxide or peroxide intermediate which would be the analogue of a monooxygenase system[25]. One oxygen atom was thought to react with ethylene to give epoxide while the second O atom would be removed from the surface through combustion of ethylene. The maximum selectivity towards epoxide would then be only 6–7. The reaction proceeds through a low valency silver state.

The alternative mechanism, which is now generally accepted, is the dioxygenase analogue. Selective reaction occurs through Ag in a high valency state, generated by the overoxidation of the Ag surface. This has an optimum selectivity at conditions where the oxygen adatom to Ag atom ratio (O_{at}/Ag_s) is equal to 2. Half of the oxygen atoms that belong to the surface-layer atoms are subsurface. Oxygen atom insertion into the ethylene bond occurs by its reaction with an AgO^+ species. In principle now the two oxygen atoms from molecular O_2 can be inserted, which gives a theoretical selectivity of 100%.

7.7 Reduction Catalysis

The catalytic reduction of a particular intermediate involves the addition of hydrogen. Just as in oxidation catalysis, there is the issue of whether one or two hydrogen atoms from H_2 are incorporated into the substrate through identical intermediate hydrogen atoms. In heterogeneous catalysis, this question translates into whether dissociation occurs homolytically or heterolytically. On a transition-metal surface H_2 dissociation generates two equivalent hydrogen atoms such as we have seen in Chapter 3. As discussed in Chapter 4, however, H_2 can dissociate heterolytically: $H_2 \rightarrow H^- + H^+$.

The presence of H^+ at the cation site of a cation-exchange zeolite will become the acidic proton on the zeolite lattice. The presence of H^-, on the other hand, will take on the form of an MH species. We discussed such a reaction for the Zn^{2+} ion-exchanged zeolite. The reaction is driven by the strong Lewis base nature of the negatively charged zeolite oxygen atoms.

In biochemistry there are two ways to hydrogenate or to dehydrogenate. In one reaction path, an H^- ion is transferred from NADH or NADPH to the substrate molecule that is to be reduced and the additional hydrogen atom is added as a proton. Such a step occurs, for instance, in the reduction of glutamate to proline. Cations such as Zn^{2+} or Mg^{2+} play a role in enzymes catalyzing such reactions.

The reactions can also be of the acid–base type induced by the electrostatics of the enzyme cavity.

A schematic description of such a concerted reaction is shown in Scheme 7.2.

An important representative of the reaction route through equivalent hydrogen atoms is that for the reduction of N_2 to NH_3 in the nitrogen fixation reaction. The nitrogenase enzyme catalyses the overall reaction:

$$N_2 + 10H^+ + 10e^- \longrightarrow H_2 + 2NH_3 + H_2$$

The reaction proceeds with co-evolution of hydrogen. Nitrogenase has been studied extensively and appears to contain three different kinds of Fe/Mo/S clusters distributed over two proteins. The actual nitrogen reduction takes place at the Fe_7MoS_8 cluster shown in Fig. 7.15.

It has recently been discovered that structure (A), which was widely believed to be the active cluster, was incorrectly identified. Structure (B), which contains a nitrogen atom attached to six Fe atoms, is the actual structure of the active nitrogen reducing cluster.

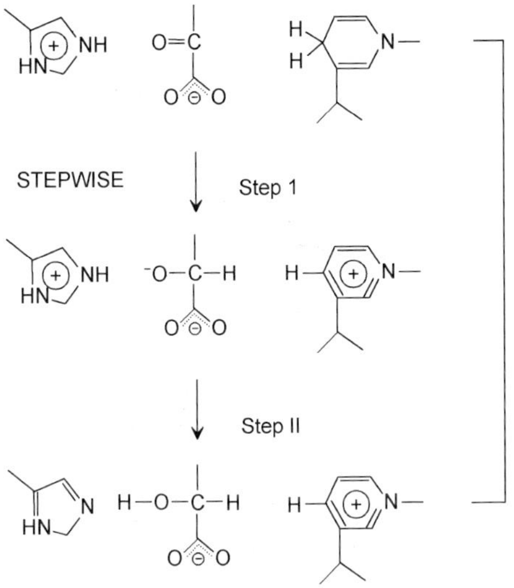

Scheme 7.2 The stepwise and concerted reactions[26] carried out by lactate hydrogenase.

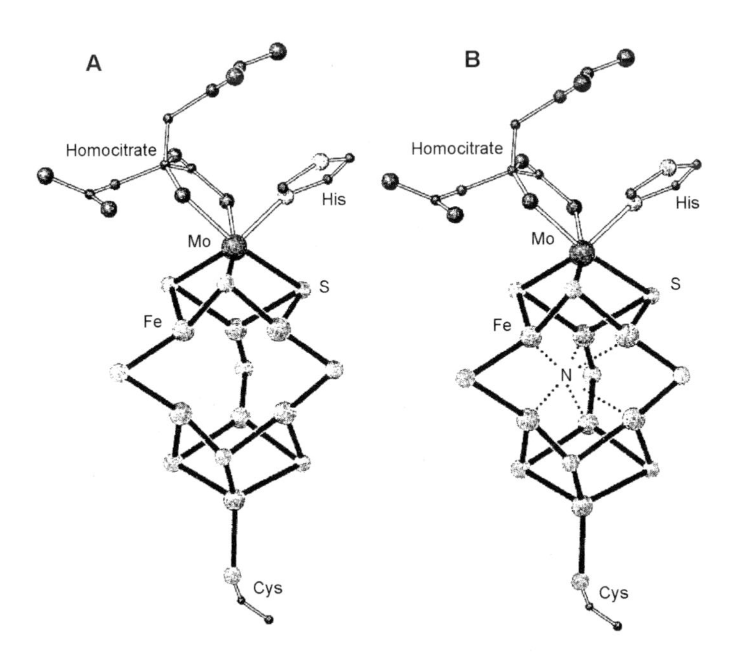

Figure 7.15. Two models of the FeMo cofactor of the nitrogen-fixing enzyme nitrogenase[27].

We will discuss available theoretical results for this reduction reaction and report on experimental studies of a recently discovered homogeneous reaction system. We subsequently conclude this section with an analysis of the heterogeneous catalyst used to synthesize ammonia industrially from N_2 and H_2 over Fe.

All theoretical models used so far to study the activation of N_2 have been based on the structure shown in Fig. 7.14A. It is not clear whether structure (B) is an intermediate or actually the catalytically reactive system that initiates the reaction by the adsorption of an N_2 molecule. It is generally believed that the seven Fe atoms provide the binding site for the N_2 molecule. We summarize here the general theoretical understanding of this system[28].

Nitrogen can initially adsorb end-on or side-on. Nitrogen binding is affected by protonation and the addition of an electron to generate a bridging SH group. Fe then experiences a weaker Fe–S interaction and, hence, can bind more strongly with the substrate N_2 molecule. Successive hydrogen transfers (protonation and electron donation) to nitrogen lead to unstable intermediates such as NNH, HNNH and $HNNH_2$ (Fig. 7.16).

Thermodynamics becomes favorable only once a three-electron transfer occurs. The first stable irreversibly bonded intermediate is hydrazine. Adsorbed NH_3 is the most stable state. It will desorb as NH_4^+.

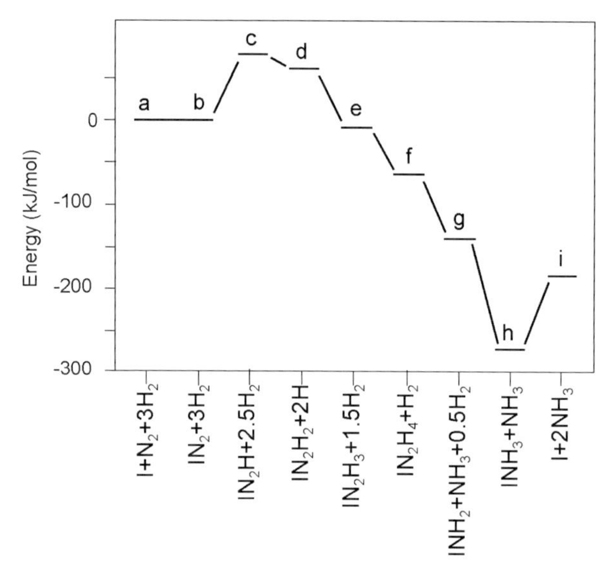

Figure 7.16. The calculated binding energies of the intermediates for hydrogenation of N_2[28b].

It is of interest to compare the reaction energy diagram for the N_2 reduction by the "Fe_7MoS_8" cluster with those for the heterogeneous Fe-based catalyst that are used industrially and operates at a reaction temperature at around 400°C. The reaction energy diagram constructed from experimental results for the industrial Fe–metal surface is shown in Fig. 7.17.

The relative energies of the reaction intermediates are shown to be completely different. On iron, the initial step is not the hydrogenation of N_2, but N_2 dissociation. Compared with the adsorbed N_{atom} state, the formation of NH_{ads}, NH_{2ads} and NH_3 are all thermodynamically endothermic. The industrial catalyst is promoted with oxides such as potassium oxides, which are thought to lower the activation energy Fe in the rate-limiting step, which is the dissociative adsorption of N_2.

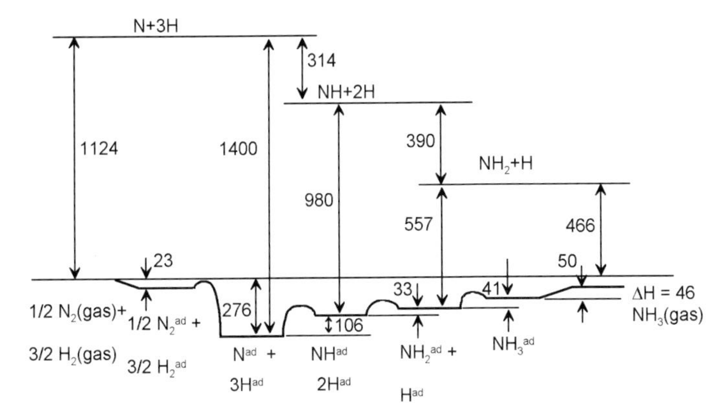

Figure 7.17. Schematic energy profiles for the ammonia synthesis on a promoted iron catalyst with energies in kJ/mol. Adapted from G. Ertl[29].

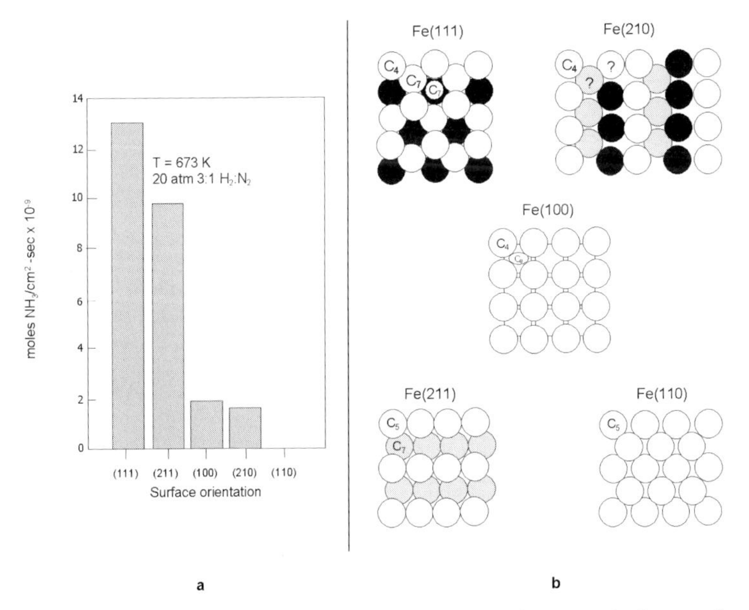

Figure 7.18. (a) Rates of ammonia synthesis over five iron single-crystal surfaces with different orientations: (111), (211), (100), (210), and (110). (b) Schematic representations of the idealized surface structures of the (111), (211), (100), (210), and (110) orientation of iron single crystals. The coordination of each surface atom is indicated. Adapted from Spencer et al.[30b].

As shown in Fig. 7.18, the ammonia synthesis reaction is highly structure sensitive. The most reactive face of iron is the Fe(111) surface plane[30]. The different phases tested in the reactivity study are shown in Fig. 7.18. The preference for this surface in the hydrogenation of nitrogen implies that the barrier for the activation of N_2 on this surface is the lowest[31]. Interestingly, on this surface, the site responsible for N_2 dissociation consists of seven Fe atoms, as in the enzyme "Fe_7MoS_8" cofactor. It illustrates that heterogeneous catalyst can also contain unique sites that are responsible for maximal substrate activation.

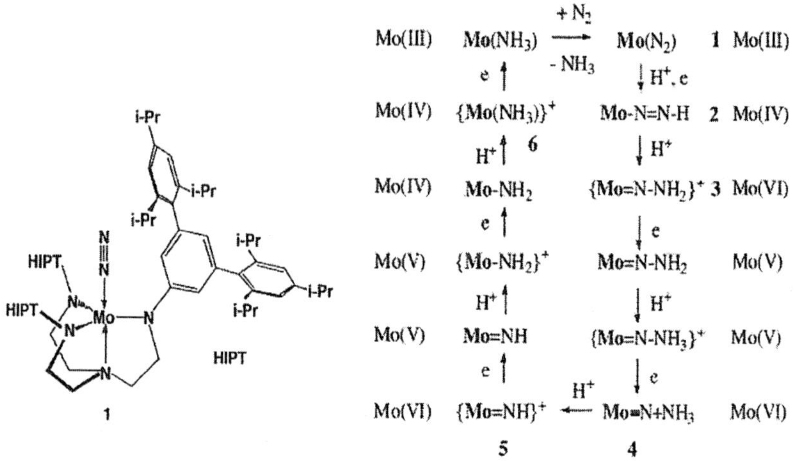

Figure 7.19. Proposed intermediates in the reduction of dinitrogen at an [HIPTN$_3$N]Mo (Mo) center through the stepwise addition of protons and electrons[32].

The concept that the low-temperature nitrogen fixation reaction requires electron transfer to nitrogen followed by subsequent protonation steps has been exploited to design biomimetic single-center homogeneous organometallic catalysts based on Mo. Yandulov and Schrock[32] designed different Mo complexes with tetradentate triamidoamine ligands that allowed for N_2 reduction using a mildly acidic molecule and reductant in heptane. The reaction sequence they propose is shown in Fig. 7.19, which is similar to earlier proposals for reduced Mo systems[33].

Unique to this biomimetic system over that of the previously reported [Fe_7MoS_8] cofactor systems is the inequivalence between the reduction of the first and second nitrogen atoms. The finding of the nitride atom in the new crystal structure of the Fe_7MoS_8 cluster may indicate that such a pathway may also occur in the nitrogenase system.

In summary, protonation and electron transfer steps in biochemical systems are typically rate limiting steps. Biomimetic chemocatalytic systems have been designed based on ideas of incorporating specific functions that could carry out similar chemistry.

7.8 Enzyme Mechanistic Action Summarized

Fischer[3] suggested at the end of the 19th century that unique activity of enzymes is related to the need for reactant molecules to fit optimally in the enzyme cavity. This is the lock and key molecular recognition model. Later Koshland[4] postulated the concept of induced fit; the enzymes assume shapes that are complementary to that of the substrate after the substrate is bound.

Pauling[34] suggested in 1948 a strategy for developing enzyme cavities that stabilize the transition state of the rate-limiting step. This can be recognized as the need for a substrate to have an optimum interaction with the enzyme. We have seen that the stabilization of pretransition-state structures is usually the essential step that stabilizes transition states.

Transition-state or pretransition-state structures must attain maximum free energy stabilization. The induced changes of the enzyme pocket, the concerted bond cleavage and the formation of bonds within the enzyme cavity, within reacting reagents and between substrate and enzyme lead to the optimum stabilization of pretransition-state structures. Enthalpy gains and entropy counteracts. This sometimes leads to anti-lock and key behavior when enthalpy advantages are too small to be overcome by entropy loss. The altered shape of the enzyme cavity after reaction decreases the optimum fit and, thus helps the desorption of the product. Hence the flexibility of the enzyme–protein framework is an intrinsic and an essential feature for its high activity.

References

1. L. Stryer, *Biochemistry*, Freeman, New York (1995)
2. (a) S. Natsch, *Thesis*, Münster, 2002;
   (b) T.A. Steitz, R.J. Flathericky, W.F. Anderson, C.M. Anderson, *J. Mol. Biol.* 104, 197 (1976);
   (c) P.R. Kuser, S. Kranchenco, O.A.C. Antunes, I. Polikarpov, *J. Biol. Chem.* 275, 20814 (2000)
3. E. Fischer, *Ber. Dtsch. Chem. Ges.* 27, 2984 (1894)
4. D.E. Koshland, in *The Enzymes*, Vol. 1, P.D. Boyer, H. Hardy (eds.), Academic Press, New York, p305 (1959)
5. W. Klaffke, personal private communications
6. I.A. Rose, *Biochemistry*, 36, 12346 (1997);
   I.A. Rose, *Biochemistry*, 37, 17651 (1998)
7. G.I. Liktenstein, *New Trends in Enzyme Catalysis and Biomimetic Chemical Reactions*, Kluwer, Dordrecht (2003);
   P.D. Boyer, *Biokhimiya*, 10, 1312 (2001);
   M. Loog, D.O. Morgan, *Nature*, 434, 104 (2005)
8. W.S. Allison, *Acc. Chem. Res.* 31, 819 (1998)
9. R.L. Cross, *Annu. Rev. Biochem.* 50, 681 (1981)
10. I. Ogilvie, R. Aggeler, R.A. Capaldi, *J. Biol. Chem.* 272, 16652 (1997)
11. F.M. Harold, *The Vital Force*, Freeman New York (1986)
12. (a) E.A. Erikson, P.M. Jones, A. Liljan, *Proteins*, 4, 283 (1980;
    (b) D.N. Silverman, S. Lindskog, *Acc. Chem. Res.* 21, 30 (1988)
13. D. Lu, G.A. Voth, *J. Am. Chem. Soc.* 120, 4006 (1998)
14. G. Wulff, *Chem. Rev.* 102, 1 (2002)
15. V.T. D'Souza, M.L. Bender, *Acc. Chem. Res.* 20, 146 (1987)
16. J.M. Lehn, *Angew. Chem. Int. Ed. Engl.* 27, 89 (1988)
17. M.W. Hisselni, J.M. Lehn, *J. Am. Chem. Soc.* 109, 7047 (1987)
18. C.J. Walter, H.L. Anderson, J.K.M. Sanders, *J. Chem. Soc. Chem. Commun.* 458 (1991);
    L.G. Mackay, R.S. Wylie, J.K. Sanders, *J. Am. Chem. Soc.* 116, 3141 (1994)
19. E. Kimura, T. Koike, in *Bioinorganic Catalysis*, J. Reedijk, E. Bouwman (eds.), Marcel Dekker, New York, p. 33 (1999)
20. A.E. Shilov, *Metal Complexes in Biomimetic Chemical Reactions,* CRC Press, Boca Raton, 1997;
    A.E. Shilov, *CaTTech*, 5, 72 (1999)

21. D.E. de Vos, B.F. Sels, P.A. Jacobs, *CaTTech*, 6, 14 (2002)
22. H.B. Dunford, in *Comprehensive Biological Catalysis, III,* M. Sinnott (ed.), Academic Press, New York p. 210 (1998)
23. IBIDEM, p. 200
24. IBIDEM, p. 381
25. R.A. van Santen, H.P.E. Kuipers, *Adv. Catal.* 35, 265 (1987)
26. G. Naray-Szabo, M. Fuxreiter, A. Warshel, in *Computational Approaches to Biological Reactivity*, G. Naray-Szabo, A. Warshel (eds.), Kluwer, Dordrecht, p. 266 (1997)
27. B.E. Smith, *Science*, 297, 1654–1655 (2002);
    O. Einsle, F. Akif Tezcan, S.L.A. Andrade, B. Schmidt, M. Yoshida, J.B. Howard, D.C. Rees, *Science*, 297, 1696 (2002)
28. (a) H. Deng, R. Hoffmann, *Angew. Chem. Int. Ed. Engl.* 32, 1062 (1993);
    (b) Th.H. Rod, J.K. Nørskov, *J. Am. Chem. Soc.* 122, 12751 (2000)
29. G. Ertl, in *Science Technology,* Vol. 4, J.R. Anderson, M. Boudart (eds.), Springer Verlag, Berlin, p. 273 (1983)
30. (a) R. Brill, E.L. Richter, E. Ruck, *Angew. Chem. Int. Ed.* 6, 882 (1967);
    (b) N.D. Spencer, R.C. Schoonmaker, G.A. Somorjai, *J. Catal.* 74, 129 (1982)
31. J.J. Mortensen, L.B. Hansen, B. Hammer, J.K. Nørskov, *J. Catal.* 182, 479 (1999)
32. D.V. Yandulov and R.R. Schrock, *Science*, 301, 76 (2003)
33. G.J. Leigh, *Science*, 301, 55 (2003)
34. L. Pauling, *Am. Sci.* 36, 51 (1948)

CHAPTER 8
Self Organization and Self Assembly of Catalytic Systems

8.1 General Introduction

In the previous chapters we predominantly considered catalysis as a molecular event, in which substrate molecules are activated by the catalyst. In this chapter and the next we will emphasize catalytic features of dimensions in space much larger than that of single catalytic centers and times much longer than those associated with the individual molecular catalytic cycles. Often mass and heat transport cause reaction cycles, which occur at different sites, to interact. Under particular conditions this gives rise to cooperative phenomena with oscillatory kinetics and temporal spatial organization. As such, interesting surface patterns such as spirals or pulsars may form. Such complex cooperative phenomena are known in physics as appearances of excitable systems. Their characteristic features are easily influenced by small variations in external conditions. Hence these systems have also features that are called adaptive.

In this chapter, we bring together several topics in catalysis that at first sight appear unrelated. However, all of them share features of complex adaptive systems. This aspect unites topics as different as the biological immuno-response systems and zeolite catalysis.

We start the next section with a discussion of self repair of the catalytic site after reaction to restore it to its initial state when the reaction cycle has been completed. Self repair in a catalytic system is the lowest level of self organization. It is an intrinsic property of a catalytic system and occurs locally at each catalytic site. In the two sections that follow we will introduce the general features of self organization, that result from collective cooperative effects, due to the interaction of catalytic reaction cycles of reactant molecules occuring at different catalytic centers. The example chosen is CO oxidation on a reconstructing Pt surface. It will appear that fundamental studies in computer science and the cellular automata have contributed in an essential way to understanding such phenomena.

As we will see in a final chapter, dynamic Monte Carlo methods, genetic algorithms and evolutionary computational strategies help to determine the optimum structure of catalysts for maximum performance.

Two intermediate sections deal with experimental systems, in which a catalyst develops in response to a templating molecule. The template can be chosen similar to the shape of a molecule in the transition state of the reaction to be catalyzed. The biological immuno response system is an important example. It can respond to different templates. Molecular recognition of different templates leads to different reorganization of the immunoglobin molecule that is key to biological reactions that follow. A response that is amplified by biological transformation within the biosystem. The reaction sequences can be physically characterized as evolutionary, recombinatorial events. Interestingly, the self assembly of siliceous oligomers around a template in zeolite synthesis shows many of the same features that can be recognized as similar but much less sophisticated as in the biological system.

We expect that the unconventional way in which the different topics have been brought together in this chapter will help to assist the reader to appreciate the many different forms in which complex organization realizes itself in catalysis.

Molecular Heterogeneous Catalysis. Rutger Anthony van Santen and Matthew Neurock

ISBN: 3-527-29662-X

8.2 Self Repair in Chemocatalysis

Self repair of chemical bonds altered within the catalyst during reaction has to occur in order for the reaction cycle to close. This is due to bonding changes on the surface or within a catalyst during its reaction cycle. Catalysis is a cyclic event, that consists of a series of elementary reaction steps in which the reaction center or a key reaction intermediate is regenerated after completing the reaction cycle. Interactions between the catalyst atoms and the atoms in the reacting molecules have to be strong, so that reactant molecular bonds are broken, bonds between reactant atoms and catalyst atoms are formed and rearrangement reactions occur within a reactant molecule or between adsorbed reactant molecules. In the process bonds between atoms within the catalyst also become weakened and there can be changes in the atomic positions. Finally, product molecules desorb, leaving behind an empty active center or an activated complex of new reactant molecules and catalyst. To regenerate the initial catalytic site, the displaced catalyst atoms must return to their original positions.

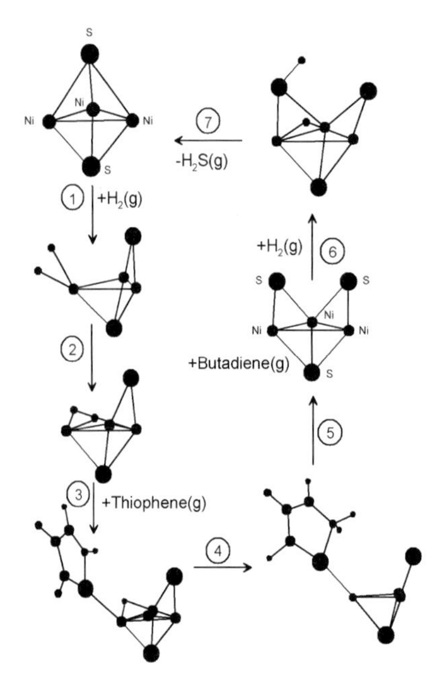

Figure 8.1. Catalytic HDS cycle via η^1-thiophene and dihydrothiophene intermediates: H_2 adsorption initiated [1].

The cyclic succession of reactions with self repair of the catalyst is illustrated for the catalytic reaction in which sulfur is removed from thiophene with hydrogen. The successive reaction steps are shown for an Ni_3S_2 cluster in Fig. 8.1. Thiophene adsorbs on the vacant Ni site of the Ni_3S_2 complex. Hydrogen can dissociate on the Ni_3S_2 cluster and weaken the Ni–S bonds in the cluster. Hydrogen can subsequently add to the carbon–sulfur bond in thiophene, thus enhancing the C–S bond-breaking reaction. When the reaction is concluded, butadiene and H_2S desorb and the Ni_3S_2 particle is restored to its original state. Owing to the strong chemical interaction between catalyst surface atoms and reactant atoms, within the catalytic complex the bonds between the catalyst atoms themselves weaken. This can be best understood on the basis of the Bond Order Principle[2] in chemical bonding, that we discussed in Chapter 3.

The Bond Order Principle is an approximate theory. It assumes a spherical distribution of electrons, hence exceptions to its rule exist. However, for many chemical systems it is found to have predictive value. According to the Bond Order Principle, the valency or bonding power of each atom is distributed over the chemical bonds in which the atom is involved. The total bonding power of an atom is considered to be a constant and this is to be distributed over all the bonds directed towards neighboring atoms. It implies that when more bonds are shared there is less bond strength per bond. As a consequence, attachment of reactant molecules to the catalyst surface atoms will weaken the bonds in the adsorbate via their interaction with the catalyst. In addition, there is a weakening of the internal catalyst molecular bonds that may lead to significant distortions and cleavage of internal catalyst chemical bonds and even to reconstruction of surfaces or clusters during catalytic reactions. In the particular example illustrated by Fig. 8.1, the Ni–Ni and Ni–S bond distances change upon adsorption and reaction. There is even a cleavage of one of the Ni–S bonds that after reaction is restored.

A second example that nicely illustrates the relevance of self repair is found in the comparison of two experimentally related systems. The first is a true catalyst and the other an unstable reactive, but non-catalytic, material. We consider the selective oxidation of an alkene to an epoxide by silica-based catalysts that contain Ti. In such catalysts, Ti is four-coordinated to the oxygen atoms. There are two important systems that are used in practice. In the first system, Ti istetrahedrally bound to a silica surface. Its state is as schematically shown in Fig. 8.2.

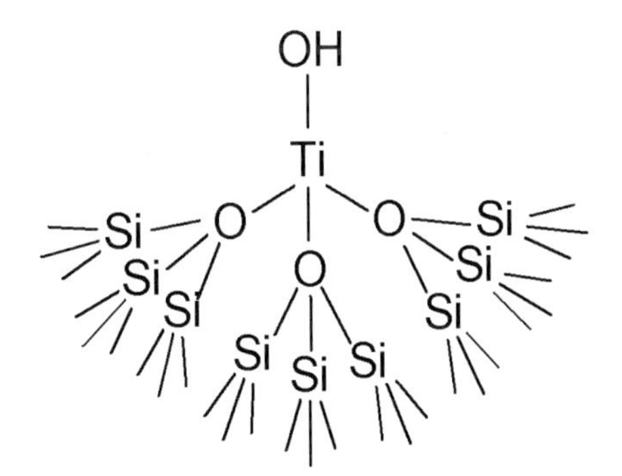

Figure 8.2. Ti–OH attached to a silica surface.

Ti is coordinated through three oxygen atoms to the silica surface and terminated by a hydroxyl group[3]. This system can epoxidize propylene with hydroperoxide to the corresponding epoxide and alcohol by a reaction path to be discussed at the end of this section. The second system is a crystalline zeolite system in which a silicon cation is tetrahedrally surrounded by four oxygen atoms. The tetrahedra form a crystalline network with a four, five or six tetrahedra-containing ring structure. Many structures can be formed, some of them containing channels of 10 or 12 rings through which molecules can diffuse. Zeolites were extensively discussed in Chapter 4. In a particular zeolitic SiO_2 polymorph, silicalite, the replacement of Si by Ti results in an active epoxidation catalyst[4], in which hydrogen peroxide can be used as the oxidation agent (see Fig. 8.3).

Whereas the above-mentioned systems are catalytically active, one can design other systems in which Ti is four-coordinated with oxygen that turn out to be catalytically unstable. Such a system is formed, for instance, when $TiCl_4$ reacts with cubic silica clusters such as silsesquioxanes, with two dangling silanol groups. Such a cluster is illustrated in Fig. 8.3 (top left).

The reaction of $TiCl_4$ with such a silsesquioxane cluster connects Ti through oxygen atoms with four silsesquioxane clusters (Fig. 8.3 , bottom left)[5]. The result is a very flexible gel. The titanium atoms become part of a rather loose network of silsesquioxane clusters connected through Ti atoms. This system appears to be catalytically active for epoxidation of alkenes by hydroperoxide. When contacted with the Ti center, the reaction sequence shown in Fig. 8.4 occurs in Ti-silicalite and in the gel.

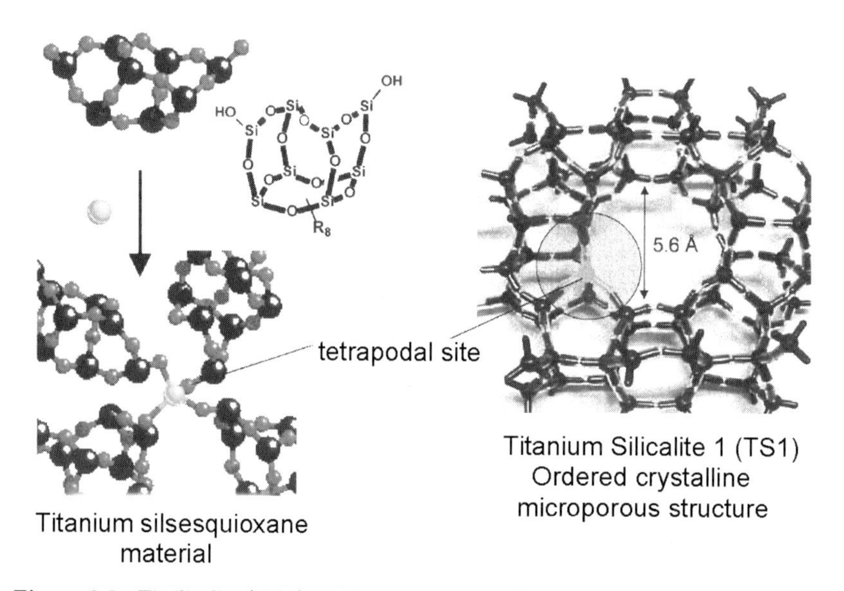

Figure 8.3. Ti-silicalite (right) and Ti-silsesquioxane (left) epoxidation systems.

Cleavage of the OH bond in the peroxide $^{-}$OOH and the catalyst Ti–O bond occurs with formation of an SiOH group. In consecutive reaction steps the alkenes reacts with the Ti-OOR group to give the epoxide and -TiOR. The catalytic center is restored when the Ti–O–Si bond is re-established through formation of the alcohol or H_2O by reaction of silanol with Ti alkoxy. The difference between zeolite and gel is that in the gel the resulting silanol group will move away from the Ti center, and, hence, restoration of the TiO bond after reaction becomes impossible. However, in the zeolitc there is only a very limited motion of the SiOH group possible. This is due to the high connectivity of the silicon units in rather tight small rings. Once in the zeolite, one of the oxygen atoms of the OOH group has been inserted into the π-bond of the alkene the OR group left on Ti will now react with the silanol proton to H_2O and the Si–O–Ti bond is restored.

Titanium that is three-coordinated as in Fig. 8.2 is also part of a catalytically stable reaction systems. This has been demonstrated by the incorporation of Ti as a corner

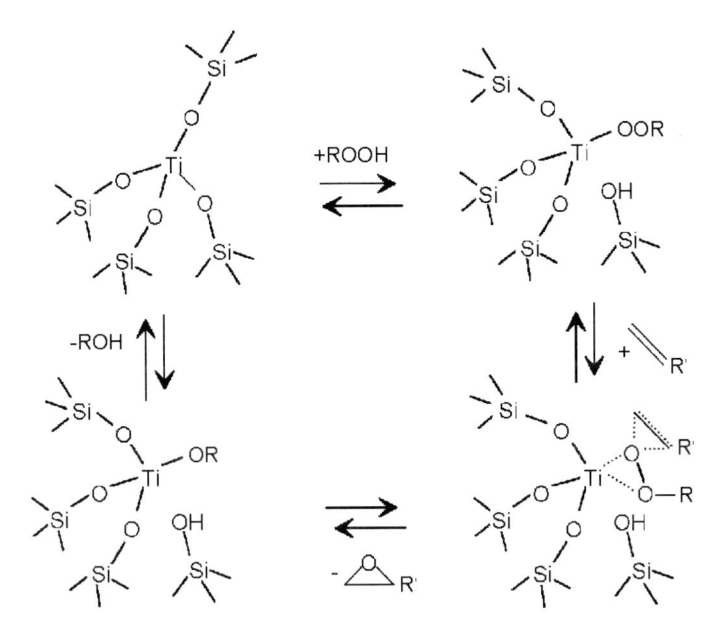

Figure 8.4. The epoxidation reaction cycle.

atom in a Ti silsesquioxane complex (see Fig. 8.3). NMR measurements of this complex during the epoxidation reaction indicate complete stability. The stability of this complex is related to the the Ti–OH group, which can react with the peroxide to form an OOR bond, so that now no Si–O–Ti bond must be activated during reaction. If the Ti–O breaks in a parallel reaction, the silsesquioxane lattice constrains the Ti–O–Si site so that the Ti–O–Si is restored after reaction as sketched in Fig. 8.4.

These examples illustrate that local changes in the catalytically reacting systems induce local stress or strain, that result from bond cleavage or bond formation as the result of the catalytic reaction. The catalytic system must therefore be flexible as well as robust. Flexibility is needed so that local volume changes can be accommodated through changes in bond angles of the surrounding bonds of the atoms around the catalytically reactive systems. This happens in the zeolitic system in which the energy needed to alter the Si–O–Si bond angles is small.

Zeolites are robust enough that local stress or strain does not disrupt their framework. This is not the case for many of the metal surfaces, as was discussed in Chapter 2.

8.3 Synchronization of Reaction Centers

The cyclic nature of the catalytic reaction usually does not typically lead to an overall cyclic time dependence or synchronization of the reaction system. Rarely is there a synchronization of the cycle reaction phases on different reaction centers. As long as the reaction conditions are close to the equilibrium condition of a system, stationary kinetics tend to rule. However, for particular systems, when reactions are performed far from their equilibrium complex oscillations in time and even spatial organization may occur. The necessary conditions for the occurrence of cooperative time dependent and spatial events are:

- the presence of autocatalytic elementary reaction steps (that enhance the rate)
- a reaction step that slows the reaction
- synchronization of the phase of the catalytic reaction cycles at the different catalyst centers by mass or heat transport over the surface or through the gas phase.

Spatial or temporal self organization effects have been observed on heterogeneous transition metals as well as oxide catalysts. We will illustrate here in detail these concepts using the CO oxidation reaction as an example [6]. We briefly referred to self organizing catalytic systems earlier in Chapters 2 and 4.

The catalytic oxidation of CO over the Pt(100) surface provides a nice illustration of these ideas. This surface has the interesting feature that in a vacuum its surface topology is different from that of the bulk terminated surface. A slightly more stable situation is obtained when the surface layer reconstructs from a configuration in which each surface atom has four surface atom neighbors (the total number of nearest neighbor atoms is to eight, implying that there are four neighbors with atoms in the layer next to the surface) to a configuration in which each surface atom has six surface atom neighbors (total number of nearest neighbor atoms between nine and twelve). This is denoted the Pt(100)hex surface. The larger average surface atom coordination number implies for the Pt(100)hex surfaces a lower degree of coordinative unsaturation of the surface atoms, which will lower the surface energy since the cost of surface generation is reduced. However, it will also decrease the reactivity of the surface atoms with respect to adsorbing gas-phase molecules. This is a consequence of the Bond Order Principle. The bond energy of the adsorbate decreases when there is an increase in the number of surface metal atom neighbors, since this dilutes the bonding power towards the adsorbed molecule as compared with bonding with a surface atom that has fewer metal atom neighbors. As a consequence, the adsorbing oxygen molecules interact so weakly with the reconstructed surface, that the activation barrier for O_2 dissociation cannot be overcome. On the other hand, CO will adsorb since no intramolecular bonds are cleaved upon chemisorption and, hence, chemisorption will always be exothermic. The adsorption of CO on the reconstructed Pt(100)hex surface, weakens the Pt–Pt bonds. The interaction with CO will increase when the number of neighboring metal atoms in the surface that bind CO decreases. This can drive the surface to reconstruct back to the less stable, more reactive bulk terminated Pt(100) surface. This tends to occur when at least five CO atoms adsorb near each other on the Pt(100) hex surface. The cost of reconstruction is now compensated for by the increased interaction with CO, that is the result of the decrease in the number of neighboring metal atoms of the surface atoms bonded to CO. Also, part of the neighboring surface uncovered by the CO molecules surface reconstructs back to the Pt(100) surface. These surface atoms are now more reactive, since they have fewer neighboring metal atoms and can now readily dissociate oxygen. The sketch of the events that occurs during CO oxidation on the Pt surface gives a beautiful illustration of the consequences that bond weakening effects can have on the structure of a reactive catalyst surface as a function of the phase of the catalytic reaction cycle. So far we reached the conclusion that dissociative adsorption of O_2 can occur if at least five CO molecules have been adsorbed on a patch of the (100)hex surface in order to reconstruct it to the more active phase. Once oxygen atoms are coadsorbed with CO, rapid recombination of adsorbed O and CO occurs to give CO_2 that rapidly desorbs into the gas phase. At the same time the surface atoms free of CO and O reconstruct back to the low-reactivity Pt(100) hex phase. As a consequence, the dissociative adsorption of O_2 stops and the reaction cycle will only start once enough vacant Pt(100)hex sites have been generated. Once again five CO molecules must adsorb in order to start the reaction cycle again. Under particular reaction conditions this catalytic system shows an oscillating time dependence and, hence the catalytic reaction cycle has an autocatalytic elementary reaction step.

Autocatalysis is strictly defined as a reaction in which the product enhances the rate of the reaction, as, for instance, in the following reaction scheme:

$$\mathrm{A + B \rightarrow 2B + C}$$

In this autocatalytic reaction scheme, B is replicated by its reaction with A. Autocatalysis is therefore an elementary form of replication, a topic we will discuss extensively in the next chapter.

The autocatalytic elementary reaction step for CO oxidation is the removal reaction of CO from the catalyst surface. The presence of adsorbed CO suppresses the dissociative adsorption of O_2, because vacant surface sites are required to accommodate the oxygen atoms that are generated by dissociated molecular oxygen. The CO_2 removal reaction, on the other hand, is actually autocatalytic in vacant sites since a molecule of oxygen can ultimately remove two adsorbed CO molecules, thus freeing up two additional sites.

$$\begin{aligned} \mathrm{O_2^{(gas)}} + \square\square\mathrm{CO_{ads}CO_{ads}} &\rightarrow \mathrm{O_{ads}O_{ads}CO_{ads}CO_{ads}} \\ &\rightarrow \square\square\square\square + \mathrm{2CO_2^{(gas)}} \end{aligned}$$

The reaction step which is autocatalytic in CO is the regeneration of the surface phase necessary to dissociate O_2. Two vacant sites generate four vacant sites. Several autocatalytic reaction steps are coupled in the CO oxidation reaction ($\square$ is schematic representation of surface vacancy).

The surface reconstruction reactions are slow and act to slow the reaction. This presence of a reaction step which slows the reaction system is a second condition for self organization. Slowing the reaction progress creates the opportunity for local synchronization. Synchronization of the reaction cycles at different reaction centers over larger surface distances occurs when surface diffusion homogenizes the surface composition at different reaction sites of the catalyst.

In the next section, we will illustrate the need for synchronization of reaction cycles that occur on different reaction centers using results from dynamic Monte Carlo simulations. As shown by Pecora and Carroll[7], oscillatory dynamics at different reaction centers have to satisfy particular conditions, so that synchronization between reaction centers can occur.

A system of coupled autocatalytic reactions that can be used to illustrate the generation of stable oscillations under non-equilibrium conditions is the Lotka–Volterra system:

$$\begin{aligned} \mathrm{A + X} &\xrightarrow{k_1} \mathrm{2X} \\ \mathrm{X + Y} &\xrightarrow{k_2} \mathrm{2Y} \\ \mathrm{Y} &\xrightarrow{k_3} \mathrm{E} \end{aligned}$$

The corresponding kinetic expressions for this system are

$$\frac{d[X]}{dt} = k_1[A][X] - k_2[X][Y]$$
$$\frac{d[Y]}{dt} = k_2[X][Y] - k_3[Y]$$

Using as ansatz for [X] and [Y] solutions

$$[X(t)] = [X_0] + xe^{i\omega t} \qquad (x \ll [X_0])$$
$$[Y(t)] = [Y_0] + ye^{i\omega t} \qquad (y \ll [Y_0])$$

$[X_0]$ and $[Y_0]$ are the steady-state solutions of the kinetic equations. They satisfy the conditions

$$k_1[A] - k_2[Y_0] = 0$$
$$k_2[X_0] - k_3 = 0$$

For the frequency ω one finds the solution

$$\omega = \sqrt{k_1 k_3 [A]}$$

Surface diffusion coupled to autocatalysis can also lead to spatial self organization of the surface, resulting in time-dependent pulsing or spiral-type overlayer patterns. Under particular conditions growing spiral patterns may split into smaller spirals, which will grow in turn. This can be considered a chemocatalytic mimicry of reproduction.

An important lesson to be learned from this exposition is that chemocatalytic systems adapt their state to the reaction mixture composition or rather the chemical potential of the gas phase to which they are exposed. To predict catalysis properly one therefore has also to be able to predict the state of the catalyst surface during reaction.

In section 8.4 we discuss in more detail self organization and synchronization continuing the analysis of the CO oxidation reaction.

8.4 The Physical Chemistry of Self Organization

Self organization is a general phenomenon that occurs in many particle systems that are defined as active media [8]. Such systems can be generally described by reaction-diffusion equations for their individual components i:

$$\dot{a}_i = g_i\left(\left\{a_j\right\}\right) + D_i \nabla a_i$$

where $a_i\left(\vec{r}, t\right)$ are the local concentrations of i and $g_i(a_i.....a_M)$ are a set of linear functions $g_i(a_i, ..., a_M)$ that describe the reaction rates for components i. D_i represents the diffusion constants of the different reacting components. ∇ is the second-order derivative operator of the spatial coordinate vector $\vec{r}$. Chemical reactions are considered to take place in small volume elements of uniform composition in which the concentrations

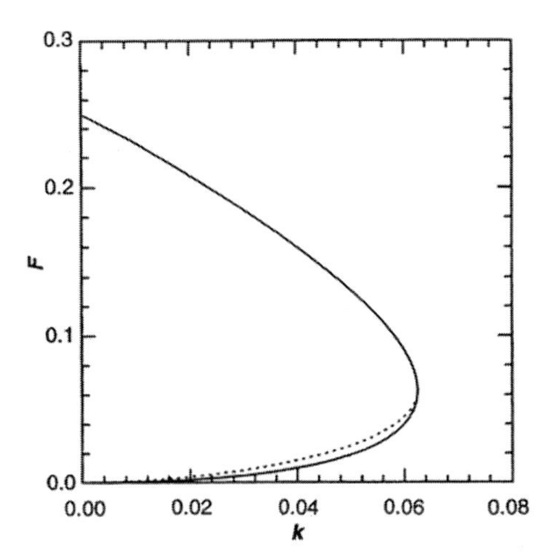

Figure 8.5. Phase diagram of the reaction kinetics for the Pearson autocatalytic model. F and k refer to the feed and the rate parameters respectively. Outside the region bounded by the solid line, there is a single spatially uniform state (called the trivial state) that is stable for all $(F,\ k)$. Inside the region bounded by the solid line, there are three spatially uniform steady states. Above the dotted line and below the solid line, the system is bistable[10].

change. Active media can be classified as bistable, excitable and oscillatory. In a bistable medium the kinetic set of equations $g_i(\{a_j\})$ has two states as its stationary solution. Large perturbations trigger transition between these states, which may result in trigger waves, typical for instance for flame propagation. They may also lead to a large variety of irregular spatio-temporal patterns.

A catalytic example is provided by the autocatalytic reaction scheme in which x multiplies:

$$\mathrm{A} + 2\mathrm{x} \underset{k_1}{\overset{k_2}{\rightleftharpoons}} 3\mathrm{x}$$

$$\mathrm{x} \underset{k_3}{\overset{k_4}{\rightleftharpoons}} \mathrm{B}$$

It provides a simple model case that illustrates formation of spatio-temporal patterns due to such finite-amplitude perturbations.

In the Gray–Scott model[9] of this system, both reactions are considered to be irreversible. This reaction scheme is a simplification of the autocatalytic model of the glycolysis cycle (see Chapter 7). A is a feed term and B an inert product. Pearson[10] has shown that as a function of kinetic and diffusion parameters this system leads to the formation of local regions of concentration defined by sharp boundaries. These local regions take on cell-like characteristics, thus undergoing multiplication and division behavior. We discuss some of the results in detail, also because of the discussion in the next chapter on self replication and the origin of protocellular systems. As a function of feed (F) and rate parameter (k), a state phase diagram can be constructed (see Fig. 8.5).

Two areas can be identified: one with three steady states and another area to be characterized as excitable. At the boundary of the two time regimes transient patterns form in response to a small disturbance. In this unstable region a particularly interesting phenomenon is observed that behaves like cell multiplication. Patterns occur with concentration profiles of cellular form that grow and replicate. When these "cells" exceed a particular dimension, the interior destabilizes (in this case because a necessary concentration gradient is not maintained) and cells divide.

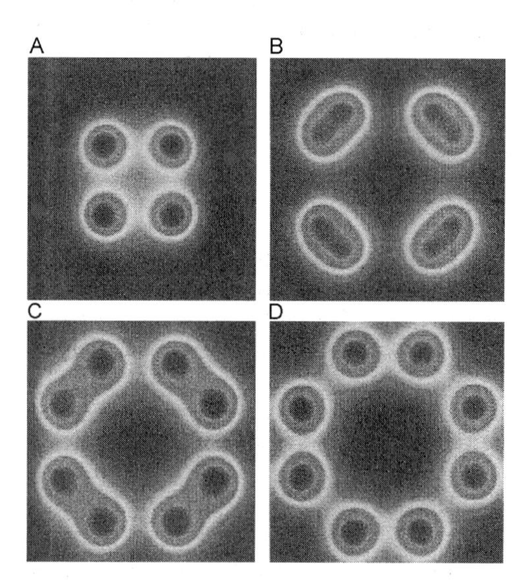

Figure 8.6. Time evolution of spot multiplication. This figure was produced in a 256 by 256 simulation with physical dimensions of 0.5 by 0.5 and a time step of 0.01. The times t at which the figures were taken are as follows: (A) $t = 0$; (B) $t = 350$; (C) $t = 510$; and (D) $t = 650$[10].

The resulting spatio-temporal pattern at a particular point in this unstable region is shown in Fig. 8.6. The compositional profile in Figure 8.6 was made just after an initial square perturbation had decayed to leave four cell-like spots. Cell growth and division are shown to succeed. Experimental verification of such self replication has been given by Lee et al. [12] in a reaction system with ferrocyanide, sulfide and iodide that has a bistable stationary solution.

An element of an excitable medium returns to its initial state of rest after being activated by an external perturbation that exceeds a particular threshold. In an excitable medium this results in a traveling excitation pulse.

A bistable medium is converted into an excitable medium by coupling a production rate with an inhibitor component. An excitable medium undergoes a fast transition into an excited state if triggered by the threshold transgressing perturbation. After this transition, the medium at the point becomes refractory, slowly recovering its excitability until it again becomes receptive. Target patterns and spiral waves are characteristic resulting features.

Oscillatory media have many characteristics of an excitable medium. They consist of a large population of self oscillating elements, which are weakly coupled. Depending on the properties of the medium, oscillations can synchronize or desynchronise with time. The os-

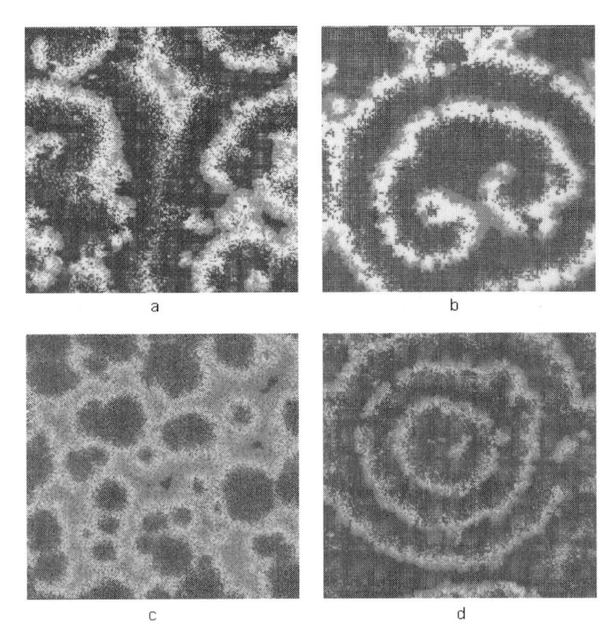

Figure 8.7. Four examples of spatio-temporal pattern formation obtained in dynamic Monte Carlo simulation of the reconstructing CO oxidation system: (a) turbulent patterns; (b) a double rotating spiral; (c) cellular structures; (d) target patterns[11].

cillations in rate and self organization of overlayer patterns observed for the CO oxidation reaction, discussed earlier, belong to this category. Diffusion of CO synchronizes the different reaction sites on the surface. Otherwise, overall catalytic time-dependent behavior would be chaotic in time, controlled by the collision moment of the individual molecules. The need to couple the reaction phases at different catalytic reaction centers is neatly illustrated by dynamic Monte Carlo simulations on grids of varying size representing the reconstructing surface during the CO oxidation reaction [11].

Different spatio-temporal patterns can emerge depending on the kinetics in this system. Figure 8.7, for example, presents turbulent double rotating spirals, cellular, and target patterns. In Fig. 8.7, reactions are represented by composition changes with a particular probability on grid positions, dependent on the occupation of neighboring grid positions. These figures illustrate self-organized surface pattern formation in the oscillatory medium of the catalytic oscillators. They appear as the result of colliding wave fronts of different local surface concentrations. The spatial dimension of these patterns depends on the square root of diffusion rate and oscillation time. When the grid is small, the grid represents only a small number of surface atoms and, hence, can only support a single reaction center which requires enough surface atoms to allow for reconstruction. This results in only a single oscillation. When the grid size is increased, the possibility of more reaction centers arises. As Fig. 8.8b shows, in the absence of diffusion there is no synchronization and the amplitude of the oscillation decreases rapidly with the dimensions of the grid system. Figure 8.8d shows, however, stable oscillations when there is diffusion that covers a large fraction of the grids within the period of a catalytic cycle. Of course, there will be a limitation when the grid size becomes too large and time scales of diffusion and local

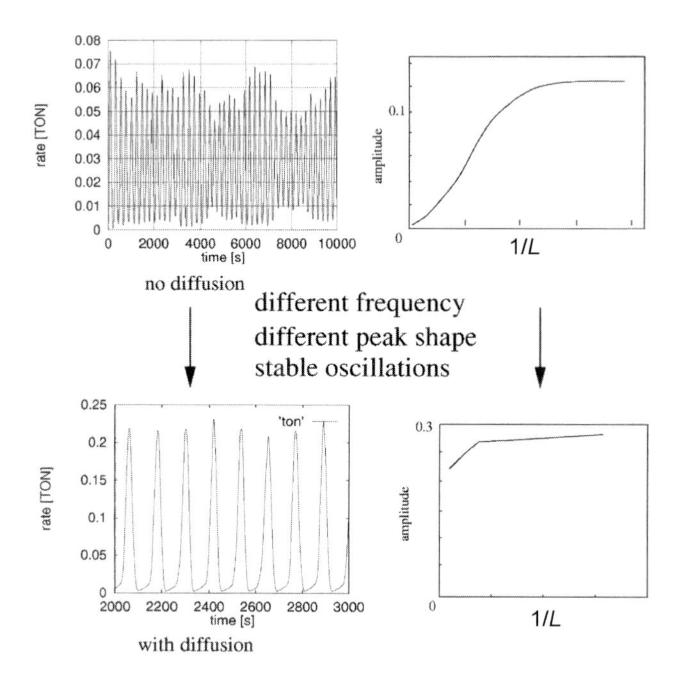

Figure 8.8. Simulation of the influence of diffusion and grid size on overall oscilatorry behavior. L is the dimension of the grid. (a) and (b), absence of diffusion. The frequency of the oscillations (a) does not change with grid size, but the amplitude of the oscillations (b) rapidly decreases with increasing grid size. (c) and (d(, with diffusion; with diffusion amplitude and shape of the oscillations change (c), but the amplitude of the oscillation is stable over a large grid size trajectory.

oscillation no longer match. Self organizing patterns will only occur in specific reaction condition regimes. For the CO oxidation reaction, such a condition arises when the temperature of the reaction is chosen just beyond the maximum of the overall reaction rate. In this temperature regime the auto catalytic reaction steps start to become rate limiting.

The assignment of reaction probabilities and different grid states as a function of the states of neighboring gridpoints can be considered as a specific case of cellular automata, on which a rich literature exists [13]. Cellular automata were originally designed by von Neuman to model self reproducing networks of cells. The von Neuman machine relied on the reproduction of a code that instructs the building of a machine, the analogue of cellular reproduction through the replication of DNA. He was able to construct such a machine in a cellular array with 29 states per cell and a five-cell neighborhood.

Later Codd [14] reduced the complexity of the Von Neuman machine inspired by insights in the physiology of the nervous system in animals. Important to our later discussion in Chapter 9 of the design of self organizing catalytic systems, he proposed a universal configuration of only eight-states per cell.

In 1997, Chou and Reggia [15] showed that it is possible to create cellular automata models in which a self replicatory structure emerges from a random density of individual components. Chou and Reggia designed the rules such that a viable structure of arbitrary size generated by a random process replicates and grows. This precedes the discovery of

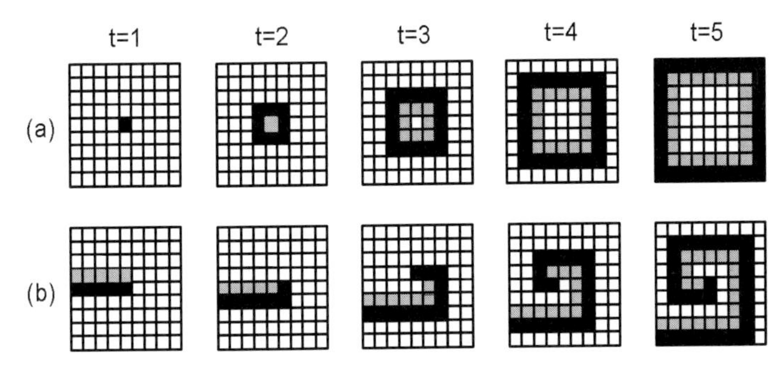

Figure 8.9. Fourfold wave-symmetry obtained by a cellular automaton with square cells: (a) target pattern; (b) spiral wave. Adapted from O. Arino et al.[16].

self organization in the grid model dynamic Monte Carlo simulations of the CO oxidation reaction shown in Figs. 8.7 and 8.8.

Cellular automata simulations on the periodic grid neatly illustrate for an excitable medium how a two-dimensional automaton leads to target patterns or spiral waves (see Fig. 8.9). It is assumed in this automaton that all eight neighbors of an excited cell (block) become excited in the next step, except for the neighboring cells that become refractory (dashed), i.e. do not change. An excited cell becomes refractory after one time step and receptive (white) after two time steps. The simple mechanism that generates the target and spiral wave patterns illustrates that the key to self organization is the rules of interaction between neighboring sites. We will return to this subject in the next chapter.

8.5 Size Dependence and Cooperative Behavior

Phase transitions such as order–disorder transitions have a well-defined critical temperature when a system is sufficiently large, so that boundary effects can be ignored. An example of a disorder–order transition for a surface overlayer system is the mixing or demixing that occurs in a surface phase when repulsive interactions between adatoms dominate. At low temperatures two different phases demix and above a critical temperature a mixed phase can be formed. In the oxidation of methanol catalyzed by Cu, overlayers consisting of separate islands of oxygen atoms and adsorbed methoxy species are formed[17]. Reaction occurs only at the boundary of the two surface overlayer phases. The oxygen islands are formed from an oxide overlayer in which the free metal surface has reconstructed resulting in effective attractive interactions between adsorbed oxygen atoms.

Surfaces of small particles have small dimensions. Such surfaces of small dimensions are present on stepped surfaces with short terrace length. With such small dimensions, no sharply defined critical temperature for phase separation exists and an intermediate surface state with partial ordering over a wide temperature interval can be formed.

In the previous section we noted that synchronization of reaction events that occur at different parts of a catalyst is a necessity for an oscillatory time dependence of reaction kinetics. The cooperative phenomenon called self organization can then take place. When particles become so small that only one catalytic cycle takes place per particle, synchronization is lost and no such self organization can occur.

A precondition for oscillatory behavior is the bistability of different surface phases. Under stationary conditions on Pd, the oxidation of CO to CO_2 can give rise to hysteresis

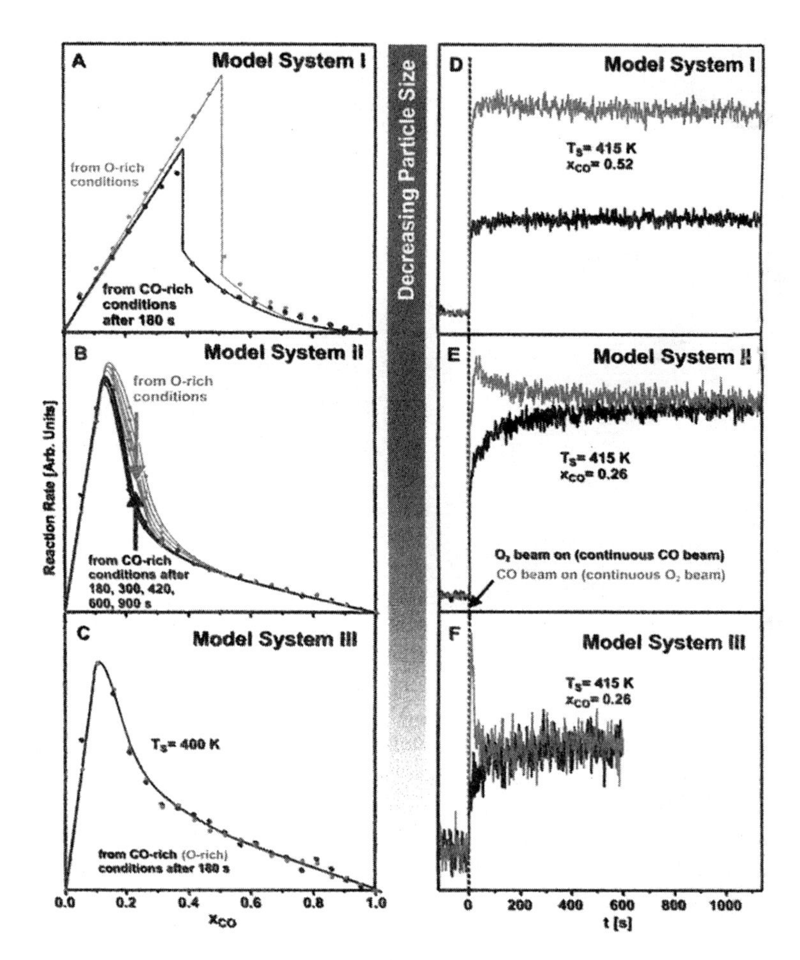

Figure 8.10. Particle size-dependent bistability and hysteresis. (A and D). On model system 1 (500 nm paricles), the CO oxidation shows a perfectly stable bistability behavior. On the time scale accessible by the experiment ($>10^3$ sec), one can arbitrarely switch between the two states by pulsing either pure CO or O_2. (B and E) For the model system II (6 nm particles), a very slow transition toward a single global state is observed in the transition region between the CO- and O_2-rich reaction regimes. This is accelerated by the presence of defect sites. (C and F). For the smallest particles of the model system III (1.8 nm), globally monostable kinetics are rapidly established under all conditions [for all experiments, the total flux of CO and O_2 beams at the sample position was equivalent to a local pressure of 10^{-4} Pa; surface temperature in (A) to (C), 400 K; (D) to (E), 415 K; the continuous curves in (A) to (C) are only a guide to the eye[18].

phenomena depending upon the O_2/CO ratio in the system. The two phases are a CO rich overlayer phase and alternatively an oxygen-rich overlayer phase. On large particles the dependence of rate on increasing O_2/CO ratio is different from the dependence found when one starts with a high O_2/CO ratio and then this ratio is decreased. This behavior has been studied as a function of Pd particle size by Johánek et al.[18]. As shown in Fig. 8.10, the hysteresis loop appears when the particle size exceeds 6 nm. When the particle size decreases there is a smaller number of surface atoms and, as a result, the surface phase composition now shows large concentration fluctuations that are non-synchronized. Phase separation is now suppressed.

The phenomenon of altered cooperative kinetic behavior as a function of surface dimension may be quite general. It will apply to those situations where surface overlayer island formation occurs with different kinetics than predicted within mean field kinetic expressions, in which ideal mixing of the surface overlayer is assumed. For instance, the electrochemical butterfly-type current–voltage diagram measured for sulfate adsorption (Chapter 3, Section 3.10.4), is significantly changed when a surface is taken with short terraces. The order–disorder transition on the extended terraces is responsible for the butterfly pattern. On shorter terraces, the well-defined features are no longer present and the sharp order–disorder transition no longer occurs.

8.6 Immunoresponse and Evolutionary Catalysis

The aim of theoretical catalysis is to predict catalytic reactivity for an arbitrary system and, hence, to direct the synthetic chemist to the exploration of new materials. A completely different approach, with important modeling consequences, would be possible if systems could be developed that not only self repair during the catalytic reaction, but also would change or adapt as a function of the product formed. The ideal catalyst would form itself from catalyst building components in a reaction mixture in response to desired products. To analyze the conditions for the chemodesign of such systems is one of the great challenges of modern catalysis. We refer further to this topic in Chapter 9.

Natural systems that are able to adapt to a desired product are the antibodies in biochemical systems. Their modular composition makes combinations in many different configurations possible. For instance, the macromolecular biological immunoglobulin system contains a great variety of compounds that vary in shape and size that can be organized to recognize a reagent by combinatorial association. More than 10^8 different antibodies can be formed.

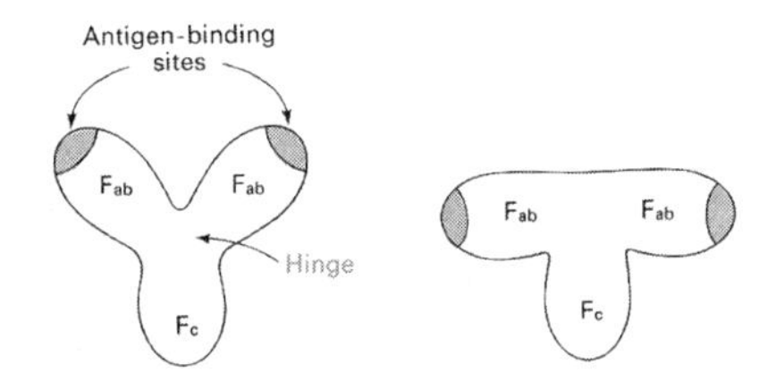

Figure 8.11a. The shape of immunoglobulin G. Adapted from C. Branden and J. Tooze[19].

Antibodies are synthesized by cells as a cellular response to a foreign molecule (antigen). Antibodies are made specific for that antigen and reproduced by triggering cell division. The bio-immunoresponse reaction has been used to produce catalytically active immunoglobulin catalyst molecules by the use of antigen molecules with a shape and charge distribution close to those of the transition state of a desired reaction. Cells of the immune system produce a pool of antibodies by genetic recombination. Antibodies discriminate themselves by their unique combination of amino acids. They form large folded polypeptides that bind virtually any natural or synthetic molecule. In the genetic recombination process the genes encoding each antibody are spliced and recombined from

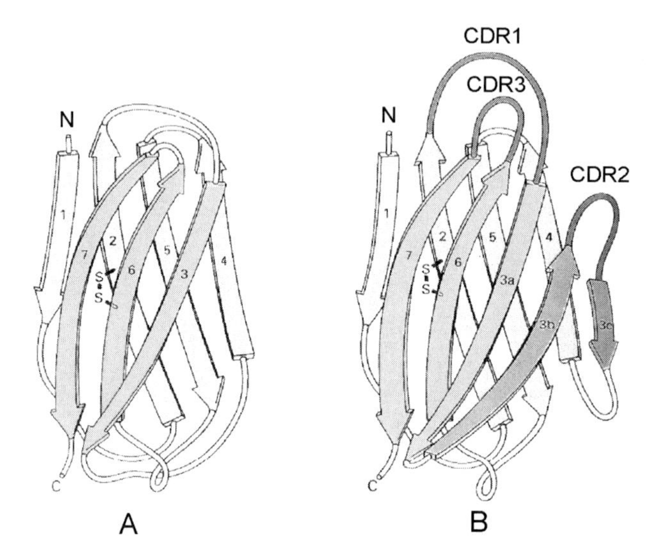

Figure 8.11b. The F_a and F_{ab} domains of immunoglobuline. Adapted from C. Branden and J.Tooze[19].

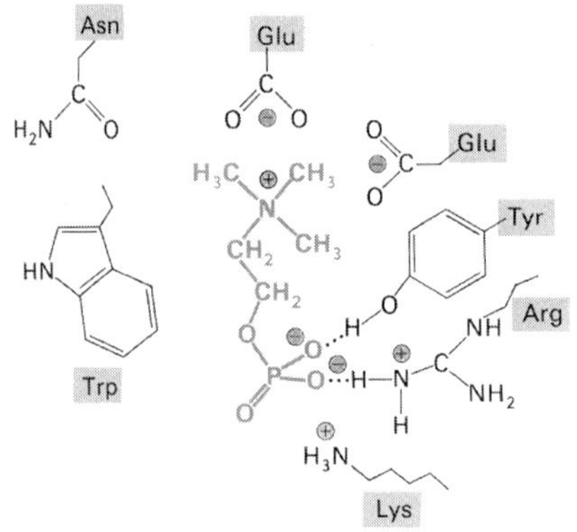

Figure 8.11c. Binding of phosphorylcholine to an antibody-combining site. Adapted from E.A. Padlan et al.[20b].

a battery of gene segments, which enable an organism to produce a multitude of different antibody molecules.

As generally described by L. Stryer [20a] and more specifically by Padlan et al.[20b], immunoglobulin has the shape of the letter Y (Fig. 8.11a). The antigen combines with the two F_{ab} units, that have so-called segmental flexibility which enhances the formation of antibody–antigen complexes. Figure 8.11b shows the immunoglobulin domains F_a and F_{ab} that consist of two sheets of antiparallel β strands. The sheets are bridged by a disulfide bond. The variable F_{ab} units contain two additional β strands. Three key loops compose the complementary-determining regions (CDR) that form part of the antigen binding site. The detailed structure of the binding of phophorylcholine to an antibody-combining site is shown in Fig. 8.11c.

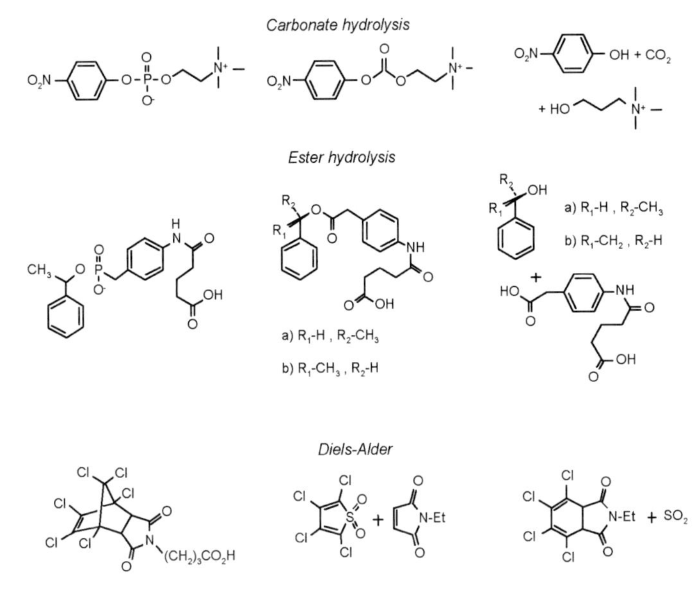

Figure 8.12. Picture of transition states, antigen with transition state analogous to transition states for the hydrolysis of carbonates and esters and Diels–Alder reaction. Adapted from R.A. Lerner et al.[25].

The structure of the antigen–antibody binding site shows a highly structure-dependent interaction that is dominated by hydrogen bonds and van der Waals and electrostatic interactions.

In response to a specific antigen, a particular antibody molecule is selected and amplified through the cellular system. Jencks[21] suggested that antibodies could be used as catalysts by selective stabilization of rate-determining transition states on a reaction pathway. Lerner et al.[22] showed that catalytic antibodies can be converted into selective catalysts induced by antigens with the shapes of transition states of a desired organic conversion reaction. They used amongst others tetrahedral, negatively charged phosphate and phosphonate transition-state analogues of the transition states for the hydrolysis of carbonates and esters. Many catalytic antibodies have been generated for a wide range of catalytic reactions.

An alternative biological approach is the use of evolutionary adaptation. An evolutionary biological implementation of the combinatorial adaptational approach towards the generation of improved and novel enzymes has been developed by Reetz and Jaeger[23] and Arnold[24]. Bacteria are used to produce catalytic proteins encoded by evolution of mutated genes. Mutations are introduced by the biochemical polymer chain reaction or other random mutagenesis methods. Recombinative techniques such as DNA shuffling are also used. Gene evolution is then carried out by selection or screening to identify a large library of potential genes, that will produce the desired biocatalyst. This directed evolution approach, for example, has been used to develop bacterial lipases with significantly enhanced enantioselectivity. In this approach to catalysis, no mechanistic information on the catalytic reaction is used to optimize the system. The desired catalyst is found by feedback of the information obtained by screening into the selection of the bacteria possessing the desired gene sequences.

These two approaches illustrate the impressive state of biomolecular catalysis, which has successfully developed approaches to direct a catalytic system towards optimum performance for a particular reaction. Molecular recognition (as in the immunosystem) and evolutionary adaptation by combinatorial self learning techniques are important principles that can be used in the design of an adaptive catalytic system. Once a preferred system has been generated, a mechanism should be in place for its amplification. In biological systems the cellular reproductive system serves this purpose.

Catalytic modeling can be used to assist identification of the proper testing molecule for the evolution of the desired catalytic system. An important goal in modeling efforts is the prediction of the shape of the transition state of a particular desired reaction step. We discussed this in some detail in previous chapters. In some instances the optimum shape of a transition state is dominated by the reaction complex itself, rather than by the interaction of the substrate with the catalyst. This happens to be the case for zeolite and enzyme systems. In contrast, for transition-metal catalysts the interaction between the substrate and the catalyst dominates the shape of the transition state.

Modeling of catalytic systems requires the ability to differentiate between changes in surface structure and surface composition as well as three-dimensional structural aspects. As we have already discussed dynamic Monte Carlo simulation and ab initio quantum mechanical methods can be used to model adsorbate surface interactions and changes in surface structure and composition. Molecular mechanism and structured Monte Carlo simulation methods can be used to model three-dimensional aspects such as shape selectivity. Taken together, these methods can be used to model catalytic performance for a host of materials including metals, zeolites, metal oxides and metal sulfides. This was discussed in Chapters 3–5. These methods allow one to predict the overall rate of a catalytic reaction as a function of catalyst composition and structure. Recent genetic algorithms or related combinatorial evolutionary techniques can be used in conjunction with "predictive simulation methods" in order to modify the catalytic model system, computationally test the system for catalytic performance and begin to "design" more active material structures and compositions. This approach will be discussed in the Section 8.8 on Evolutionary Computation Methods.

Inorganic systems are much less sophisticated than the immunoresponse system, but similar in that, in principle, information on the transition state through the correct choice of template can be incorporated into the catalytic system such as in the case of zeolite synthesis. The mechanism of the zeolite synthesis reaction, discussed in the next, section has combinatorial evolutionary characteristics.

In Chapter 9, we summarize the current understanding of the origin of living cell system early in the evolution of life as an orientation on the question of, whether an artificial cell-type reaction that self assembles can be designed to optimize the rate and selectivity of catalytic reactions by an adaptive evolutionary process.

8.7 Inorganic Self Assembly Processes; Zeolite Synthesis

8.7.1 General Aspects

In zeolite synthesis, silicate and aluminate ions are reacted with base under hydrothermal conditions. Basic components such as inorganic alkali or alkaline earth metal cations can also be used in the synthesis, in addition to organic cations. The organic cations can act as a structure-directing template.

The Al/Si ratio of the zeolite framework and the structure of zeolite are quite sensitive to the base cation used in the synthesis. The use of organic templates tends to decrease the Al/Si ratio in the zeolite. Their bulkiness restricts the number of cationic template molecules adsorbed per unit micropore surface area and, hence, the compensating negative charge density on the micropore zeolite wall that relates to the Al/Si ratio of the zeolite framework.

The interaction between template and zeolite lattice also controls to a significant extent the shape of the zeolite micropores and, hence, the zeolite structure. Similarly, to catalyst design of antibodies in response to a templating antigen, one can select the organic base template as an analogue of a transition state or key intermediate in the reaction sought to be catalyzed by the zeolite. In this way, the size and shape of the micropore that forms are selected to maximize interaction with the desired reaction intermediate. This will bias the reaction channels that pass through such intermediates. In zeolite formation a particular aluminosilicate cluster organizes around the template.

The zeolite synthesis solution provides a multitude of small oligomeric molecules that can display very different interactions with a selected template molecule. A specific complex with template molecule will have unique stability. Because of the equilibria between the oligomeric units, crystallization will consume all molecules from the mother liquid to form this particular complex through the recombination of particular oligomers formed.

Hence zeolite synthesis shares some of the same combinatorial self organization and self learning aspects as seen in the antibody system. During zeolite crystallization template molecules are incorporated in unique positions. Molecular mechanics techniques have been developed[26] that allow the prediction of optimum template zeolite interaction useful to select template molecules to synthesize preselected zeolite structures.

An interesting advance in the synthesis of reactant-directed solid-state catalysts that are not zeolitic is catalyst design of an alkene epoxidation catalyst[27]. They showed self assembly of a catalytic polyanion cluster $\alpha\text{-}\left[(\mathrm{Co^{II}})\mathrm{PW_{11}O_{39}}\right]^{5-}$ during catalysis from components Co^{2+}, $H_2PO_4^-$, ${WO_4}^{2-}$ and protons. The components themselves are not catalytically reactive. The catalytic activity increases with time until it saturates. Most likely the formation of the catalytically active cluster is catalyzed by the oxidized complex itself and we will deal with an example of autocatalysis.

8.7.2 Mechanism of Zeolite Synthesis

Zeolite synthesis occurs at temperatures between 100 and 200°C in a basic medium under hydrothermal conditions. Usually high concentrations of silica and alumina sources are used and synthesis occurs from a state where initially a gel is formed. Upon heating, silicate and aluminate species dissolve from the gel and chemistry relevant to crystallization occurs in the solvent phase. Nucleation itself is assisted by nucleation sites offered by the gel material.

The most detailed picture of zeolite synthesis available to us is for siliceous silicalite. We will discuss this here in an attempt to illustrate molecular recognition of template, evolutionary recombination of reaction intermediate oligomers and self replication. Detailed information became possible once homogeneous conditions for zeolite formation in silicate synthesis were found in the absence of gel formation. For a concise and excellent review on general mechanistic aspects of zeolite synthesis we refer to Cundy an Cox[37].

A basic silicate solution consists of a mixture of many different monomers, illustrated in Fig. 8.13. Monomers, dimers, trimers and higher oligomers are formed, in ratios that

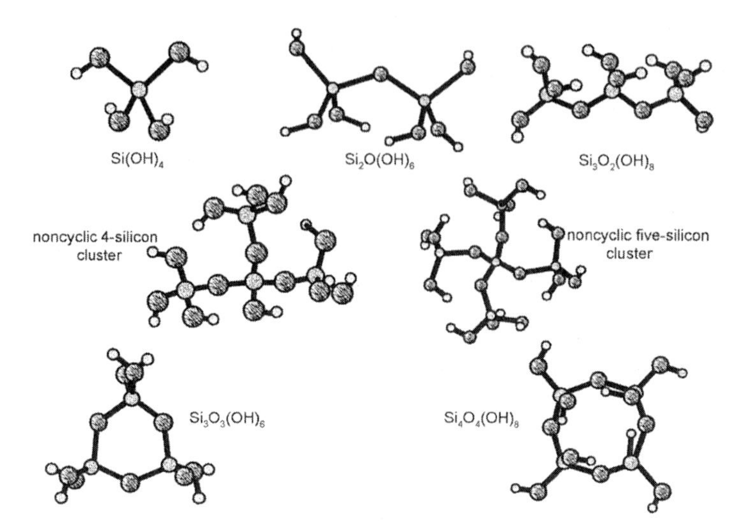

Figure 8.13. Silicate oligomers as identified in waterglass solution. Neutral oligomers are shown.

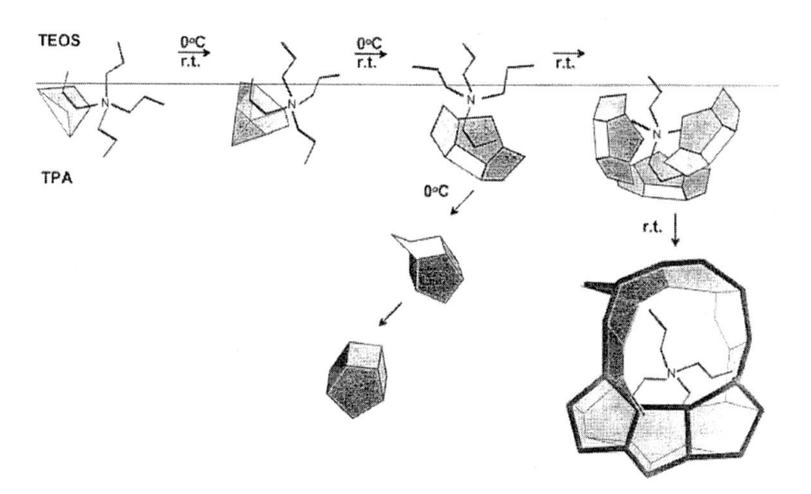

Figure 8.14. The oligomers formed in the presence of tetrapropylammonium cation[28].

vary with the concentration of the basic cation and the reaction conditions. When organic cations are present, the interaction of the hydrophobic molecule with silicate oligomer can become fairly strong, especially when the oligomer dimensions and template size become comparable. The tetrapropyl cation is the preferred cation in silicalite synthesis. The tetrapropyl cation is shown schematically attached to various silicate oligomers in Fig. 8.14. A special cluster appears to be the Si_{33} cluster shown in Fig. 8.15.

The clusters in Figs. 8.14 and 8.15 have been assembled from oligomers as shown in Fig. 8.13, by a process which is driven by the specific stabilizing interaction of the tetrapropylammonium cation and silicate cluster. The Si_{33} cluster apparently has an optimum interaction between the tetrapropyl ion and the silicalite cage. It is the result of self assembly

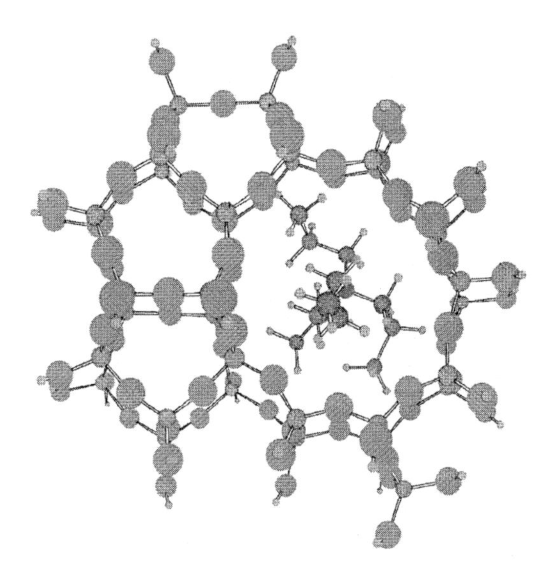

Figure 8.15. The Si_{33}-tetrapropylammonium precursor complex.

by a molecular recognition process between silicate oligomers and tetrapropylammonium cation. The equilibria that exist between the different solution oligomers shift towards the Si_{33} cluster. Formation of this intermediate already occurs under relatively mild reaction conditions. Its formation process can be considered evolutionary, in the sense that interaction between template and silicate oligomers selects the desired oligomers and rejects those that do not give the preferred interaction. Under zeolite synthesis conditions the Si_{33} precursor molecules, formed around the tetrapropylammonium ion, dimerize and condense into partially organized precursor species that can be considered as the nanoblocks shown in Fig. 8.16.

Interestingly, the local environments of the tetrapropylammonium ion in the Si_{33} precursor and of tetrapropy ammonium in silicalite are slightly different. The Si_{33} environment resembles that of a channel, whereas the tetrapropylammonium ion occupies a channel cross-section in silicalite. The need to reorient the tetrapropylammonium ions when the zeolite crystal is formed may necessitate the intermediate formation of nanoslabs. The dimensions of these nanoslabs depend on the reaction conditions. Their size is typically 4 x 4 x 1.3 nm or 4 x 2 x 1.3 nm. The number of 33 cluster rows varies in this example by a factor 2.

In situ X-ray scattering studies have convincingly shown (Fig. 8.17) that only when nanoblocks reach a significant concentration does zeolite crystallization occur. Zeolite crystallization, hence, does not occur directly from the monomer, but its complex unit cell structure requires formation through stages. Initially a specific precursor molecule is to be formed. In the case of silicalite, this is the Si_{33} cluster. In a second stage, the precursor molecule self assembles into nanoblocks of approximately the unit cell dimension of the silicate, developing its key geometric motif. The size of the nanoblocks is controlled by colloid-chemical properties. It depends on the double potential of the negatively charged silicate clusters with counteracting adsorbed and dissolved cations[31]. A higher temperature is needed to overcome the double potential barrier so that the nanoblocks can

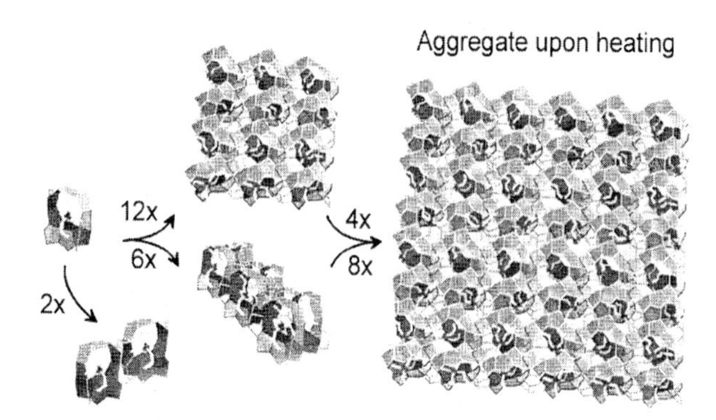

Figure 8.16. Proposed schematic structures for the silicalite MFI-type zeosil nanoslabs. The Si_{33} precursor can self assemble to form discrete and organic–inorganic hybrid nanoslabs with dimensions depending on synthesis conditions[29].

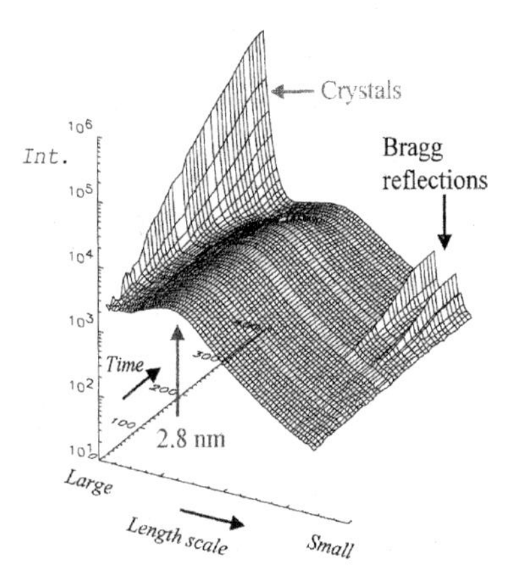

Figure 8.17. Time-dependent scattering curves of silicate formation in the homogeneous phase, followed by small angle and wide angle scattering[30].

crystallize.

The type of precursor molecule formed depends on the templating molecule. This key step is described as a molecular recognition event. When the templating molecule is chosen to resemble a key intermediate of a reaction that the zeolite should selectively catalyze, a zeolite structure crystallizes that contains cavities with optimum fit for that reaction intermediate.

8.8 Evolutionary Computational Methods

In this section, we will begin with a short exposition of different evolutionary computational approaches. Then we will apply one of these methods to the design an alloy catalyst with optimal performance for a particular dissociation reaction. The performance of the catalyst is theoretically tested using the dynamic Monte Carlo method to predict the kinetics for a surface reaction.

Evolutionarycomputation (EC) is a term used to denote a group of methods from computer science that mimic natural evolution [32,33]. These methods are mainly used in optimization problems. All EC methods have some common properties. They usually work with a set of objects that represent the object that one wants to optimize (parameters, a structure, or a process). The set of objects is called a population and the objects are individuals. An individual is represented in some coded form. The encoding is often the genotype and the object itself the phenotype. Working with a (large) number of objects at the same time helps in obtaining a global instead of a local optimum.

EC improves a population by three genetic operations: selection, mutation, and crossover. Selection does what its name implies. It picks out those individuals that are allowed to reproduce and to make a new population. Other individuals are discarded. Reproduction proceeds via the combination of two individuals by a crossover. Parts of the genetic materials of the individuals are chosen and put together to form one or two new individuals. The idea is that the good parts of the genetic material of different individuals can be combined in this way. Finally, the genetic materials can be more or less arbitrarily changed. This is called mutation and it is used to prevent premature convergence to a local optimum. The advantages of EC methods are that they provide more of a global optimization rather than a local optimization. In addition, they can also be used to optimize non-numerical objects. They are also very good at scanning large search spaces. This means that they can be employed to optimize objects with many components. They are also able to handle so-called NP-complete problems (NP stands for non-polynomial). These are problems that scale faster than a power of the size of the problem. An example of such a problem would be the optimization of the structure of a bimetallic catalyst. Each atom can be either one of two metals. With N atoms in the unit cell this give 2^N possible structures, so the problem scales exponentially with the size of the unit cell.

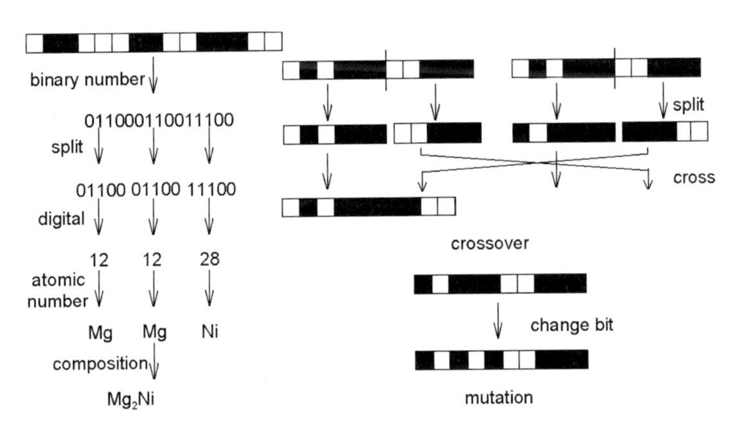

Figure 8.18. Example of a string of bits representing the composition of a bimetallic catalyst on the left. The standard crossover and mutation of genetic algorithms is shown on the right.

A drawback of all EC methods is that they can be computationally intensive. Important are the differences between different EC methods. The EC method that has been used most in chemistry and chemical engineering is genetic algorithms (GA). There are other methods, however, that can be much more efficient. The most important difference between the EC methods is the way in which an object is encoded. GA uses a string of bits. Figure 8.18 shows how this can be used to encode the composition of an alloy.

The figure also shows how crossover and mutation can be done with such an encoding. Theoretical work on GA has shown that the method will find global optima and why it is capable of searching in a very large search space. Computer experiments have shown that an encoding and genetic operators that reflect properties of the objects that are to be optimized can lead to much faster convergence. In genetic programming, the encoding can use data structures (usually trees or graphs) that can adapt themselves. The idea is that an encoding evolves that is best adapted to the object to be optimized.

Another difference between EC methods is the order in which the genetic operators are applied. Figure 8.19 shows the two main variants. In GA there is first a selection. The number of individuals is then increased, by taking the best individuals more than once so that the same number of individuals is obtained as in a full population. Finally, a new population is constructed by crossover and mutation.

In evolution strategies, a method that is well suited for parameter optimization, crossover and mutation is used first to generate a large number of offspring. Selection is then used to reduce the offspring to a new population. The selection can include the old population or not, and there are many different methods to make the selection.

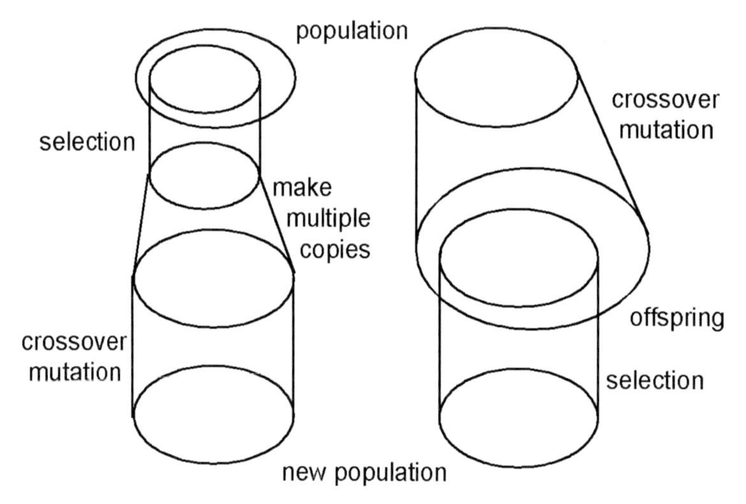

Figure 8.19. Ordering of the genetic operators as in genetic algorithms (left) and in evolution strategies (right).

The selection process is the place where the optimization in EC methods really takes place. This is where methods such as dynamic or kinetic Monte Carlo (DMC) simulations become important. They are used to compute the properties of a system or process. These properties are then converted to a fitness value. This fitness value is for satisfaction of a particular requirement of performance which is then operated on by the EC methods. The conversion is different for each system and property and also determines how effective the selection is. Dynamic Monte Carlo simulation, as we have already discussed, is a method to simulate elementary processes along with the actual rate. The method uses each individual reaction as an elementary event, which means that time scales comparable to actual experiments can be simulated. The reaction rate constants that it needs as input can be calculated using quantum chemical methods such as density functional theory, which results in what has been termed ab initio kinetics (see Chapter 3.10.4).

The application of such methods to the modeling of overall kinetics has been described in detail in Chapter 3. EC and DFT can be combined to investigate the effect of replacing atoms or chemical groups by others. The simplest application would be to have the EC determine only the composition. This might be done with a straightforward GA. The coding can be done more or less as shown in Fig. 8.18.

The genotype consists of a number of parts each of which correspond to an atom (or molecule). In principle, standard crossover and mutation can then be used; e.g., this has been the procedure followed by the group of Nørskov to find new super-strong alloys [34]. Computer experiments have shown that adapting crossover and mutation to the problem can speed up that optimization. The combination EC plus DMC can also be used to carry out structural optimizations.

We will illustrate here the generation of an optimum alloy surface configuration for the reaction of 2A + 2B → 2AB[35].

There is a bimetallic surface with two types of sites. As will adsorb on only one type (α), and Bs only on the other type (β). The specific application of such a system might be CO oxidation, where A refers to CO and B_2 refers to O_2. The adsorption of Bs is a dissociative one. The reactions in the model therefore are the following.

$$\mathrm{A(gas)} + \alpha \longrightarrow \mathrm{A(ads)}$$
$$\mathrm{B_2(gas)} + 2\beta \longrightarrow \mathrm{2B(ads)}$$

and

$$\mathrm{A(ads)} + \mathrm{B(ads)} \longrightarrow \mathrm{AB(gas)} + \alpha + \beta$$

or if we only look at how the site occupation changes:

$$\alpha \longrightarrow \mathrm{A}$$
$$2\beta \longrightarrow \mathrm{2B}$$

and

$$\mathrm{A} + \mathrm{B} \longrightarrow \alpha + \beta$$

Here α and β are both vacant sites, but of different type. A and B in the last reaction should be nearest neighbors. The rate constants of the adsorption of A, B and the surface reaction are $W_{\mathrm{ads}}^{(\mathrm{A})}$, $W_{\mathrm{ads}}^{(\mathrm{B})}$ and W_{rx}, respectively. We consider the case here where the adsorbate does not diffuse. For $W_{\mathrm{ads}}^{(\mathrm{A})}/W_{\mathrm{rx}} = 0.2$ and $W_{\mathrm{ads}}^{(\mathrm{B})}/W_{\mathrm{rx}} = 1$ the optimal structure was determined using a genetic algorithm. The structure of the surface was coded as a string of bits (0s and 1s) as follows. A 0 represented an α site, a 1 a β site. A string represented all sites in a unit cell. The size and shape of the unit cell was fixed for each optimization run with the genetic algorithm. Runs were done for square unit cells with 2 x 2 to 8 x 8 adsorption sites. The structure of the whole surface was obtained by periodic repetition of the unit cell. Figure 8.18 shows how we got from the string of bits to the structure cell. The string of length 25 represents a 5 x 5 unit cell. The 25 bits are first split into five groups of five. These groups are put into a square, and each bit is replaced

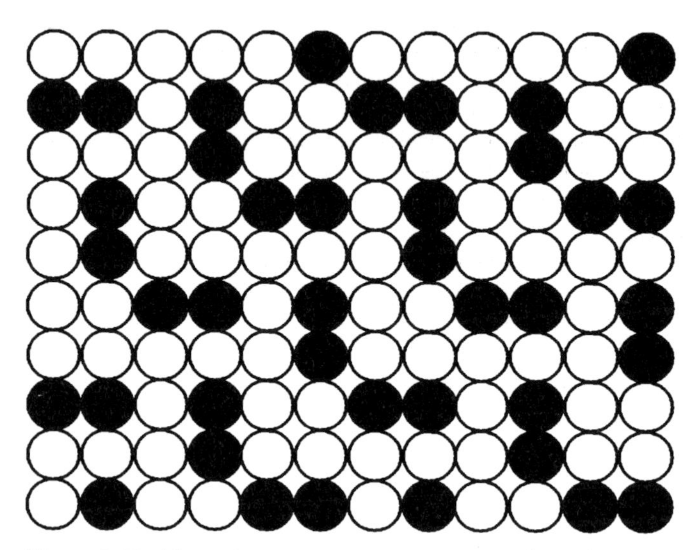

Figure 8.20. The optimum topology of AB alloy for catalysis reaction $2A + 2B \rightarrow 2AB$[35].

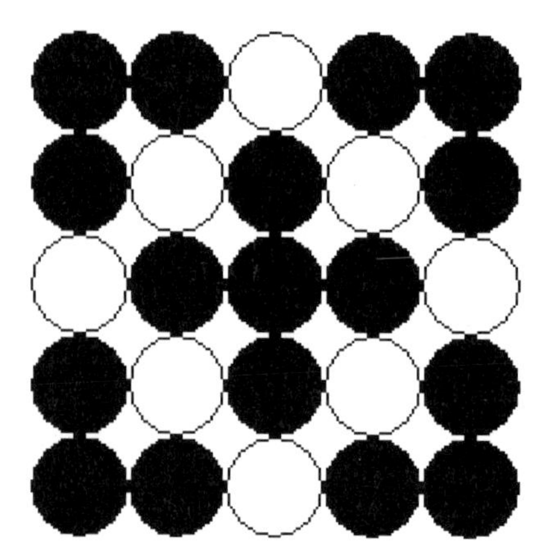

Figure 8.21. Representation of Pt–Au(100) alloy surface optimum for selective decomposition of NO in presence of excess oxygen[36].

by its adsorption site. This gives us the unit cell, which is repeated periodically to obtain the whole surface.

A typical optimization run used 64 strings per generation and had 50 generations. New generations were obtained from old ones by the usual selection, crossover and mutation. Typical crossover probabilities were 0.02 per bit. The rate of AB formation was used to make the selection, but it was scaled linearly so that the best string in a generation had a probability of being selected that was 3.5 times the probability for an average string. Selection was done using remainder stochastic sampling. The resulting optimum structure is shown in Fig. 8.20

One third of the sites are β sites and two-thirds are α sites. All β sites occur in pairs. (Isolated β sites are not found, because B_2 needs at least two neighboring sites for adsorption). Half of the α sites have one β neighbor and the other half have two β neighbors in both structures. The paired β sites now have an optimum contact with α sites.

Interestingly, detailed dynamic Monte Carlo studies of realistic systems indeed indicate the relevance of such detailed type of surface topologies. Kieken et al.[36]. studied the decomposition of NO to N_2 in excess O_2 over Pt/Au alloys. They deduced that an alloy which is comprised of a Pt ensemble assembled into a "+" structure surrounded by Au atoms as is shown in Fig. 8.21 is more active. The cluster ensemble of Pt atoms is made up of bridging Pt sites that will adsorb NO but help prevent O_2 dissociations. The bridge sites share metal atoms, which makes it very unfavorable for it to accommodate the two strongly bound O atoms that would result. O_2, therefore, is difficult to dissociate since two oxygen atoms cannot be accommodated. NO dissociation is also hindered. However, two NO molecules can adsorb and recombine in an associative reaction to give N_2O and adsorbed oxygen. N_2O can readily decompose in a consecutive reaction step to give N_2 and adsorbed oxygen. The latter atoms will recombine to give molecular oxygen by surface diffusion. By patterning the surface we can help the surface to separate the product species which can poison active sites.

8.9 Summary

The complexity of the catalytic event has been analyzed in this chapter in great detail.

First elementary reaction steps at an isolated reaction center have been considered and then the increasing complexity of the catalytic system when several reaction centers operate in parallel and communicate. This situation is common in heterogeneous catalysis. On the isolated reaction center, the key step is the self repair of the weakened or disrupted bonds of the catalyst once the catalytic cycle has been concluded. Catalytic systems which are comprised of autocatalytic elementary reaction steps and communication paths between different reaction centers, mediated through either mass or heat transfer, may show self-organizing features that result in oscillatory kinetics and spatial organization. Theory as well as experiment show that such self-organizing phenomena depend sensitively on the size of the catalytic system. When the system is too small, collective behavior is shut down.

Complexity features of catalytic phenomena have obtained a firm basis. All the ingredients to predict catalytic reactivity as a function of catalyst composition and structure are available. A strategy to optimize the catalyst reactivity is in place.

The outline of an adaptive approach has been given based on evolutionary combinatorial principles. The theoretical combinatorial approach at the end of the chapter is introduced with a summary of the biological immunoresponse system. This is of great interest because it exemplifies evolutionary dynamics of an adaptive system consisting of many components, that can self assemble a catalytic system optimized for the catalytic reaction of choice. Since the mechanism of zeolite synthesis can be viewed as being mechanistically related, this topic is discussed here. Molecular recognition and self assembly of different catalyst precursor aggregates directed by template molecules are present in such a system, but the amplification shown by the immunoresponse system is lacking.

Evolutionary adaptation has been used experimentally in biological systems to optimize enzyme performance. The use of combinatorial self learning computational approaches has been illustrated for the optimization of the composition of alloy catalysts by simulating catalytic performance using the dynamic Monte Carlo approaches.

References

1. M. Neurock, R.A. van Santen, *J. Am. Chem. Soc.* 116, 4427 (1994)
2. E.M. Shustorovitch, *Surf. Sci. Rep.* 6, 1 (1986)
3. J.M. Thomas, C.R.A. Catlow, G. Sankar, *Chem. Commun.* 24, 2921 (2002)
4. (a) G. Belusi, M.S. Rigutto, in J.C. Jansen, M. Stöcker, H.G. Karger, J. Weitkamp (eds.), *Advanced Zeolite Science and Applications, Stud. Surf. Sci. Catal.* 85, 177 (1994)
   (b) P. Ratnasamy, D. Srinivas, H. Knözinger, *Adv. Catal.* 48, 1 (2004)
5. H.C.L. Abbenhuis, S. Krijnen, R.A. van Santen, *Chem. Commun.*, 331 (1997)
6. R. Imbihl, G. Ertl, *Chem. Rev.* 95, 697 (1995)
7. L.M. Pecora, T.L. Carroll, *Phys. Rev. Lett.* 64, 84 (1990)
8. A.S. Mikhailov, *Foundations of Synergetics I*, Springer, Berlin (1994)
9. P. Gray, S.K. Scott, *Chem. Eng. Sci.* 38, 29 (1983); 39, 1087 (1984)
10. J. Pearson, *Science*, 261, 189 (1993)
11. R.J. Gelten, R.A. van Santen, A.P.J. Jansen, in *Molecular Dynamics*, P.B. Balbuena, J.M. Seminario (eds.), Elsevier, Amsterdam, p.737 (1999)
12. K-J.L. Lee, W.D. McCormick, J.E. Pearson, H.L. Swinney, *Nature,* 369, 215 (1994)
13. C.G. Langton, *Physica D*, 10, 135 (1984)
14. E.F. Codd,*Cellular Automata,* Academic Press, New York (1968)
15. H.M. Chou, J.A. Reggia, *Physica* D110, 252 (1997)
16. O. Arino et al., *Mathematical Population Dynamics*, Marcel Dekker, New York (1900)
17. M. Bowker, H. Houghton, R.A. Hadden, J.N.K. Hyland, K.C. Waygh, *J. Catal.* 109, 263 (1988):
    K.C. Waugh, *Catal Today*, 15, 51 (1992)
18. V. Johánek, M. Laurin, A.W. Grant, B. Kasemo, C.R. Henry, J. Libuda, *Science* 304, 1639 (2004)
19. C. Branden, J. Tooze, *Introduction to Protein Structure*, Garland, p. 185 (1991)
20. (a) L. Stryer, *Biochemistry*, Freeman, New York (1995);
    (b) E.A. Padlan, D.R. Davies, S. Rudikoff, M. Potter, *Immunochemistry*, 13, 945 (1976)
21. W. Jencks, *Catalysis in Chemistry and Enzymology*, McGraw-Hill, New York (1969)
22. R.A. Lerner, S.J. Benkovic, P.G. Schulz, *Science*, 252, 659 (1991)
23. M.F. Reetz, K.E. Jaeger, *Top. Curr. Chem.* 200, 31 (1999);
    M.T. Reetz, *Proc. Natl. Acad. Sci. USA*, 101, 5716 (2004)
24. F.H. Arnold, *Acc. Chem. Res.* 31, 125 (1998)
25. R.A. Lerner, S.J. Benkovic, P.G. Schultz, *Science*, 252, 659 (1991)
26. D.W. Lewis, D.J. Willock, C.R.A. Catlow, J.M. Thomas, G.J. Hutchings, *Nature*, 382, 604 (1996)
27. C.L. Hill, X. Zhang, *Nature*, 373, 324 (1995)
28. C.E.A. Kirschhock, R. Ravishankar, F. Verspeurt, P.J. Grobet, P.A. Jacobs, J.A. Martens, *J. Phys. Chem. B*, 103, 4965 (1999)
29. C.E.A. Kirschhock, V. Buschmann, S. Kremer, R. Ravishankar, C.J. Houssin, R.A. van Santen, B.J. Mojet, P.J. Grobet, P.A. Jacobs, J.A. Martens, *Angew. Chem. Int. Ed.* 40, 2637 (2001)
30. P.P.E.A. de Moor, T.P.M. Beelen, B.U. Komanschek, L.W. Beck, P. Wagner, M.E. Davis, R.A. van Santen, *Chem. Eur. J.* 5, 2083 (1999)

31. C.E.A. Kirschhock, R. Ravishankar, P.A. Jacobs, J.A. Martens, *J. Phys. Chem. B*, 103, 11021 (1999)
32. K. Mainzer, *Thinking in Complexity*, Springer, Berlin (2003)
33. J.M. Holland, *Hidden Order*, Addison-Wesley, Reading, Mass. (1995)
34. G.H. Johannesson, T. Bligaard, A.V. Ruban, H.L. Skriver, K.W. Jacobsen, J.K. Nørskov, *Phys. Rev. Lett.* 86, 25506 (2002)
35. A.P.J. Jansen, personal communication;
A.P.J. Jansen, C.G.M. Hermse, *Phys. Rev. Lett.* 83, 3673 (1999)
C.G.M. Hermse, A.P.J. Jansen, *Surf. Sci.* 461, 168 (2000)
36. L.D. Kieken, M. Neurock, D. Mei, *J. Phys. Chem.* 109, 2234 (2005)
37. C.S. Cundy, P.A. Cox, *Microporous Mesoporous Mater.* 82, 1 (2005)

CHAPTER 9
Heterogeneous Catalysis and the Origin of Life, Biomineralization

9.1 General Introduction

The ultimate goal in catalysis science is the a priori design of a catalyst that selectively produces a desired product from a specific set of reactants at a high rate. We have argued that such a design requires, in addition to insights into material properties, which are not the primary focus of this book, especially insights on the molecular level of the elementary reaction steps that support the catalytic reaction cycle. We have learned in earlier chapters that predictive design requires knowledge of the pretransition state along with the transition state of the intermediates that participate in the reaction cycle. This requires the prediction of the catalyst surface or topology of the catalytic complex under reaction conditions also. The interaction between reactants and catalyst should be at an optimum. The interaction needs to be strong enough for reaction to proceed. On the other hand, to prevent catalyst poisoning, the rate of product desorption should be in balance with that of adsorption. Particular attention must also be paid to non-selective or non-desired side reactions that deactivate the catalyst system. The actual catalyst phase that is present as the catalytically active state during catalysis is often only formed in situ during the catalytic reaction.

We learned in the previous chapter that in biological systems the immune system can generate catalytic activity for a desired reaction by triggering a selected process of cell replication through a process of self recognition and amplification. Many options for pretransition-state structure recognition are created by the combinatorial possibilities of the genetic system. We have also noted that zeolite synthesis, representative of an important class of heterogeneous catalytic materials, is a self assembly process in which aluminosilicate building blocks are formed by a template recognition process. Based on the principle of molecular recognition, both inorganic and organic self -assembled supramolecular systems[1] have been designed that have found application as sensors or separation agents. A conceptual approach to design a catalyst based on supramolecular ideas is illustrated in Fig. 9.1.

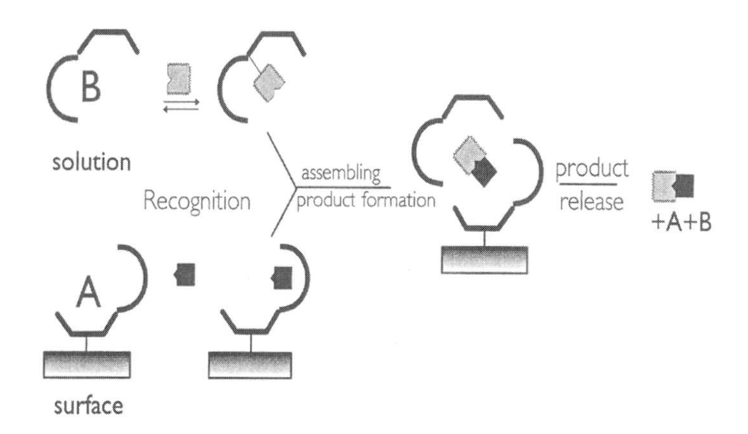

Figure 9.1. Catalysis proceeding through supramolecular assembly (schematic).

Molecular Heterogeneous Catalysis. Rutger Anthony van Santen and Matthew Neurock

ISBN: 3-527-29662-X

In this scheme, catalysis occurs through two molecular recognition steps. One reactant is trapped by a molecular catalytic component which resides in solution and the other reactant is trapped by an immobilized component of the catalyst. The reaction between the two reagents occurs in a self assembled system of the two components, kept together by supramolecular interactions such as hydrogen bonds and van der Waals interactions. The complex decomposes, releasing the product after reaction. Since three molecular recognition events take place, this reaction scheme should be highly selective.

The ultimate catalyst synthesis process is a self organization process in which the catalyst system organizes in situ from catalyst components. Such a process is more complex than proposed in Fig. 9.1, which is designed for a specific reaction with specified reactants. Here we aim at a process analogous to that for zeolite synthesis. In the solution phase there are many different oligomers that are able to form well-defined molecular clusters with a particular template. Different templates give rise to different clusters. The ultimate self organizing catalytic system forms different catalyst assemblies depending on the reactant used. In the process in Fig. 9.1, this means that catalyst components A and B self assemble differently depending on the catalytic reagents. This implies that in addition to molecular recognition, self assembly must also occur. To mimic the biological system, the self assembled catalyst should be able to replicate. In its more elementary form an autocatalytic reaction cycle should follow in which the self assembled system reproduces itself. More generally, this would require a reproductive and self organizing type of reaction system. If as a template for self assembly a molecule is used that is analogous to the reaction transition state, the system should not only self assemble around a template, but should also be able to replicate the template. As discussed in the previous chapter, in the biological combinatorial evolutionary immuno system reaction such a process is actually realized.

The design of catalytic self–organizing cell type systems can be helped by the answers found for the evolutionary origin of metabolic cellular living systems. We will review here our current understanding of the origin of protocellular systems, with a focus on the chemical and physical aspects that relate to catalysis. Such theories imply the evolution of life-like cellular systems from lifeless chemosystems. In the first five sections of this chapter, we will review theories on the generation of protocellular systems. As a follow-up in later sections we will introduce the application of biomineralization towards the synthesis of mesoporous siliceous systems with a variety of cellular structures related to the siliceous skeleton of diatoms. Biomineralization of catalytic systems is essential chemistry to convert preorganized catalyst precursor assemblies into robust, solid, heterogeneous catalysts.

The history on the theories of the origin of life is extensive. A major advance early in the last century was Pasteur's demonstration that no life originates spontaneously from lifeless materials. In the same period, Wöhler synthesized urea from inorganic components, which can be considered the start of organic chemistry. Urea is a molecule that only occurs in living systems. Its synthesis implies that molecules of living systems can be created from the lifeless world. Synthesis from inorganic components of a cell type system mimicking biological cells is a great challenge and would have not only important technological consequences, but would also alter Pasteur's paradigm. The succesful synthesis of a cellular "living" system would imply that lifeless material can be designed to behave life like. The acceptance that processes exist that spontaneously generate replicating, protocellular systems increases the probability of some kind of life on a planet in another solar system. Several theories and also experiments indicate that the generation of life-like systems from lifeless material should be possible.

First, there are the physical models of self organization and reproduction, originating from irreversible thermodynamics as proposed by Prigogine[2]. He discovered the rules for the generation of stable[1] self organized systems cyclic in time and patterned in space far out of equilibrium. A precondition[2] for the stability of such states is mass and energy flow through the system.

Then there are the theories of complex adaptable systems heralded especially by Kauffmann[3], that propose that reproductive living cellular systems can be generated once autocatalytic systems have exceeded a particular limit of complexity. These theories refer to reproducing systems, without the need for a template system that act as a code to be replicated for reproduction. The above theories consider self organization and reproduction to be a consequence of complexity. The properties of active media discussed in the previous chapter are related. It has been discussed there that under particular, unique conditions self reproduction and self organization features emerge. In order for a living system to reproduce and convert matter and energy, von Neumann[4] (see also Chapter 8, page 348) proposed the necessity for an algorithmic program that instructs cell operation, as well as the need for hardware in order to execute the program. A reproducing system has to replicate both the instructing code and the hardware. As instructing code we recognize in the biological system the DNA genetic code that replicates in cell multiplication. The hardware of the cell is the proteins that act as the enzymes and thus determine which chemical reaction in a cell is executed.

In addition to the physical theories, there are also several chemical approaches. Here one can broadly distinguish two ways of thinking, that are not only basic to the chemistry one proposes, but also to the physical models one intends to explore. The so-called RNA world view proposes that life has coincided with the origin of the genetic apparatus. The genetic apparatus is the DNA template that controls cell architecture. Eigen and Schuster[5] and also Kuhn and Försterling[6] have developed evolutionary models based on these premises. The alternative view is the Oparin[7] view, which proposes that the origin of first cellular systems coincided with the development of reproducing, self organized metabolic systems, that convert feed molecules to cellular material and waste. Such systems would be applicable as catalytic systems, when feed molecules are also converted into a particular product. This view is also consistent with the models proposed by Prigogine and Kauffmann. Reproduction takes place by cell multiplication. In life, reproduction is a necessity because no living system has an infinite existence. We know this well from material science, because materials tend to age and disintegrate with time. This provides a natural reason for the existence of reproduction and, hence, of evolution[8]. Since reproduction is never faultless, some systems will start to reproduce faster than others, depending on system conditions. A process that aims to develop self assembled catalytic systems that are able to adapt to different reaction requirements can also exploit evolutionary development by enabling different growth rates for particular mutants. As long as the concentrations in the cell system are homogeneously distributed, there is no need for genetic instruction. Genetic reproduction becomes necessary once cellular organization has reached a level of organizational complexity such that it cannot be reproduced directly by cell multiplication.

Evolution within the RNA world stems from copying errors of DNA, that are created upon the reproduction of different systems. Several options have been proposed to prevent the so-called error catastrophe that might occur. The error catastrophe is due to the accumulation of errors that gives a progressive deterioration of the system until it is totally disorganized. To resolve this, Eigen proposed the existence of hypercycles. These are part of a metabolic system coupled to the replicative system that initially exists as quasi-species of the RNA type. The quasi-species undergo a Darwinian process of selection. In a hypercycle several such quasi–species chemically associate with protein enzymes. As Dyson[9] describes, the enzymes associated with one species are supposed to assist the replication of a second quasi-species, and vice versa. The linked populations then become locked into a stable equilibrium. There appear to be additional catastrophes so that the system may yet collapse. Short circuits may occur minimizing the cycle, or a single RNA may multiply too efficiently and become a parasite, choking the rest of the population to death. There is also the possibility of statistical collapse, when one of the components of a cycle disappears. There appears to be a narrow range of oligomer population size for which the hypercycle acquires an ample, but finite, lifetime.

We will focus on the chemical proposals here in addition to computational models for evolutionary formation of metabolic cellular systems that do not yet have a genetic apparatus. Because of its very general nature, however, we will summarize first an important model of Kuhn on the evolutionary development of the genetic apparatus. We have selected this more generally applicable model because it highlights some key chemical principles for evolutionary reproduction. According to Kuhn, design principles for fabricating supramolecular systems are:

- Lock and key molecular recognition related concepts.
- Programmed environment change concept.

Supramolecular systems are aggregates held together by weak chemical interactions such as hydrogen bonds or van der Waals interactions. We recognize them in cellular systems as the aggregates formed by macromolecules. Supramolecular systems are also fundamental to self assembly and molecular recognition systems as we discussed above. The lock and key concept, and variations on it as we discussed in this book, are the basis of molecular recognition and catalytic selectivity. The need for a programmed environment refers to identifying the conditions that lead to the evolution of life-like systems at the end of a long sequence of consecutive steps.

For the formation of macromolecular chains from a limited number of monomers, Kuhn proposed the following aspects as important for the environment of the developing system, aspects that we also will recognize in chemical model systems:

- A spatial and temporal structure has to be present in special locations. For instance, a microporous spatial structure in a rock structure helps by maintaining a high concentration of molecules, to enable the initial start of a cycle of multiplication of simple strands of oligomer molecules. Energy-rich building blocks should be available. Self organization decreases entropy, which is only possible when energy is consumed.
- Microdiversity of the environment serves as an evolutionary gradient. Neighborhood regions with slightly different structural properties cannot be populated in the beginning but later by casually occurring slightly improved chemical systems.

The evolution of early life according to Kuhn can be summarized as follows. Initially under prebiotic conditions, the building blocks amino acids, nucleotides and lipids were formed. An autocatalytic replication mechanism of oligomers is proposed. The above-mentioned environmental changes lead to supramolecular engineering of simple living cells, through self-learning, evolutionary, adaptive processes.

Computational methods such as the genetic algorithms discussed in Chapter 8 have

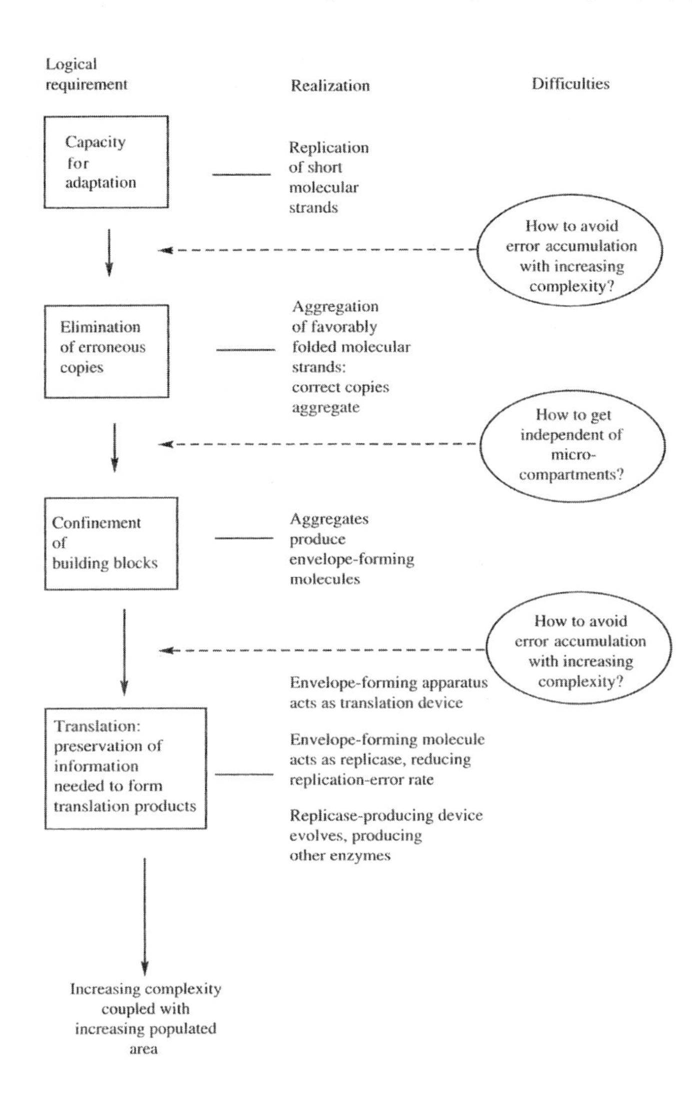

Figure 9.2. Modeling of proto life: logical requirements, their realization, and barriers to overcome[6]

been developed to simulate such processes. The scheme shown in Fig. 9.2 summarizes processes identified with such computer simulations. Four stages are distinguished:

(a) Possibility of adaptation. Several different monomers are required to form short molecular strands that replicate with some error rate.

(b) For sustainable replication, erroneous copies have to be eliminated. Kuhn now makes the proposal that only favorably folded molecular strands form aggregates. Those copies that do not aggregate are rejected.

This would agree with the observation that often prior to crystallization of complex (inorganic or organic) systems, precursor species are formed of the order of a few nanometers. Interesting in this context is the example of zeolite crystallization that we described in

Chapter 8. We argued earlier that this can be considered a combinatorial process based on template recognition. Prior to zeolite crystallization aggregates are formed, of the size of a few nanometers, specific for each zeolite. These intermediate aggregates may serve the same role as Kuhn's oligomeric aggregates. They may enhance the formation of these siliceous oligomers that are the elementary building units from which the crystal is to be formed.

(c) In order to create an interior separate from the environment, an envelope has to develop around the aggregates, leading to confinement of the building blocks. Liposomes, that are bounded by membranes, can be considered examples.

(d) In order to minimize error accumulation in a replicating system with increasing complexity, a translation device, that relates membrane development to the replicating molecules, has to develop, as part of the envelope-forming apparatus. This, according to Kuhn, is the RNA–DNA machinery. However, as we will see for a homogeneous system of limited complexity, such a machinery may not be necessarily needed.

Russell et al.[11] and Wächtershäuser[10] have proposed a very important chemical realization of a system close to the Kuhn model. Differently from the Kuhn model, Russell and Wächtershäuser focus on the generation of a self assembled, reproducing metabolic cellular system, that does not yet contain a genetic reproducing hereditary system. They propose that the initial protocell system was inorganic and prebiotic reactions were of a heterogeneous catalytic nature. The generation of enzymes and organic membranes is a later system evolutionary step. Both authors proposed that protocellular life emerged initially at sulfidic submarine springs, in a stage of the evolution of the Earth where the atmosphere was largely reducing. Typical conditions are an atmosphere of $\sim$10 bar, the presence of CO_2 and CO, with some CH_4 and NH_3 and of course water. Energy-building materials proposed by Wächtershäuser's scheme are those reagents which drive pyrite-forming reactions such as:

$$(n+1)CO_2 + 3nFeS + 3nH_2S \longrightarrow HO(CH_2)_nCOOH + 3nFeS_2 + (2n-1)H_2O$$

Alternatively, photochemical reactions may have been important;

$$4H^+ + 2Fe(OH)^+ \xrightarrow{h\nu} 2Fe^{3+} + 2H_2O + H_2$$

The hydrogen produced is used to form lipophylic molecules in catalytic chain growth reactions. Wächtershäuser[12] proposed that chain growth would occur initially through a so-called archaic reductive citric acid cycle in which SH groups coexist with OH groups and carbonyl groups coexist with this derivative. The archaic citric acid is autocatalytic in the succinate intermediate that is formed by the incorporation of four CO_2 molecules. Hydrogen and energy are produced by the ferrosulfide reaction with H_2O to give pyrite.

Cairns-Smith and Walker[13] propose reaction networks in which formaldehyde and glycolaldehyde are key intermediates. The different networks are considered phenotypes, formed by catalytic contact with clay minerals. Clays are proposed to play a role similar to DNA in replication. Replicating clays are thought to contain the "genetic information" (cation composition, distribution, imperfections, etc.) for pre-life metabolic systems.

Membrane growth at sulfide mounds can occur at the interface between iron-deficient, HS-bearing and thiolate $(RS)^-$-bearing alkaline (pH $\sim$8) reduced hot spring water and iron-bearing, mildly oxidized and acidic (pH $\sim$5) ocean[14]. Traces of tungsten and molybdenum may be present. Wächtershäuser also pointed out that initial reactions should have occurred as surface reactions on the iron–sulfide surface, since a closed membrane would not allow for the diffusion of reactant molecules into the quasi-cellular system. Attachment to the surface also concentrates surface reaction intermediates so that long-chain molecules can be formed in a chain-growth type process. Once lipophilic molecules are formed they will envelop the pyrite particles. In the process, the sulfide particles may decompose and actually become incorporated into the lipophilic membrane. This provides an interesting evolutionary path to the generation of enzymes, from the initially heterogeneous sulfide system. The membrane molecules may become ligands of the metal sulfide cationic clusters. This will begin to introduce catalytic selectivity and now peptide molecules may be formed from reactions with NH_3 produced by catalytic reactions with N_2 and H_2. Hydrothermal conditions have been identified under which NH_3 is produced from N_2[15].

Initial pre-life reproducing cellular systems emerge as a membrane-bonded vesicle incorporating autocatalytic enzyme-type catalysts. There is an ongoing metabolic process that lead to cell growth and cell multiplication. Evolutionary adaptive systematics leads to cell growth and cell death, depending on the relative value of different reaction parameters. Computational models of such processes are discussed in Sections 9.3 and 9.4.

9.2 The Origin of Chirality

There are many speculations on the origin of chirality of biosystems. Most interesting for the self assembly of reproducing catalytic systems are theories on the amplification of enantiomeric excess. Frank[16] proposed a general mechanism for spontaneous asymmetric synthesis. He showed that if the production of living molecules of life is rare and, hence, slow compared with their rate of multiplication, the whole Earth is likely to be extensively populated with the progeny of the first event before another appears. A living entity is defined as one able to reproduce its own kind. Frank showed that a simple and sufficient life model is a chemical substance which is a catalyst for its own production (hence, autocatalytic) and an anticatalyst for the production of its optical enantiomers.

Rate events are fluctuations and statistical averaging requires a large number of them. If the time scale of averaging is long compared with the amplification of the fluctuations, symmetry breaking occurs and one enantiomer dominates. This view is in line with mathematical analysis[17] which shows that macroscopic behavior derived from collective dynamics of microscopic components cannot be modeled using spatially continuous density functions. One needs to take into account the actual individual/discrete character of the microscopic components of the system.

Bonner[18] concluded that efficient polymerization mechanisms that involve enantioselective enrichment via α-helix or β-sheet secondary structures during polypeptide growth, are applicable to prebiotic environments. Total spontaneous resolution of racemates during crystallization involving secondary asymmetric transformations can also be important. The polymerization amplification concept involves the partial polymerization of a slightly enriched amino acid mixture, followed by an autocatalytic sequence of additional partial hydrolysis and polymerization steps. These amplification reactions occur because reactions of one enantiomer with another to form two diastereomeric products occur at different rates for each diastereomer.

Wynberg[19] suggested that the enantiomeric form of the chiral autocatalytic products might be able to form semi-stable dimer complexes, resulting in enrichment of the uncomplexed catalytic product. Experimental proof was given[20] that enantiomerically enriched alkali metal alkoxides, which can give aggregates in solution with both products and reactants, can act as chiral catalysts for their own formation from achiral reactants, yielding a product with enhanced excess enantiomeric selectivity of the same chirality.

Breaking of symmetry has been reported in stirred[21] crystallization. $NaClO_3$ crystallization in an unstirred solution produces a statistically equal number of *l*- or *d*-crystals, but crystallization in a stirred achiral solution can produce 99% crystal enantiomeric excess. This is due to a secondary nucleation phenomenon. Dendritic or needle-like structures on the surface of a crystal break off in a stirred solution. The result is an amplification of the corresponding enantiomeric phase.

9.3 Artificial Catalytic Chemistry

Kauffmann defined a living system as a physical cell able to self reproduce and at least able to complete a single thermodynamic cycle that executes work. A minimal model of primitive self maintaining cells named chemoton was defined by Ganti [22]. It is composed of:

(1) a metabolic system of autocatalytic molecules;
(2) self replicating molecules that inherit genetic information;
(3) a self organizing membrane molecule to enclose the system.

Since the production of a membrane costs energy, a reaction cycle is required that generates energy and allows its use in the production cycle. External resources are used as an energy source and as building materials of the protocell and partially converted to waste. The artificial catalytic cell would be selective in its waste production and produce instead desired products.

The different reaction cycles often require different conditions. This generates a need for compartmentalization with communication between compartments.In modern cells, electrocatalytic processes with the consumption and generation of electrons and protons and their transport through membranes play an important role in this respect. The system operates far from equilibrium, where cyclic behavior can be maintained.

Kauffmann demonstrated that chemical reaction networks that exceed a minimum requirement of complexity convert to a state of self reproduction coupled to a production cycle. With increasing complexity, the system undergoes a phase transition from a disordered, non-reproducing state to a self-organizing and self-reproducing state. It follows from the previous paragraph that there is also an upper limit to this complexity, beyond which a genetic apparatus becomes necessary for reproduction.

Several computational models have been designed to analyze the evolution of self organization of the first protocell. The supporting growth processes are evolutionary, combinatorial process sequences based on a selection principle. In in Chapter 8, we discussed active media that show under particular conditions self organization or chaotic behavior. An evolutionary, combinatorial process leads to the formation or selection of a particular functional material, as a catalyst, by a response reaction with the template. Adaptation implies that the multiplying system develops altered properties due to a Darwinian selection process. Combinatorial processes make adaptation possible. Once a replication principle is operational, mutation by errors occurs and self-correction mechanisms also have to be present. We will summarize these concepts by discussing here the Graded Autocatalysis Replication Domain (GARD) model and Lattice Artificial Chemistry model.

9.3.1 Graded Autocatalysis Replication Domain Model

An important question is whether a rudimentary genetic memory may emerge based on statistical rules of mutually interacting catalytic reaction networks. The basic idea of such a compositional genome is that in a mixture of relatively simple chemicals, the array of relevant concentrations may be viewed as a vehicle of information storage. According to Morowitz[23a], memory can also exist without specific macromolecules, but may be initiated in a chemical network with catalytic loops and reflexive autocatalysis in which the same catalyst participates in different networks.

In biological systems, memory [the genetic apparatus) and operating system (the catalytic enzymes organized into (auto)catalytic networks], are distinguished. According to the Oparin model, in very early life these may not have been separated. Morowitz defined a boundary for molecular assembly to ensure its compositional inheritance (see Fig. 9.3). If an assembly contains m molecular species and if each is present with an average copy number $2r$ per assembly, then the probability that all molecular types are present in the progeny in at least one copy is given by $Pb = (1 - e^{-r})^m$. r is to be interpreted as the number of (autocatalytic) reaction paths that interconnect the different molecular species. When the number of molecules increases, r scales with a power of m.

The Morowitz boundary is defined as $Pb = 0.5$. At large values of r, Pb approximates 1 and the system reproduces. One can consider this to be due to the redundancy of the system, so that upon replication the key autocatalytic cycles are easily transmitted.

The model systems described so far belong to the class of the so-called Replicative–Homeostatic Early Assemblies (RHEA). The concept of homeostasis and the idea of self replication stems from Oparin: "..... the stationary drop of a coacervate (particle enclosed by a membrane), or any other open system, may be preserved as a whole for a certain time while changing continually in regard to both its composition and the network of processes taking place within it, always assuming that these changes do not disturb its dynamic stability".

The GARD model assumes combinatorial or random chemistry with random emergence of diverse organic molecules. The transition from random chemistry to self-replicating entities occurs because of intrinsic statistical factors. A key step is the definition of a matrix for random catalytic interactions, which may be based on molecular recognition within random receptor assemblies. The GARD model assumes a finite enclosure (e.g. an amphiphilic vesicle), containing the catalytic set of members, and absorbing energy-rich chemical precursors from the external environment. Computational implementation occurs by numerical solution of differential equations or by Monte Carlo simulations.

In the Lattice Artificial Chemistry model and also in later extensions of the GARD model, formation of the membrane itself is explicitly included. In GARD every species A_j may, in principle, catalyze every possible reaction in which another species A_i is formed/decomposed, with a catalytic probability β_{ij}. This β matrix defines the chemical structure of the (auto)catalytic networks. The role and formation of the amphiphilic assembly that becomes the enclosing membrane can be incorporated into the catalytic network by assuming that the same molecules that form the assembly are also responsible for the mutually catalytic functions. This is a model that contains similarities with the sulfide protocell systems discussed earlier. The results of numerical simulations based on extended GARD model versions demonstrated the spontaneous emergence of catalytical assemblies that tend to lie below the Morowitz boundary.

In the above models, interest was focused on evolutionary system development to

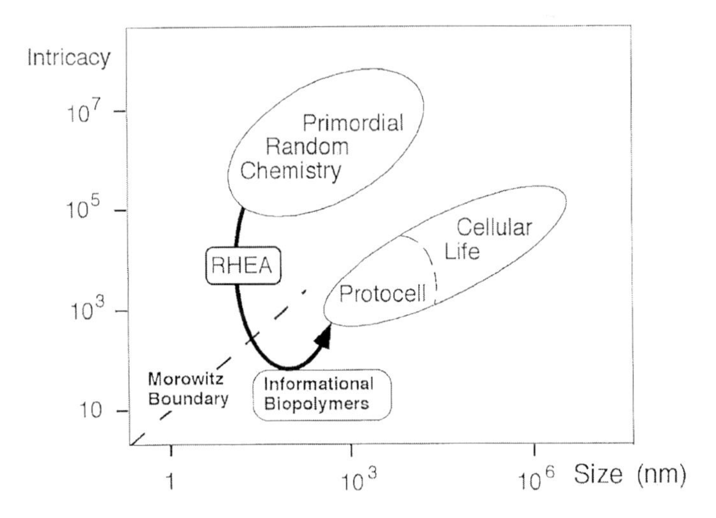

Figure 9.3. A compositional space diagram illustrating conjectural rough relationships between assembly composition and size in early prebiotic events. Intricacy, the number of different types of molecular species in an assembly, is plotted against assembly size. Early prebiotic chemistry is assumed to have generated many millions of different organic molecules in the size range of a few hundred daltons. These may have formed small assemblies (e.g. micelles) in the 0.1–10 μm range. Such primordial assemblies might lie in a region of the diagram (upper oval) where intricacy is related to assembly size as dictated by a multinomial distribution statistics. A hypothetical path is shown (thick arrow) from prebiotic random chemistry to more biased, life-like assemblies, capable of transmitting compositional information. It is proposed that mutual catalysis-based, "metabolisms first scenarios", provide a likely path for such assemblies to cross the Morowitz boundary (broken line, representing a probability = 0.5 for successful division), by selecting subsets of molecules. Below this boundary, those assemblies that grow and split can propagate their compositional information with some fidelity. This consequent decreased intricacy, potentially associated with only minor changes of assembly size, means the generation of assemblies with a relatively small number of types of low molecular weight chemical species ("monomers"), thus paving the way for informational biopolymers. Only then may intricacy rise again, as the complex attributes leading to cellular life begin to emerge. Protocells may have been characterized by an augmented size, and the appearance of structural complexity, with only modest increase of intricacy. This trend is clearly manifested in present day cells, which may be up to 1 mm in size, contain trillions of molecules, but may have only a few tens of thousands different kinds of molecular species, including proteins. Adapted from D. Segré and D. Lancet[23b].

reach homeostasis. This implies replication with continued growth. The latter is ensured by autocatalytic events. In the two-dimensional Lattice Artificial Chemistry approach proposed by Ono[24a] and Ikegami[24b], the emergence of protocells is actually followed by the reproduction of cells. Selective growth occurs of those cells that have the higher activity of membrane production. Chemicals are represented by particles on reaction sites arranged at a two-dimensional triangular lattice. Chemical reactions are expressed by the probabilistic transition between different particle types. There are five types of particles: A, M, X, Y, and W. A particle A replicates itself by consuming a resource particle X using it self as template:

$$\mathrm{X} + \mathrm{A}_{\mathrm{template}} + \mathrm{A}_{\mathrm{catalyst}} \longrightarrow 2\mathrm{A}_{\mathrm{template}} + \mathrm{A}_{\mathrm{catalyst}}$$

It can also catalyze the production of membrane particle M by consuming resource particle X. All particles can decay into waste particles Y, which are recycled into X by an external

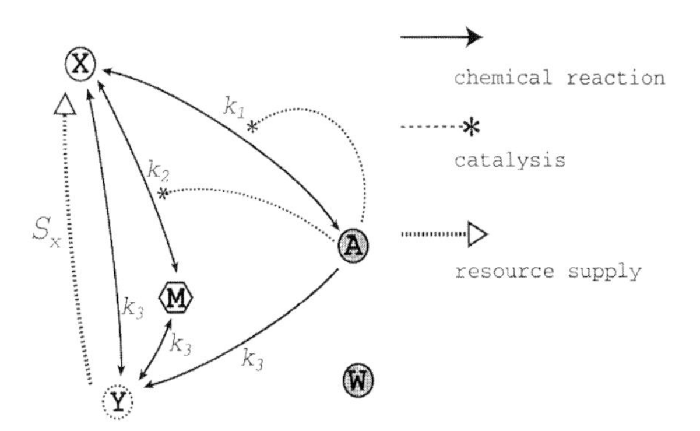

Figure 9.4. Reaction paths. Particle A produce A or M from X. All particles decay into Y. There is a source which supplies X.

source. The number of particles is, hence, constant. Particles W play the role of water. Particles A and W are hydrophilic, particle M is amphiphylic and particles X and Y have neutral interactions. The reaction is represented schematically in Fig. 9.4. As in conventional lattice Monte Carlo methods, diffusion is expressed by random walks of particles on the sites. Autocatalytic and membrane particles are assumed to be larger molecules so that their diffusion coefficients are smaller than those of other particles.

Ono and Ikegami found that the evolution of the system is roughly divided into three characteristic stages:

1. chemical evolution;
2. emergence of protocells;
3. cellular evolution.

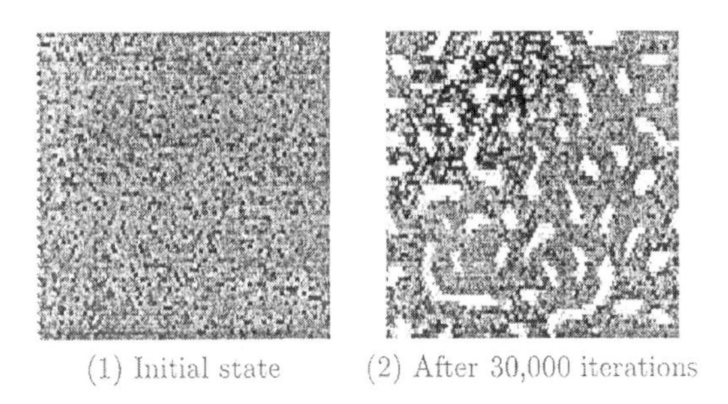

Figure 9.5a. Chemical evolution. The white regions are dominated by particle M. The depth of gray shade represents the total population of the autocatalysts ($\sum A_i$). The black regions are dominated by particle W. Resource and waste particles are not displayed. Pieces of membranes are produced by the catalysts which emerged through mutations[24a].

In chemical evolution small pieces of membrane are formed. Once membranes are formed, there is a restriction in the diffusion of molecules. A difference in concentration or different sites within the membrane emerges (Fig. 9.5a).

Most regions become inactive, while there remain some active regions in which protocells help reproduction. The autocatalysts inside them reproduce themselves and me-

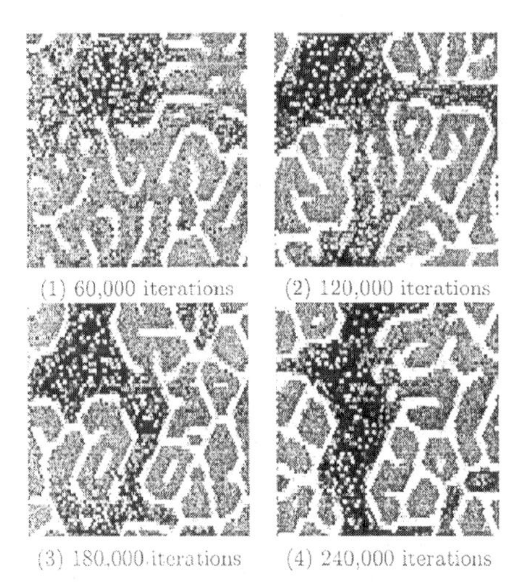

Figure 9.5b. Emergence of protocell structures. As the membranes grow, the competition for resources between regions separated by membranes takes place. The regions differentiate into two states. In some regions that are enclosed by membranes, the density of autocatalysts stays high. In the other regions, their density becomes almost zero[24a].

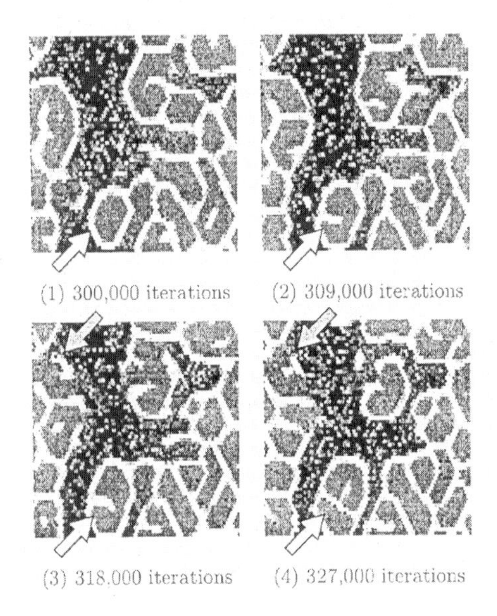

Figure 9.5c. Reproduction of protocell. Snapshots from 300,000 iterations to 327,000 iterations. The protocell indicated by the white arrow grows gradually in size. When it becomes too large, another membrane appears inside it. Finally, it divides the cell into daughter cells. On the other hand, the cell indicated by the gray arrows could not produce enough membrane particles to keep its membrane. Catalysts in the cell diffuse away through the defect of the membrane[24a].

tabolize the membrane particles to maintain their membranes. These assemblies are the protocells (Fig. 9.5b).

When a protocell grows inside, it starts to produce more membrane particles than it needs to maintain its membrane. When it reaches a particular size, surplus membrane particles begin to form another membrane within the cell. This divides the mother cell into a few daughter cells (Fig. 9.5c). A Darwinian selection principle appears to operate. Owing to cellular selection the population of catalysts is biased towards higher membrane production activity.

The model, as presented, can be changed into an artificial catalytic cell by introducing the selection principle that only particular waste molecules are desirable products and other cause cell death, by poisoning (membrane) catalytic sites. The cell can, in principle, be trained to produce the desired molecules as waste. For instance, shape-selective diffusion can be incorporated for molecules moving through the cell membrane, allowing desired molecules of particular shape to leave the cell. Cells without such micropores, not allowing the molecules to escape, can be made to die by poisoning. This would incorporate additional autocatalytic networks into cell metabolism. Alternatively, within the cell wall catalytic centers can be generated as a response to an external stimulus as a templating molecule. Again, a competitive reaction chain has to be designed so that cells would be induced to die, when such molecules were not produced or accumulated. The developing protocells can in principle be trained to produce the desired waste product molecule, by autocatalytic generation of micropores or selective reaction centers as a response to an external stimulus as a templating molecule.

9.4 Control Parameters and the Emergence of Artificial Life

9.4.1 The Logistic Map

The design of a system that adapts its function by an evolutionary process, in our case the development of a cellular catalytically behaving system, raises an interesting question that was originally formulated by Descartes[25]: how can a designer build a device which outperforms the designer specifications? If specifications are followed too closely, there is no way that improvements can be made. On the other hand, if they are followed too loosely, the device will not operate according to specifications.

For most efficient evolutionary design, this implies that there will exist an optimum condition, at which the compromise of the two conflicting requirements is found. Langton defined a complexity parameter λ that determines such a condition[26].

We learned in Chapter 8 of excitable systems and active media and in the previous sections of this chapter of complex autocatalytic reaction networks. Computational systems have been designed[27] that provide insights into the general condition by which complex behavior emerges. A more general understanding of the dynamic features that determine complex behavior is obtained from analysis of the the so-called logistic map[28a]and Wolfram's two-dimensional cellular automata studies[28b].

It will appear especially from Wolfram's work that a priori prediction of macroscopic behavior even for many particle systems that follow simple interaction rules is often not possible. The behavior can be sensitive to initial conditions and disturbances. These are, of course, conditions that are optimum for a learning system, where microscopic rules have to be adapted to macroscopic requirements. The optimum condition for emergence of life-like multiplication appears close to conditions where the system behavior becomes unpredictable. An important feature of a living cell is its finite lifetime. Sustained existence of a living collective system of cells is maintained by self reproduction.

In order to introduce the concept of a control parameter, we will analyze first the logistic map to illustrate more precisely the different phases of self organization that one can distinguish. We will follow closely the work by Nicolis[28a]. The logistic map is a generalized equation that describes the development of a population x due to growth and decline by competition between the species:

$$x_{n+1} = 4\mu x_n(1 - x_n) \quad \begin{cases} 0 \leq \mu \leq 1 \\ 0 \leq x \leq 1 \end{cases}$$

where μ is a control parameter. The solutions x of the corresponding set of equations vary qualitatively depending on the value of μ. When $\mu < \frac{1}{4}$, only the trivial solution $\overline{x} = 0$ exists, and no growth is possible. In the interval $\frac{1}{4} < \mu < \frac{3}{4}$, the solution converges to a unique value:

$$\overline{x} = 1 - \frac{1}{4}\mu$$

Its path towards $\overline{x} = \sum_{n \to \infty} x_n$ is illustrated in Fig. 9.6. The solution is at the cross-section of diagonal $f(x) = x$ and the inverse parabola. The subsequent values of x_n are the cross-section of the respective vertical lines with the parabola. After an initial onset, the numbers x_n spiral to $\overline{x}$, with increasing value of n.

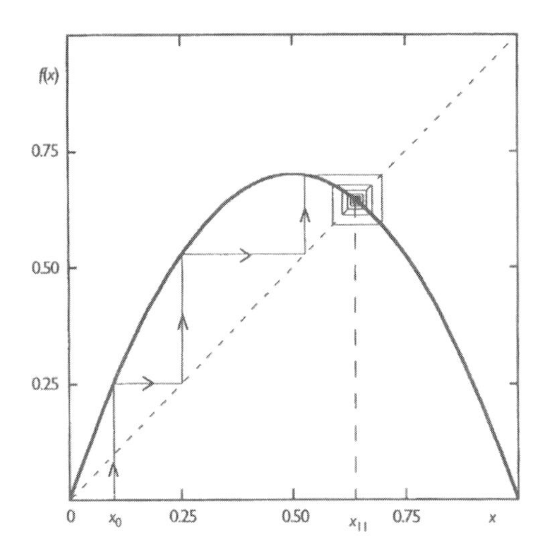

Figure 9.6. Evolution toward the stable fixed point in the logistic map for $\mu = 0.7$[28a].

A qualitativily very different behavior is observed when $\mu > \frac{3}{4}$.

An example of such behavior is sketched in Fig. 9.7, for $\mu = 0.775$. Now two solutions exist for $\overline{x}$:

$$x_{2^+} = 4\mu\, x_{2^-}(1 - x_{2^-})$$

$$x_{2^-} = 4\mu\, x_{x^{2+}}(1 - x_{2^+})$$

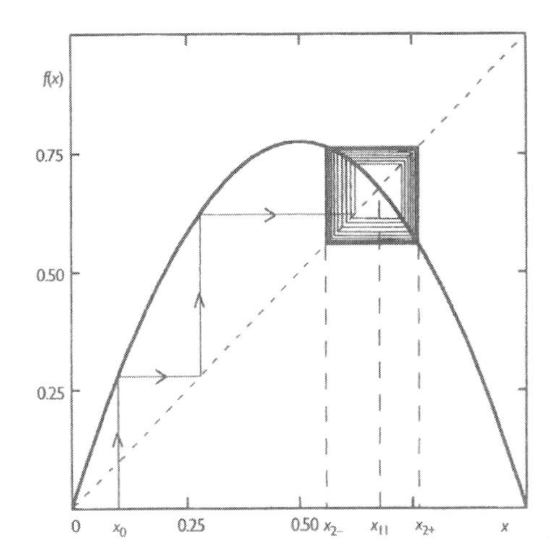

Figure 9.7. Evolution toward a stable cycle of order two in the logistic map for $\mu = 0.775$[28a].

with solutions

$$x_{2\pm} = \frac{1}{8\mu}\left[1 + 4\mu \pm (16\mu^2 - 8\mu - 3)^{1/2}\right]$$

After an initial period, the solutions at infinite n iterate between x_{2+} and x_{2-}. This phenomenon is called period doubling.

In the regime $\frac{3}{4} < \mu < 1$ an infinite sequence of successive period doublings occurs at increasing values of the control parameter $\mu(\mu_1 = \frac{3}{4}, \mu_2, \ldots.., \mu_n)$ culminating at a well-defined value $\mu_\infty < 1$.

The logistic map at $\mu = 1$ generates different chaotic behavior (see Fig. 9.8).

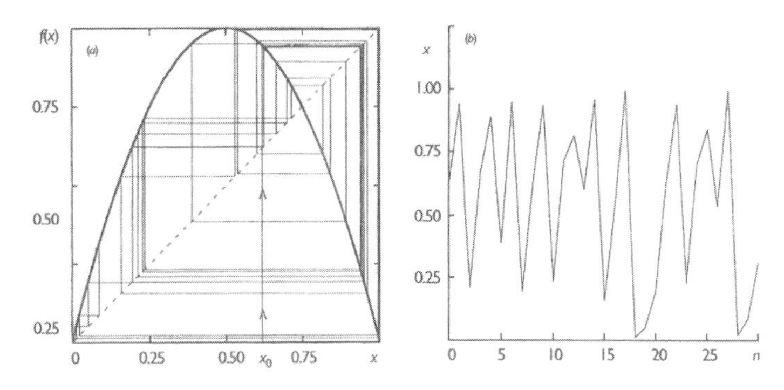

Figure 9.8. Fully developed chaos in the logistic map for $\mu = 1$: (a) successive iterations starting from $x_0 = (\sqrt{5} - 1)/2$; (b) time series generated by these iterations[28a].

In some systems an additional interesting phenomenon, called intermittency, appears when the functions $f(x)$ as in Fig. 9.6 are close to curve at $f(x) = x$. This is illustrated in Fig. 9.9 for a particular function $f(x)$. After a short transient in the narrow region between the graph and bisector, the process resembles convergence to a fixed point that

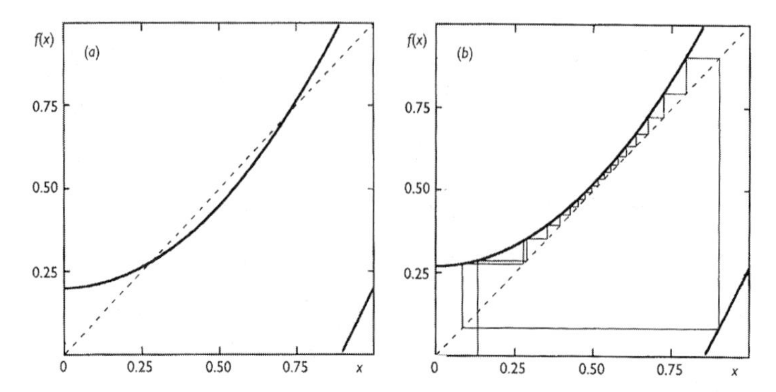

Figure 9.9. Generation of intermittent behavior through limit point bifurcation in the map $f(x) = 0.25 + \varepsilon + x^2 \bmod 1$: (*a*) $\varepsilon = 0.55$: the system possesses one stable and one unstable fixed point; (*b*) $\varepsilon = 0.02$: the fixed points have been destroyed and the system undergoes chaotic bahavior of the intermittent type[28a].

is nearly stable. It therefore has a long time scale. However, the iterations will eventually leave this region and evolve to another branch of the graph of $f(x)$ and be reinjected back to the region of near tangency.

This will appear as a series of long periods of quiescence interrupted by short-lived bursts. This phenomenon is called intermittency and is a signature of turbulence. It is a well-known phenomenon that occurs, for instance, for mass flow in trickle flow reactors at flow velocities where flow becomes turbulent.

The change in dynamic behavior as a function of control parameter can be considered analogous to the phase behavior of a material. For instance, as a function of temperature, water will go through a highly ordered state, ice, to an intermediate disordered state water, and finally to a gas state of complete disorder.

9.4.2 Life at the Edge of Chaos

Emergent behavior occurs in many systems. For catalysis in porous systems bond or site percolation is an important example. It describes the probability that in a statistical network a particle can move through the system. It models the mass transport through a porous medium. This is illustrated in Fig. 9.10[29], where a two-dimensional network is shown with all points connected. At some specific point, the number of bonds that are broken is high enough that the statistical probability of finding a network that completely spans the lattice, P_c, is enough small that there are only disconnected clusters. The overall connectivity at this point is lost and percolation is no longer possible.

Wolfram[28b] discovered analogous behavior from simulation studies with cellular automata. His work shows that, notwithstanding well-defined short-range interaction rules between components on a microscopic level, macroscopic dynamic behavior can become unpredictable. This implies that external disturbances can have an important outcome on both temporal and structural events. This is consistent with a condition of life, where there is change due to evolutionary response. Examples of the four basic classes of behavior Wolfram discovered are shown in Fig. 9.11.

In Wolfram's computational work pattern formation is studied on a two-dimensional lattice. Interaction rules are defined between lattice squares for the evolution of blackness

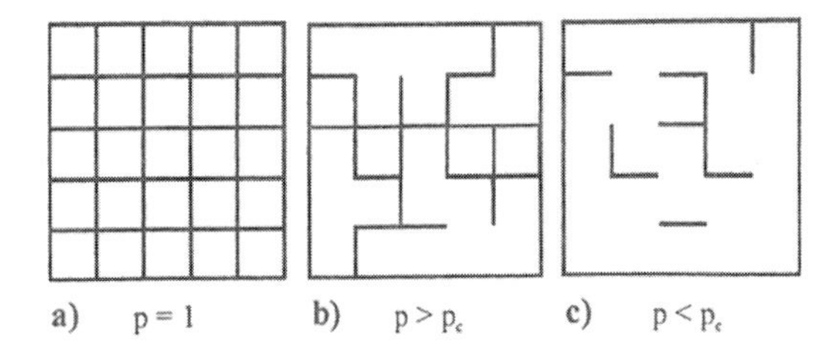

Figure 9.10. Bond percolation: (a) all bonds are closed; (b) part of the bonds are open, but there is still a cluster of infinite size; (c) there exist only clusters of finite size.

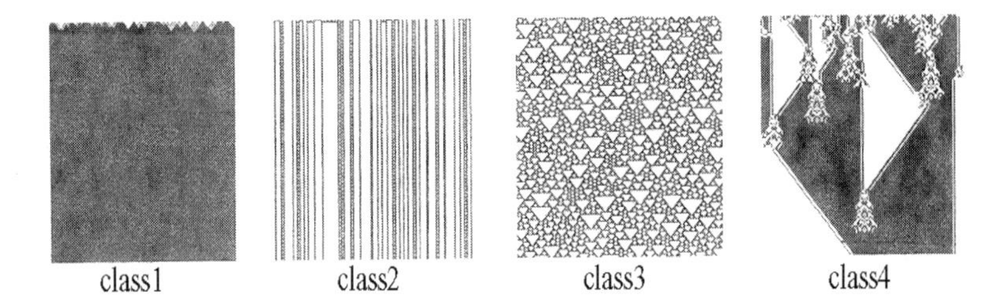

Figure 9.11. Examples of the four basic classes of behavior seen in the evolution of cellular automata from from random conditions. Adapted from S. Wolfram[28b].

or whiteness of a square on the lattice starting with an initial condition in the first row at the top of the system defined by a particular sequence of black and white squares. One then moves downwards in successive rows. The coloring of an element in each successive row depends on the coloring of neighboring and next neighboring elements in the previous row.

Hence rules have been set up in which the color of the cell in a row n depends on the coloring of the three neighboring cells in row $n-1$. Each system is determined by eight different rules. In total, 256 different choises can be made, generating an amazing variety in pattern behavior. Wolfram distinguishes four classes of behavior.

- *Class 1*: All initial conditions lead to exactly the same result. Initial differences die out rapidly.
- *Class 2*: There are many different final states, but all of these consist of simple structures, that remain the same for ever or repeat. Changes due to different initial conditions may persist but remain localized.
- *Class 3*: behavior is complicated and seems, in many, respects random, although nested structures such as triangles and other small structures always tend to appear. There is a rapid increase in network complexity. Changes in initial conditions spread everywhere, as in a percolating system. There is long range communication but there is no repetitive behavior. Changes are irreversible.
- *Class 4*: There is a mixture of order and randomness. Localized structures are formed, but move around and interact in complicated ways. Structures persist but are floating, they change sporadic. Long range communication is lost. The patterns formed are extremely sensitive to initial conditions.

Representative structures from all four of these classes are shown in Fig. 9.11. We recognize the overall changes in behavior from complete order to disorder and intermediate stages with several stationary solutions. What is new are the stages with intermediate structure formation, decay and reappearance. The first three catagories of Wolfram can be identified with the three types of behavior we have met in the logistic map model of a continuous dynamic system. Class IV behavior with long transient and complex patterns of localized structures has no direct analog. The scene is now set for a return to the theories of the origin of life.

Following the ideas of Anderson[27], Langton[26], who is one of the founders of artificial life research, coined the phrase "life at the edge of chaos". To identify the condition of life for a cellular automaton system, he defined a control parameter λ. For a cellular automaton with K states and N neighbors the λ parameter is defined as follows. An arbitrary state S_q is chosen as a quiescent state, as for instance, the solution zero, with $\mu < \frac{1}{4}$ for the logistic map. If there are n_q transitions defined to state S_q, the remaining $K^N - n_q$ can be filled randomly and uniformly over the other $K - 1$ states, where

$$\lambda = \frac{K^N - n_q}{K^N}$$

If $\lambda = 0$, all the transitions in the system create the quiscent state S_q. This implies full order. When $\lambda = 1$ as the point where all of the states are filled randomly and there is complete disorder.

Langton considers that in a living system the dynamics of information has gained control over the dynamics of energy. Langton suggests that information can dominate, the dynamics of physical systems in the vicinity of a second-order (or critical) phase transition.

At intermediate values of λ, a phase transition can occur between periodic and chaotic dynamics. At either end of the λ spectrum, behavior seems simple and predictable. It is the behavior in the vicinity of this phase transition that is complex and unpredictable. There are long transients, intermediate and sometimes periodic structures. They are generated and later collapse, very similar to the class 4 system of Wolfram.

The dependence of the different phase space regimes on λ is illustrated in Fig. 9.12, in which the four Wolfram class of behavior are indicated.

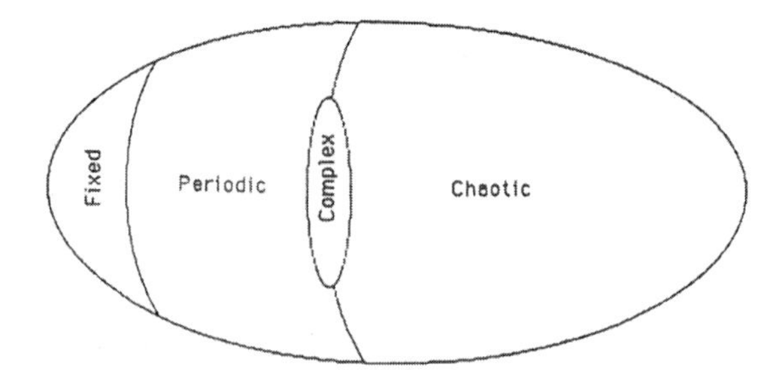

Figure 9.12a. Schematic drawing of cellular automaton rule spaces indicating relative location of periodic, chaotic and complex regimes[26].

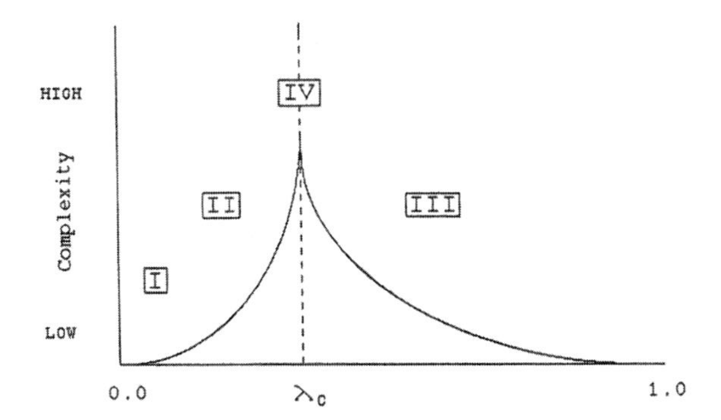

Figure 9.12b. Schematic drawing of complexity versus λ over CA rule space, showing the relationship between the Wolfram classes and the underlying phase-transition structure[26].

Approaching the transition from "below", one progresses from simple fixed-point to simple periodic behavior, accompanied by larger and longer transients showing more and more sensitivity to the array size, until the transition regime is reached. The slow growth and occasional collapse of complex dynamics to periodic behavior makes the outcome of a particular rule operating on a particular configuration impossible to predict in the general case. Just slightly past the transition regime, the ultimate collapse to periodic behavior becomes extremely rare for finite arrays. Ultimately, the behavior becomes maximally chaotic and, hence, again predictable.

Living systems, according to Langton, can probably best be characterized as avoiding attractors, as the values $\overline{x}_i$ in the logistic map, towards which the solution converge or oscillate. In the periodic system such attractors exist and are characterized as limit cycles or fixed-point attractors towards which the system converges with time. In the chaotic regime attractors can also exist, and are known as strange attractors typically of a high dimension. Living systems have learned to maintain themselves on extended transients, near a critical transition, where the state remains undecided. This is also illustrated by the simulations discussed in the Chapter 8, page 345, on replication near system phase boundaries, when the fluctuations in concentration in time and space are large.

When one exceeds a particular complexity threshold, the autocatalytic networks must give rise to the emergence of a self sustaining network. A metabolic system crystallizes that is able to replicate by division. There is, however, a limit to such complexity for reproducing systems. As we discussed in the previous section, once complexity passes another critical parameter, replication requires coding of the instructions in a program, as von Neumann proposed for his self-reproducing computer and life realized through the DNA genetic system.

9.5 Different Levels of Self Organization in Catalysis; a Summary

In Chapter 8, we described how combinatorial evolutionary processes optimize the system by adaptation to a template. In the have been formed they can be reproduced.

In the previous sections we have analyzed computational models of artificial chemistry that indicate that, in principle, the chemistry can be designed so as to create an artificial catalytic system, that optimizes its selectivity by evolutionary adaptation.

The ultimate catalytic design would be the generation of such an artificial catalytic cell. A few essential ingredients have to be part of the design. The catalytic system has to self organize, replicate and adapt itself. The system should not only self assemble around a template, but should also be able to replicate the template or the template structure. It should also behave as a reproductive and self-organizing reactor.

The property of reproduction and self organization will only exist for sufficiently complex systems. Towards the design of the artificial catalytic cells a combinatorial, evolutionary process will have to be used. The realization of such an evolutionary adaptive process requires system conditions far from equilibrium and at a state near a critical point, such as those near a dynamic phase transition, where the developments in time and space are undecidable and hence, very sensitive to the choice of initial conditions.

9.6 Biomineralization, the Synthesis of Mesoporous Silicas

In the previous sections, we focused on chemical theories for the evolutionary origin of life-like systems. We learned that an important condition in the initiation of early life is a microporous or cell-type enclosed environment that can sustain local concentration variations.

Biomineralization is a biological synthesis process that exploits the unique interactions at the interface of the inorganic and organic material to produce composite materials with unique properties and inorganic materials with a unique structure, regular over nanometer length scales to micrometers and millimeters[33a]. The essence of biomineralization processes is that mineral is deposited around or within a mould, prestructured by organic building blocks that may consist of various types of molecules such as sugars or fatty esters. The key feature that characterizes these molecule is that they are amphiphilic. Such molecules can organize as double layers, as presented in cell membranes, and can have the shape of spheres, tubes or three-dimensional structures. These phases are known as liquid crystals. In the biomineralization process, minerals are deposited at the interfaces of the preorganized structures formed by the liquid crystal-type amphiphilic systems. This process that can be seen as a fixation of a fluidous system.

An extensive research activity in biomineralization has developed that investigates the structure-directing principles described above, but now chemically without the use of biological systems. This is of great interest in the context of designing the microporous cell-type enclosed conditions for protocell systems. Second, conversion of protocell-type catalytic systems into heterogeneous catalytic systems requires mineralization chemistry similar to biomineralization.

In this section and also in the next sections, we will highlight aspects of the chemistry that biomimics biomineralization. The treatment of the topic material will be much more detailed than the previous sections dealing with origin of life-like systems. We need the detail to compare the synthesis of ordered materials prepared in biomineralization with the disordered, often fractal, materials with high surface areas synthesized with more conventional chemistry. This comparison will be done in later sections. Again, processes in such materials are important to strenghten the walls of such systems. Stochastic modeling of such systems is possible, as will be shown in final sections. The chapter will be concluded with a summarizing section, which is less detailed. Some readers may prefer to skip the more technical sections that follow and read the final summarizing section of this chapter.

Zeolites are well-defined microporous systems that we introduced earlier. They have micropores ≤ 1 nm. Their synthesis requires the use of organic template molecules. As we described in Section 8.7.2, these template molecules organize silicate oligomers into precursor molecules for crystallization of the zeolite. These precursor molecules differ depending of the templating molecule or cation used. In these systems there is no preorganization of the template molecules before crystallization. Using biomineralization synthesis techniques, ordered mesoporous and microporous systems can be made with micropore dimensions that vary over many length scales. In contrast to the zeolites, with channel walls of coordinatively saturated oxygen atoms, the channel walls of those systems are hydroxylated.

Biological systems that realize such ordered nano- and micrometer porosity include diatoms, which are monocellular algae. Figure 9.13 compares the micropore structure in a diatom with that of synthesized mesoporous materials.

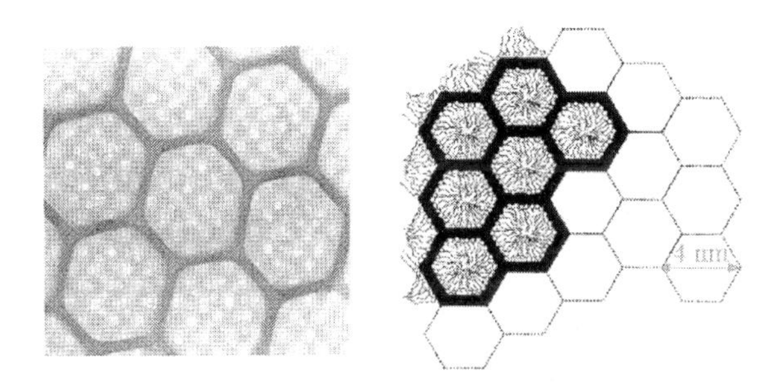

Figure 9.13. Schematic drawings of (left) diatoms and (right) synthetic mesoporous MCM-41.

Here we will present the basic physicochemical principles on which the formation of such ordered microporosity is based. We limit ourselves here and in the following sections to the example of silica formation since this has been most extensively studied. The introduction in 1992 of the MCM-type periodic nanometer pore size materials[30], using surfactants in the synthesis instead of structure-directing template molecules, brought forward the development of methods to prepare mesoporous silicas with a great leap. In this approach, different lyotropic liquid crystal phases of a large variety of surfactants and amphiphilic polymers that can be organized in numerous mesophases are used to structure the developing silica phase. The silica phase is grown around the self assembled organic polymers, which act as a template. Surfactant molecules or amphiphilic polymers form micelles or liposome vesicles with a membrane that consists of a double layer of amphiphilic molecules. In an aqueous solution, their hydrophobic features point inwards, whereas their hydrophilic features are directed into the water phase. At increased concentration the micelles can become converted into tubular structures or even networks. Silicas with a well-defined network of uniform pore dimensions are formed by introducing silica precursors that oligomerize and subsequently solidify around the micelles. The pore sizes of as-made porous silica range from a few to tens of nanometers. Crystalline siliceous materials with larger pore dimensions (~7.4 nm) can be made using non-ionic triblock copolymers of ethylene oxide and propylene oxide as templates under acidic conditions. The crystallinity of such materials can sometimes be directly deduced from the shape of the crystal particles. Figure 9.14 shows the rhombodecahedron shape of the corresponding crystals.

According to Zhao et al.[33a], a major breakthrough in biomineralization was the discovery of how to synthesize such materials at different pH. Kresge et al.[30] originally

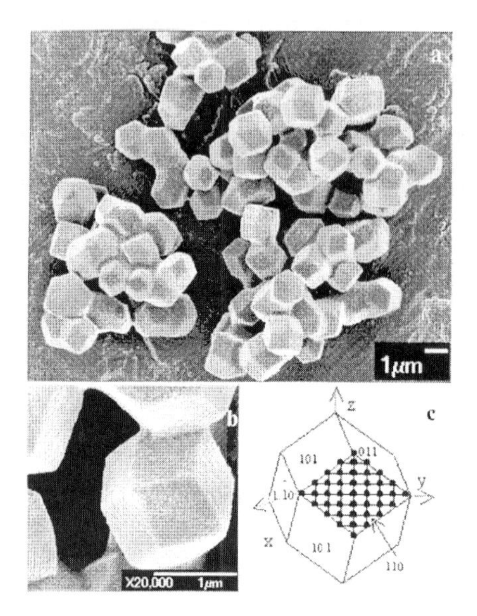

Figure 9.14. SEM images of mesoporous single crystal synthesized with F108 at 38 °C with 0.5 mol/L K_2SO_4: a) at low magnification, b) at high magnification, c) index of a rhombdodecahedron model, the mesopore array in (110) plane is demonstrated. [32].

synthesized mesoporous materials in basic media with anionic silica species. The use of nearly neutral pH conditions, the same as in biological systems, was accomplished by use of $[SiF_6]^{2-}$ and Triton X-100 as non-ionic surfactant, complementary to the use of conventional ethyltrimethylammonium cations, similar to those used in zeolite synthesis, as structure-directing agent[34]. The halide ion associates with the surfactant molecule. There is no need in this system to charge balance the surfactant charge with the negatively charged silica wall. The silica wall remains neutral. Tanev and Pinnavaia[35] were the first to use non-ionic surfactants such as poly(ethylene oxide) and poly(alkylene oxide) block copolymers with hydrophilic and hydrophobic parts. At neutral pH, disordered, worm-like pores are formed. If, on the other hand, low molecular weight block copolymer solutions of relative high concentration are used at low pH, ordered mesoporous silica phases are synthesized[36,37].

9.6.1 Biomimetic Approaches for Amorphous Silica Synthesis

Diatom biosilica formation can be mimicked by using bio-analogous reagents for silica synthesis. The diatom contains ordered micropores in a silica matrix which is amorphous. In this section, we show that amorphous silica can be considered as structures built for elementary particles. The packing of the elementary particles and their internal structure depend sensitively on the amphiphilic structure-directing molecules. The pH controls the preorganization of the structure-directing molecules. In this section we discuss materials made without preorganization of the structure-directing molecules.

Two types of enzymes have been identified as playing a role in the formation of biogenous silica. In a sponge, such enzymes are the silicateins[38] that catalyze the formation of Si–O–Si bonds from the corresponding monomer. Mechanistically the reaction sequences in the enzyme fold are similar to the elementary reaction step in the papain-like hydrolysis reaction discussed in Chapter 4, section 4.4. The other enzymes, identified in diatoms, are the silaffins[39]. They are small peptides that consist of 15–18 amino acid residues linked to linear polyamines consisting of 5–11 *N*-methylamidpropylamine units. The silaffins induce the precipitation of silica under nearly neutral pH conditions. Synthetic polyamines such as polyethylenimine (PEI) along with natural species found in diatoms induce silica precipitation very efficiently. The reaction is generally too fast in the laboratory to investigate silica transformations and structure-directing processes. Polyethylene oxides (PEOs) and derivatives thereof are good alternatives, as they induce silica polymerization at a slower reaction rate. Consequently, it has been posible to perform and monitor time-resolved silica transformations, which eventually can be manipulated by temperature-controlled aging[40].

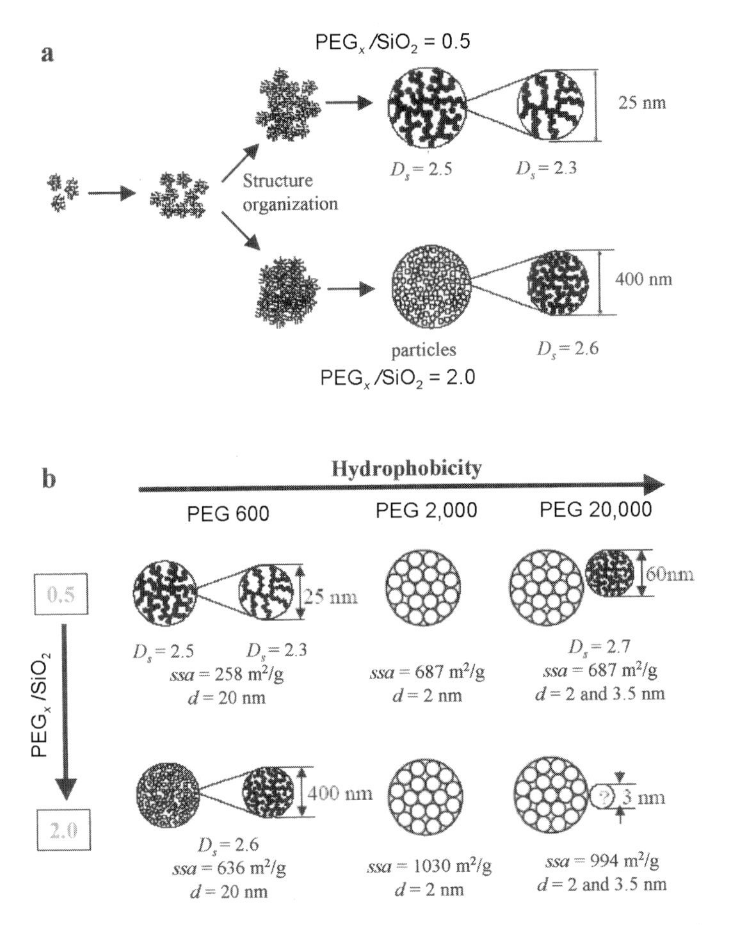

Figure 9.15. (a) Aggregation model for PEG_{600}-mediated silica at PEO/ silica ratios 0.5 and 2.0. (b) Schematic presentation of the relation of PEO/silica ratio and hydrophobicity (which increases with size of the PEO monomer) and the physicochemical characteristics of the silica formed: specific surface area (ssa in m^2g^{-1}) determined by N_2 sorption–desorption, fractal dimension (D_s) determined by SAXS, and pores diameter (d) determined from SAXS and pore volume contributions (derived from sorption–desorption isotherm)[41].

When different hydrophylic homopolymers such as PEO_{600}, PEO_{2000}, $PEO_{20,000}$ and $PEI_{1,000,000}$, or triblock copolymers such as PEO_{76}–PPO_{29}–PEO_{76} or mixtures of them are used under mildly acidic conditions as additives in the biomimetic synthesis of silica from waterglass, well-defined spherical and smooth non-fractal and fractal silica particles (for definition fractals see Section 9.7.2) are formed. The silica aggregation behavior and pore formation behavior which are controlled by PEO homopolymers at various molecular weights and PEO/silica ratios have been studied with in situ (ultra) small-angle X-ray scattering [(U)SAXS]. The experimental results clearly indicate that PEO plays three different but important roles in silica aggregation. First, it serves as a flocculation agent in the formation of silica sols, similarly to PEI and the (analogues of) diatomaceous polyamines. Second, the hydrophobic silica–PEO interactions steer the silica polymerization. Finally, the presence of PEO induces phase separation during the reaction, in which PEO-rich and silica-rich phases are formed in the formerly homogeneous aqueous solution. As we will see later, phase separation processes play an important role in micromorphogenesis of the diatom cell wall.

From these studies, it becomes clear that the length of the PEO chain and the PEG/silica ratio affect the physicochemical properties of the silica formed with respect to a variety of pore dimensions from 2 up to 20 nm, their connectivity (determined as fractal dimensions with SAXS) and their specific surface areas. Consequently, different silicas can be created by choice (see also Section 9.7). A fractal dimension of <3 implies that a gel-type disordered, chaotic and open structure will form due to cluster–cluster aggregation processes that are diffusion or reaction rate limited. A fractal packing of particles implies self similar scaling of the porosity distribution over several length scales. A fractal dimension of the order of 2.5 implies diffusion-limited aggregation due to high sticking probabilities when reaction clusters meet. Figure 9.15a shows as a function of PEG[PEG = (poly ethylene oxide glycol)]$_x$/SiO_2, ratio that initially open or more dense gel-type structures are formed. They become converted into particles that themselves are built from smaller elementary particles that are fractally ($PEG_x/SiO_2 = 0.5$) or regularly packed. The elementary particles themselves can exist again with a fractal gel-inner structure as is the case for the silicas made with PEG_{600}.Figure 9.15b summarizes the different types of end products as a function of polymer hydrophobicity and polymer/SiO_2 ratio.

9.6.2 Micro-Emulsion Mediated Silica Formation

The diversified porous patterns of diatomaceous silicas are on the nano- to submicrometer scale (~10–300 nm) and these meso- and macropores cannot be mediated by single macromolecules, not even proteins. To mimic these meso- and macroporous structures, a different approach can be applied based on a phase separation process as in the vesicle-mediated macromorphogenesis processes extensivily reviewed in Pickett-Heaps et al.[42]. In this case oil-in-water (O/W) emulsions are applied as a model system. O/W emulsions are isotropic and thermodynamically stable liquid media with a continuous water domain and an oil domain, which are thermodynamically stabilized by a surfactant as micrometer-sized liquid entities.

Amphiphilic PEO–PPO–PEO triblock copolymers can form a variety of aggregates depending on the length of the respective polymer blocks. In such aggregates, the more hydrophobic PPO blocks are shielded from the aqueous envirement by hydrophilic PEO blocks that protrude into the aqueous phase. The triblock copolymer poly(ethylene oxide)–poly(propylene oxide)–poly(ethylene oxide) ($PEO_{76}/PPO_{29}/PEO_{76}$: see Fig. 9.16a) can

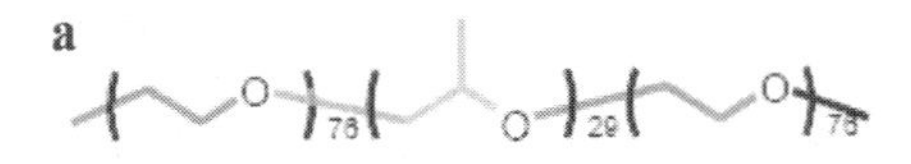

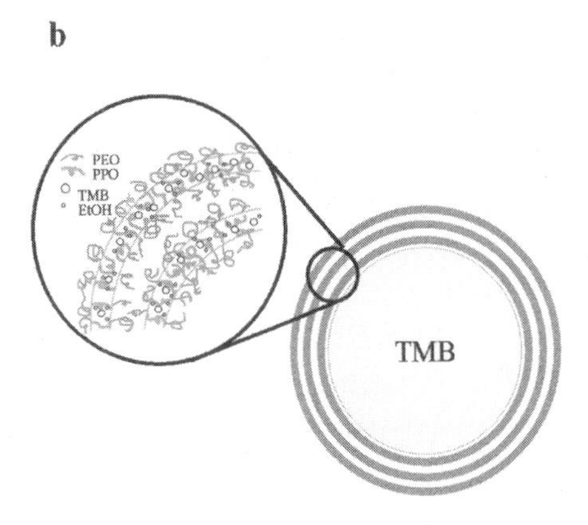

Figure 9.16. (a) the chemical structure of the triblock copolymer poly(ethylene oxide)–poly(propylene oxide)–poly(ethylene oxide) (PEO_{76}–PPO_{29}–PEO_{76}) and (b) the proposed structure of the emulsion droplets. TMB = 1,3,5-trimethylbenzene.

also be used with emulsions of 1,3,5-trimethylbenzene (TMB). These emulsion droplets act as templates in the biomimetic formation of silica (Fig. 9.16b), thereby exploiting the ability of the PEO chains on the aggregate surface to induce the precipitation of silica. After removal of the organics by high-temperature calcination, cavities of micrometer-size dimensions should remain in the as-made silicas.

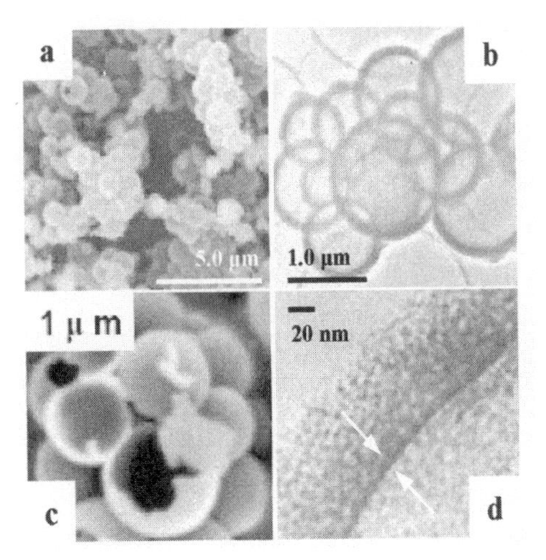

Figure 9.17. Scanning (a, c) and high-resolution (b, d) micrographs of hollow silica spheres synthesized with the TMB/PEO–PPO–PEO/silica emulsion system at 80 °C[43].

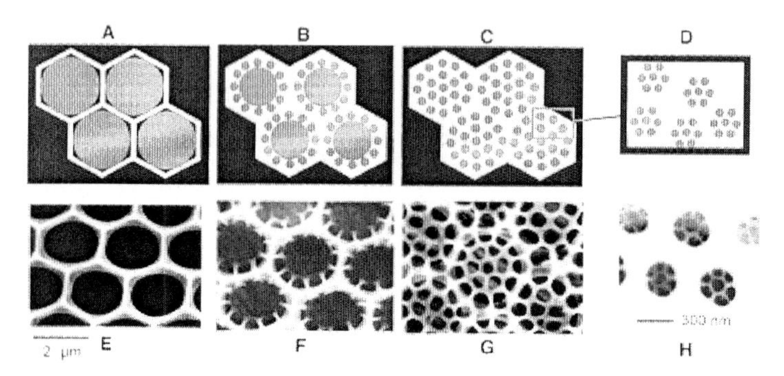

Figure 9.18. Schematic drawing of the templating mechanism by a downscaling phase separation model [(A) to (D)] in comparison with scanning electron micrographs of the valves from diatom *Coscinodiscus wailesii* [(E) to (H)]. (A) The monolayer of polyamine-containing droplets in close-packed arrangement guides silica deposition. (B and C) Consecutive segregations of smaller (about 300 nm) droplets open new routes for silica precipitation. (D) Dispersion of 300 nm droplets into 50 nm droplets guides the final stage of silica deposition. Silica precipitation occurs only within the water phase (white areas). The repeated phase separations produce a hierarchy of self-similar patterns. (E to H) SEM images of valves at the corresponding stages of development[44].

At pH 5.2, hollow spheres are formed with a uniform size distribution and an average diameter of approximately 1 μm. These silica spheres are thermally very stable and hollow (Fig. 9.17), possessing a relatively thin shell with a very regular thickness of 80–130 nm composed of 7–8 nm thick layers of silica with an interlamellar spacing of ~3–4 nm[43] .

The biomimetic studies of biomineralization have also provided inspiration to explain biopolymer-mediated silica formation in the diatoms. As mentioned in the previous section, biogenic silica contain small proteins and polyamines such as the silaffins.

The silaffin-1A proteins are small (2.4–3.1 kDa) polycationic proteins with highly modified amino acid residues. A particular well-characterized peptide contain seven phosphorylated serine residues and one phosphorylated trimethylhydroxylysine moiety.

The zwitterionic structure of the native silaffins leads to the self assembly of these molecules, which explains their extremely efficient induction of silica precipitation. Following up on this model, Sumper[44] proposed, on the basis of simulations, a model that consists of multiple steps in which phase separation processes occur. At the initial stage phase separation of protein phase and silica permits the formation of the large, honeycomb structures, followed by several intermittent steps of silica formation – each mediated by phase separation processes – to create smaller structures (see Fig. 9.18).

In this *"downscaling"* model, the largest structures are formed at the start (Fig. 9.18a) followed by the formation of the smaller ones (Figs. 9.19b,c) and finally the more delicate details (Fig. 9.18d).

In contrast, Vrieling et al.[45] derived an *"upscaling"* model based on experimental data obtained from in situ time-resolved ultra-small angle X-ray scattering analysis of silica transformations mediated by synthetic polymers [poly(ethylene glycol), polyethylenimine)] and protiens (myoglobin, horseradish peroxidase). In the course of the polymerization and transformation processes (the latter of which is induced by aging), phase separation occurs. Silica-rich and structure-directing containing macromolecule phases appear, which continuously interact during the transformation of silica from smaller aggregates to the larger structures (see Fig. 9.19).

Ultimately, removal of the structure-directing-rich phase leads to the creation of a

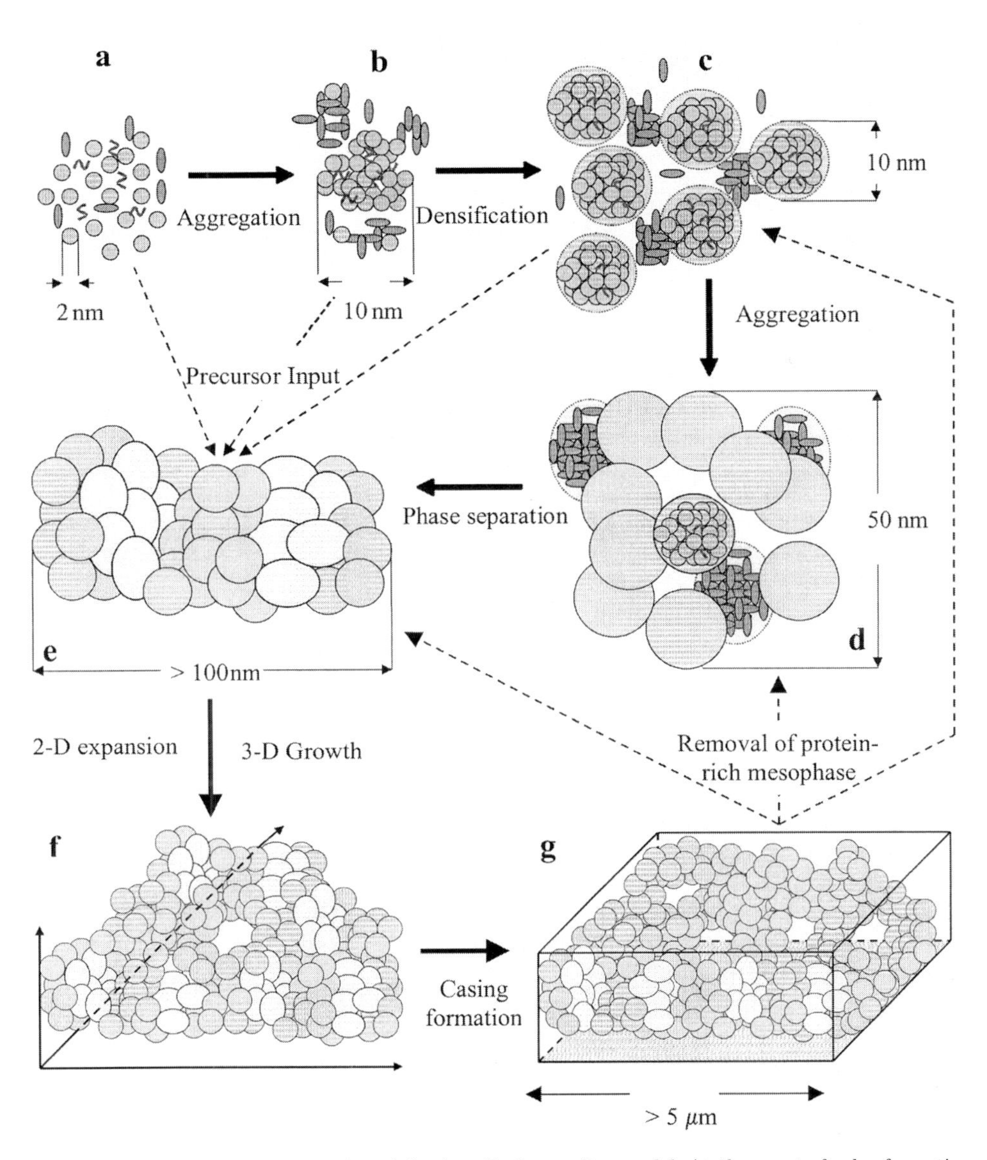

Figure 9.19. Schematic presentation of the described *upscaling* model. At the onset of valve formation (a) silica precursors and peptides are imported into the silica deposition vesicle (SDV), where precipitation of silica is induced by small organic molecules (silaffins and/or polyamines) to form silica sols (b). These sols further densify and grow to larger silica particles, while larger peptides start to interact with the silica (c). When aggregation continues (d), silica forms larger particles (up to 50 nm). At this stage the silica and protein aggregates become transferred to silica- and protein-rich mesophases by phase separation (e). This process proceeds until the SDV has reached it final two and three-dimensional size (f) and the protective casing has been formed (g). In order to leave the pores, the protein-rich mesophase is somehow removed prior to assembly of a protective casing before the wall leaves the cell[45].

porous network ordered over several length scales. The two models can be reconciled by viewing the phase separation process proposed by Vrieling et al. as step A in the diatom formation model according to Sumper.

9.7 Aging of Silica Gels

9.7.1 Silica Gel Synthesis

After their initial formation, conventional silicas, formed by acidifying silicate solutions, can be significantly influenced by consecutive aging processes. When the fragile, low-density silicas as prepared in solution are dried, they often collapse. The surface tension of the water droplets disrupts the silica structure. Aging processes are important because in the process the walls of the micropores thicken and, hence, become resistant to drying processes.

Here we will summarize the reaction steps that lead to silica gel formation, and focus on the physical chemistry of the aging processes. NMR proton spin–latice relaxation measurements[46] can be used to follow the morphological changes that occur in a silica gel as a function of time. Owing to the restricted movements and, hence, increased correlation times of the water molecules near to the silica–water interface, protons of water included in a porous solid have longer relaxation times than water in the bulk. A decrease in the spin–spin relaxation time is to be interpreted as an increase in the number of water molecules restricted in their movements due to the proximity of the silica structure. Figure 9.20 shows the spin–spin relaxation T_2 for a silica system as a function of relaxation time. The decrease in T_2 in the first 20 min of the reaction indicates the formation of increasing silica surfaces in contact with water.

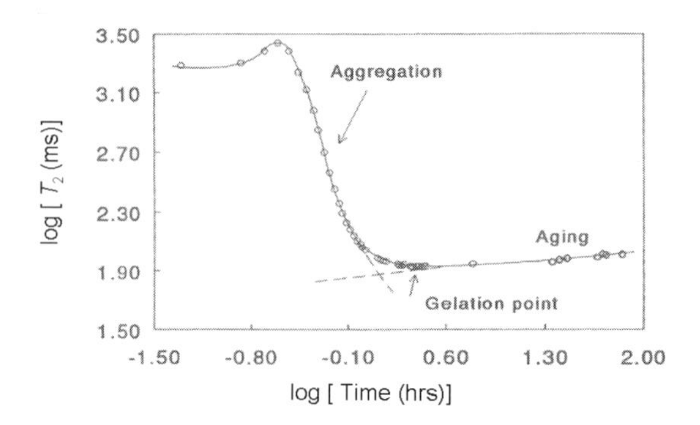

Figure 9.20. Spin–spin relaxation behavior as a function of reaction time for a silica system. C_{SiO_2} = 0.73 M, pH = 4 , T = 40 °C, F/Si = 0. The small initial increase is caused by initial warming up of the sample[46].

This surface formation is due to the polymerization of silica monomers and oligomers and subsequent aggregation of the particles formed into clusters. When the minimum value of T_2 is reached, the beginning of gelation starts. After gelation, aging processes will become dominant. Silica dissolution leads to the formulation of convex soluble surfaces and redeposition of this material in the crevices and necks (concave surfaces) decreases the specific surface area and the average correlation time of the interacting water molecules increases slightly.

In acidified wate glass at pH = 4, the following different reaction steps may be discerned:

1. particle growth
2. aggregation
3. gelation
4. aging.

Steps 1, 2 and 4 are mass transformations successively in the reaction mixture. When gelation occus, a percolating network will be observed only above a certain silica concentration. First the monomers and oligomers present in the reaction medium will start to combine according to the condensation reaction

$$\mathrm{W \sim Si{-}O^- + HO{-}Si \sim W \rightleftharpoons W \sim Si{-}O{-}Si \sim W + OH^-} \tag{9.1}$$

resulting in a mixture of monomers, oligomers and small polymers.

Particle growth is favored at high pH, leading to larger sphere-like particles. At low pH, however, gel-type structures evolve.

As soon as particles are formed, also particle–particle bonds may be formed according to Eq. (9.1), thus leading to open aggregates (Fig. 9.22) which may exhibit fractal properties, to be discussed in Section 9.7.2.

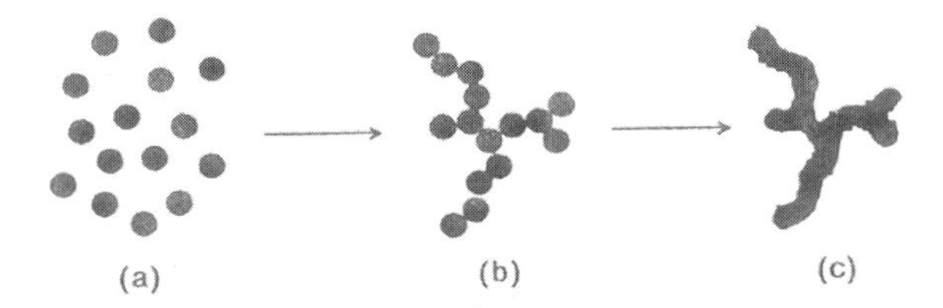

Figure 9.21. Aggregation of single particles (a) into agglomerates (b) and subsequent aging (c).

The structure of the aggregates formed is dependent on reaction parameters used during preparation of the silica gel. Diffusion rates are inversely proportional to the size of the diffusing particles or aggregates.

There are two limiting cases of aggregation, namely reaction-limited and diffusion-limited cluster–cluster aggregation. In the case of diffusion-limited aggregation, the reaction between particles occurs upon collision, resulting in open, tenuous structures. For reaction-limited systems, particles will collide more often before reaction occurs and the primary particles are therefore able to penetrate deeper into the already existing aggregates, thus resulting in more compact, denser aggregates.

Aggregation continues until all primary particles have been incorporated into aggregates or, if the concentration is sufficient, when a percolating network is formed throughout the entire reaction vessel. Gelation is characterized by the fact that the viscosity increases rapidly from the viscosity of the liquid to the viscosity of a solid.

Particles that are not part of the aggregates or network can continue to diffuse freely through the liquid and condense on this gel backbone, changing the properties of the gel. This is a form of aging (see Fig. 9.21), which may result in a substantial growth of the branches of the network long after aggregation and gelation.

Another form of aging may be due to crosslinking processes, where adjacent hydroxyl groups or hydroxyl groups from adjacent gel branches will polymerize into siloxane bonds. During this process the gel may contract and with shrinkage of the gel will take place expulsion of reaction liquid from the pores. This form of aging is called syneresis.

Coarsening or ripening is a third form of aging where dissolution and reprecipitation of silica particles occur, driven by the differences in solubility between surfaces with different radii of curvature. Redissolution will occur from surfaces with positive curvature and will precipitate in regions with a negative curvature, causing growth of necks between aggregated particles, increasing the strength and stiffness of the entire gel.

Aging results in a reinforcement of the structure, originated during aggregation (and gelation). The average pore radius of the gel will increase by these reorganization reactions and the specific surface area will decrease owing to the transport of material to energetic more favorable positions. If the gels are not aged, the pores are very likely to collapse during drying owing to the minimal strength of the backbone and strain forces due to the surface tension of liquid droplets. Aging of the gels, prepared as described above, is a necessity in order to obtain mesoporous materials from them.

9.7.2 Fractals

The aggregates formed during silica gel preparation often tend to exhibit fractal properties. This implies a powerlaw distribution of particle or micropore sizes. Power law kinetics appear to be the consequence of processes where objects grow and/or decompose. The growth rates are typically random and independent of the particle size.

A large number of computer growth models which produce fractal structures have been published[47]. Six basic models can be distinguished and are differentiated by two main factors: transport or diffusion and reaction or sticking probability. In Fig. 9.22 a survey of these six basic models is presented.

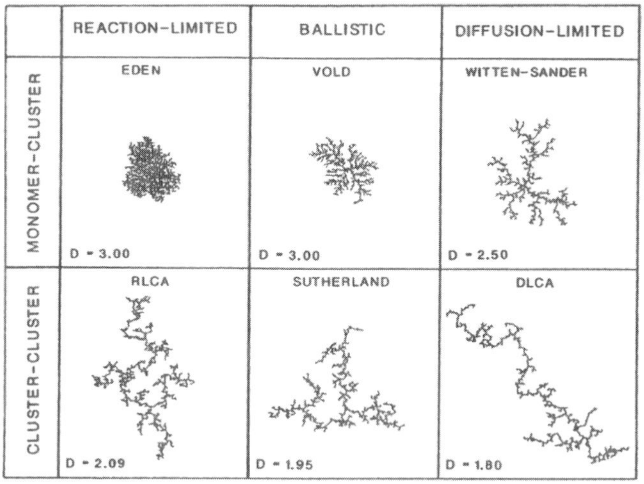

	REACTION-LIMITED	BALLISTIC	DIFFUSION-LIMITED
MONOMER-CLUSTER	EDEN D = 3.00	VOLD D = 3.00	WITTEN-SANDER D = 2.50
CLUSTER-CLUSTER	RLCA D = 2.09	SUTHERLAND D = 1.95	DLCA D = 1.80

Figure 9.22. Models of kinetic growth. The mass fractal dimensions of the three-dimensional simulations are shown but for clarity reasons the corresponding two-dimensional clusters are shown. Adapted from R. Julien and R. Botet[47].

Clusters approach each other by Brownian motion in the diffusion-imited regime. Because of the diffusive nature of the transport, few particles ever penetrate into the interior of the structure that is formed and most growth takes place at the tips of the cluster. In ballistic growth, the trajectories of the particles are linear and large compared with the size of the growing clusters. As a result, the monomers are more likely to penetrate into the interior of a cluster than in diffusion-limited growth, resulting in more compact structures having a larger fractal dimension. In the case of reaction-limited growth, transport is considered facile with respect to reaction. The sticking probability between particles becomes important here. Numerous contacts are required before a bond is succesfully formed, permitting monomers to penetrate deeper into the interior of the growing clusters, resulting in compact aggregates. As is illustrated in Fig. 9.22, it is important to distinguish aggregation processes with growth through monomer–cluster reaction from systems that aggregate through cluster–cluster reactions. The interpretation of dimension D is given in the next paragraph.

Mass fractal structures are characterized by a mass gradient in the cluster according to: $M \sim R^{D_f}$ where M the mass of the cluster, R the radius of the cluster and D_f the mass fractal dimension. This relation represents self similarity. Self similarity means that a similar dimensionality, that reflects the connectivity of a network, is found when different length scales of a material are probed. The self similarity of a cluster is defined between R_g, the radius of gyration of the cluster, and R_0, the radius of gyration of the primary building unit. Systems with Euclidean, non-fractal dimensions such as a rod, a disc and a sphere-like structure, have mass fractal dimensions of $D_f = 1$, 2 and 3, respectively, consistent with the common characterization of these objects. For mass fractal objects the fractal dimension D_f is a non-integer and varies between 1 and 2 in two-dimensional space and between 1 and 3 in three-dimensional space. The fractal dimension is defined by the formula $M \sim R^{D_f}$, implying

$$D_f = \frac{\log \frac{M_{i+1}}{M_i}}{\log \frac{R_{i+1}}{R_i}} \tag{9.2}$$

In a particle growing by iteration, in each iteration step, when increasing the radius of the object from R_i to R_{i+1}, the mass of the object will increase from M_i to M_{i+1}. This is illustrated in Fig. 9.23.

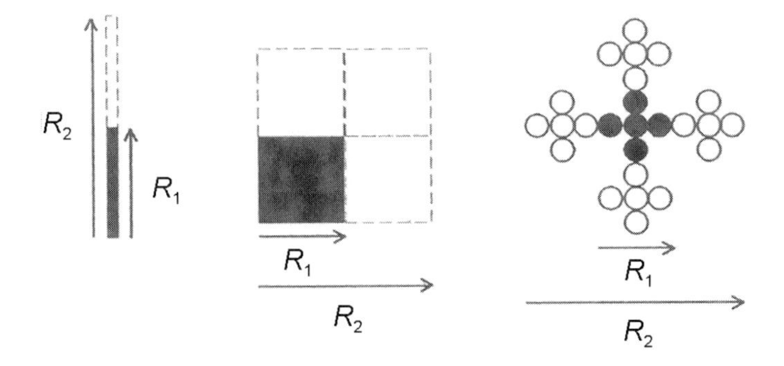

Figure 9.23. Increase in mass of some common objects and a fractal object with increasing radius. The mass fractal dimensions are discussed in the text.

Rod:	M increases by a factor of 2 if R increases by a factor of 2
	$M \sim R^{\log 2/\log 2} \sim R^1$
Plane:	M increases by a factor of 4 if R increases by a factor of 2
	$M \sim R^{\log 4/\log 2} \sim R^2$
Fractal object:	M increases by a factor of 5 if R increases by a factor of 3
	$M \sim R^{\log 5/\log 3} \sim R^{1.465}$

In principle, the same approach is followed when dealing with other objects in three-dimensional space. In this way, the dimensionality of the structure of silica gels is determined. As illustrated in Fig. 9.24, when zooming in on smaller length scales, initially the internal structure is determined by the primary building. Further magnification measures the surface structure or internal structures of the primary building unit. The dimensionalities can now be different.

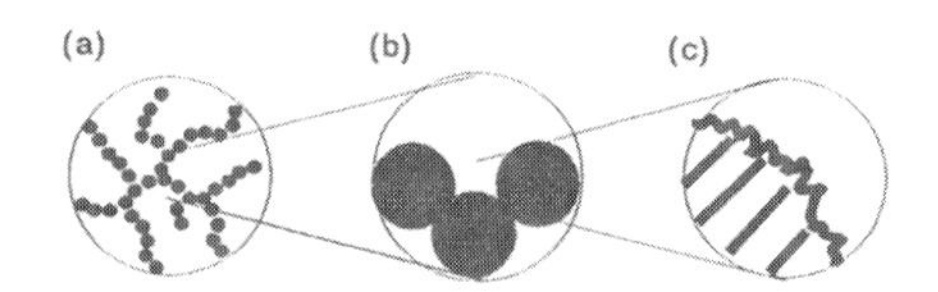

Figure 9.24. The structure of a mass fractal object on several length scales.

The surface roughness can be described in a fashion analogous to that for the mass fractal structure according to

$$S \sim a^{D_s} \tag{9.3}$$

The variable a can be considererd a measure of the surface of the rough particle. It is a ruler. If a ruler is large a lot of small cavities will not be noticed and the resulting surface area is small. When the length of a ruler decreases, more and more surface irregularities will be found, contributing to an exponential increase in surface area. The power-law exponent of the relation between length of the ruler and the corresponding surface is termed the surface fractal dimension (D_s).

For $D_s = 2$, the particles have a smooth surface and the surface is two-dimensional. When the surface roughness increases, D_s can vary between 2 and 3. The surface fractal dimension will approach three for an exceedingly rough surface that also may be described as a structure with a homogeneous distribution of mass and voids. The surface area will then be proportional to the mass: $D \to 3$. It has become three-dimensional.

9.7.3 Simulation of Aggregation Processes

The concentration of monomers plays a very important role in aggregation models. We discussed this earlier in Section 9.4.2, where we noted that percolation through a network becomes realized only when a limit concentration of connections is exceeded.

Here we will analyze the behavior of cluster–cluster aggregation as a function of the momomer volume fraction[48]. We will recognize features that relate to the behavior of percolation. A particular aim of the simulations is to analyze not only the aggregation but also the aging processes. Simulations on aging will be presented in Section 9.7.4.

The interesting aspect of concentration-dependent modeling is that at high concentrations through aging processes a microporous network is formed. The two-dimensional simulations are analyzed with an analytical model that permits the determination of the critical volume fraction ϕ_c for crossover from a non–percolating to a percolating system. Beyond the percolation point the concentration of the fractal–dimensional structure becomes initially local, and at high concentration the local fractal structures disappear.

The results of off-lattice diffusion limited cluster–cluster aggregation (DLCA) simulations in two dimensions are presented in Fig. 9.26. Simulations are limited to two dimensions because it enables one to use large unit cells with many particles. Brownian trajectories are followed. Larger aggregates are generated as a result of bond formation between overlapping aggregates. One finds crossover from fractal to Euclidean behavior when a particular monomer concentration is exceeded.

In the DLCA, reaction limitations are ignored and instantaneous bond formation is assumed. Figure 9.25 depicts resulting snapshots of the aggregation process for a low monomer concentration at successive moments in time. The onset of the formation of a fractal aggregate is visible after 729 moves (Fig. 9.25b). Both small and large aggregates are present, the latter being more anisotropic (less circular). Figure 9.25c shows continuing growth after 2187 moves, which has now resulted in the formation of some large fractal aggregates. The mass is now much less homogeneously distributed over the available space. The snapshot for move 6561 in Fig. 9.25d shows the system just before final aggregate formation. The final aggregate formed is shown in Fig. 9.26a. Only a few large aggregates remain. One of these will soon dominate the aggregate mass distribution. The aggregation process stops when one aggregate has been formed.

The aggregation process depicted in Fig. 9.25 clearly shows a wide variety of aggregates. The aggregate mass distribution, n, as found in aggregation models such as the DLCA model used here, is polydisperse. For a theoretical description that includes ζ, the correlation length associated with the length scale over which the structures exhibit fractal properties, it is convenient to approximate n as a delta function. This process defines the hierachical DLCA model. N aggregates are assigned mass M, while $NM = N_0M_0$ to ensure mass conservation during aggregation. The aggregates formed are fractal on a length scale between L_0 and L_1. The mass M of the aggregates as a function of length scale L between L_0 and L_1 is

$$\frac{M}{M_0} = \left(\frac{L}{L_0}\right)^{D_f} \tag{9.4}$$

The lower limit of fractality L_0 is assumed to be the monomer radius R_0. For distances L shorter than L_0, the density ρ_0 of the initial primary scatterer is set to unity. The density ρ of the fractal region can be expressed in analogy to Eq. (9.4) as

$$\frac{\rho}{\rho_0} = \left(\frac{L}{L_0}\right)^{D_f - D} \qquad (D_f \leq D) \tag{9.5}$$

Hence, for the effective volume fraction ϕ_e, one finds

$$\frac{\phi_e}{\phi_0} = \left(\frac{L}{L_0}\right)^{D - D_f} \tag{9.6}$$

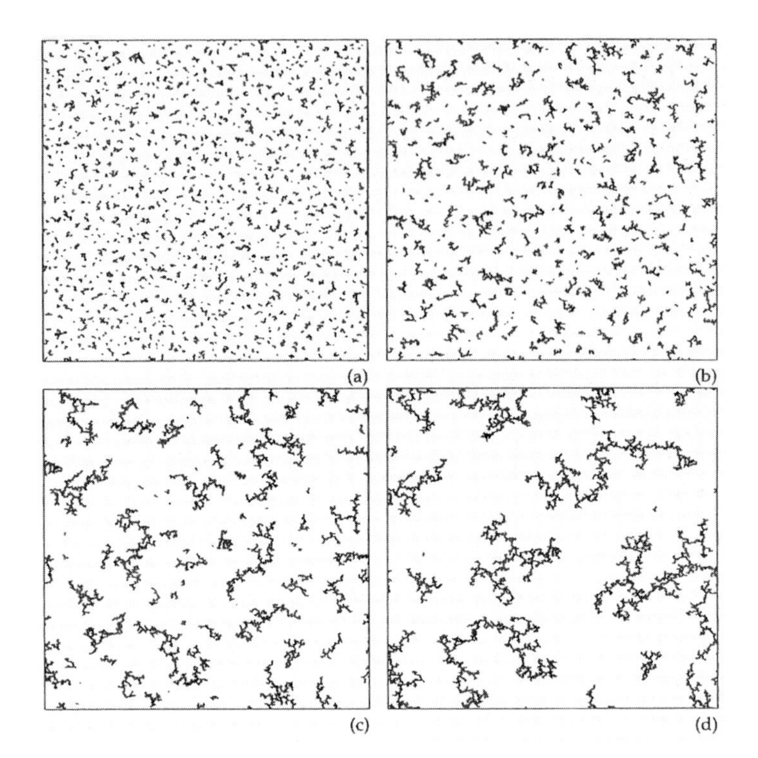

Figure 9.25. Various stages of growth processes of DLCA aggregates in $2D$ at $\phi_0 = 0.05$: (a) move 243, (b) 729, (c) 2187 and (d) 6561.

In this idealized model, the aggregates grow in time until finally one aggregate is formed. It is necessary to distinguish between two situations for the formation of the final single aggregate and the upper length scale L_1 of fractality. In the first case (situation a), for high volume fractions, the growing aggregates touch at a certain stage of the simulation (see Fig. 9.26b). At this moment the total effective volume of the aggregates has become equal to L_s^D, which is the volume of the simulation space where the aggregates can no longer perform their Brownian movements independently. The effective volume fraction of the aggregates, ϕ_e, has reached 1.0 (percolation) before the final aggregate has been formed (gelation). For these high volume fractions one might say that percolation is faster than gelation: $\phi_e^a = 1$ and $N_a > 1$. This limits the development of fractal properties, leading to a change in L_1. It then follows from Eq. (9.6) that for $L_1 = L_a$

$$\frac{L_a}{L_0} = \phi_0^{1/(D_f - D)} \tag{9.7}$$

The upper length scale of fractality depends on ϕ_0 and is smaller for high volume fractions.

In the second case (see Fig. 9.26a), at low ϕ_0, even the final aggregate is too small to span the box. The system does not percolate and that only gelation occurs: $\phi_e^b < 1$ and $N_b = 1$. It then follows from Eq. 9.4

$$\frac{L_b}{L_0} = N_0^{1/D_f} \tag{9.8}$$

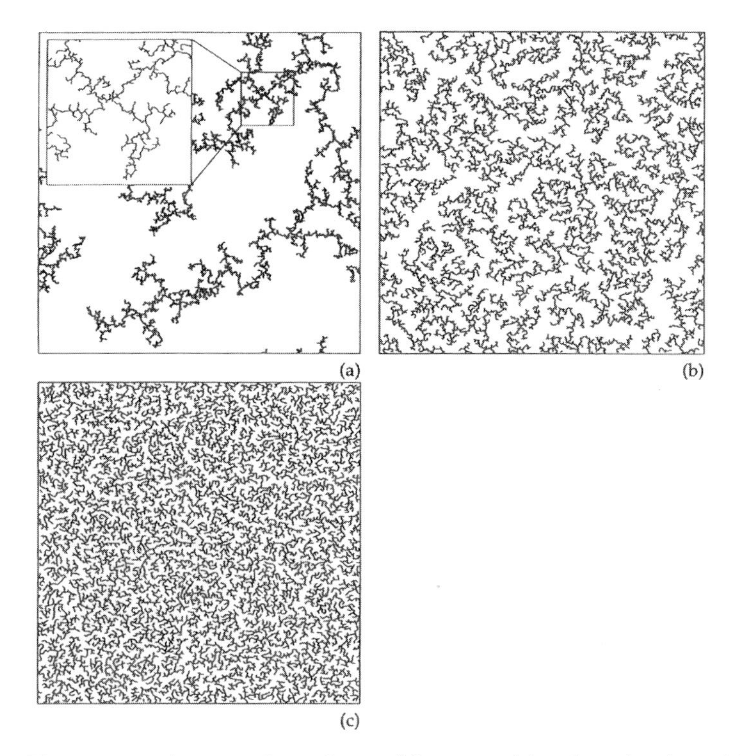

Figure 9.26. Aggregate formation at different particle volume fractions: (a) $\phi_0 = 0.05$, (b) 0.25 and (c) 0.50. The aggregates are built using periodic boundary conditions and consequently some of the connecting bonds reach over the boundaries.

For low volume fractions the aggregate growth is not limited due to percolation. Intrinsic fractal bahavior is observed so that L_b depends only on the number of monomers in the system and the fractal dimension of the aggregatin process.

The critical volume fraction ϕ_c for the crossover situation can be calculated by combining Eqs. (9.7) and (9.8):

$$\frac{L_c}{L_0} = \phi_c^{1/(D_f - D)} = N_0^{1/D_f} \Leftrightarrow \phi_c = N_0^{(D_f - D)/D_f} \tag{9.9}$$

The results are typical for percolating systems[29]. Equation (9.9) shows that the occurrence of percolation depends on the value of N_0 used in the simulation. Equation (9.7) indicates that the upper length scale of fractal behavior does not depend on N_0 for $\phi_0 > \phi_c$.

It should be realized that the upper length limit of fractal behavior, L_1, is related to, but not identical with, the correlation length, ζ. L_1 is the radius of the fractal regions while ζ must be related to the radius of gyration of the fractal regions. The correlation length can be calculated by combining the definition of the radius of gyration and eq. (9.5):

$$\zeta^2 = \frac{\int^{L_1} \rho(L) L^2 \mathrm{d}L}{\int^{L_1} \rho(L) \mathrm{d}L} = \frac{2 - D + D_f}{4 - D + D_f} L_1^2 \tag{9.10}$$

L_1 can be calculated using Eqs. (9.7)–(9.9) depending on ϕ_0. For $D_f = 1.45$ and $D = 2$, the following relation between L_1 and ζ is found:

$$\zeta = \sqrt{\frac{D_f}{2 + D_f}} L_1 \simeq 0.65 L_1 \tag{9.11}$$

The essential features of cluster–cluster aggregation as a function of concentration are shown in Fig. 9.26 at volume fractions 0.05 (a), 0.25 (b) and 0.50 (c). At $\phi_0 = 0.05$, the mass-fractal properties are well developed and the aggregate has a very ramified structure. At $\phi_0 = 0.25$, the mass-fractal properties are only present at short length scales. The long–range structure seems almost homogeneous. At $\phi_0 = 0.50$, no mass-fractal structure is visible and the morphology resembles a high volume fraction $2D$ random network. The structure looks homogeneous at all length scales.

9.7.4 Expressions for Aging of Fractal Systems

Results are reviewed here that describe morphology changes by aging of aggregates formed by the diffusion-limited aggregation process as described in the previous section. As we noted before in Section 9.7.1, the aging process creates larger, stable micropores in initially fragile and narrow pore gels. We saw in the previous section for a high monomer content (high volume fraction) that the development of fractal properties is limited during percolation and consequently the correlation length at high volume fractions is smaller than the size of the aggregate. For low volume fraction the aggregates can grow unrestricted by the size of the space constraints and the correlation length can be identified as the radius of gyration of the resulting aggregate. A crossover between these situations is present at the critical volume fraction. For this volume fraction the final aggregate size equals the size of the space (either the simulation space or the silica gel container) in which it is grown in. Aging processes are responsible for the formation of stable and larger micropores in systems prepared through sol–gel processes.

Aging processes such as random bond breaking and ring formation induce changes in the (fractal) properties of the system such as:

- Formation of new primary particles of radius R'_0 with density ρ'_0.
- Shrinking of the aggregates, due to sintering or resolvation, thereby inducing changes in the correlation length ζ, the system size and the aggregate density ρ and the volume fraction ϕ_0.
- Changes in the fractal dimension D_f.

Often, several changes in the fractal properties appear simultaneously. Using the modified system parameters, an expression equivalent to Eq. (9.5) can be written down for the aged system:

$$\frac{\rho'}{\rho'_0} = \left(\frac{L'}{L'_0}\right)^{D'_f - D} \tag{9.12}$$

Equations (9.5) and (9.12) allow us to relate the fractal properties of non-aged and aged-aggregates.

When a fractal system shrinks homogeneously, the upper length scale of fractality L will decrease to L'. The density of the aggregate will increase from ρ to ρ'. The lower limit of fractal behavior, L_0, and consequently the density, ρ_0, of the primary building unit of the system are not affected by the homogeneous shrinking process.

We will analyze simulation results of one particular kind of aging, called shaking. Bond breaking and monomer fusion are not allowed for. The monomers within an aggregate are moved (shaken) with respect to each other. When monomers overlap, an additional bond is formed.

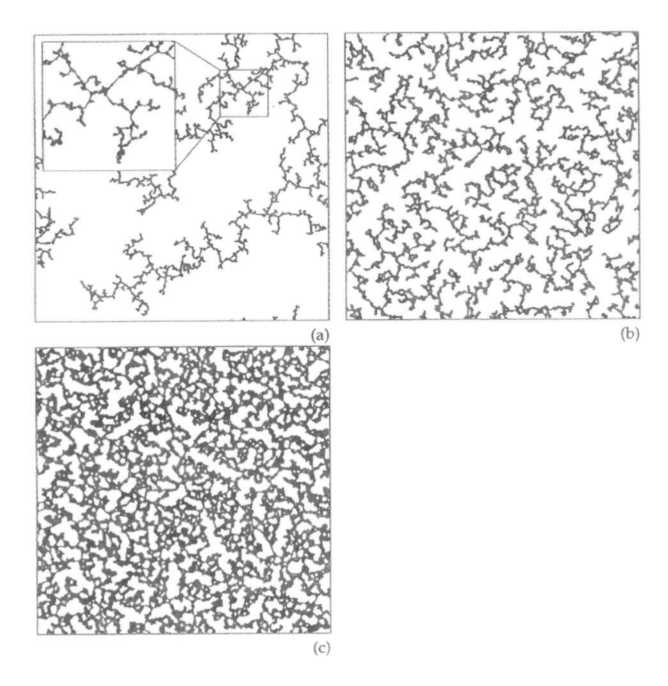

Figure 9.27. The same aggregates as in Fig. 9.27 after 25,000 shaking moves. (a) $\phi_0 = 0.05$, (b) 0.25 and (c) 0.50.

The effect of shaking on the aggregates of Fig. 9.26 is shown in Fig. 9.27. The short side-chains of the aggregate of Fig. 9.27a have folded on to the backbone forming chains of 3-monomer rings. The disappearance of side-chains causes a more sparse appearance. Long chains have been deformed only slightly so that the long-range mass-fractal structure has been preserved. In the aggregate of Fig. 9.27c the short side-chains have also disappeared, but because of the absence of mass-fractal behavior the system has become more compact instead of open. For the aggregate of Fig. 9.27b the two effects are visible at low and high ϕ_0 are combined. Both sparse and dense regions have been formed. So for all volume fractions short side chains have been folded into chains of 3-monomer rings or larger structures built of 3-monomer rings.

One notes especially from the comparison of Figs. 9.26c and 9.27c, above the critical volume fraction ϕ_c the creation of large micropores due to the aging processes.

9.8 In Conclusion; Self Organization and Self Assembly

The design of catalytic systems requires the ability to synthesize specific material architectures and the atomic control of the catalytically active reaction centers. The biomineralization part of this chapter illustrated how preorganization of lipophilic molecules in an ordered phase controls the synthesis of crystalline microporous crystals with dimensions on the order of several nanometers. Polymers with hydrophilic and hydrophobic subunits can be used to synthesize a wide variety of structured materials. The polymer–silica ratio, temperature and solution phase conditions can be tuned in order to create mesoporous systems with ordered pore structures, fractal particles or hollow particles of large dimensions. The membrane of the latter contains well-defined nanopores. A fractal pore surface optimizes the surface area and at the same time contains wide channels. The great challenge in the design of the catalyst architecture is to design materials that allow for ready access of catalytic centers in high-density material. Ready access requires wide pore channels of high connectivity, implying low density. A high concentration of catalytic sites, on the other hand, requires a high-density material. Materials with channels that could feed the catalytic centers independently are highly desirable.

For example, in oxidation catalysis the introduction of separate channels for the oxidant, the reactant and the product would result in significantly improved selectivity. The catalytically active component should reside in the wall of one of the channels and the pore dimensions of the channel should be such as to allow only migration of one of the reaction components.

When liquids are used, the channels should be at least 1 μm in diameter; the catalytic centers can still exist as nanometer-size cavities, to be connected through thin walls with well-defined molecular-sized micropores. Channels or micro cavities could be hydrophilic or hydrophobic and thus tuned by the needs of the particular reaction studied. The architecture of a catalytic system aims to reduce mass transport limitations and optimize the mixing of components at least near catalytically active centers. Second, for highly exothermic or endothermic reactions, the design should also include heat transfer considerations. Clearly this is a topic where there is a need for extensive catalytic process modeling and innovative catalyst synthesis.

In this chapter, the siliceous framework of the diatoms has been introduced as an example of a system that has micropores and channels that vary over many dimensions. Synthetic design approaches that allow for the self assembly of amphiphilic molecules into large units and phase separation provide a way to make such multi-scaled ordered porous materials.

In the biomimetic biomineralization approach, such materials with ordered porosity of varying size and connectivity are made through conversion of structure-directing organic phases converted into a solid catalyst by fixation through hydrolysis of co-mixed or attached siloxy fragments (see review by Zhao et al.[33a]). This approach can be generalized to produce microporous materials of other elements such as Al, Ti and V with widely varying composition[33a].

The first part of the chapter dealing with the chemistry of evolutionary protocell formation discussed new strategies for the design of self assembling catalytic systems that can self adapt in order to control activity or selectivity for a particular conversion reaction. Biomimetic biomineralization processes are successful catalyst synthesis strategies to produce structured microporous systems. Mineral fixation as exemplified in this chemistry can also be of use to convert the protocell-type catalytic system into a practical heterogeneous catalyst. The aim of the exposition of theories of artificial life and chemistry was to establish the fundamental elements necessary to synthesize a catalytically active center by evolutionary synthesis design. In Chapter 7 we discussed that the catalytically active site in an enzyme provides an optimum arrangement of functional groups around the reacting molecules where matching the shape and size of a reaction cavity are key to generating high selectivity and activity. The evolutionary adaptive systems, discussed in the earlier part of the chapter on the origin of life, were based on the philosophy that the predictive capabilities of the relation between catalyst structure and function are still imperfect. In addition, well-defined synthesis strategies are often not available or possible. Hence rational catalyst design should employ an approach in which the catalytic center configures itself. Molecular recognition chemistry of a template, that relates to the transition state of a kinetically important catalytic reaction intermediate, is the link between the catalytic reaction and the catalytic system. Evolutionary adaptive synthesis the implies formation of a catalytic system that proceeds in stages. After each synthesis stage the system is tested. Systems survive and multiply when the test is positive. Intrinsic to the evolutionary process after each stage, protocells grow and multiply. In the multiplication process there is a possibility of mutation of building or replication units.

A protocellular system requires a membrane to separate it from the environment. The chemical machinery has to be in place that can synthesize the membrane and the catalytically active system. We discussed especially in detail the Wächtershäuser proposal in which FeS, which acts as a catalyst, is coupled with an energy source to provide for the evolutionary formation of protocells. The energy-providing source in this system is the pyrite formation reaction from FeS itself. A necessity for multiplication and self organization is an autocatalytic reaction system that can be induced to reproduce the cell machinery when the cell membrane can no longer support internal growth and starts to divide. Once the optimum catalytic system has evolved, a fixation step as in biomineralization has to be added to convert the preorganized protocellular system to a heterogeneous solid catalyst. Succesful protocellular systems are becoming within reach. Rasmussen et al.[49] and Szostak et al.[50] have discussed liposome systems with self dividing walls. In their system the liposome interior contains a replicating chemical reaction system and employs light as an energy source. Once a molecular recognition system has been added and the chemistry of its replication has been implemented, an evolutionary adaptive design process can be made operational. Interestingly, adaptation implies, as in the immunoresponse system, that the molecular recognition system has to improve in the different synthesis stages and should be able to operate against a range of test molecules. The molecular recognition site therefore has to assemble itself from molecular building units, that can mutate.

Although so far the ultimate goal of evolutionary adaptive catalyst self assembly has not been realized, there are many uses of parts of the system that are within close reach. For instance, there is a need to assemble catalysts based on the replication of atomic rearrangement patterns in two dimensions. The predicted alloy configurations, discussed in previous chapters, which are optimal for particular catalytic reactions have to be translated into three-dimensional layered materials. Molecular recognition, combined with chemical amplification, could be the basis of the design of sensors. Adaptive design could make them biocompatible. Rational design will likely require the close combination of theory, modeling and experimental exploration.

In this book we have provided an overview of the physical chemistry of catalytic systems with a focus on molecular mechanistic aspects. Theories, concepts and techniques to describe the relevant chemistry have advanced to a state that the reaction mechanism in heterogeneous catalysis is becoming well understood. Clearly, similar advances relevant to the practice of catalyst synthesis and design are now on the near horizon. The foundation of theoretical molecular heterogeneous catalysis provides a firm basis towards these future endeavors.

References

1. J.M Lehn, *Angew. Chem. Int. Ed. Engl.* 27, 89–112 (1988)
2. I. Prigogine, *From Being to Becoming*, Freeman, New York (1980)
3. S.A Kauffmann, *The Origin of Order*, Oxford Univereity Press, Oxford (1993)
4. J. von Neumann, *Theory of Self-Reproducing Automata*, University of Illinois Press, Urbana (1966)
5. M. Eigen, P. Schuster, *Naturwissenschaften*, 64, 541 (1997); 65, 341 (1978)
6. H. Kuhn, H.D. Försterling, *Principles of Physical Chemistry*, Wiley, New York 2000
7. A.E. Oparin, *The Chemical Origin of Life,* Thomas, Springfield (1964))
8. D.E. Koshland, *Science*, 295, 2215 (2004)
9. F. Dyson, *Origins of life*, Cambridge University Press, Cambrideg (1999)
10. G. Wächtershäuser, *Proc. Natl. Acad. Sci. USA*, 87, 200 (1990)
11. M.J. Russell, R.M. Daniel, A.J. Hall, J. Sherringham, *J. Mol. Evol.* 39, 231 (1994)
12. G. Wächtershäuser, *Prog. Biophys. Mol. Biol.* 58, 85 (1992)
13. A.G. Cairns-Smith, G.L. Walker, *Biosystems*, 5, 173 (1974)
14. J.P. Grotinger, J.F. Kastring, *J. Geol.* 101, 235 (1993)
15. M. Dörr, J. Kässbohrer, R. Grunert, G. Kreisel, W.A. Brand, R.A. Werner, H. Geimann, C. Apfel, C. Robl, W. Weignand, *Angew. Chem. Int. Ed.* 42, 1540 (2003)
16. F.C. Frank, *Biochem. Acta*, 11, 459 (1953)
17. N.M. Shnert, Y. Louzain, E. Bettelheim, S. Solomon, *Proc. N. Y. Acad. Sci.* 97, 10322 (2000)
18. W.A. Bonner, *Origins Life Evol. Biosphere*, 21, 59 (1991)
19. H. Wynberg, *Chimia* 43, 150 (1989)
20. A.H. Alberts, H. Wynberg, *J. Am. Chem. Soc.* 111, 7265 (1989)
21. D.R. Kondepudi, C. Sabanayam, *Chem. Phys. Lett.* 217, 364 (1994)
22. T. Ganti, *J. Theor. Biol.* 187, 583 (1997)
23. (a) H.J. Morowitz, *Beginning of Cellular Life*, Yale University Press, London (1992): (b) D. Segré, D. Lancet, *Biochem. Mol. Biol.* 12, 382 (1999).
24. (a) N. Ono, Artificial Chemistry, *Thesis*, University of Tokyo, (2001);
    (b) T. Ikegami, *Artif. Life Robot.* 3, 242 (1999)
25. P. Cariani, in *Artificial Life II, FFI studies in the Science of Complexity,* Vol X, C.G. Langton, C. Taylor, J. D. Farenerand, S. Rasmussen (eds.), Addison-Wesley, Redwood City, p. 776 (1991)
26. C.G. Langton, Life at the Edge of Chaos, in C.G. Langton, C.Taylor, J.D. Farmer, S. Rasmussen (eds.); *Artificial Life II,* Addison-Wesley, Redwood City, p. 41 (1992)
27. P.W. Anderson, *Proc. Math. Acad. Sci.* 80, 3386 (1983)
28. (a) G. Nicolis, *Introduction to Nonlinear Science*, Cambridge University Press, Cambridge p. 180 (1995);
    (b) S. Wolfram, *A New Kind of Science,* Media Publisher, Campaign, Illinois (2002)
29. D. Stauffer, A. Aharong, *Introduction to Percolation Theory*, Routledge, New York (1994)
30. C.T. Kresge, M.E. Leonowicz, W.J. Roth, J.C. Vartuli, J.S. Beck, *Nature*, 359, 710 (1992)
31. J.M. Kim, S.K. Kim, R. Ryoo, *Chem. Commun.*, 259 (1998)
32. C. Yu, B. Tiau, J. Fan, G.D. Stucky, D. Zhao, *J. Am. Chem. Soc.* 124, 4556 (2002)
33. (a) D. Zhao, P. Yang, Q. Hus, B.F. Chemelka, G. D. Stucky, *Curr. Opin. Solid State Mater*, 3, 111 (1998);

(b)P.Yang, D. Zhao, D.I. Margolese, P.F. Chmelka, G.D. Stuckly *Nature*, 396, 152 (1998);
(b)M.R. Fisch, *Liquid Crystals, Laptops and Life*, World Scientific, Singapore 2004
34. A.C. Voegtin, F. Ruch, J.L. Guth, J. Patarin, L. Huol, *Microporous Mater.* 9, 95 (1997)
35. P.T. Tanev, T.J. Pinnavaia, *Science*, 271, 1267 (1996)
36. G.S. Attard, J.C. Glyde, C.G. Göltner, *Nature*, 378, 366 (1995)
37. D. Zhao, J. Feng, Q. Huo, N. Melosh, G.H. Frederikson, B.F. Chemelka, G.D. Stucky, *Science*, 279, 548 (1998)
38. K. Shimizu, J. Cha, G.D. Stucky, D.E. Morse, *Proc. Math. Acad. Sci. USA*, 95, 6234 (1998)
39. N. Kröger, R. Deutzman, M. Sumper, *J. Biol. Chem.* 276, 26066 (2001)
40. E.G. Vrieling, Q. Sun, T.P.M. Beelen, S. Hazelaar, W.C. Gieskes, R.A. van Santen, *J. Nanosc. Technol.* 5, 63 (2005)
41. Q. Sun, T.P.M. Beelen, R.A. van Santen, S. Hazelaar, E.G. Vrieling, W.W.C. Gieskes, *J. Phys. Chem. B*, 106, 11539 (2003)
42. J. Pickett-Heaps, A.M. Schmid, L.A. Edgar, *Prog. Phycol. Res.* 7, 1 (1990)
43. Q.Sun, P.J. Kooymann, J.G. Grossmann, P.H.H. Bomans, P.M. Frederik, P.C.M.M. Magusin, T.P.M. Beelen, R.A. van Santen, N. A. J. M. Sommerdijk, *Adv. Mater.* 15, 1097 (2002)
44. M. Sumper, *Science*, 295, 2430 (2002)
45. E.G. Vrieling, T.P.M. Beelen, R.A. van Santen, W.W.C. Gieskes, *Angew. Chem. Int. Ed.* 41, 1543 (2002)
46. W.H. Dokter, H.F. van Garderen, T.P.M. Beelen, J.W. de Haan, L.J.M. van de Ven, R.A. van Santen, *Colloids Surf. A*, 72, 1165 (1993)
47. R. Julien, R. Botet, *Aggregation and Fractal Aggregates* World Scientific, Singapore (1987)
48. H.F. van Garderen, W.H. Dokter, T.P.M. Beelen, R.A. van Santen, E. Pantos, M.A.J. Michels, P.A.J. Hilbers, *J. Chem. Phys.* 102, 480 (1995)
49. S. Rasmussen, L. Chen, D. Deamer, D.C. Krahauer, N.H. Packard, P.F. Stadler, M.P. Bedau, *Science*, 303, 963 (2004)
50. J.W. Szostak, D.P. Bartel, P.L. Luisi, *Nature*, 409, 387 (2001)

CHAPTER 10
Postscript

In this chapter we summarize the catalytic principles highlighted in this book. This provides the reader with an umbrella of important theoretical catalytic concepts and founding laws. Each concept is listed along with the chapter(s) in which it was introduced.

Loose and Tight Transition States

Chapter 2

A transition state is considered loose when its entropy is high thus implying that it has a significant mobility. A transition state is considered tight, when its entropy is low, thus implying that it has little mobility. Tight transition states with low entropy occur when there are strong covalent interactions between the substrate and the catalyst surface. The strong intercations help to activate bonds in the reacting molecule or to form bonds with it. This is typical for the activation of molecules on metal surfaces. Loose transition states with high entropy occur when the reagent and the reactant form strong covalent bonds in the transition-state complex. Their interaction with the catalyst surface is simultaneously weakened, thus creating a more mobile transition-state complex. This is usually the case in proton-activated reactions that occur in zeolites and other solid acids. Bond formation between the zeolitic proton and the substrate occurs together with the cleavage of the bond between the proton and the zeolite. The energy is offset by the strong electrostatic interaction between the protonated transition-state complex and the negatively charged zeolite lattice, which is weakly directional.

Sabatier Principle

Chapter 2

Catalysis involves a cycle of elementary physicochemical steps which includes adsorption, surface reaction, desorption and surface diffusion processes. These steps require both bond making and bond breaking between the substrate molecule and the catalyst surface. Hence, there is an optimal interaction between the molecular substrate and the catalyst surface that leads to a maximum catalytic activity. This is known as Sabatier's principle. A plot of the reaction rate verses the substrate–catalyst interaction energy thus goes through a maximum which is described as Sabatier's maximum. To the left of this maximum, the reaction order is positive in the reactant concentration and the surface coverage is low. The surface is suboptimal in that it does not readily active the substrate molecule. To the right of the Sabatier maximum, the reaction order is negative in the reactant concentration, which indicates that the overall rate is impeded by high surface coverage. While the intrinsic rate of reaction here may be higher, the number of vacant surface sites is too low.

Surface Topology and Geometry Effects

Chapters 2 and 3

The reactivity of a coordinatively unsaturated surface atom tends to increase with decreasing coordination number. Therefore, high-index surfaces tend to be more reactive than low-index surfaces. Steps and kinks are often the preferred sites for dissociation to occur. If dissociation leads to fragments that preferentially require a specific site or sites

Molecular Heterogeneous Catalysis. Rutger Anthony van Santen and Matthew Neurock

ISBN: 3-527-29662-X

that have more than one surface atom, dissociation will only occur when the specific ensemble and arrangement of surface atoms is available for chemical bond activation. In such a case, the reaction is structure sensitive for both bond cleavage and bond association. The activation energy of the surface reaction is lowered in both directions since the reactant and the product states become more stabilized. For reactions that occur over a single metal atom, bond activation is enhanced when the metal surface atom is less coordinatively saturated. This lowers the barrier for bond activation but increases the reverse barrier for recombination. For a late transition state, the change in the activation energy for recombination will be less than the change in the activation energy of the bond-breaking reaction, which then will be most surface sensitive.

Brønsted–Evans–Polanyi Relation

Early and Late Transition States. Chapters 2 and 3

The Brønsted–Evans–Polanyi (BEP) relationship relates the changes in the activation energy for a particular reaction over different catalysts to the changes in their corresponding heat of reaction (or overall reaction energy). Similar relationships have also been developed for changes in the reactant molecule rather than changes in catalyst. The BEP equation is a simple empirical relationship, in which a non-equilibrium property, such as the activation energy, is related to an equilibrium property such as the reaction energy. The proportionality constant, α, in the BEP relationship is close to one when the transition-state conformation is close to product geometry (late transition state). The value of α is close to zero when the transition state is near the reactant state configuration (early transition state). For simplicity, the value of α in the literature is often approximated as 0.5.

Promotion

Chapters 2, 3 and 5

- Alloys: Mixed metal surfaces can promote the selectivity of a reaction, by offering reduced ensemble sizes for adsorption, by poisoning reactions at surface edges and/or by offering coadsorption sites. In addition, bi- and multi-metallic systems can also enhance the activity of the surface by changing the strength of the metal–adsorbate surface bond. These changes can occur through an "ensemble" or geometric effect where the adsorbates on the surface bind preferentially to sites which maximize their surface adsorbate bond strength. The preference for one metal type over the second, in some cases, is even stronger than its coordination site preference. Therefore, alloying the surface can drive adsorbates to specific sites which can either promote or poison the site's reactivity. The changes in activity can also occur through electronic or ligand effects, which refer to the changes in the electronic structure at the active site due to changes in its nearest-neighbor ligands. Pseudomorphic metal overlayers and near-surface alloys demonstrate this point rather nicely in that the reactivity on these surfaces can be quite different even though the surface metal layer is the same. The changes in their reactivity can be explained by changes in the electronic properties of the surface that result from charge transfer or changes in the lattice spacing. More explicitly, the changes in binding energies and activation barriers are nicely correlated with shifts in the d-band center of the metal substrate.
- Mixed oxides: Mixed oxide surfaces can help to promote reactions by utilizing a combination of different types of sites with different functionality to aid the reaction.

These include oxidation sites with different oxidation states, as well as Brønsted and Lewis acid and base sites. In addition, the sites responsible for O_2 activation may be different from those for oxygen insertion or dehydrogenation. In order to stabilize particular reaction centers or catalyst phases, the formation of compounds with non-reactive cations is sometimes desirable.
- Mixed sulfides: Create enhanced activity by stabilizing sulfide vacancies
- Coadsorbed molecular moderators: Much of the chemistry on metal surfaces is two-dimensional in nature with no participation of the weaker stabilizing van der Waal's interactions from organic functional groups such as those which make up the enzyme cavity and no influence of the pore wall as one may have in zeolites. Certain heterogeneous catalyzed reactions carried out on metal surfaces, however, have shown significant improvement on the addition of molecular modifiers that can coadsorb along with the reactant. Upon adsorption, they can begin to create partial three-dimensional pockets on the metal substrate. The well-known example here is the addition of *Cinchona* alkaloids to modify Pt and promote its ability to carry out hydrogenation of α-keto esters with high enantioselectivity.

Transient Reactive Intermediates
Chapter 2

Much of what we know about the reactive intermediates for a particular reaction has been established from either in situ or ex situ spectroscopic analyses of the reaction surface. What is usually measured, however, is the most stable species on the surface and not necessarily the most reactive species. There are a growing number of examples which have shown, through either experiment or theory, that the reaction may be controlled by species that are very reactive on the surface. They have very short lifetimes, thus making it difficult to catch them "in action". Some of the notable examples include the π-bound ethylene species on Pt and Pd which are more weakly bound than their di-σ-bound intermediate but also tend to be the more predominant reaction channels. Similarly, transient O_2^- surface intermediates on Cu and La_2O_3 and also O^- on different metal surfaces have been identified.

Cluster Size Effects
Chapters 2 and 5

The reactivity of supported metal, metal oxide and metal sulfide nanoparticles can change owing to changes in the size of the particles and changes in the support. In this section, we refer only to changes that are the direct result of particle size. Indirect effects that arise from the particle support interactions are described in the next section. In moving from bulk particles down to nanometer size dimensions, metal, metal oxide and metal sulfide particles all show significant changes in their electronic structure as we move from their band structures for the bulk materials to distinct molecular states with unique electronic properties for the nanoparticles. Therefore, systems that were once conducting, insulating or semiconducting may now have very different properties. The unique electronic structures of these molecular states govern the chemical properties and hence catalytic reactivity. It is well established that metal clusters of with fewer than 20 atoms in the gas phase show distinct electronic properties, chemisorption characteristics and surface reactivity. In addition, the band gap for metal oxide particles is increased significantly on moving down to very small nanoparticles, which are on the order of 40 atoms or less.

A second predominant effect that occurs on shrinking the cluster size is related to the changes that occur in the relative composition of different exposed facets and defect sites. The percentage of exposed corner and edge sites increases substantially as the cluster size is decreased down to molecular scales. This change in site composition density can dramatically influence the reactivity at the particle surface.

Support Effects

Chapters 2 and 5

The support that is used to anchor the active metal, metal oxide, or metal sulfide particle can significantly influence the reactivity of the particle depending upon its structure, morphology and chemical properties at its interface with the particle along with its local environment. The acid/base characteristics, the composition of surface species such as hydroxyl groups and cation or anion defect can all act to change the properties of the nanoparticle. First, the surface properties of the oxide can control the size, shape and morphology of the nanoparticle that forms. Second, the support can act as a ligand and control the electronic properties of catalytic particle via charge transfer. The electronic properties of the naked gas-phase particles can be significantly altered through electron transfer either from the support to the particle or from the particle to the support, thus leading to particles that are electron-rich or electron-poor. This will depend upon the bonding between the metal and the support and the presence of defects sites. Third, the interface formed between the nanoparticle and the support creates uniquely active sites such as in the formation of new mixed metal oxide sites (M_1–O–M_2) where the metal or metal oxide nanoparticles attach to their support, or new mixed metal sulfide (M_1–S–M_2) sites where the metal sulfide attaches to its support. In addition, new sites can be created by local charge transfer at the nanoparticle/support interface, the creation of bifunctional sites that cooperate across the interface and the formation of Brønsted acid sites to accommodate bridges between cations of different charges. The properties of the particles that form can be very different from those of the naked gas-phase particles. The factors that control the catalytic activity of nanometer metal and nanometer metal oxide particles on different supports are still a subject of great debate.

The Material and Pressure Gaps

Chapters 2, 3, 4, 5 and 6

The material and the pressure gaps refer to the two greatest differences between experiments performed under ultrahigh vacuum conditions over single crystal surfaces and those carried out over actual catalysts run at near operating conditions. These same issues relate to the extrapolation of ideal theoretical results to catalysis over supported particles under industrial conditions. The ability to bridge the materials gap requires a more complete understanding of the active sites and models to represent them. For systems with well-defined structures such as organometallic clusters, heterpolyacids and many zeolites, there is no real materials gap since we know the location of the atoms and we can develop reasonable models to capture the primary structure. Simulating supported metals, on the other hand, requires modeling, or at least understanding, how changes in (1) metal-support interactions, (2) surface structure, (3) defect sites and (4) particle shape, size and composition all influence catalytic reactivity.

The pressure gap refers to the difference in pressure between reactions run under ultrahigh vacuum conditions and those run under actual industrial conditions. This difference in pressure can be of the order 10^{10}-fold, which will significantly influence the surface coverages and can affect the surface reactivity and selectivity. We have shown that the most stable surface structures for oxides, sulfides and metals are strongly dependent on the reaction conditions and can readily change throughout the course of reaction. This requires the application of phase diagrams to map out the lowest energy surface structures as a function of chemical potential. The dynamic changes that occur as the result of reaction can be simulated using kinetic Monte Carlo simulations, provided that the elementary processes have been measured experimentally or calculated quantum mechanically. The simulations of catalytic surface reactions on both metals and oxides show that the system is strongly dependent upon the total pressure, partial pressures and the temperature as they significantly influence the total surface coverage, the relative surface compositions and the potential reactivity.

Lock and Key Related Molecular Recognition Concepts

Chapters 2, 4 and 7

The concept of pretransition-state orientation is general to molecular catalysis. A catalyst has to provide the optimum opportunity to stabilize reactants in a configuration that, with minimum movement, can lead to the respective transition state. Maximum stabilization of the transition-state free energy has to occur for the transition state. Optimum stabilization of the pretransition-state structure requires a steric match between the shape of the pretransition-state configuration and the catalyst cavity or the catalyst surface. This involves a molecular recognition process.

The flexibility of the catalyst framework helps to accommodate the different steric requirements of reactant, transition and product states. The steric match of the pretransition state and the transition state should not be so tight that it prevents the entropic movement of substrate. The product that forms must be unfavorably bound so that it will desorb from the active site once it forms.

The induced lock and key principle refers to a flexible catalyst lattice that adapts its shape to that of the substrate. The anti-lock and key principle refers to enantioselective catalytic systems where the state of most unfavorable binding yields the preferred product. The entropy difference here determines the selectivity.

Selectivity

Chapters 2, 3, 4, 5, 6

There are many factors that determine the rate of a reaction sequence that lead to a particular product. Within the same catalytic system, reaction sequences leading to different products may compete. The two key parameters, which are important to the selectivity of a catalytic reaction, are the difference of the rate constants of elementary reaction steps controlled by electronic, geometric or steric parameters and the overlayer composition of the reactive catalytic surface or occupancy of complex or cavity. This affects the relative probability for product molecule formation from the recombination or dissociation of reaction intermediates generated during the catalytic cycle. The relative stability of the fragment molecules determines their concentration and, hence the probability that they are present at high enough concentration to result in a finite quantity for recombination. Site occupancy controls also the probability of surface vacancies necessary for dissociation. The last, for instance, is an important parameter that discriminates between associative and dissociative reaction steps. In addition, the differences in elementary rate constants determine also the selectivity.

Structure-Sensitive and -Insensitive Reactions

Ensemble Effect. Chapter 3

The surface structure can affect surface reactivity electronically by changes in the degree of coordinative unsaturation of the surface atoms, and geometrically by creating ensembles of surface sites with different topologies.

Elementary steps which proceed through transition states in which the product atoms remain bonded to the same metal atom are typically insensitive to the binding site topology. This occurs for the activation of C–H and N–H bonds in molecules. Elementary steps, however, that generate atoms or fragments that demand higher fold coordination sites as products are typically much more sensitive to surface binding site configurations, and hence structure sensitive. For instance, this tends to occur for reactions involving the activation of diatomics that are strongly bound to the surface such as CO and NO. Alloying a reactive metal with an inert metal decreases the size of the reactive metal surface ensembles that form. The activation of the adsorbed molecule is therefore suppressed.

There is an important difference in the activation barriers that result when the adatoms that initially form upon dissociation share bonding to one or more surface metal atoms and when they do not share bonding with the same metal surface atoms. The transition-state energy is substantially lowered for the forward reaction when product fragments do not bind to the same metal atoms. The barrier, however, is also reduced for the reverse recombination reaction. The reactivity of the (100) surface as compared with the (111) surface and the reaction at step or kink sites as compared with those on terrace sites two good examples of this. Reactions that proceed by activation over a single metal atom tend to be less structure sensitive. The activation barrier will change when the surface metal atom changes coordinative unsaturation. When the reactivity of the metal atom varies, the activation barriers for the forward and the reverse reaction move in different directions.

Lateral Interaction Effects and Surface Reconstruction

Chapter 3

When two (or more) adsorbed atoms bond to the same surface atom(s), they experience a repulsive interaction. When two adsorbed atoms bond to two different neighboring metal atoms that share a metal–metal bond, they tend to experience attractive interactions. These two rules can readily be deduced from the Bond Order Conservation principle which indicates that the atom-surface bond strength decreases with an increase in the number of adatoms bonded to the same surface metal atom. This change does not occur linearly with the number of neighboring atoms or molecules, but instead tends to vary exponentially.

The formation of an overlayer of adatoms or molecules can lead to reconstruction of the surface metal layers. This will reduce strain in the surface layer due to the altered metal–metal atom interactions. Often ordered surface phases are formed, in which the adatoms have reduced reactivity, because of the increased interaction with the reconstructed surface atom overlayer. The reactivity of the adsorbate overlayer is then limited to the boundary atoms of the overlayer surface islands. Once ordered overlayers are formed and the surface concentration of adatoms or molecules is further increased, bonding in the surface overlayer becomes weakened because more unfavorable bonding sites are occupied. The catalytically active surface species sometimes are the more weakly bonded species in the surface overlayer.

Orbital Symmetry Control

Chapter 3

Substrate bond cleavage reactions occur with low activation energies when the unoccupied adsorbate antibonding molecular orbitals become populated with a finite electron density in the transition state. This is the result of electron transfer from highest occupied states within the catalyst to the lowest unoccupied states of the reactant. Substrate bond cleavage and associative reaction steps often occur along a reaction coordinate that is nearly parallel to the surface so as to stabilize the interactions between the two fragment and the catalyst surface in the transition state. The corresponding antibonding orbital is therefore usually antisymmetric with respect to the surface normal. Electron transfer between this substrate orbital and the surface requires an interaction between this state and surface orbitals of the same symmetry. At a local level, the symmetry of substrate orbitals and local orbital fragments have to match. The corresponding local surface orbital fragments are called group orbitals. For a substrate chemical bond crossing a surface metal atom, the d_{xz} or d_{yz} atomic orbitals which are antisymmetric with respect to surface normal can interact with the antibonding unoccupied substrate orbital. The surface atomic d_{z^2}, s and p_z orbitals are symmetric. Hence, atop s-atomic orbitals do not interact with the antibonding substrate orbital fragment. This is only possible in a valley or bridging position, where the interaction occurs now with group orbitals, which are linear combinations of atomic orbitals. On metals with small d-orbital extensions, crossing over valleys tends to be preferred. The interaction with the antisymmetric s-atomic orbital combinations then dominates.

Universal Relationships

Chapter 3

Diatomic molecules adsorbed to transition-metal surfaces dissociate through tight transition states. A general Brønsted–Evans–Polanyi relationship can then be defined which is valid for nearly all diatomic molecules that dissociate along a similar reaction path:

$$E_{\mathrm{act}} = E_0 + \alpha E_{\mathrm{react}}$$

The parameters E_0 and α are adjustable.

Carbenium and Carbonium Ions in Zeolites and Solid Acids

Chapters 4 and 5

Transition states in proton-catalyzed reactions that occur in zeolites and other solid acids proceed through activated intermediates close to the carbenium or carbonium ions found in superacid solutions. A carbenium ion is a positively charged ion, that can be formed by the protonation of an alkene. The positive charge is localized on an sp^2-hybridized C atom. A carbonium ion is a protonated saturated alkane, that forms non-classical valencies such as a protonated σ C-C bond or a five-coordinated C atom. In zeolites, the positively charged intermediates are compensated for by the negatively charged zeolite framework and are therefore intrinsically unstable. Carbonium ions decompose into carbenium ions and neutral molecules. The carbenium ions prefer either to back-donate a proton to the zeolite lattice or adsorb as alkoxy species.

Lewis Acidity in Zeolites

Chapter 4

One can distinguish the reactivity of soft cations such as Zn^{2+} or Ga^{+} from that of hard cations such as Na^{+} or Ca^{2+} through the different mix of covalent and electrostatic interactions with these cations. Whereas probe molecules such as CO bind to the hard cations by induced polarization, their binding to soft cations is stronger as it involves additional covalent interaction with d-electrons of the cation. Hard cations can influence the reactivity by creating strong electrostatic fields which stabilize ionic transition states that form as the result of charge-transfer reactions. This is seen in photochemical radical reactions. Soft cations can activate C–H bonds heterolytically or homolytically. Zn^{2+}, for example, activates alkanes in a heterolytic manner, thus producing a $[Zn\text{-alkyl}]^{+}$ intermediate and a zeolitic proton. Ga^{+}, on the other hand, activates alkanes by oxidative addition, thus forming an $[HGa\text{-alkyl}]^{+}$ intermediate. The adsorbed alkyl intermediate can undergo a subsequent β-C–H scission reaction to form the alkene product.

Pretransition-State Configuration

Chapters 4 and 7

The most stable adsorption state of a reactant molecule on the surface of a catalyst, in the cavity of a zeolite, or within the cleft of an enzyme, typically does not have the appropriate configuration of the reactants necessary to allow them to react directly. Therefore, they typically have to reorient themselves before the reaction can proceed. This requires energy. On a metal surface, this is the energy required to bring fragments close together. This requires overcoming the repulsive lateral interaction energy, especially if they have to share surface metal atoms. In the zeolite cavity, the adsorbed molecules may have to rotate partially to reorient themselves, or else they can first form an intermediate with partially broken hydrogen bonds. In the enzyme, there is reorientation of the reactants to maximize their interaction with peptide functional groups, as well as an adaptation of the framework to optimize the interaction geometry. The small energetic cost of this pretransition-state orientation can have a large effect on the actual transition-state energies. The transition state are typically lowered substantially as a result of these more favorable reactant-catalyst structures.

Associative Versus Direct Mechanisms

Chapter 4.

The creation of a new bond between two different molecules or molecular fragments through consecutive reaction steps is typically called a direct reaction mechanism. In each individual reaction step, a chemical bond between the molecule or molecular fragment and the catalyst surface is formed. The molecular fragments bonded to the catalyst surface can subsequently react with a second adsorbed molecule. In an associative reaction mechanism, a cluster of at least two molecules adsorbs at the reaction center. Bond formation and cleavage reaction, now occur as a single event within the adsorbate cluster consisting of several molecules. This is assisted by transient chemical bond formation with the catalyst surface.

Diffusion in Zeolites
Chapter 4

Diffusion in most zeolites cannot be accurately described as Knudsen diffusion since the number of collisions between molecules is large compared with that with the micropore wall. Only in the wide-pore zeolite faujasite does the diffusion constant show Knudsen-type behavior which follows an $m^{-1/2}$ dependence. In the narrow and medium pore size zeolites, the diameter of the pore is on the order of the diameter of the molecules that we wish to study. The diffusion is then dominated by the interactions between molecule and zeolite wall. Once inside the zeolite, the molecules move in the shallow potential of the micropore, hence their rate of diffusion becomes mass independent. Diffusion inhibition occurs as the result of very narrow pore-size openings. The motion of the zeolite framework atoms controls the diffusion barrier.

If molecules can pass one another in the pore, then an increased pore occupation may increase the diffusion rates since molecule–molecule interactions may be weaker than molecule–zeolite channel interactions. When molecules cannot pass one another, the phenomenon known as single file diffusion may arise. In single file diffusion, which tends to occur in one dimensional porous systems, the large reduction in diffusion rate is represented by an effective diffusion constant proportional to the center of mass of the file of molecules occluded in the zeolite channel.

When a reaction becomes diffusion limited, the product molecules of a zeolite-catalyzed reaction equilibrate. No equilibration, however, can take place between the inner and outer parts of the zeolite. Equilibration within the zeolite occurs with different chemical potentials (due to the confined space) than in the gas or liquid phase. The product selectivity in such cases is controlled by inner zeolite equilibration.

Zeolite Medium Effects
Chapter 4

Activation of molecules by protonation depends on zeolite cavity shape and size. Charge separation is screened by the polarization of zeolite lattice oxygen atoms. The stability of intermediate cations strongly depends upon the steric inhibition due to zeolite curvature. The framework of cation-exchanged zeolites behaves as a Lewis base. Heterolytic reactions are promoted in which a proton is accepted by the basic zeolite framework.

Structure Dependence of Zeolite Reactivity
Chapter 4

The dominant interactions between the zeolite framework and the hydrocarbon are van der Waals dispersion interactions that take place between the polarizable zeolite oxygen atoms and the adsorbed hydrocarbon. These interactins are in addition to those between the hydrocarbon and the protons or cations which lead to the activation of the hydrocarbon. The overall van der Waals interaction between hydrocarbon and zeolite cavity depends strongly on the match of hydrocarbon size and shape and that of the zeolite cavity. As a consequence, at the same partial pressure and temperature, the micropore occupation of different zeolites may vary significantly for the same adsorbents. This has an important consequence on zeolite catalysis that depends on the concentration of reactant molecules adsorbed at reaction centers. Second, it will strongly affect the rates of diffusion which are strongly micropore occupation dependent.

Oxygen-atom Reactivity
Chapters 4 and 5

In oxides such as V_2O_5 or MoO_3, the M=O bond, which has a bond order of three, is strong, and can help to abstract hydrogen atoms from the reactants. The M=O bond, however, is too strong to be able to insert its oxygen atom into the reactant substrate. In contrast, cation-bridging oxygen atoms are readily inserted into hydrocarbon reactants. The bridging oxygen atom is more weakly bonded with a lower bond order. In addition, energy is gained because of oxide reconstruction upon the oxygen transfer. This can take on various forms. Locally, the octahedrally coordinated cations can be converted into tetrahedrally coordinated cations. In some instances there are more dramatic changes such as those which occur as a result of glide shear planes.

In addition, the bridging oxygen atoms of cationic clusters are highly reactive. In zeolites, the protonation energy of the oxygen atom bridging a cationic cluster occluded in the zeolite is higher than the protonation energy of the basic zeolite-lattice oxygen atoms. Oxygen atoms coordinated to cations of high valency are highly electrophilic and reactive (e.g. $[AgO]^+$). Oxygen atoms bonded to atoms in low-valency states are usually less reactive (e.g. AgOAg).

Proton Transfer Mechanism
Chapters 4 and 6

Since H_2O will readily form a hydronium H_3O^+ in the presence of a proton, the coadsorption of H_2O near a reaction complex in which heterolytic bond splitting occurs with generation of a proton can significantly lower the activation energy of such a reaction.

Instead of direct proton transfer to a basic Lewis oxygen atom, coadsorbed H_2O will more readily accept the proton and then transfer one of its other hydrogen atoms as a proton to the accepting oxygen atom. This two-step path can also be carried out simultaneously ,thus leading to more direct hydrogen shuttling path.

Hybridization at Transition-Metal Oxide/Sulfur Surface
Chapter 5

In transition-metal compounds such as metal oxides and metal sulfides, the chemical bonding of a transition-metal cation can be rationalized using concepts derived from organometallic chemistry such as atomic hybridization. A cation in a bulk metal compound which is surrounded octahedrally by six anions can be described by six hybridized d^2sp^3 oriented orbitals. A cation which is tetrahedrally coordinated is described by four sp^3 orbitals whereas that which is in a planar configuration is described by four sp^2d-hybridized orbitals. An approximate electronic structure diagram for the oxide can be constructed by considering that each oxygen atom contributes two electrons in an s-type atomic orbital. In an oxide, six bonding and six antibonding orbitals are then formed between an oxygen and a d^2sp^3-hybridized cation. When the hybridization is sp^2d or sp^3, four bonding and four antibonding orbitals are formed, respectively. The bonding orbitals can be considered doubly occupied. Then the d-orbitals are occupied with as many electrons as required for charge balance. In octahedral symmetry, three nonbonding d-orbitals are available. In tetrahedral and planar coordination, five and four d-orbitals, respectively, are available. At the surface, an oxygen anion vacancy creates an empty lone pair orbital which is directed towards the vacancy. On the surface oxygen atoms, occupied lone pair orbitals are formed. This creates surface Lewis acid and base sites.

Pauling Charge Excess
Chapter 5

Within the ionic chemical-bonding description, the bond order of a chemical bond is determined by the cation or anion charge divided by the number of nearest-neighbor atoms. In a stable system, the sum of the bond orders of the bonds from the coordinating counter ions should not differ by more than $\pm\frac{1}{6}$ from the cation or anion charge. This is the Pauling stability criterion. At a surface, the charge excess is much larger. A high positive charge excess implies Brønsted or Lewis acidity and a large negative charge excess implies Brønsted or Lewis basicity.

Synchronized Action
Chapter 7

Pretransition-state stabilization in enzyme catalysis occurs by a synchronized adjustment of multiple atom positions in the enzyme. The resulting multipoint adsorption of the reactant maximizes its interaction with the activating enzyme atoms or substituents. It also results in an overall entropy loss. This loss becomes larger as the fit between the reactant and the cavity created by the enzyme becomes tighter. Hence, pretransition stabilization occurs only with an optimum motion of the enzyme atoms so that the free energy is maximized

Influence of Aqueous Media
Chapter 6

Solvents can play an active role in promoting the reaction chemistry for a wide range of different reactions carried out in solution. Similar effects can also be found in surface-catalyzed reactions. In particular, protic solvents such as water help to stabilize the heterolytic dissociation of molecules into solution by stabilizing the charge on the anion and the cation that form. Water is both a H^+ acceptor and an OH^- donor. Water can act in a classical way to help stabilize the transition state of charged complexes over the uncharged reactant states in catalyzing a solution-phase reaction. More interestingly, water can actually take part in the chemistry by offering a conduit to conduct protons.

For reactions that occur over a surface carried out in a protic solvent, both homolytic and heterolytic bond activation steps are possible. Which route prevails depends upon various factors, which include the ability for the anion and the proton that form to migrate into solution, while thermodynamics would suggest that both the anion and the proton prefer to reside in the solution phase. This, however, is difficult to accomplish since an anion that forms has a relatively strong interaction via charge transfer from the anion into unfilled states of the metal. In addition, because of its size, the anion would also pay a high energy penalty for desorption into solution because of the cost associated with solvent reorganization. The protons, on the other hand, typically do not bond very strongly to a neutral surface. In addition, they are much more mobile than the anions and can therefore more readily shuttle off into solution.

A second alternative for the heterolytic activation path involves anion adsorption and proton dissolution. This results in the well-known double layer seen electrochemically. In order for this heterolytic path to proceed, the surface must be able to accommodate the extra electron and therefore the work function of the metal must be higher than 4.8 eV. The water here facilitates proton transfer.

Electrocatalysis
Chapter 6

The potential that results at an electrochemical interface can significantly influence the reactivity of the electrode surface. More negative potentials are more reducing whereas more positive potentials are more oxidizing. The influence of the potential can be accounted for by simply calculating the overall surface energies for homolytic reactions and then shifting the energies by the potential required to match the standard hydrogen electrode. A more rigorous analysis of the influence of the electrochemical potential, however, indicates that the applied potential can significantly polarize the surface in the presence of solution and thus lead to more enhanced changes in the chemical bonding to the surface as well as surface reactivity.

The reactions for C–H, O–H and S–H bond activation typically involve metal atom insertion reactions. In the presence of solution, the hydrogen that forms can be directly transferred into solution as proton. The site dependence for these reactions, which are at the heart of many electrochemical processes, may not be very strong. The reactivity of terraces, steps and kinks may be quite similar. This is different to the activation of the molecules over a metal in the gas phase, which is structure sensitive. The electrochemical behavior will, of course, be strongly dependent upon the potential.

An alcohol or acid can adsorb at an electrode as an alkoxide or a carboxylate anion respectively, whereas the proton is accepted by H_2O. One can consider this as H_2O assisted chemisorption. On the other hand, if H_2O dissociation is rate limiting, as is the case for oxidizing conditions, OH surface species are found. The reactivity will then strongly depend on the presence of steps or kinks.

Inner- and Outer-Sphere Reactions
Chapter 6

Coordination complexes in a solvent tend to have two reaction modes:

- Inner-sphere reactions:
  Chemical reaction occurs between molecules or ions, coordinated to the central metal atom, that are in the first coordination shell of the complex. Reacting molecules are in direct contact with the cationic center. Bond formation or cleavage can occur by direct (non-cation intermediated) reactions between ligands (such as H^+ transfer) or are activated by the redox center.
- Outer-sphere reactions:
  Reactions occur by the interaction of a molecule or in the solvent with a molecule ion that is a ligand of the coordination complex.

Evolutionary Adaptive Synthesis Processes
Chapter 8

The molecular recognition of a template by a self assembling system leads to preferential assembly of a unique synthesis reaction intermediate. Through nucleation processes, the intermediate undergoes self assembly with other intermediates and results in the formation of cavities or channels in a substance uniquely related to the template. When the template resembles a particular transition state, the material may have unique selectivity with respect to the corresponding reaction.

Self Assembly
Chapter 8

Self assembly involves the organization of molecules in a cluster or ordered system. The structure of the self assembled state is ordered and molecules interact through weak hydrophobic or hydrophilic interactions. Structural aspects are often important. Self assembly is a process that is driven by thermodynamics. The final state is a local or absolute minimum free energy state.

Self Organization
Chapter 8

Stationary oscillating states arise far from equilibrium when autocatalytic elementary reaction steps are part of the catalytic cycle. Synchronizing the phase of catalytic reaction cycles in different parts of the catalyst is assisted by delaying events such as self organization. Overall reaction rates may show temporally varying phenomena such as oscillations or chaotic time dependence. On the surface of the catalyst, patterns in the form of spirals or replicating pulsars can occur. It is a property of so-called complex systems, in which the components interact in a specific way. Excitable systems are built from autocatalytic reactions that amplify signals and contain an inhibition reaction. When activated beyond a particular threshold, the system may show complicated time-dependent behavior. When the components have different diffusion rates, the system can self organize in complicated patterns that may show replicating features.

Self Repair
Chapter 8

In addition to the changes in the chemical bonds of the adsorbates that occur during a catalytic cycle, there are also changes in the chemical bonds of the catalyst that take place. Some bonds are weakened whereas others are broken. These bonds must be restored upon completing the catalytic cycle in order for the cycle to continue. This catalytic process then is one which must contain self repair.

The Complexity of the Catalytically Reactive Phase
Chapters 2, 3 and 8

Whereas defects such as kinks or steps are often sites of unique reactivity, the reactive phase of a heterogeneous catalyst is often locally disordered and transient in character. In self organizing systems, these local events can be ordered in time and space. During the course of the catalytic cycle, adsorbed molecules are converted into fragments and subsequently transferred into product molecules which ultimately desorb. In addition to these molecular rearrangements and transformations, the atoms on the surface of the catalyst can also become displaced as a result of local and long-range reconstruction processes. Diffusion, adsorption, desorption, dissociation and recombination reactions can all occur with quite different demands on the local arrangement of atoms about the reactive center where a specific reaction occurs and its environment.

Templated Catalyst Synthesis
Chapters 8 and 9

Organic cations are able to complex silicate anions to form silicate complexes. These complexes can condense to form ultimately microporous zeolite structures. Different templating molecules may form different microcavities. Hence zeolite synthesis is an example of template-directed synthesis. Mesoporous materials can be prepared by using nanometer-sized micelles preorganized as liquid crystals as a template rather than a single molecule template.

Artificial Catalytic Chemistry
Chapter 9

Computational models that use cellular automata to simulate the reproduction of primitive cells, are often composed of:

- A metabolic system of autocatlytic molecules
- Self replicating molecules that inherit genetic information
- A self organizing membrane molecule to close the system.

Such models can be developed for the computational design of catalytic systems that self organize and adapt themselves for optimum catalytic performance. Adaptation occurs in the reproduction process with mutation of the self-replicating molecules coupled to the metabolic system. The metabolic system acts as the bio-immune molecular recognition and response system. The conditions for the emergence of such a system are far from equilibrium in the complex regime. The behavior as a function of time is unpredictable, similar to the class 4 system proposed by Wolfram.

Biomineralization
Chapter 9

Biomineralizatoin involves the exploitation of liquid crystal or self ordering properties of amphiphilic molecules or polymers to design inorganic materials with porous structures ordered over several dimensional length scales.

Aggregation Kinetics
Chapter 9

The non-template-controlled synthesis of amorphous porous materials occurs through aggregation processes in which the relative rate of chemical bond formation and rate of component diffusion compete. The aggregates initially formed often have fractal properties that depend on the ratio of these two parameters. Ultimate pore formation occurs via a secondary aging process in which the walls densify.

APPENDICES
Computational Methods

Introduction

A comprehensive understanding of the electronic, molecular, micro- and meso-scale issues associated with modeling catalytic processes requires a multiscale approach in order to integrate:

1. the electronic and structural changes that govern the intrinsic reaction steps
2. the dynamics of adsorbates in the adlayer and the atoms in the catalytic surface
3. the kinetics for the physicochemical adsorption, reaction and diffusion processes
4. the fluid dynamics along with heat and mass transfer in the reactor
5. the deactivation to the catalyst over time.

This would cover changes in time-scale which range from 10^{-15} sec for electronic transitions to months and years for deactivation phenomena. In addition, this spans a range of length scales that cover ångstroms to meters. These same multiscale issues are apparent in a number of other engineering systems where chemistry is important. Multiscale modeling has, therefore, been a subject of much interest. Our focus in this book is predominantly on the intrinsic physical chemistry and the operative catalytic kinetics. This limits the scope to understanding structure and dynamics of the electronic, atomistic and microscopic scales. This requires the integration of electronic structure methods to establish the catalytic reaction steps and the influence of the local environment, along with atomistic methods in order to simulate kinetics, dynamics and equilibrium.

In the three Appendices A–C, we provide a broad overview of electronic structure, atomistic, and kinetic simulation methods along with references for readers who are interested in more detailed discussions. In general, electronic structure methods are used to solve the Schrödinger equation subject to a series of fundamental approximations. Schrödinger's equation describes the state of the many body N-electron system and its corresponding energy. The solution thus provides information on the electron states in the system, along with the population of these states with electrons. Schrödinger's equation can ultimately be used to calculate a wide range of different properties that are based on the electronic structure of the system including relative energies, geometric structure, spectroscopic signatures, and reactivity. The fundamental entity in electronic structure methods is the electron and is based on quantum mechanics. The system is described by the fundamental forces that act upon the electron. Quantum mechanical methods can typically be characterized as either semiempirical or ab initio. In general, ab initio methods can be subdivided into wavefunction methods and density functional theory. Quantum mechanical methods are necessary for modeling electron transfer processes or chemical reaction steps since they require modeling of the changes that occur to electronic structure. Electronic structure calculations, however, are only practical for systems which have fewer than 10^3 atoms for the highest level computing systems or a few hundred atoms on more conventional computing systems.

The fundamental entity in atomistic or molecular simulations is the atom or the molecule and is based on the fundamentals of statistical mechanics. The detailed electronic structure is no longer present, thus preventing the treatment of electron transfer, bond breaking and bond making processes. The system is instead described by the forces that act upon the atoms or molecule. This significantly lowers the CPU cost on a per atom basis, thus allowing the simulation of 10^6–10^7 atoms. Atomistic simulation can be

Molecular Heterogeneous Catalysis. Rutger Anthony van Santen and Matthew Neurock

ISBN: 3-527-29662-X

divided into methods for simulating equilibrium properties including structure, sorption and phase behavior, dynamic properties such as diffusivity and thermal conductivity, and kinetic properties such as reactivity.

A: ELECTRONIC STRUCTURE METHODS

1. General overview

The goal of quantum mechanical methods is to predict the structure, energy and properties for an N-particle system, where N refers to both the electrons and the nuclei. The energy of the system is a direct function of the exact position of all of the atoms and the forces that act upon the electrons and the nuclei of each atom. In order to calculate the electronic states of the system and their energy levels, quantum mechanical methods attempt to solve Schrödinger's equation. While most of the work that is relevant to catalysis deals with the solution of the time-independent Schrödinger equation, more recent advances in the development of time-dependent density functional theory will be discussed owing to its relevance to excited-state predictions.

The following discussion on electronic structure methods is rather general in order to provide a simple overview. More in-depth discussions can be found in a number of very good references, including [1–11].

The time-independent Schrödinger equation is

$$\hat{H}\Psi = E\Psi \tag{A1}$$

where Ψ is the wavefunction and E is the energy of the N-particle system. $\hat{H}$ is the Hamiltonian operator, which is comprised of the kinetic and potential energy operators which act on the overall system wavefunction Ψ. The wavefunction can extend between $+\infty$ and $-\infty$ and depends upon the positions of the atoms in the system along with the spin of each electron. The square of the wavefunction (Ψ^2) describes the probability distribution for the N-particle system. The Schrödinger equation is actually nothing more than a force balance on the electrons and the nuclei of the system. The Hamiltonian is comprised of two essential terms, the kinetic energy operator, $\hat{T}$, and the potential energy operator, $\hat{V}$. For an N-particle system, these operators can be written as

$$\hat{T} = -\sum_{i=1}^{N} T_i = -\sum_{i=1}^{N} \frac{\hbar}{2m_1}\nabla_i^2 = -\sum_{i=1}^{N} \frac{\hbar}{2m_i}\left(\frac{\partial^2}{\partial x^2} + \frac{\partial^2}{\partial y^2} + \frac{\partial^2}{\partial z^2}\right) \tag{A2}$$

$$\hat{V} = \sum_{i=1}^{N}\sum_{j>i}^{N} V_{ij} \tag{A3}$$

where T_i is the kinetic energy of particle i and V_i and V_j refer to the potential energy terms for electronic interactions between electron–electron, electron–nuclei and, nuclei–nuclei interactions.

A number of simplifying approximations are required to solve this N-particle system, as will be discussed later. The first is the Born–Oppenheimer approximation, which indicates that since the mass of an electron is nearly 2000 times smaller than the mass of a proton, the electrons move many orders of magnitude faster than nuclear motion. Therefore, the electronic motion can be strictly decoupled from the nuclear motion. The electronic wavefunction can then be solved separately for a fixed set of nuclear positions (R). The hamiltonian for the electronic system now becomes

$$\hat{H} = -\sum_{i}^{n} \frac{\hbar}{2m_i} \nabla_i^2 - \sum_{i}^{n}\sum_{a}^{N} \frac{Z_a}{|r_i - R_a|} e^2 + \sum_{i}^{n}\sum_{j>i}^{n} \frac{1}{|r_i - r_j|} e^2$$

Kinetic Nuclear–Electron Attraction Electron–Electron Repulsion

$$+\sum_{a}^{n}\sum_{b>a}^{n} \frac{Z_a Z_b}{|R_a - R_b|} e^2 \qquad \text{(A4)}$$

Nuclear–Nuclear Repulsion

The first term of the hamiltonian describes the kinetic energy of the electron. The second term describes the attractive interaction between the electron and the nuclei where r_i and R_a refer to the positions of electron i and atom a. The number of electrons is defined as n and the number of nuclei as N. The third term describes electron–electron repulsion. The final term refers to the nuclear–nuclear repulsive interactions. Since the nuclear charges are decoupled from the electronic wavefunction, this summation can be computed in a straight forward manner and does not change upon the solution to the electronic structure.

The Born–Oppenheimer approximation[1] is usually a very good approximation since the nuclear mass is so much greater than the electronmass[2,3]. Uncoupling the electronic motion from the nuclear motion enables one to solve for the electronic structure for a fixed set of nuclei. The final term, which describes electron–electron repulsion, prevents the direct solution to the electronic structure. The solution requires convergence of the electronic structure via an iterative scheme. This is known as the self-consistent field approximation, which is discussed later[3,4].

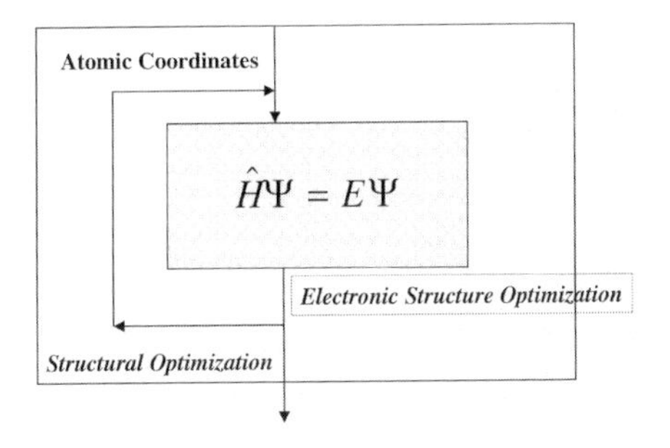

Figure A1. The lowest level hierarchical structure for most quantum mechanical computational algorithms. The inner loop is used to converge the self-consistent field in order to establish the electronic structure to within a user-defined tolerance. The outer loop is used to optimize the structure to within a defined geometric tolerance.

Nearly all quantum mechanical codes are comprised of the basic structure shown in Fig. A1, where there are two primary nested loops. The inner loop requires the convergence of the electronic structure for a fixed set of atomic positions. The outer loop moves the atoms in order minimize the forces upon each atom and to converge upon the lowest energy of geometric structure for the system.

The initial atomic positions are necessary to describe the starting structure. The structure studied can be that of a simple molecule, a macromolecule, a bulk metal or metal oxide, the unit cell of a zeolite, a complex system comprised of catalytic surface along with adsorbates, solution molecules and ions, etc. Structures can be described in terms of Cartesian coordinates, direct coordinates, or z-coordinates[5]. The electronic structure can be mathematically represented by an infinite number of basis functions. More practically, these functions are truncated and described by a finite number of basis sets. A wide range of different basis sets currently exist and depend on the solution method used, the type of problem considered and the degree of accuracy required for solution. These functions can take on one of several forms, including Slater-type functions, Gaussian functions, and plane waves[6]. An infinite number of basis functions would be required for true accuracy and the complete electronic structure. This can be relaxed to a finite number of basis functions with some potential loss in accuracy. The number of basis functions used is then based on the relative degree of accuracy that one desires along with the CPU expenditures required to calculate.

The structural positions of the atoms and their basis functions are the only chemically specific input. Of course, there are typically a number of additional variables which describe the state of the system (the number of electrons, orbital occupations and the charge of the system), the type of calculation performed (a single point calculation, geometry optimization, a transition-state search, a molecular dynamics simulation), instructions on how the calculation is performed (the electronic and geometric convergence criterion, the density mixing scheme), the relative accuracy of the calculation (expanded basis sets, increased energy cutoffs, etc.), and the specifications on output that are requested (orbitals, population analyses, density of states, frequencies, thermodynamic properties, etc.) that are all required.

2. The Potential Energy Surface

The output from the simulation of an optimized molecular structure includes the optimized atomic coordinates, which define the optimal structure, the optimized electronic structure for these specific coordinates and the total energy for this system. A schematic of the optimization of the O_2 in the gas phase is shown in Fig. A2.

The energy for the O_2 molecule is plotted at its initial starting geometry (i.e. $\lambda =$ O–O distance $= 2$ Å) The electronic structure is calculated and subsequently used to determine the forces on each atom for the particular state that is being probed. These forces are then used to determine the new positions of the atoms in the system. This process is repeated in order to converge upon the energy for the optimized geometric structure. The results can be used to determine a host of different molecular properties, including electron density, electron affinity, ionization potential, relative energies for reaction processes, vibrational spectroscopy and a wide range of other chemical properties.

As we move from the one-dimensional O_2 example shown in Fig. A2, the potential energy surface becomes much more complicated.
Figure A3, which was taken from Foresman and Frisch[5], depicts the presence of local and global minima as well as local and global maxima. The local, and a global minima occur when the derivative of the energy with respect to the structural degree of freedom λ_i is zero for all degrees of freedom $\lambda_i(\mathrm{d}E/\mathrm{d}\lambda_i = 0)$. Transition states occur at saddle points

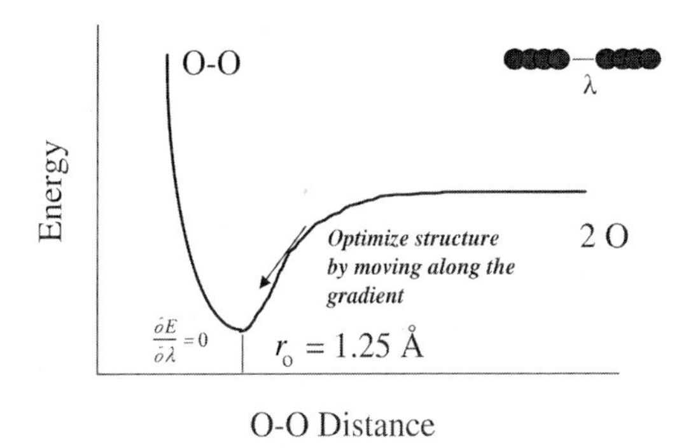

Figure A2. A schematic which shows the one-dimensional potential energy surface for O_2. The single defining internal coordinate, λ_i, is the distance between oxygen atoms. The energy is minimized when its derivative, with respect to changes in its Cartesian or its internal atomic coordinates, is zero ($dE/d\lambda_i = 0$) and the second derivative of energy, with respect to changes in its Cartesian or internal atomic coordinates, is greater than 0 ($d^2E/d\lambda_i^2 > 0$).

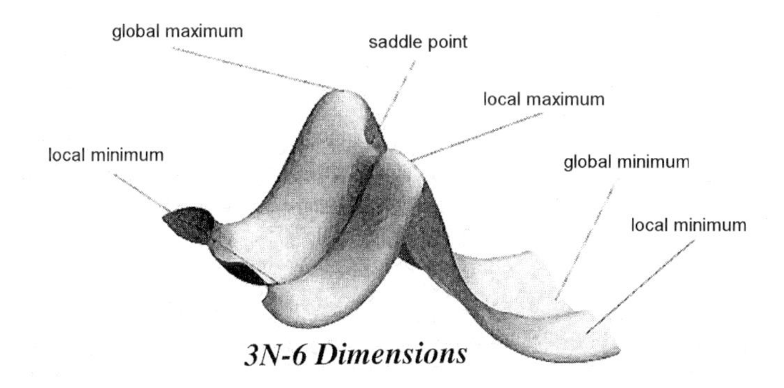

Figure A3. A more complex three-dimensional potential energy surface. The surface displays a global maximum and minimum ($dE/d\lambda_i = 0$) and transition (or saddle) points $d^2E/d\lambda_i{}^2 > 0$ for all modes, λ_i, except the reaction trajectory, which instead is defined as $d^2E/d\lambda_i{}^2 < 0$. The graph is reprinted from reference [10].

along the potential energy surface[2,4,5]. The derivative of the energy with respect to the degree of freedom λ_i is zero for all degrees of freedom for transition-state structures. In addition, the second derivative of the energy with respect to the degree of freedom λ_i is equal to zero for all degrees of freedom λ_i except for the mode which corresponds to the reaction coordinate.

3. General Electronic Structure Methods

Electronic structure methods can be categorized as ab initio wavefunction-based, ab initio density functional theoretical, or semiempirical methods. Wavefunction methods start with the Hartree–Fock (HF) solution and have a well-prescribed methods that can be used to increase its accuracy. One of the deficiencies of HF theory is that it does not treat electron correlation. Electron correlation is defined as the difference in the energy between the HF solution and the lowest possible energy for the particular basis set that is used. Electron correlation refers to the fact that the electrons in a system correlate their motion so as to avoid one another. This physical picture then points out the deficiency of describing electrons in fixed orbital states. The electrons in reality should be further apart than predicted by HF theory. An exact solution of the Schrödinger equation requires the full treatment of electron correlation along with complete basis sets. Although this is unachievable, the breakdown of the inaccuracies into correlation and basis set expansion provides for a well-prescribed way in which to improve continually the accuracy and approach the exact wavefunction for the N-particle system.

Density functional theory is also derived from first principles but is fundamentally different in that it is not based on the wavefunction but instead on the electron density of the N-particle system[7]. Hohenberg and Kohn[8] showed that the energy for a system is a unique functional of its electron density. The true exchange-correlation functional necessary to provide the exact DFT solution, however, is unknown. The accuracy of density functional theory (DFT) is then limited to quality of the exchange-correlation functional that is used.

Semi-empirical methods avoid the solution of multicenter integrals that describe electron–electron interactions and instead fit these interactions to match experimental data [4,9,10]. We will only discuss ab initio wavefunction and DFT methods here as they are more reliable for calculations concerning heterogeneous catalytic surfaces.

4. Ab Initio Wavefunction Methods

A series of general approximations are necessary in order to solve the Schrödinger equation. We have already introduced the Born–Oppenheimer and the time-independence approximations, which indicate that the energy of the system can be determined by solving for the electronic wavefunction.

4.a. Hartree–Fock Self Consistent Field Approximation

The self-consistent field approximation, which was briefly introduced earlier, is used to reduce the N-electron problem into the solution of n-single-electron systems. It reduces a $3n$ variable problem into n single electron functions that depend on three variables each. The individual electron–electron repulsive interactions shown in Eq. (A4) are replaced by the the repulsive interactions between individual electrons and an electronic field described by the spatially dependent electron density, $\rho(r)$. This avoids trying to solve the difficult multicenter integrals that describe electron–electron interactions. The only trouble is that the electron density depends upon how each electron interacts with it. At the same time, the electron interaction with the field depends upon the density. A solution to this dilemma is to iterate upon the density until it convergences. The electron density that is used as the input to calculate the electron-field interactions must be equivalent, to within some tolerance, to that which results from the convergence of the electronic structure calculation. This is termed the self-consistent field (SCF).

This approach used in solving for n molecular orbitals within a self-consistent field is known as the Hartree–Fock solution[11,12]. The molecular orbitals are the individual electronic states that describe the spatial part of the molecular spin orbital[3]. Electrons are

Multibody electronic interactions cannot be calculated

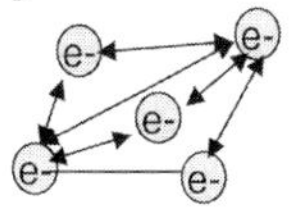

Guess the resultant field (electron density)

Calculate the interactions between individual electrons and the field

Convergence on the electron density

Figure A4. The general self-consistent technique for solving the electronic structure via an iterative approach.

fermions and have non-integral spin. The wavefunction must therefore be antisymmetric with respect to the exchange of spin, that is, $\Psi = -\Psi$. The Slater determinant shown in Eq. (A5) provides the simplest wave function with the correct antisymmetry.

$$\Psi(x_1, x_2, \ldots\ldots x_n) = \frac{1}{\sqrt{N!}} \begin{vmatrix} \psi_1(x_1) & \psi_2(x_1) & \ldots & \psi_n(x_1) \\ \psi_1(x_2) & \psi_2(x_2) & \ldots & \psi_n(x_2) \\ \ldots & \ldots & \ldots & \ldots \\ \psi_1(x_n) & \psi_2(x_n) & \ldots & \psi_n(x_n) \end{vmatrix} \tag{A5}$$

The N-electron Schrödinger equation is now reduced to n single-electron problems that take the following form:

$$\hat{h}_i \psi_i = \varepsilon_i \psi_i \tag{A6}$$

$$\left[-\frac{1}{2} \nabla^2 + V_C(r) + \mu_x^i \right] \psi = \varepsilon_i \psi_i \tag{A7}$$

where ψ_i refer to the individual molecular orbitals. The single-electron Hamiltonian, which is shown here in brackets, depends upon the electronic distribution within the molecular orbitals Ψ_i. This is what leads to the SCF solution scheme where the electrons simply interact with an average potential. In this solution scheme, electron correlation which describes the interactions between electrons is not included. This results in much of the errors associated with the Hartree–Fock solution.

4.b. Basis Set Approximation

The molecular orbitals can be described by a linear combination of atomic orbitals (X_i) as follows

$$\varphi_i(r) = \sum_{j=1}^{N_{basis}} C_{ij} \chi_j(r) \tag{A8}$$

where C_{ij} is a coefficient which relates the atomic orbital j to molecular orbital i. This is known as the basis set approximation[14]. More generally, the basis functions presented in Eq. (A8) do not have to be atomic orbitals but can simply be a series of basis functions used to describe the molecular orbitals. Atomic orbitals tend to be the most natural choice of basis functions for molecular-based systems. Gaussian- or Slater-type basis functions are often used because they are easier to solve for computationally[2,4]. Solid-state systems described by periodic methods, on the other hand, are more naturally represented by using periodic plane wave basis functions[13]. In theory, the most accurate solution would require an infinite number of basis functions. Instead, the number of basis functions is truncated to a smaller set which is still able to capture the essential features of the wavefunction. The accuracy can improve by increasing the number and extent of the basis orbitals[14]. In the atomic basis scheme, for example, one can increase the number of basis functions on each atom to increase the size and spatial extent. In addition, polarization and diffuse functions can also be added to improve the displacement of electron density away from an atom in a particular environment as well as its spatial extent[2,4]. In periodic systems, the number of plane waves must be expanded[13].

Full ab initio treatments for complex transition metal systems are difficult owing to the expense of accurately simulating all of the electronic states of the metal. Much of the chemistry that we are interested in, however, is localized around the valence band. The basis functions used to describe the core electronic states can thus be reduced in order to save on CPU time. The two approximations that are typically used to simplify the basis functions are the frozen core and the pseudopotential approximations. In the frozen core approximations, the electrons which reside in the core states are combined with the nuclei and frozen in the SCF. Only the valence states are optimized. The assumption here is that the chemistry predominantly takes place through interactions with the valence states. The pseudopotential approach is similar.

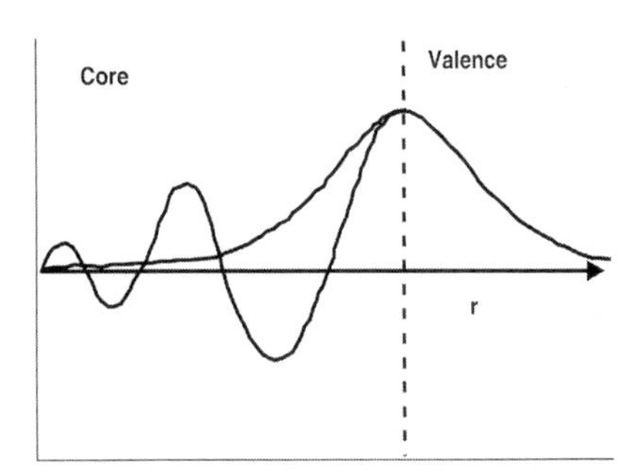

Figure A5. A schematic showing the comparison of the full electron wavefunction and the pseudopotential-derived wave function. The strongly bound core electrons are replaced by a smoother analytical function. This schematic was adapted from Payne et al.13].

The valence electrons oscillate in the core region as is shown in Fig. A5, which is difficult to treat using plane wave basis functions. Since the core electrons are typically insensitive to the environment, they are replaced by a simpler smooth analytical function inside the core region. This core can also now include possible scalar relativistic effects. Both the frozen core and pseudopotential approximations[13,15,16] can lead to significant reductions in the CPU requirements but one should always test the accuracy of such approximations.

4.c. Hartree–Fock Solution Strategy

The single-electron wave equations from Eq. (A6) can be written in a more compact matrix form as the following equation:

$$F^t C^t = SC^t \varepsilon \tag{A9}$$

where F^t, C^t and S refer to the Fock, orbital coefficient and orbital overall intergral matrices, respectively. ε is a diagonal matrix which is comprised of the molecular orbital energies[5,16,17]. Hall[17] and Roothaan[18], simultaneously, proposed a solution strategy to solve the Hartree–Fock system based on the following secular equations:

$$\sum_{\nu=1}^{N} \left(F_{\mu\nu} - \varepsilon_i S_{\mu\nu}\right) c_{\nu i} = 0 \tag{A10}$$

where $F_{\mu\nu}$ refers to the Fock operator elements, $H_{\mu\nu}$ are the Hamiltonian elements, $S_{\mu\nu}$ are the overlap integrals for electrons in orbitals μ and ν, and $C_{\nu i}$ are the molecular orbital coefficients. These matrix elements are defined by the following equations:

$$F_{\mu\nu} = H_{\mu\nu}^{core} + \sum_{i}^{N/2} \sum_{\lambda}^{N_{basis}} \sum_{\sigma}^{N_{basis}} C_{\nu i} C_{\sigma i} \Big[2(\mu\nu|\lambda\sigma) - (\mu\sigma|\lambda\nu)\Big] \tag{A11}$$

$$H_{\mu\nu}^{core} = \int \mathrm{d}\bar{r}_1 \chi_\mu(\bar{r}_1) h(\bar{r}_1) \chi_\nu^*(\bar{r}_2) \tag{A12}$$

$$S_{\mu\nu} = \int \mathrm{d}\bar{r}_1 \chi_\mu(\bar{r}_1) \chi_\nu^*(\bar{r}_1) \tag{A13}$$

The terms $(\mu\nu|\lambda\sigma)$ and $(\mu\sigma|\lambda\nu)$ are electron repulsion integrals:

$$J_{ij} = (\mu\nu|\lambda\sigma) = \int\int \mathrm{d}\bar{r}_1\, \mathrm{d}\bar{r}_2\, \chi_\mu(\bar{r}_1)\chi_\nu^*(\bar{r}_1) \frac{1}{r_{12}} \chi_\lambda(\bar{r}_2)\chi_\sigma^*(\bar{r}_2) \tag{A14}$$

$$K_{ij} = (\mu\sigma|\lambda\nu) = \int\int \mathrm{d}\bar{r}_1\, \mathrm{d}\bar{r}_2\, \chi_\mu(\bar{r}_1)\chi_\sigma^*(\bar{r}_1) \frac{1}{r_{12}} \chi_\lambda(\bar{r}_2)\chi_\nu^*(\bar{r}_2) \tag{A15}$$

which more specifically refer to the Coulomb (J_{ij}) and exchange (K_{ij}) interaction between an electron and other electrons in the system.

The specific orbital energy levels can be written in terms of core Hamiltonian elements along with the Coulomb and exchange energies as follows:

$$\varepsilon_i = H_{ij}^{core} + \sum_{j=1}^{N/2} (2J_{ij} - K_{ij}) \tag{A16}$$

The total energy of the ground-state system can then be written as:

$$E_{HF} = \frac{1}{2} \sum_{\mu=1}^{N} \sum_{\nu=1}^{N} P_{\mu\nu} \left(F_{\mu\nu} + H_{\mu\nu}^{core}\right) + \sum_{a \neq b}^{nucl} \frac{q_a q_b}{|R_a - R_b|} \tag{A17}$$

where P is the charge density matrix which is made up of the elements, $P_{\lambda\sigma}$, which are comprised of the orbital coefficients $C_{\lambda i}$ and σi evaluated over all occupied orbitals:

$$P_{\lambda\sigma} = 2 \sum_{i=1}^{occupied} C_{\lambda i} C^{*}_{\sigma i} \tag{A18}$$

The spatial electron density $\rho(r)$ is defined by the density matrix elements as follows:

$$\rho(r) = \sum_{\mu=1} \sum_{\nu=1} P_{\mu\nu}\, \phi_{\mu}(r) \phi_{\upsilon}(r) \tag{A19}$$

The solution strategy for solving for the self-consistent field and the final energy in the basic Hartree–Fock theory is shown in Fig. A6. The user starts with a simple guess for the initial $\rho(r)$ density or the orbital coefficient matrix, C. The coefficient matrix can then be used to calculate the Fock elements. The Hamiltonian and overlap elements are also computed and used to solve the Roothan–Hall equations. This results in a new set of orbital coefficients along with the overall energy for the system. The new orbital coefficients are used to calculate new density and Fock matrix elements along with a new system energy. The procedure continues until the calculated density (orbital coefficients) is the same as that which was used in the input to the problem. The results ultimately provide the electron density, orbital overlap and the final energy state levels of the system.

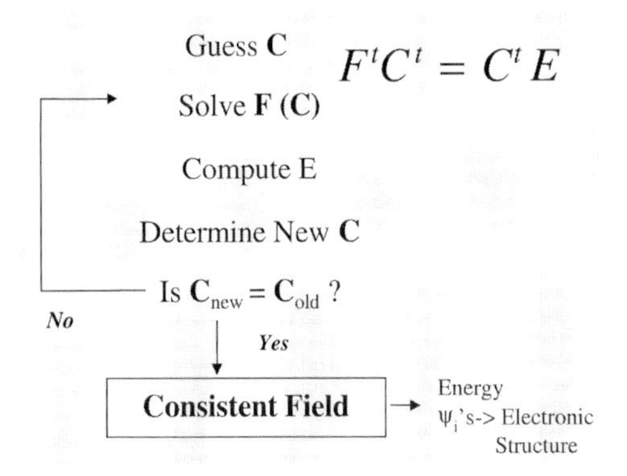

Figure A6. Schematic illustration of the basic solution strategy for solving for the self–consistent field and the final energy in Hartree–Fock methods.

For more in-depth discussions the reader is referred to texts by Jensen [2], Szabo and Ostlund[18], Levine[3], and Leach [4]

5. Advanced Ab Initio Methods

The Hartree–Fock solution strategy avoids the direct solution of electron–electron interactions but instead replaces these interactions by a mean field approach. This ignores the fact that the motion of individual electrons may be correlated. The schematic shown in Fig. A7 indicates that as the electron from point 1 moves toward point 2 , the electron at point 2 would likely move due to repulsive interactions between the two. The electrons therefore should have correlated motion. By definition, the difference in the energy calculated by Hartree–Fock theory for a specific basis set which treats the systems as a mean field (without correlation) and the exact energy is the correlation energy. There are two primary strategies for treating correlated motion between electrons. Electrons with the same spin behave differently to electrons with opposite spins. The basic Hartree–Fock theory already includes the treatment of electrons with the same spin by virtue of the fact that the wavefunction is required to be antisymmetric. Hartree–Fock theory, however, does not treat appropriately the interaction of electrons which have opposite spins. The wavefunction of the system, Ψ, cannot be described by a single determinant.

Three general approaches have been developed to treat electron correlation:

1. Configurational Interaction (CI),
2. Many Body Perturbation Theory
3. Couple Cluster (CC) theory.

The methods are briefly described below. More detailed discussions on each of these methods can be found elsewhere[2,14].

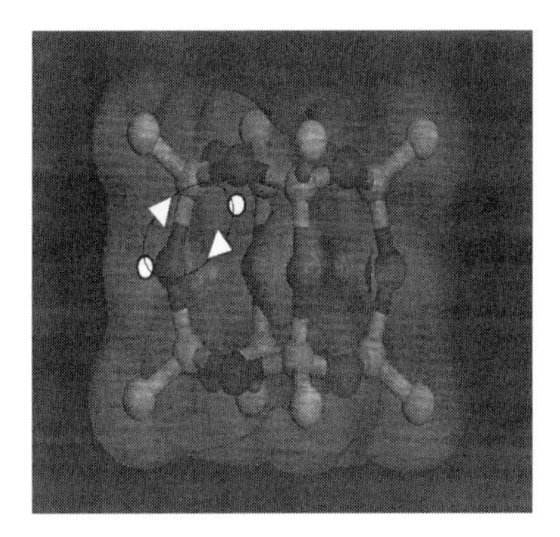

Figure A7. A schematic cartoon illustrating the basic idea behind electron correlation. The movement or position of electron 1 should be correlated with the movement or position of electron 2 owing to the repulsive interactions that exist between the two.

5.a. CI Methods

The general solution strategy for CI methods is to construct a trial wavefunction that is comprised of a linear combination of the ground- state wavefunction Ψ_0 and excited-state wavefunctions Ψ_1, Ψ_2, etc. The trial wavefunction is shown in Eq. (A20), along with possible excited states Ψ_1 and Ψ_2:

$$\Psi = C_0\Psi_0 + C_1\Psi_1 + C_2\Psi_2 + \dots \tag{A20}$$

The trial wavefunction can include the exchange of 1, 2 or 3 electrons from the valence band into unoccupied orbitals; these are known as CI singles (CIS), CI doubles (CID) and CI triples (CIT), respectively. CIS, CISD, and CISDT are methods configurational

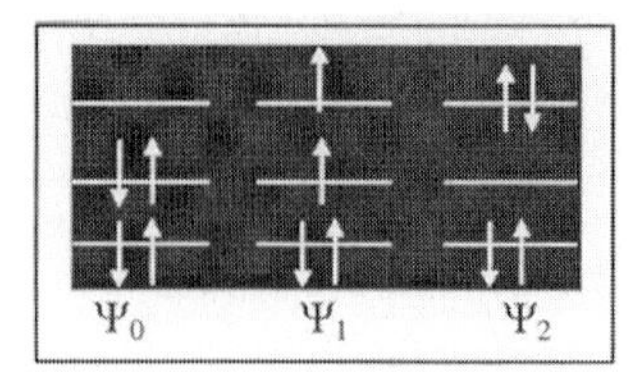

Scheme A1. Groundstate Ψ_0, single-excited state Ψ_1, double-excited state Ψ_2

interaction methods which allow for single, single/double and single/double /triple excitations (see Scheme A1). All of these methods are based on the variational principle, which allows one to optimize the coefficients before each of the trial determinants shown in Eq. (A20). A full configurational interaction exchange (Full-CI) would involve all possible electron substitutions into the full manifold of occupied and unoccupied states. The Full-CI expression is Eq. (A21)

$$\Psi_{CI} = C_0\Psi_{SCF} + \sum_{s} C_S\Psi_S + \sum_{D} C_D\Psi_D + \sum_{T} C_T\Psi_T + \ldots \tag{A21}$$

where C_S, C_D, and C_T refer to the coefficients for the singly, doubly or triply excited states. The actual wavefunction now contains contributions from the ground state wave function Ψ_0 and all of the other possible determinants. In addition to expanding the number of potential states, the coefficient multipliers for each state, C_i, can be optimized by variationally minimizing the energy. Full-CI calculations are computationallydemanding, and therefore, full-CI is possible only on very small systems. The multireference framework, however, provides a well-defined scheme for systematically improving the level of accuracy. Since these calculations are variationally optimized, the solutions should approach an accurate solution as the number of excitations increase. The full-CI then should begin to provide exact solutions within the limit of the basis set expansion. This will overcome the mean field approximation that is introduced in using Hartree–Fock.

All of the CI methods described so far are considered single determinant wavefunctions. Multiconfigurational SCF methods use multiple determinants[2,14]. In these methods, the coeffficients that multiply each state in Eq. (A21) and also the molecular orbital coefficients used to construct the determinants must also be optimized. This involves an iterative SCF-like approach.

The second critical approximation that needs to be improved in order to improve accuracy is that of the limited basis functions used. Expanding the number of wavefunctions will help increase the resolution and accuracy. Figure A8 suggests that the most efficient improvement in accuracy come from increasing the CI treatment and basis set expansion together. The crudest HF basis set is that of a single valence. This can be improved by going to split valence, double valence, and triple valence along with adding polarization and diffuse functions. These increase the number of basis functions considerably, allowing for a more complete mathematical treatment of the wavefunction of the system.

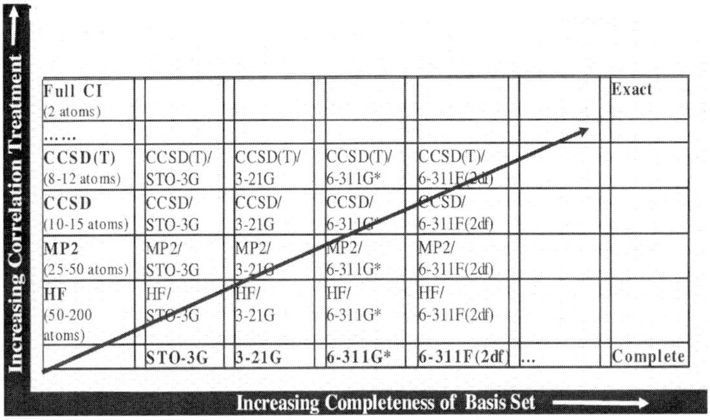

Increasing Correlation Treatment →

Full CI (2 atoms)						Exact
......						
CCSD(T) (8-12 atoms)	CCSD(T)/ STO-3G	CCSD(T)/ 3-21G	CCSD(T)/ 6-311G*	CCSD(T)/ 6-311F(2df)		
CCSD (10-15 atoms)	CCSD/ STO-3G	CCSD/ 3-21G	CCSD/ 6-311G*	CCSD/ 6-311F(2df)		
MP2 (25-50 atoms)	MP2/ STO-3G	MP2/ 3-21G	MP2/ 6-311G*	MP2/ 6-311F(2df)		
HF (50-200 atoms)	HF/ STO-3G	HF/ 3-21G	HF/ 6-311G*	HF/ 6-311F(2df)		
	STO-3G	3-21G	6-311G*	6-311F(2df)	...	Complete

Increasing Completeness of Basis Set →

Figure A8. A comparison of model chemistries and their consistent improvement in accuracy as one increases both the correlation treatments and the completeness of the basis set. The optimal approaches for given CPU resources lie along the diagonal in that both the correlation and basis set are at optimal positions. Adapted from Head-Gordon [14] and Foresman and Frisch [5].

5.b. Many-Body Perturbation Theory/Møller–Plesset (MP) Perturbation Theory

Many-body perturbation theory is based on the premise that the Hamiltonian from HF theory provides the basic foundation for the solution of the electronic structure and that configurational interactions can be treated as small perturbations to the Hamiltonian. The Hamiltonian is, therefore, written as the sum of the reference (HF) Hamiltonian (HO) and a small perturbation H′:

$$H = H_0 + \lambda H' \tag{A22}$$

where λ is a variable which describes the relative degree of perturbation. The perturbation is derived from the constructs of the true Hamiltonian and is equal to nuclear attraction and electron repulsion terms.

The wavefunction and the energy can then be written as a Taylor series expansion.

$$E = \lambda^0 E_0 + \lambda^1 E_1 + \lambda^2 E_2 + \lambda^3 E_3 + \dots \tag{A23}$$

$$\Psi = \lambda^0 \Psi_0 + \lambda^1 \Psi_1 + \lambda^2 \Psi_2 + \lambda^3 \Psi_3 + \dots \tag{A24}$$

The terms E_1, E_2, E_3, Ψ_1, Ψ_2, Ψ_3, etc, are the higher order corrections to the energy and the wavefunction. The higher order corrections are solved subsequent to the unperturbed solution of Ψ_0 and E_0 from the H_0 Hamiltonian.

The solution mechanism described is quite general. The most common choice for the reference of the unperturbed Hamiltonian operator is the sum over Fock operators. This is known explicitly as Møller–Plesset (MP) perturbation theory[2,19].

Most systems can be solved using relatively low perturbation orders, i.e. MP2 or MP4. MP2 can typically recover 80–90% of the correlation energy[2]. MP4 usually provides a reliably accurate solution to most systems. Nearly all of the studies where MP methods have been used to examine catalysis, however, have been performed at the MP2 level owing to the size and complexity of the systems modeled.

5.c. Couple Cluster Methods

Couple cluster methods differ from perturbation theory in that they include specific corrections to the wavefunction for a particular type to an infinite order. Couple cluster theory therefore must be truncated. The exponential series of functions that operate on the wavefunction can be written in terms of single, double and triple excited states in the determinant[2,14]. The lowest level of truncation is usually at double excitations since the single excitations do not extend the HF solution. The addition of singles along with doubles improves the solution (CCSD). Expansion out to the quadruple excitations has been performed but only for very small systems. Couple cluster theory can improve the accuracy for thermochemical calculations to within 1 kcal/mol. They scale, however, with increases in the number of basis functions (or electrons) as N^7. This makes calculations on anything over 10 atoms or transition-metal clusters prohibitive.

Returning to Fig. A8, there is a well-prescribed way of improving the accuracy for ab initio-based wavefunction calculations[2,14]. This involves an increase in the level of CI from:

$$\text{HF} < \text{MP2} < \text{CCSD} < \text{CCSD(T)} < \text{CCSDT(Q)} \ll \text{Full CI}$$

and also an increase in the relative extent of the basis set from:

single zeta (SZ) < double zeta (DZ) < double zeta with polarization (DZP) < triple zeta with double polarization (TZ2P)

Higher level CI calculations provide the most accurate predictions of properties including structures which can be determined to within 1%, reaction energies and enthalpies to within 1 kcal/mol, free energies to 2 kcal/mol and acid strengths to less than 2 pK_a units.

This increase in accuracy, however, comes with a significant price in terms of CPU. Hartree–Fock formally scales as N^4 but most current methods can bring this down to N. The scaling for different CI methods, however, follows

$$\text{MP2}(N^5) < \text{CISD}, \text{MP3 and CCSD}(N^6) < \text{MP4 and CCSD(T)}(N^7).$$

The application of these methods has been limited in the area of catalysis since the sizes of the systems of interest are typically too big to handle at any level above MP2.

6. Density Functional Theory

6.a. Theory

The development of density functional theory (DFT) has had a tremendous impact on modeling heterogeneous catalytic systems. There are now a number of reviews which describe the application and impact of DFT on catalysis. The relative accuracy of DFT, along with the size of the systems that it can handle, makes it attractive for modeling heterogeneous catalytic systems[20–22,34]. Density functional theory is "ab initio" in the sense that it is derived from first-principles and does not require adjustable parameters. DFT methods formally scale as N^3 and thus permit more realistic models of the intrinsic reaction than can be afforded by higherlevel wavefunction-based methods. The theoretical accuracy of DFT, however, is not as high as the higher level ab initio CI wave function methods.

Density functional theory can be traced to the developments by Thomas[23], Fermi[24] and Dirac[25] in which electron correlation was treated as a functional of the electron gas. The practical application of DFT theory, however, is attributed to work of Hohenberg and Kohn, who formally proved that the ground-state energy for a system is a unique functional of its electron density[8]. Kohn and Sham extended the theory to practice by showing how the energy could be partitioned into kinetic energy for the motion of the electrons, potential energy for the nuclear–electron attraction, electron–electron repulsion which involves with Coulomb as well as self interactions and exchange correlation which covers all other electron–electron interactions[26]. The energy of an N-particle system can then be written as

$$E[\rho] = T[\rho] + U[\rho] + E_{XC}[\rho] \tag{A25}$$

Kohn and Sham demonstrated that the N-particle system could be rewritten as a set of n-electron problems (similar to the molecular orbitals in wavefunction methods) that could be solved self-consistently in a manner which was similar to the SCF wavefunction methods[26]. Namely,

$$\hat{H}\psi_i = \varepsilon_i \psi_i \tag{A26}$$

or more specifically

$$\left[-\frac{\hbar^2}{2m} \nabla^2 + V_{ion}(r) + \frac{e^2}{2} \int \frac{\rho(r)\rho(r')}{|r - r'|} \mathrm{d}^3 r + V_{XC}(r) \right] \psi_i(r) = \varepsilon_i \psi_i(r) \tag{A27}$$

The first three terms are similar to HF theory, thus corresponding to the kinetic energy of the electron, the potential for nuclear–electron attractive interactions and the Hartree repulsive interactions between electrons. The final term, $V_{XC}(r)$, corresponds to the exchange correlation potential which is the derivative of the exchange correlation energy with respect to the density. This is more formally recognized as the chemical potential and written as

$$\mu_{XC}(r) = \frac{\delta E_{XC}[\rho(r)]}{\delta \rho[r]} \tag{A28}$$

The total energy of the system is then defined as

$$\begin{aligned} E[\{\psi\}] = 2 \sum_i \psi_i \left[-\frac{\hbar^2}{2m} \right] \nabla^2 \psi \mathrm{d}^3 r + \int V_{ion}(r) \rho(r) \mathrm{d}^3 r + \\ \frac{e^2}{2} \int \frac{\rho(r)\rho(r'}{|r - r'|} \mathrm{d}^3 r \mathrm{d}^3 r' + E_{XC} \left[\rho(r) \right] + E_{ion}(\{R_i\}) \end{aligned} \tag{A29}$$

The energy is formally a function of the density. The density of the system is still written as the sum of squares of the Kohn–Sham orbitals:

$$\rho(r) = \sum_{OCC} |\psi_i(r)|^2 \tag{A30}$$

The Kohn–Sham equations are solved in a very similar manner to that used to solve the Hartree–Fock system in that one iterates on the correct spatial distribution of the electron density.

In the theory presented thus far, DFT can be considered as an exact approach. Unfortunately, the exchange correlation energy is not known. It is at this point where approximations must be introduced in order to solve the electronic structure problem.

The most basic solution to Eq. (A27) is to invoke the local density approximation which assumes that exchange-correlation per electron is equivalent to the exchange correlation per electron in a *homogeneous electron gas* which has the same electron density at a specific point r. This is typically written as:

$$E_{XC}(r) = \int \rho(r)\varepsilon_{XC}\left[\rho(r)\right] \mathrm{d}r \tag{A31}$$

The local density approximation (LDA) is valid only in the region of slowly varying electron density. The LDA approximation is obviously an oversimplification of the actual density distribution and is well-known to lead to calculated bond and binding energies that are over-predicted[27].

One of the primary shortcomings of the local density approximation is that the exchange correlation charge distribution is not spherically homogeneous. Non-local gradient corrections are introduced to allow for non-spherical electron density distributions [2,7,27]. As such, the correlation and exchange energies are functionals of both the density and the gradient with respect to the density. These gradient corrections take on various different functional forms which include the BP86 (Becke[28] and Perdew[29] corrections), PW91 (Perdew–Wang exchange functional)[30], PBE (Perdew–Burke–Ernzerhof)[31] or RPBE (Revised PBE functional)[32]. By way of example, the widely used Becke (B88) correction to the Local Spin Density Approximation to the exchange is given by

$$E_x^g = b\sum_\sigma \frac{\rho_\sigma x_\sigma^2}{|1 + 6bx_\sigma \sin h^{-1} x_\sigma} \mathrm{d}r \tag{A32}$$

where

$$x_\sigma \equiv \frac{\nabla\rho}{\rho_\sigma^{4/3}} \tag{A33}$$

The functional takes on the correct r^{-1} asymptote behavior. x_σ is the dimensionless density gradient shown in Eq. (A33) and ρ_σ is the density. The term b is simply a fitting parameter for the energy that is regressed against atomic data. Despite the importance of the exchange correlation functional, there is no formal path toward the development of more accurate functionals. The accuracy of DFT is therefore typically less than what can be expected from higher level ab initio methods such as coupled-cluster theory[2,5]. More recent developments in functionals attempt to couple an exchange component derived from Hartree–Fock theory which provides for a more exact match of the exchange energy for single determinant systems along with the correlation (and exchange) calculated from LDA theory in "hybrid" functionals. The most notable is the B3LYP functional, which is a combination of the Lee, Yang and Parr functional and the three-parameter model by Becke[33]:

$$E_X^{B3LYP} = a_0 H_x^{HF} + (1 - a_0)E_x^{LDA} + a_x E_x^{B88} + (1 - a_c)E_C^{VWN} + a_c E_c^{LYP} \tag{A34}$$

The theoretical chemistry community developed density functional theory for finite molecular systems which involve molecules and cluster models that describe the catalytic systems. They use the same constructs used in many ab initio wavefunction methods, i.e. Gaussian or Slater basis sets. The solid-state physics community, on the other hand, developed density functional theory to describe bulk solid-state systems and infinite surfaces by using a supercell approach along with periodic basis functions, i.e. plane waves[9]. Nearly all of our discussion has focused on finite molecular systems. In the next section we will describe in more detail infinite periodic systems.

6.b. Periodic Density Functional Theory Algorithms

The cluster approach described so far can nicely begin to capture the local surface chemistry but is limited in terms of describing metals or metal oxides that take on more bulk-like characteristics including electronic, optical, and magnetic properties. In addition, there are also more practical considerations for solid-state periodic calculations which include the ability to examine readily surface relaxation and reconstruction effects, higher surface coverages, and the degree of adsorbate ordering. The calculations for surfaces then are likely more easily modeled using a supercell approach along with plane wave basis functions[13]. The supercell is defined by three lattice vectors as well as the length along these vectors, thus providing a 3-D unit cell. The supercell is used to replicate the system infinitely along all three vectors using periodic boundary conditions, thus simulating the solid state. This is shown for the bulk structure of Pd in Fig. A9A. For three-dimensional bulk systems this is straightforward. The simulation of surfaces, however, requires that the metal atoms be truncated along the vector perpendicular to the surface and replaced with a vacuum region[13,21,34]. The unit cell is still repeated periodically along all three vectors. In this case, however, the result is a set of periodic slabs of some metal thickness sandwiched between two vacuum regions as shown in Fig. A9B.

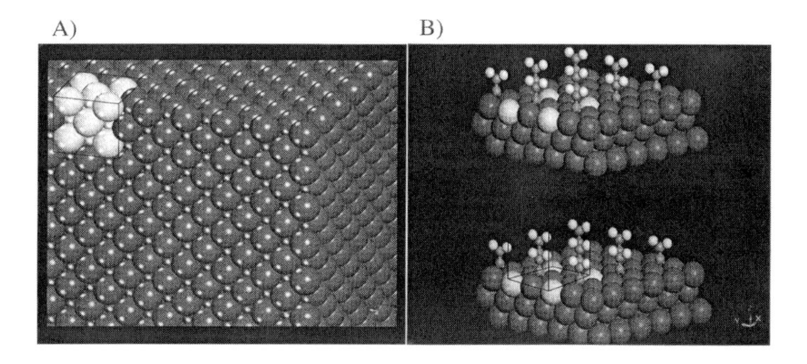

Figure A9. Three-dimensional supercells used to replicate the bulk Pd metal and the Pd(111) surface with adsorbed ethylidyne.

The wavefunction, according to Bloch's theorem, is one which contains a wave-like portion [the exponential term in Eq. (A35)] and a periodic cell portion [the $f_i(r)$ term in Eq. (A35) and (A36)][13]. The wavefunction is expanded so as to take on the same periodicity of the lattice, where G defined in $Gl = 2\pi m$ is the reciprocal lattice vector and m is an integer. The wavefunction is described by the summation of plane waves expanded out to a chosen cutoff energy. k is the symmetry label in the first Brillouin zone.

$$\psi_i(r) = \sum_G c_{i,k+G} \exp\left[i(k+G)r\right] f_i(r) \tag{A35}$$

$$f_i(r) = \sum_G c_{i,G} e[iGr] \tag{A36}$$

The choice of the cutoff energy dictates the expansion of the wavefunction. Increasing the cutoff energy is, therefore, similar to increasing the number of orbitals in a molecular calculation in that in increases the accuracy by allowing for more expansive wavefunction[13]

The numerical integration for periodic solid-state systems is typically carried out in reciprocal space where the first Brillouin zone is divided and described by a finite number of k-points. The k-points describe the sampling of the electronic wavefunction. Observables such as the energy and the density are integrated over all k-points within the first Brillouin zone. Chadi–Cohen[35] and Monkhorst–Pack[36] are two particular approaches that have been developed to provide an optimal division of special k-points so as to provide a reasonably accurate description of the electronic potential. The total energy of the system should converge with increasing number of k-points since the increase in the number provides for a more dense k-point mesh and finer sampling of the Brillouin zone. The single particle wavefunctions are then described by plane wave basis sets that obey Bloch's theorem[13].

Although plane waves are the natural choice for periodic systems, they pose difficulties in accurately solving for the wavefunction near the core of the nuclei. The orbitals near the nuclei core are tightly bound and have significant oscillations, both of which make it difficult to model using expanded plane waves. They require an extensive number of plane waves, which is CPU intensive. Since most of the chemistry occurs via the valence electrons, the detailed electronic structure of the core can be avoided by using pseudopotentials. The pseudopotential approach, which was discussed earlier, substitutes the strong ionic potential and valence wavefunction with a weaker pseudopotential along with peudo wavefunctions. The pseudopotential removes the radial nodes in the core region and matches the valence electron wavefunction outside of the core region. This was shown in Fig. A5.

To summarize, the main differences between the molecular and periodic DFT calculations involve:

1. the periodic replication of the unit cell which is described in a plane wave formalism as opposed to the local finite molecular system.
2. the basis set expansion is controlled by the cutoff energy in the periodic system rather than by the number orbitals as it is in the molecular system.
3. the description of the system in reciprocal space verses real space for molecular systems.

Other than these differences, the solution strategy is essentially the same for solving both the molecular and periodic systems.

We have outlined here some of the main features in density functional theory. A number of other developments have occurred since its inception. Wimmer[6] elegantly captured some of the different approaches and how they differ with respect to the fundamental Kohn–Sham equations. An adaptation of one of his figures is give here in Fig. A10.

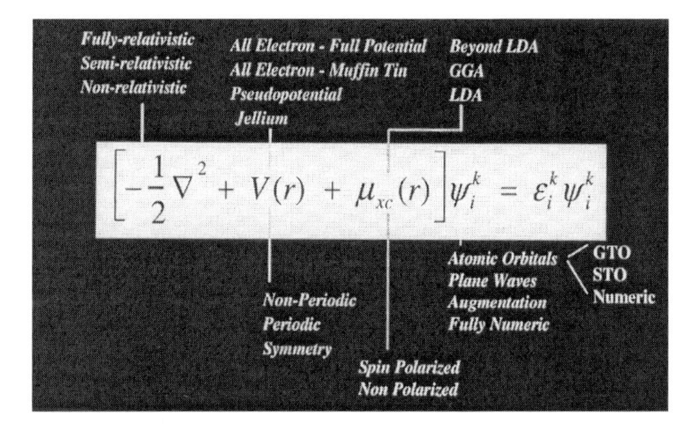

Figure A10. A suite of options available in the solution of the Kohn–Sham equations for density functional theory. This figure nicely captures some of the essential differences that exist between the full spectrum of different DFT codes that currently exists. Adapted from Wimmer [6].

7. Model Versus Method Accuracy

Method accuracy describes the relative accuracy of a quantum mechanical method based upon how well the quantum mechanical method can accurately predict electronic structure and energetics for an exact molecular structure or system[5,14]. In modeling heterogeneous catalytic systems, we seldomly can model all of the atoms in the system and must make a choice on how many atoms to include in the model of the active site and its environment. The size of the model used to describe the reaction environment can be critical to obtaining reliable numbers. While the calculations of small molecules such as NO, CO, ethylene, and NH_3 on a single metal atom can be performed accurately by fairly using couple cluster theory, its relationship to the adsorption of these molecules on a metal surface is rather poor since a single metal atom cannot accurately represent the electronic structure of the metal surface. We describe this as the "model" accuracy as opposed to the method accuracy, which has been the focus to this point. In order to model catalytic systems, both the method and model accuracies need to be balanced. Most of the current efforts with heterogeneous catalytic systems have used density functional theory primarily to improve the model accuracy since DFT scales as N^3 and can thus allow the simulation of hundreds of atoms rather than 20 atoms or less. The scaling for some of the methods and the number of atoms that can be treated is shown in Fig. A11.
The relative accuracy of density functional theory methods for catalytic systems suggest, that the structures can typically be optimized to within 0.05 Åand 1° in terms of bond lengths and angles, respectively[27]. Spectral properties such as infrared, Raman, and NMR can be predicted to within about 5% of the known experimental values. Optical properties which require excitation energies are less reliable for standard DFT methods. Time-dependent density functional theory, however, provides for a better estimation of these properties. The binding energies and overall reaction energies from DFT methods are typically on the order of 5–8 kcal/mol in terms of accuracy[22,27,37]. This is not the 1 kcal/mol level of accuracy required for engineering purposes. In addition, there are also known outliers which are beyond the accuracy ranges cited. The accuracy of DFT methods is currently controlled by the exchange correlation energy functionals that are

Method	Number of basis functions	Estimates for the maximum number of atoms that can currently be handled
DFT	N^3-N^4	50-100
HF	N^4	
MP2	N^5	25-50
MP4	$<N^5$	
CISD	N^7	10-15
QCISD	N^7	

Figure A11. A comparison of the scaling and the approximate number of heavy atoms that can be handled by different theoretical treatments in the general order of increasing scaling and increasing accuracy. While the CISD and QCISD methods provide much more accuracy, the systems that can be examined are significantly smaller.

employed. These functionals, as discussed earlier, are somewhat arbitrary and therefore there is no systematic way of improving their accuracy. Despite these limitations, there is a wealth of current activity by various groups around the world towards the development of more accurate exchange correlation potentials.

8. Advancing to Larger Systems

There are various approaches that have been used to begin to simulate larger systems. The two most notable are the quantum-mechanical embedding and linear scaling methods.

8.a. (Quantum-Mechanical) Embedding

Quantum mechanical embedding involves dividing the reaction environment into different regions which are each treated using a different level of theory. The atoms directly connected to the actual reaction site are treated with a higher level of theory whereas those atoms which are further removed are treated with a lower level. The greatest difficulty comes in linking two different regions together. The link region is usually defined in order to provide an adequate transfer of information between the inner and outer regions. The different regions are shown schematically in Fig. A12 for the adsorption site in a zeolite.

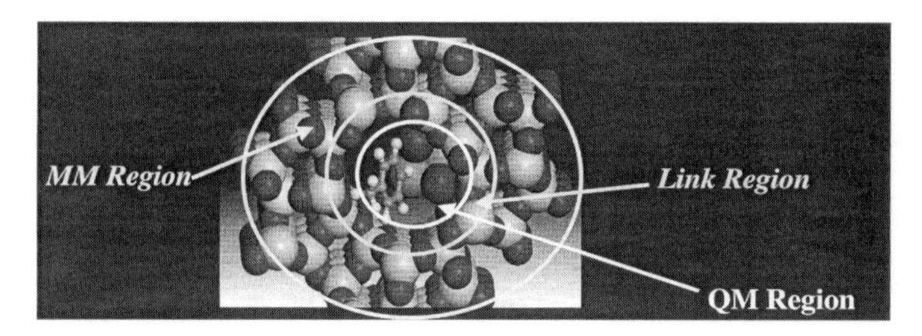

Figure A12. A schematic picture of the process of embedding for benzene adsorption at a metal oxide cluster fixed into the pore of a zeolite. The inner region which defines the adsorption site can be treated with a higher level of theory whereas the outer region is defined with a significantly lower level of theory such as force field. An overlap or "link" region is defined between the two in order to transfer information effectively between the two models.

The QM energy for this system is then calculated by the following equation:

$$E_{\mathrm{QM}}(\mathrm{System}) = E_{\mathrm{QM}}(\mathrm{Core}) + E_{\mathrm{MM}}(\mathrm{System}) - E_{\mathrm{MM}}(\mathrm{Core}) \tag{A37}$$

where $E_{\mathrm{QM}}(\mathrm{Core})$ refers to the QM energy calculated for the inner core region only. $E_{\mathrm{MM}}(\mathrm{System})$ – $\mathrm{E}_{\mathrm{MM}}(\mathrm{Core})$ refers to the difference in energy between the full system calculated using the lower level of theory and the core region using the lower level of theory. Although MM refers to molecular mechanics, it is not restricted to molecular mechanics. It can be any method which is a lower level method which is faster than the QM region of the core.

The ability to divide a system into separate regions and solve the inner system with a rigorous method and the outer region with a much faster method is fairly powerful. It allows for the simulation of systems comprised of O (10^4) atoms. In addition, it enables one to increase the accuracy of a particular calculation by using high-level CI calculations to describe the central QM core region.

The earliest efforts in this area are ascribed to work by Warshel and Karplus[38]. It was not until the 1990s when it began to take on a much more active following. Perhaps the most widely used scheme is the ONIOM method, which was developed by Vreven and Morokuma[5,39]. This is a general approach which is now part of the Gaussian suite of codes[123]. In the ONIOM method implemented into Gaussian, the user can choose between various methods for both the core and exterior regions. Homogeneous catalytic systems, zeolites, and enzymes have all been modeled by using DFT or higher level quantum mechanical treatments for the core region and force field models to describe the external region. They can therefore capture the local reaction chemistry and also begin to describe the longer range effects.

For supported metal and metal oxide systems, one typically has to resort to using two different QM methods owing to the lack of accurate force fields or empirical potentials to describe these systems. Both Whitten and Yang[40] and Govind et al.[41] have developed schemes which embed more accurate CI wavefunction methods into lower level QM methods in order to provide for more accurate descriptions than DFT. Sauer's group has used standard ab initio methods along with shell models to describe the oxide environment for zeolite systems[42,43].

8.b. Linear Scaling

A number of recent efforts have been focused on improving DFT's N^3–N^4 scaling down to N. Linear scaling would thus enable one to examine much larger heterogeneous catalytic systems as well as biocatalytic systems. One of the inherent difficulties in developing linear scaling methods resides in the Coulomb electron–electron repulsion integrals shown in Eq. (A38) which formally scale as N^4.

$$\left(\mu\nu|\lambda\sigma\right) = \iint \mu(1)\nu^*(1)\frac{1}{r_{12}}\lambda(2)\sigma^*(2)\mathrm{d}r_1\mathrm{d}r_2 \tag{A38}$$

Very fast multipole methods have been developed in order to calculate these electron repulsion integrals[44,45]. The near field is determined by analytical Gaussian calculations. The far field is calculated using multipole expansions to treat the distant charges and their interactions. The scaling for this approach has been reduced to $N^{1.35}$. Fast quadrature methods for calculating the exchange correlation potentials have also been developed to improve scaling. In addition, there have been developments to provide linear-scaling approaches to the diagonalization of the density matrix. Traditional quantum mechanical methods are based on wavefunctions which characterize eigenstates for discrete energy levels. Orthogonality requirements on the wavefunction thus lead the system to scale as N^3. As the system grows larger, the wavefunction must extend over a much larger volume, thus increasing in larger basis sets[46]. In addition, more wavefunctions must be orthoganolized with respect to one another. These issues lead to N^3 scaling. The newer methods provide novel means of diagonalizing the density matrix to preserve linear scaling.

Siesta (the Spanish Initiative for Electronic Structure of Thousands of Atoms) is a self-consistent DFT method that demonstrates linear scaling for very large systems[47]. They do so by using flexible numerical atomic basis sets and localized linear occupations of orbitals. In addition, they project the electronic wavefunction and density onto real space grid in order to calculate Hartree and exchange correlation potential and matrix elements. The long-range features of the potential are eliminated by screening with local atomic electron density. Various other techniques are also employed to reduce the order dependence. Siesta has now been used to simulate various different systems. Simulation system sizes can begin to approach 10,000 atoms.

9. Ab Initio Molecular Dynamics

Classical molecular dynamics simulations have proven to be invaluable in determining the structure, sorption and diffusion of organic molecules in various systems, including vapor, liquid, and mesoporous solid systems where accurate force fields and interatomic potentials have been derived. They fail, however, for systems which are not parameterized, including transition metals and transition metal oxides and sulfides. In addition, MD simulations can not be used to simulate chemical reactions or systems where electron transfer is important since the force fields are based on interatomic interactions with no treatment of the electronic structure. Simulating the dynamics of the electronic structure of a system requires the ability to follow the changes to the electronic structure as a function of time. Full quantum dynamic simulations present a significant challenge even for the simplest of systems. Fortunately, most systems obey the Born–Oppenheimer approximation, thus allowing us to separate out changes in the electronic and nuclear structure. As such, one can propagate the electronic structure with changes in the nuclear positions that result from molecular dynamics. In this way, one can do away with the necessity for an empirical force field. Instead, ab initio calculations are performed "on-the-fly" during the MD simulation to provide the forces on all of the atoms. These forces are then used to integrate the classical Newton's equation of motion to find the new positions of the ions at the next point in time. In order to ensure accuracy, the time step used in the dynamics must be significantly shorter than that of the fastest processes. For bond making and breaking reactions, this is typically on the order of about 0.05 fs[48,55,57].

Ab initio molecular dynamics methods can roughly be divided into two classifications: Born–Oppenheimer Molecular Dynamics and Car–Parrinello Molecular Dynamics[55,56]. In both simulations, the wavefunction is propagated with the changes in the nuclear coordinates. In the Born–Oppenheimer MD approach, the forces on each of ions are explicitly calculated at each MD time step. As such, the system directly follows the Born–Oppenheimer surface. The primary drawback of the Born–Oppenheimer MD approach relates to the fact that time-intensive electronic structure calculations must be converged at each time step throughout the simulation. In a landmark paper in 1985, Car and Parrinello demonstrated that the electronic structure could be propagated directly with the nuclear structure by treating the electron wavefunction as a particle with a fictitious mass[50]. This saves significantly on CPU efforts since the electronic wavefunction need not be calculated for each time step. The details of both the Born–Oppenheimer and Car–Parrinello methods are given in excellent reviews by Marx and Hutter[55a], Iftime et al.[55b], and Trout[57]. We simply try to cover some of the salient features of both methods below.

9.a. Born–Oppenheimer Ab Initio Molecular Dynamics.

As was discussed in the previous section, Newton's equations of motion result in the following expression:

$$M_I \ddot{R}_I(t) = -\nabla_I V_e^{approx}\Big(\{R_I(t)\}\Big) \tag{A39}$$

where M_I is the mass of ion I, $\ddot{R}$ is the acceleration of ion I and V_e^{approx} is the effective potential energy which is typically determined by empirically derived two- and three-body interaction potentials. The quality of the simulation results resides in the accuracy of the parameterized force field defined by V_e^{approx} .

In the ab initio Born–Oppenheimer molecular dynamics approach, the force field is defined "on-the-fly". A static electronic structure optimization is carried out at every time step within the molecular dynamics simulations. MD provides the positions for ions at each step in time. These coordinates are subsequently used as the input to the QM calculation which provides the energy and the forces which act upon each ion. Newton's equation of motion can then be described for the ground-state system as:

$$M_I\ \ddot{R}_I\ (t) = -\nabla \min_{\Psi_0}\Big\{\langle \Psi_0|\hat{H}_e|\Psi_0\rangle\Big\} \tag{A40}$$

$$E_0\Psi_0 = \hat{H}_e\Psi_0 \tag{A41}$$

where the last term in Eq. (A40) refers to the minimum total electronic energy. Equation (A41) can subsequently be integrated to solve for the position of the ions at each time step. In order to do so, the electronic structure must be optimized at each point to determine the forces on each ion and thus the right-hand side of Eq. (A41).

The solution of the effective one-particle Hamiltonians is subject to the constraint that the orbitals are orthonormal. This leads to the constraint that

$$\langle \psi_i|\psi_j\rangle = \delta_{ij} \tag{A42}$$

which can be redefined via Lagrange multipliers. The Lagrangian for this system can be defined by

$$L = -\langle \Psi_0|H_e|\Psi_0\rangle + \sum_{i,j} \wedge_{ij}\Big(\langle \psi_i|\psi_j\rangle - \delta_{ij}\Big) \tag{A43}$$

where $\wedge_{ij}$ refers to the Lagrangian multipliers[55 – 57]. The constraint can then be defined as the following for DFT methods:

$$0 = -H_e^{KS}\psi_i + \sum_j \wedge_{ij}\psi_j \tag{A44}$$

9.b. Car–Parrinello Ab Initio Molecular Dynamics

In their landmark paper, Car and Parrinello demonstrated that the electronic structure does not have to be converged to the Born–Oppenheimer surface at every time step throughout an ab initio MD simulation[50]. Instead, the orbitals can be propagated together with the atomic nuclei by assigning a fictitious mass to each electron. An important practical point in making this work is establishing the optimal step size to propagate the wavefunction. The electronic motion is still considered to be so much faster than that of the nuclei. The electrons can therefore be optimized with respect to changes in the nuclear positions. The Lagrangian for the Car–Parrinello algorithm is defined as follows:

$$L_{CP} = \frac{1}{2}\sum_I M_I \dot{R}_I^2 \frac{1}{2}\sum_i \mu_i \langle \dot{\psi}_i | \dot{\psi}_i \rangle - E[\psi_i, R_I] + \sum_{i,j} \wedge_{ij} \Big(\langle \psi_i | \psi_j \rangle - \delta_{ij} \Big) \tag{A45}$$

The first term and second terms in Eq. (A45) correspond to the kinetic energy of the nuclei and the fictitious kinetic energy of the electrons in the system, respectively. The third term reports the overall electronic energy which corresponds to the potential energy of the nuclei. The last term represents the constraints that the orbitals must be orthonormal[55].

The equations of motion for the nuclei and the electrons are then given as

$$M_I \ddot{R}_I (t) = -\frac{\partial}{\partial R_I}(E[\psi_i(\mathrm{r}, t), R_I]) \tag{A46}$$

$$\mu_i \ddot{\psi}_i (t) = -\frac{\partial}{\delta \psi_i^*}(E[\psi_i(\mathrm{r}, t), R_I]) + \sum_i \wedge_{ij} \psi_j(\mathrm{r}, t) \tag{A47}$$

respectively.

In this approach, the nuclei are simulated at some finite temperature, T, which ultimately dictates the kinetic energy of the nuclei. The electronic structure, however, is kept close to the Born–Oppenheimer surface. The fictitious temperature of the electrons must therefore be close to zero. In simulating the dynamics for a specific system, the electrons must remain "cold" while the atoms must remain "hot" and thus maintain a nearly adiabatic system. The fictitious mass of the electron and the time steps for the dynamics must be carefully structured so as to prevent energy transfer from the hot nuclei into the cold electrons. The Verlet algorithm is typically used to integrate these equations.

In order to solve the equations of motions defined above, the forces on each of the ions must be defined. This can be done by using the Hellman–Feynman theorem, whereby the force is defined as the derivative of the total energy with respect to the positions of the ions:

$$f_I = -\frac{\mathrm{d}E[\psi_i(\mathrm{r},t), R_I]}{\mathrm{d}R_I} \tag{A48}$$

The wavefunction, however, also has to change with changes in the coordinates for each particle. The total derivative of the energy with respect to the changes in the positions of the ions can therefore be written as

$$f_I = \frac{\partial E}{\partial R_I} - \sum_i \frac{\partial E}{\partial \psi_i}\frac{\partial \psi_i}{\partial R_I} - \sum_i \frac{\partial E}{\partial \psi_i^*}\frac{\partial \psi_i^*}{\partial R_I} \tag{A49}$$

The force defined in the Lagrangian is therefore not a physical force due to the second and third terms in Eq. (A45). If the wavefunction is an actual eigen state of the electronic Hamiltonian, then these last two terms are zero and so the forces calculated from Eq. (A48) are actual forces. This is known as the Hellman–Feynman theorem[55,57].

9.c. Applications

Ab initio molecular dynamic simulations have been used to study a wide range of problems including homogeneous and heterogeneous catalytic systems[49,53], reactions in solution [52], materials surface chemistry [54,55,57], biochemistry and biocatalysis. Its unique strength is its ability to follow both the dynamic changes of the nuclei along with the electronic structure. The rather limited time and length scales that can be reliably simulated are clearly limitations of this approach. Most ab initio methods are currently based on density functional theory. The foundation of these simulations is molecular dynamics, which is based firmly on statistical mechanics. Ab initio MD simulations can therefore be carried out within any of the different ensembles available to traditional MD simulations such as NVT, NVE and NPT. This enables one to calculate the structure, diffusivities and the full range of thermodynamic properties.

The tracking of the electronic structure also provides for the ability to calculate activation barriers in addition to kinetics. One can simulate about a few hundred atoms out to about 5 ps on current multiprocessor clusters. The typical time step for most catalytic purposes is on the order of about 1/20 fs. Blochl et al.[48] nicely illustrated that for a reaction that has an activation barrier of 10 kJ/mol would require simulations on the order of 1.4 ps. Simulating reactions that have an activation barrier of 50 kJ/mol, however, would require up to 10^7 ps.

Two approaches can be taken to simulate reaction systems with such high barriers. The first involves raising the temperature of the simulation in order to access higher energy states. The second approach is to carry out a sequence of a constrained AIMD simulations along a specific reaction coordinate[51].

B: ATOMIC/MOLECULAR SIMULATION

The ability to model chemical reactions required the full accounting of the electronic structure of the system and the changes to the electronic structure upon reaction. The basic building blocks for describing the electronic structure are the electrons and the nuclei. Schrödinger's equation then is simply just a force balance that operates on them to provide the total energy and the energy states of the system for a specific configuration. The ability to model the atomic structure in microporous materials, siting of sorbates, sorption isotherms and sorbate diffusion requires the ability to simulate much larger systems and longer time scales. The changes in the electronic structure are not, however, germane to simulating the structural properties or dynamic responses of the structure for systems where electron transfer is not critical. The fundamental building blocks for these systems are the atoms and molecules from which they are comprised. Atomistic scale simulations must track the forces that occur between individual atoms. Systems which contain molecular entities track both the intra- and inter-molecular forces that arise. In many of the simulations for catalysis, we are interested in modeling physisorption or diffusion processes whereby the dominant forces that control these steps are weak van der Waal's interactions which are usually very difficult to treat quantum mechanically. Atomic and molecular simulations which are based on force fields, however, are typically much better suited for modeling these weak interactions since they have been parameterized to handle such systems. Schrödinger's equation provided the framework for formulating and simulating the forces on the electrons. The simulation of the forces that act on atoms and molecules is strongly rooted in and governed by statistical mechanics and classical dynamics.[58] This allows for the rigorous simulation of a wide range of thermodynamic and dynamic properties for the system of interest. There are a number of elegant reviews on different atomistic and molecular simulations and their application to catalysis, we would refer the interested reader to the several references[58–62] and the books by Frenkel and Smit[75], Allen and Tildesley[63], Leach[64] and Rapp and Casewit[65].

Atomic and molecular simulation methods can generally be categorized as either equilibrated or dynamic. Static simulations attempt to determine the structural and thermodynamic properties such as crystal structure, sorption isotherms, and sorbate binding. Structural simulations are often carried out using energy minimization schemes that are similar to molecular mechanics. Equilibrium properties, on the other hand, are based on thermodynamics and thus rely on statistical mechanics and simulating the system state function. Monte Carlo methods are then used to simulate these systems stochastically.

Following the dynamics for the system can effectively be divided into three different categories: dynamic simulation of the system structure, dynamic simulation of both atomic and electronic structure, and the longer scale simulation of kinetics for a reaction system. In Appendix A, we described the ab initio molecular dynamics mehtod which is used to simulate the dynamics of the atomic structure along with the electronic structure. In the following section, we describe the formulation and solution to molecular or lattice dynamics. The simulation of kinetics is more involved and will be described in Appendix C.

1. Force Fields

At the heart of nearly all atomistic simulations is the force field used to describe the interaction between atoms or molecules. The accuracy of most atomistic simulations is highly dependent on the accuracy an applicability of the force field that has been developed. The force field contains both intra- and interermolecular interactions. The contributions of the intra–molecular interactions to the potential energy are the result of changes in the bond length, bond angle and torsion angle from their standard positions. The bond length potential, for example, is expressed by a parabolic equation based on Hooke's law that relates the potential energy to the differences that result in the optimized bond length (r_i) and a universal bond length for that specific type of bond (r_0), as is shown in Eq. (B1). The terms for bond angle (θ_i for the calculated and θ_0 for the universal) and torsion angle (ϕ_i for the calculated and ϕ_0 for the universal) are similar and shown in Eqs. (B2) and (B3), respectively. Intramolecular forces are fairly standard for most force fields[75]

Bond length:

$$V_r = \sum_{i=1}^{N_m-1} \frac{1}{2} K_B (r_i - r_0)^2 \tag{B1}$$

Bond angle:

$$V_\theta = \sum_{i=1}^{N_m-2} \frac{1}{2} K_\theta (\theta_i - \theta_0)^2 \tag{B2}$$

Torsion angle:

$$V_\phi = \sum_{i=1}^{N_m-3} \sum_{j=0}^{p} C_i(\cos\phi_i)^j \tag{B3}$$

The terms K_B, K_θ and C_i are simply the empirical coefficients for specific types, fit between experimental bond lengths, bond angles and torsion angles, respectively. The intermolecular potential energy terms attempt to capture different types of intermolecular interactions including electrostatic and dispersive forces. The intermolecular forces have been treated in various ways. The van der Waals interactions, for example, have been modeled via Lennard–Jones6-12, Morse and Buckingham type potentials[61]. In some cases these interactions are even neglected.

Coulombic:

$$V_C = \sum_{i=1}^{N}\sum_{j=1}^{N} \frac{q_i q_j}{4\pi\varepsilon_0 r_{ij}} \tag{B4}$$

van der Waals:

$$V_{\nu DW} = \frac{A_0}{r^{12}} - \frac{B_0}{r^6} \quad \text{(Lennard–Jones6–12)} \tag{B5}$$

$$V_{\nu DW} = A\mathrm{e}^{-Br} \quad \text{(Exponential)} \tag{B6}$$

$$V_{\nu DW} = A\mathrm{e}^{-Br} - \frac{C_6}{r^6} \quad \text{(Buckingham)} \tag{B7}$$

where q_i refers to the charge on atom i, ε is the dielectric constant of the medium, and A_0, B_O, A_1 and C are the fitting coefficients.

The total potential energy (V_T) of the system can then be described by adding in all of the contributions from intra- and intermolecular forces:

$$V_T = V_r + V_\theta + V_\phi + V_C + V_{\nu DW} \tag{B8}$$

These equations comprise the *"force field"* and provide the foundation for nearly all atomistic and molecular simulations. The force field provides the potential energy which is used to carry out energy minimization to identify the most stable structures, Monte Carlo simulations to determine the properties of equilibrated systems and molecular dynamics to follow the dynamics of the system.

2. Energy Minimization Methods

Elucidating catalyst structure is important to understanding its potential reactivity. A great deal of work has been done to derive structure from atomistic simulations. Considerable progress has been made in the development of potentials that carry out energy minimizations in order to find the most stable structures for different metal, zeolite and metal oxide systems.

2.a. Metals

The shape, morphology and composition of metal particles and thin films for systems without the presence of a reacting gas alone can be simulated with a reasonable degree of accuracy since the potentials for these systems are typically fairly good. Metals have been described by using

1. embedded atom methods(EAM) [66]
2. modified embedded atom methods (MEAM) [67,68]
3. effective medium theory (EMT) [69].

Since we talk very little about metal particle simulations in this book, we provide only a very general overview here. More detailed discussions of these methods can be found in the articles cited above. EAM, MEAM and EMT methods have been used quite effectively with molecular dynamics in simulating physical vapor deposition processes used in thin film growth. These methods have also been effective in understanding the lowest energy structures of the metal particles present for heterogeneous catalytic systems. These studies, however, have been limited predominantly to simulations in vacuum. Simulating the particle shape, morphology and composition under reaction conditions, however, has yet to be accomplished since these potentials typically only account for metal–metal bonding. Under reaction conditions the surface can be covered with strongly bound intermediates which can weaken metal–metal bonding and lead to significant changes in the surface structure as well as particle morphology. For a number of systems, the metal surface changes dynamically with reaction conditions. The ability to simulate these changes would require accurate adsorbate–metal potentials for all of the intermediates that could form. This is a significant challenge owing to the difficulty in developing metal–adsorbate force fields. Recent progress by van Beurden et al.[70] on the development of MEAM potentials to describe CO on Pt for the simulation of Pt reconstruction[71] provide hope. The development of potentials that treat the complex and dynamically changing background composition in a reacting system, however, will be considerably more difficult.

2.b. Metal Oxides

Lattice energy minimization techniques have been used fairly successfully to simulate the lowest energy structures for various metal oxides and zeolites[61]. In this approach, the total potential energy of the lattice is defined based on the summation of the potential interactions between ions in the lattice and the remaining lattice:

$$U = \frac{1}{2}\sum_{i=1}^{N} V_i \tag{B9}$$

In theory, the potential energy for the interactions between ion i and the remaining lattice can be calculated by calculating the summation of all pair, triplet, quartet, and many-body interactions:

$$\underbrace{V_i(r_1, r_2, \ldots r_n)}_{\textit{Potential energy}} = \underbrace{\sum_{i-1}^{n}\sum_{j>i}^{N} U_{ij}(r_1, r_j)}_{\textit{Pair interactions}} + \underbrace{\sum_{i=1}^{N}\sum_{j>i}^{N}\sum_{j>i}^{N} U_{ijk}(r_1, r_j, r_k)}_{\textit{Triplet interactions}} + \ldots \tag{B10}$$

These terms are nearly always truncated after accounting for only pairwise interactions. Extensions to triplet systems typically does not significantly alter the qualitative trends established from following only the binary interactions.

The binary pair interactions can be modeled by using a force field to describe the system. Various different force fields for the simulation of oxides have been developed and employed. The most basic force field would employ both Coulombic and non-Coulombic interactions such as

$$U = \underbrace{\frac{q_i q_j}{r_{ij}}}_{\textit{Coulombic}} + \underbrace{\phi_{ij}(r_{ij})}_{\textit{Non-Coulombic}} \tag{B11}$$

Ewald summation techniques are necessary for calculating Coulombic interactions. The non-Coulombic terms contain both attractive and repulsive components and can typically be modeled by using Lennard–Jones, Morse or Buckingham potentials from Eqs. (B5), (B6), and (B7), respectively.

Potentials that treat the polarization and ionization are important for modeling a number of metal oxide systems. This is difficult since polarization in solids is a many-body effect with various components and depends strongly upon changes in the electronic structure as a function of structure and forces on the ions. One of the most widely used approaches to simulate polarizability effects is that of the Shell model which uses a massless shell of charge (electron density)[61].

The simulation of the optimized oxide structure requires the minimization of the energy with respect to the changes in the atomic structure of the oxide. One can use a variety of different numerical schemes that optimize the structure with respect to the structure of the lattice. Simulating annealing is a fairly robust numerical method that can be used to find the most stable structures. Simulated annealing attempts to mimic computationally how nature forms low-energy structures. The system is started at a higher temperature which allows it to sample various states along the potential energy surface. The system is then very gradually cooled to some final state. The simulation samples random moves for all of the atoms at each temperature. The total system potential is calculated after each trial move to determine whether or not the move is accepted. If the trial move leads to a lower energy system, the move is accepted. If the system energy is higher, the probability that the move is accepted and follows a Boltzmann probability distribution:

$$P_{Accept} = \exp(-\Delta U \,/\, k_B T) \tag{B12}$$

where ΔU is the change in energy between the system at its initial state and the trial state. This is accomplished by comparing a random number between 0 and 1 with the calculated probability, P_{Accept}. If the random number is lower than P_{Accept}, the move is accepted. This is known as Metropolis sampling.

Gale developed the General Utility Lattice Program (GULP), which is a general method towards simulating the structure and energetics for 3D molecular and ionic solids, gas-phase clusters, and defect structures[72]. GULP is based on the Shell model described earlier.

It allows for the calculation of a range of structural, mechanical, and thermodynamic properties including relative energetics, sorbate siting, bulk modulus, Young's modulus, dielectric constant, refractive index, piezoelectric constants, phonon frequencies, entropy, heat capacity, Helmholtz free energy, and other properties. The approach has been used to simulate a wide range of different oxide materials including zeolites, silicates, aluminophosphates, ceramic glasses and transition-metal oxides.

3. Monte Carlo Simulation–Equilibrium Systems

The thermodynamic properties for a system of N molecules (or N atoms) can be rigorously accounted for using statistical mechanics. Monte Carlo simulation methods provide the foundation for numerically simulating the configurational integral shown in Eq. (B13) that arise from the statistical mechanics treatment.

$$Z = \int \mathrm{d}r^N \exp\Big[-U(r^N/k_BT)\Big] \tag{B13}$$

where r^N refers to the set of generalized coordinates for the N-particle system.

Monte Carlo integration allows the integral to be calculated by stochastically sampling a large discrete set of random configurations defined here as the number of MC sample steps ($N_{MCsteps}$). The configurational integral can then be calculated using

$$Z = \frac{V}{N_{MCsteps}} \sum_{i=1}^{N_{MCsteps}} \exp\Big[(-U(r^N/k_BT)\Big] \tag{B14}$$

A full range of thermodynamic properties can then be calculated via statistical mechanics. The average of some property $<A>$ for the system is then defined as the average of A over all of the different configurations generated from Monte Carlo sampling:

$$\langle A\rangle = \frac{\int A(r^N)\exp\Big(-\frac{U(r^N)}{k_BT}\Big)\mathrm{d}r^N}{\int \exp\Big(-\frac{U(r^N)}{k_BT}\Big)\mathrm{d}r^N} = \frac{\sum_{i=1}^{M} A_m}{\sum_{i=1}^{M} 1} = \frac{1}{M}\sum_{i=1}^{M} A_m \tag{B15}$$

Various methods have been used in the literature based upon the properties one wishes to simulate and thermodynamic considerations of the system being studied. A number of these methods are described below.

3.a. Canonical Ensemble (NVT) MC Simulation

In the canonical ensemble, the number of molecules (or atoms), the volume and the temperature all remain constant[4,58]. Simulations then attempt to minimize the Helmholtz free energy of the system. These simulations are used to determine the pressure in the system, lowest energy states and the optimized structures. The simulations are performed by stochastically sampling a large number of different configurations for the system where the system is restricted to obey constant number of molecules, volume and temperature. The Metropolis sampling scheme presented earlier in Section 2.b is then used in order to accept or reject each trial move. The simulation proceeds by simulating millions of different trials in order equilibrate the system. In many of the simulations of sorbates in microporous media or on surfaces, the simulations are coarse-grained so that the position of each specific atom is foregone in order to speed up the simulations. Instead, the system is described using the "United" atom approach whereby only the heavy atoms are explicitly treated. For example, the hydrogen atoms in CH_3, CH_2, or CH are collapsed into the description of the united C atom. The united atom method is also used in the subsequent methods that will be described as well.

Isothermal–isobaric simulations are performed by holding the number of molecules (atoms), pressure and temperature constant. The simulation can then be used to determine the corresponding volume in the simulation. The Gibbs free energy in this system is minimized.

3.b. Grand Canonical Ensemble (μ, V, T) MC Simulation

The grand canonical ensemble simulations model systems in which the chemical potential (μ), the volume and temperature are held fixed while the number of particles changes. The approach is very useful for simulating phase behavior which requires a constant chemical potential. Grand Canonical Monte Carlo simulation has been used to calculate sorption isotherms for a number of different microporous silicate systems. The simulations are used to model the equilibrium between zeolite and sorbate phases and, as such, it provides a natural way of simulating sorption isotherms[59,62]

The simulations proceeds by first using a gas-phase equation of state to determine the pressure and the fugacity for the gas phase. The simulation then follows a series of trial moves which involve particle displacement, particle insertion and particle removal in order to establish equilibrium. The particles (molecules) in the simulation box are allowed to move, rotate or rearrange their configuration based upon the Boltzmann-weighted Metropolis sampling probability described earlier in Eq. (B12). In order to establish a constant volume, temperature and chemical potential, the number of molecules in the box can increase or decrease. In addition to the displacement moves described already, particle insertions and particle removals are also present. A new particle or molecule can be inserted into the system at a randomly chosen point based on the following probability:

$$P_{Accept} = \frac{fV}{k_B T(N+1)} \exp(-\Delta U/kBT) \tag{B16}$$

where f is the gas-phase fugacity, V is the volume, N is the number of particles (molecules) before the insertion and ΔU is the change in potential energy due to insertion.

The removal of particles from the systems is governed by the following probability equation.

$$P_{Accept} = \frac{N k_B T}{fV} \exp(-\Delta U/k_B T) \tag{B17}$$

Simulations typically require millions of displacement, insertion and removal moves in order to equilibrate the system. The result is an adsorption equilibrium between the sorbate molecules in the gas phase and those adsorbed on the zeolite at the specific gas-phase fugacity. This would represent a single point on an adsorption isotherm. The remainder of the isotherm curve can be generated to determine the amount of gas adsorbed at various other pressures[59,62].

Grand Canonical simulations have been used fairly successfully in simulating single-component systems. More recent papers show that the method can also be used to simulate binary systems and also mixtures[73,74].

Grand Canonical MC simulation tends to work fairly well for small-molecule systems but fails for larger molecules owing to the very low acceptance probabilities for insertion moves into the system owing to the interactions between the sorbate and the zeolite or other sorbate molecules. The molecule has a difficult time taking on the preferred configuration for it to fit into the system. Configurational biasing, as discussed next, helps to overcome this problem.

3.c. Configurationally Biased Monte Carlo Simulation (CBMC)

Configurationally biased methods can be used within the simulation to avoid the difficulties that result from the low probability of insertion [58–60,62,75]. This is accomplished by allowing molecule to insert sequentially atom-by-atom. This avoids the difficulty of having the molecule adapt to limited number of configurations before it can insert. It now guides the adsorbate atom-by-atom to adapt the appropriate configuration. This, however, biases the statistical likelihood of insertion. The acceptance rules must therefore be changed in order to correct for the bias. Configurationally biased methods can lead to significant enhancements in CPU expenditure, thus allowing for simulations of systems that are typically not possible without the bias[59]. Configurational biasing is most widely adopted in Grand Canonical and Gibbs Ensemble Monte Carlo methods.

3.d. Gibbs Ensemble Monte Carlo simulation

Gibbs ensemble Monte Carlo simulation is predominantly used to simulate phase equilibrium for fluids and mixtures. Two fluid phases are simulated simultaneously allowing for particle moves between each phase[58,76,77].

3.e. Applications of Monte Carlo Simulation

Monte Carlo simulation has been used to simulate the optimized structures for zeolites, metal oxides and metals. In addition, it has been used to simulate the siting of sorbates, Henry's Law constants, heat capacities, isosteric sorption isotherms and other thermodynamic properties.

4. Molecular Dynamics

The simulation of dynamic properties such as diffusivities requires the use of dynamic methods. Molecular dynamics methods integrate Newton's laws of motion in order to follow the dynamic behavior of a system. Individual molecule or particle trajectories are obtained by solving Newton's second law to establish the positions for all of the molecules or particles at some new time $t + \mathrm{d}t$. The velocities of each particle along a specific vector can be determined by integrating the following equation with respect to time. A second integration similarly leads to the positions for each particle over time[59,63,64].

$$\frac{\mathrm{d}^2 \vec{r}_i}{\mathrm{d}t^2} = \frac{F_{\vec{r}\,i}}{m_i} \tag{B18}$$

The new position is dependent on the forces $F_{\vec{r}\,i}$ which act upon particle i along the vector $\vec{r}$. The forces are determined from the force field for the system. Similarly, one can solve for the forces acting on all of the particles along all specific vectors. The forces can be calculated from the potential energy, u, with the equation

$$F_i = -\nabla_{ri}\, u \tag{B19}$$

These equations are integrated simultaneously via finite differenc methods. A Verlet algorithm is typically used in carrying out the integration whereby the new positions and velocities, and accelerations for the particles are calculated from the previous positions[59,63,64]:

$$r(t + \delta t) = r(t) + \delta t \nu(t) + \frac{1}{2}\delta t^2 a(t) + \frac{1}{6}\delta t^3 b(t) + \ldots \tag{B20}$$

$$\nu(t+\delta t) = \nu(t) + \delta t a(t) + \frac{1}{2}\delta t^2 b(t) + \dots \tag{B21}$$

$$b(t+\delta t) = b(t) + \delta t c(t) + \dots \tag{B22}$$

The simulation of the dynamic processes of atoms and molecules requires time steps which are on the order of 10^{-15} s in order to follow intermolecular forces accurately. This results in substantial CPU requirements that ultimately limit the length of real time that can be simulated to nanosecond range.

The molecular dynamics approach outlined so far is formally for NVE systems. Many of the problems in catalysis require a constant temperature rather than a constant energy. The temperature, however, is related to the time-average kinetic energy of the system through the equation

$$\langle \kappa \rangle_{NVT} = \frac{3}{2} N k_B T \tag{B23}$$

Equation (B23) suggests that the temperature could be controlled by scaling the velocities. The scaling factor can be calculated from the following expression, which is simply derived from Eq. (B23):

$$\Delta T = \frac{1}{2}\sum_{i=1}^{N} \frac{2}{3} \frac{m_i(\lambda \nu_i)^2}{N k_B} - \frac{1}{2}\sum_{i=1}^{N} \frac{2}{3} \frac{m_i \nu_1^2}{N k_B} = (\lambda^2 - 1)T(t) \tag{B24}$$

where $\lambda = \sqrt{T_{NEW}/T(t)}$.

A second approach to controlling the temperature would be to include a heat bath whereby the bath can supply or remove energy from the system[59,63,64]. These approaches, however, do not rigorously conform to canonical averages.

Two methods that have been developed that do maintain correct canonical averaging are the stochastic collision and the extended systems approaches. Both are covered in detail elsewhere[59,63,64,78,79]. We report here only on some of the salient features from the extended systems approach since this approach is used primarily for constant temperature MD simulations for heterogeneous catalytic materials.

The extended system method was developed by Nose[78] and subsequently by Hoover [79], who considered the thermal reservoir to be an integral part of the system. The inclusion of the reservoir requires an additional degree of freedom, defined as s, be added to the system. The potential energy for this additional degree of freedom is calculated as

$$u = (f+1)k_B T \ln(s) \tag{B25}$$

where f is defined as the number of degrees of freedom.

The kinetic energy for this additional degree of freedom is calculated as

$$K.E. = \frac{Q}{2}\left(\frac{\mathrm{d}s}{\mathrm{d}t}\right)^2 \tag{B26}$$

where Q is fictitious mass defined for the additional degree of freedom. Q determines the energy flow between the extended and the real system.

Each state in the extended system is a unique state in the real system. The real velocity is then calculated as

$$\vec{\nu}_i = s\frac{\mathrm{d}\,\vec{r}_i}{\mathrm{d}t} \tag{B27}$$

Equation (B18) can then be modified to

$$\frac{\mathrm{d}^2\,\vec{r}_i}{\mathrm{d}t^2} = \frac{F_{\vec{r}_i}}{m_i s^2} - \frac{\mathrm{d}s}{\mathrm{d}t}\frac{\mathrm{d}\,\vec{r}_i}{\mathrm{d}t}\frac{1}{s} \tag{B28}$$

This modified system of equations can then be integrated in order to follow the time-dependent changes in the positions of the particles (molecules) and the state of the system.

Molecular dynamics can then be used to simulate the molecular trajectories for molecules in zeolites, the diffusivities for sorbates in zeolites, temporal changes in the pore structure due to changes in system variables (T, P) or sorption of molecules[59,62]. As was described earlier, the integration time steps (10^{-15} sec) limit the time scales that can actually be simulated to nanosecond behavior and thus preclude the simulation of longer time processes.

C. SIMULATING KINETICS

The methods discussed so far provide the ability to simulate the surface structure, micro- and mesoporous structure, physisorption and chemisorption at surface sites, lateral interactions between surface intermediates, activation barriers and overall reaction energies for elementary surface reactions, diffusion on surfaces and within porous media, and a host of other elementary or equilibrated processes important toward understanding catalysis. Simulating catalysis, however, requires following the dynamics of the entire surface adlayer and its structure as a function to changes in process conditions, including temperature, pressure and conversion. As was discussed in Chapter 2, catalysis is driven by kinetics. The ability to follow the kinetics for catalytic processes requires the ability to simulate the myriad of physicochemical elementary processes including adsorption, surface reaction, diffusion and desorption that occur simultaneously and make up the full catalytic cycle.

The kinetics for catalytic systems can be modeled by one of two general methods. The first is based on continuum concentrations and uses deterministic kinetics whereas the second approach follows the temporal fate of individual molecules over the surface via stochastic kinetics. Both approaches have known advantages and disadvantages, as will be discussed. B These methods provide the constructs for simulating the elementary kinetics. However, in order to do so, they require an accurate and comprehensive initial kinetic database that contains parameters for the full spectra of elementary surface processes that make up the catalytic cycle. The ultimate goal for both approaches would be to call upon quantum mechanics calculations in situ in order to establish the potential energy surface as the simulation proceeds. This, however, is still well beyond our computational capabilities.

Currently, the most straightforward way to bridge electronic structure and surface kinetics requires a decoupling of the time scales that govern electronic transfer processes that control elementary surface reaction steps from the overall catalytic cycle which proceeds at much longer times. Ab initio calculations are used first to calculate the kinetics, energetics and potential mechanisms necessary for an external database. The database could then be called “in situ” within the simulation algorithm.

1. Deterministic Kinetic Modeling

The kinetics for most catalytic systems are described using determinstic models rather than stochastic simulations. Deterministic models are straightforward to develop and to program. The temporal concentrations for all species/intermediates in the network are tracked by solving the full set of differential equations that describe the rate of formation and disappearance for each component in the system. The rate is defined in terms of concentrations, partial pressures or surface coverages. As such, the deterministic models average the intrinsic kinetics over the atomic structure in order to define the concentrations of reaction intermediates and are therefore considered an early averaging method. This approach ignores the features of the local structure and composition near the active site and their influence on the kinetics. By averaging out over the surface structure, the problem becomes one of solving N-differential equations. These equations can subsequently be used to solve for the reaction rate which can be plugged into various different reactor models which would follow not only the changes in the rate but also the spatiotemporal changes throughout the reactor[80].

Deterministic kinetic modeling approaches are mean-field approaches, whereby the molecules experience only an averaged interaction[81] of the others. These models are reasonable if the lateral interactions between reactant molecules, reagents or products are absent or if diffusional effects maintain a state of ideal mixing. In the latter case, the kinetic parameters will also be concentration dependent.

The input to most of the microkinetic modeling studies has been experimentally derived rate constants. This is for two reasons. The first has been the lack of available first principles kinetic data and the difficulty in simulating them. The second is due to the fact that the accuracy is not within 1 kcal/mol. Despite this drawback, there have been some very interesting studies performed even in the absence of very accurate kinetics.

2. Stochastic Methods

Stochastic methods simulate the dynamic changes that occur in the structure of the adlayer of catalytic surface and thus model the elementary surface kinetics [82–100]. The temporal changes of a system can be followed by solving the stochastic master equation which simulates the dynamic changes in the system as it moves from one state (i) to another state (j). The master equation, which can written as

$$\frac{\mathrm{d}P_i}{\mathrm{d}t} = [W_{ji}P_j - W_{ij}P_i] \tag{C1}$$

is nothing more than a balance on the kinetic "forces that drive a system from one state to another as a function of time. P_i is the probability that the system is in state i at time t and P_j is the probability that the system is in state j. W_{ij} and W_{ji} are the transition probabilities which denote the probable rate of transition from state i to state j or j to i, respectively, for the system. The self transition probabilities, W_{ii} and W_{jj}, are equal to zero. For catalytic systems, the changes in system state can be any elementary surface process that changes the nature of the adlayer including surface diffusion, surface reaction, desorption, adsorption and surface reconstruction. In order for the master equation to hold, the system must obey a detailed balance, that is

$$W_{ji}P_i = W_{ij}P_j \tag{C2}$$

The requirements for a detailed balance, however, do not control the kinetics.

The master equation, however, can only be solved analytically for very simple systems such as the gas-phase reaction A→B. The analysis of these systems typically requires numerical simulation of a lattice-based kinetic Monte Carlo model. The lattice gas model can then be used to formulate the respective transition probabilities in order to solve the master equation[81]. The groups of both Zhdanov[97–99,102–108] and Kreuzer[109–113] have been instrumental in demonstrating the application of lattice gas models to solve adsorption and desorption processed from surfaces. Once a lattice model has been formulated there are three types of solution:

1. the cluster approximation,
2. the transfer matrix technique,
3.) Monte Carlo simulation.

Zhadanov's group[97–99,102–108], as well as others, have shown that the cluster approximation is useful for understanding surface structure and surface kinetics. It is limited, however, in its ability to describe the formation of ordered surface structures. Kreuzer's group[109–113] has demonstrated the utility of the transfer matrix approach[109–110]. Mathematically, the equations become cumbersome for solution, but this technique provides an important insight into the surface physics that control the kinetics. The final approach involves the numerical simulation of the master equation by Monte Carlo algorithms. MC simulation opens up the possibilities of simulating much more complicated surface-catalyzed systems, as will be discussed.

There are two basic methods that have been used to simulated kinetics via Monte Carlo approaches[114–116]. The first is termed the fixed-time approach, in which every site on the surface has a set of probabilities associated with the different kinetic events that can occur at these sites. This could include diffusion, reaction, adsorption, and desorption processes. The state of the system then moves in *fixed incremental steps of time* and subsequently surveys all of the physicochemical steps to determine which of them can take place within the given (short) time step. This is accomplished by sampling every site and determining whether it changes due to the occurrence of a kinetic process. This is determined by drawing a random number for each potential step and comparing it with the transition probability P_{rs} for that particular kinetic step (r) at site (s):

$$P_{rs} = 1 - \exp(-k_{rs}\Delta t) \tag{C3}$$

where k_{rs} is the rate constant for reaction (r) associated with the specific environment (s)[114].

The fixed-time approach has proven effective in modeling well-defined reaction systems which have a sequence of known steps. The benefits of this approach are that the user is able to specify the time step at which the simulation proceeds. This helps to overcome some of the difficulties associated with disparate time scales. Very fast processes can be treated as pseudo-equilibrated, thus enabling the simulation to move to the time scales of interest.

One of the drawbacks associated with the fixed-time approach, however, is that at any give point in time one needs to have all of the possible future pathways worked out in order to calculate the probability that within that particular time step a sequence of events occurs. This becomes challenging and expensive computationally as the network of surface processes is rather complex.

A second drawback of the fixed-time approach is that it is mathematically not exact. The accuracy of the simulation is governed by the choice of the time step used. Only in the limit of an infinitesimal time step does the method becomes mathematically exact. For simulations performed at very small time steps, accuracy is not an issue. Although the fixed time algorithm has proven to be fairly effective for certain gas-phase reaction systems[114,117,118], nearly all of the published studies on surfaces use what is known as the variable time-step approach[82–122].

In the variable time-step approach, the system moves in event space, thus simulating the elementary kinetic processes event-by-event whereby the time is updated in variable time increments[115,116,122]. At any instant in time, t_i, the rates for all possible events are added together in order to determine the total rate, $R = \sum r_i$. The probability that some event in the entire system will occur is then defined as

$$P_i = 1 - exp(-\sum r_i \Delta t_i) \tag{C4}$$

By rearranging, we can to solve for the time at which the next event in the system will occur:

$$\Delta t_\nu = \frac{-\ln(RN)}{\sum_i r_i} \tag{C5}$$

where RN is a random number between 0 and 1 which is chosen by the computer as it defines the random probability of the time for the next event.

The time step chosen is variable and changes throughout the simulation. It can be infinitely small or infinitely large and depends both on the random probability and the overall calculated rate. The variable-time method is mathematically an exact approach and there are no concerns about accuracy due to time step size. Systems which contain fast events have very small time steps and are thus dominated by the time scale scales of the fastest processes. Faster rates of reaction lead to higher probabilities that these steps are chosen. This leads to problems for systems with disparate rates since the simulation will spend nearly all of its time simulating the faster rates without ever simulating the slower processes. This is especially a problem for systems where diffusion is fast and reaction is slow and systems which contain fast processes which are nearly equilibrated together with slow processes.

The simulation of surfaces typically requires defining an appropriate lattice. These can either be a simple lattice model or off-lattice simulations which attempt to treat sites more explicitly. The simulation proceeds in essentially the same manner as described with the one exception that we explicitly follow the surface of the lattice. At any given instant in time the entire surface is surveyed in order to construct a detailed list of all possible surface events that can occur, including adsorption, desorption, surface reaction, and diffusion. Each possible event is assigned a rate (or rate constant) based on the nature of the event and the explicit molecular environment around each species. The rates (rate constants) for each of these possible events are added together to determine the cumulative probability for that particular event. The computer draws a random number which is then used in Eq. (C5) [a modified version of Eq. (C3)] in order to establish the time step of the next event.

One of the most important features of the Monte Carlo approach is its ability to monitor explicitly of atoms on the surface and within the adlayer. This allows for the direct accounting of specific surface sites and the local reaction environment at these sites. The specific arrangement and orientation of surface ad-species can significantly influence the surface kinetics. These interactions are at the heart of coverage effects. The ability to model lateral interactions between ad-species on the surface and their influence on catalytic performance was discussed in some detail in Chapter 3 and is therefore not repeated again here.

REFERENCES

1. B. Born, J.R. Oppenheimer, *Ann. Phys.* 79, 361 (1927)
2. F. Jensen, *Introduction to Computational Chemistry,* New York, Wiley (1999)
3. I.N. Levine, *Quantum Chemistry,* Allyn and Bacon, Boston (1983)
4. A.R. Leach, *Molecular Modeling: Principles and Applications*, Pearson Education, Harlow (1996)
5. J.B.Foresman, A. Frisch, *Exploring Chemistry with Electronic Structure*, 2nd edn, Pittsburgh, PA Gaussian (1996)
6. E. Wimmer, *J. Comput-Aided Mater. Des.* 1, 215 (1993)
7. R.G. Parr, W. Yang, *Density Functional Theory of Atoms and Molecules*, Oxford University Press, New York (1989)
8. P. Hohenberg, W. Kohn, *Phys. Rev.* 136, B864 (1964)
9. M.C. Zerner, in *Reviews in Computational Chemistry,* K.B. Lipkowitz, D.B. Boyd (eds.), VCH , New York pp. 313–366 (1991)
10. J.J.P. Stewart, in *Reviews in Computational Chemistry,* K.B. Lipkowitz, D.B. Boyd (eds.), VCH, New York. pp. 45–82 (1990)
11. D.R. Hartree, *Proc. Cambridge Philos. Soc.* 24, 328 (1928)
12. V.A. Fock, *Phys.* 15, 126 (1930)
13. M.C. Payne, M.P. Teter, D.C. Allan, T.A. Arias, J.D. Joannopoulos, *Rev. Mod. Phys.* 64, 1045 (1992)
14. M. Head-Gordon, *J. Phys. Chem.* 100, 13213 (1996).
15. F.M. Bickelhaupt, E.J. Baerends, in *Reviews in Computational Chemistry,* D.B. Boyd, K.B. Lipkowitz (eds.), Wiley-VCH, New York. pp. 1-86 (2000)
16. G.C. Hall, *Proc. R. Soc. London,* A205, 541 (1951)
17. C.C. Roothaan, *Rev. Mod. Phys.* 23, 69 (1951)
18. A. Szabo, N.S. Ostlund, *Modern Quantum Chemistry,* McGraw-Hill, New York (1982)
19. C. Møller, M.S. Plesset, *Phys. Rev.* 46, 618 (1934).
20. B. Hammer, J.K. Nørskov, in *Chemisorption and Reactivity on Supported Clusters and Thin Films,* R.M.L.a.G. Pacchioni (ed.), Kluwer, Dordrecht, pp.285–351 (1997)
21. B. Hammer, J.K. Nørskov, *Adv. Catal.* 45, (2000)
22. M. Neurock, R.A. van Santen, *Catal. Rev. Sci. Eng.* 37, 557 (1995)
23. L.H. Thomas, *Proc. Cambridge Philos. Soc.* 23, 542 (1927)
24. E.Z. Fermi, *Z. Phys.* 48, 73 (1928)
25. P.A.M. Dirac, *Proc. Cambridge Philos. Soc.* 26, 376 (1930)
26. W. Kohn, L. Sham, *Phys. Rev. A*, 140, 1133 (1965)
27. T. Ziegler, *Chem. Rev.* 91, 651 (1991)
28. A.D. Becke, *Phys. Rev. B*, 38, 3098 (1986)
29. J.D. Perdew, Y. Wang, *Phys. Rev.*, B33, 8800 (1986)
30. J.P. Perdew, J.A. Chevary, S.H. Vosko, K.A. Jackson, M.R. Pederson, D.J. Singh, C. Fiolhais, *Phys. Rev.* B46, 6671 (1992)

31. J.P. Perdew, K. Burke, M. Ernzerhof, *Phys. Rev. Lett.* 77, 3865 (1996)
32. B. Hammer, L.B. Hansen, J.K. Nørskov, *Phys. Rev. B* 59, 7413 (1999)
33. A.D. Becke, *Chem. Phys.* 98, 5648 (1993)
34. C. Stampfl, M.V. Ganduglia-Pirovano, K.R. Scheffler, M. Sc heffler, *Surf. Sci.* 500 (2001)
35. D.J. Chadi, M.L. Cohen, *Phys. Rev. B*, 8, 5747 (1973)
36. H.J. Monkorst, J.D. Pack, *Phys. Rev. B*, 13, 5188 (1976)
37. M Neurock, in *Applying Molecular and Materials Modeling*, P.R. Westmoreland, (ed.) Kluwer, Dordrecht, pp. 107–144 (2002)
38. A. Warshel, M. Karplus, *J. Am. Chem. Soc.* 94, 5612 (1972)
39. T. Vreven, K. Morokuma, *J. Chem. Phys.* 113, 2969 (2000)
40. J. Whitten, H. Yang, *Surf. Sci. Rep.* 24, 55 (1996)
41. N. Govind, Y.A. Wang, E.A. Carter, *Chem. Phys. Lett.* 305, 419 (1999)
42. (a) J. Sauer, *Stud. Surf. Sci. Catal.* 84, 2039 (1994);
(b) M. Brandle, J. Sauer, *J. Mol. Catal.* 119, 19 (1997).
43. L.A. Clark, M. Sierka, J. Sauer, *Stud. Surf. Sci. Catal.* 142, 643 (2002)
44. G.E. Scuseria, *J. Phys. Chem.* A103, 4783 (1999)
45. M.C. Strain, G.E. Scuseria, M.J. Frisch, *Science*, 271, 51 (1996)
46. S. Goedecker, *Rev. Mod. Phys.* 71, 1085 (1999)
47. J.M. Soler, E. Artache, J.D. Gale, A. Garcia, J. Junquera, P. Ordegón, D. Sánchez-Portal, *J. Phys. Condens. Matter.* 14, 2745 (2002)
48. P.E. Blochl, H.M. Senn, A. Togni, *Transitions State Modeling for Catalysis*, D.G. Truhlar, K. Morokuma (eds.), American Chemical Society, Washington DC., p. 88 (1999)
49. M. Boero, M. Parrinello, K. Terakura, *J. Am. Chem. Soc.* 120, 2746 (1998)
50. R. Car, M. Parrinello, *Phys. Rev. Lett.* 55, 2471 (1985)
51. E.A. Carter, G. Ciccotti, J.T. Hynes, R. Kapral, *Chem. Phys. Lett.* 156, 472 (1989)
52. I. Feng, W. Kuo, C. Mundy et al., *J. Phys. Chem. B*, 108, (2004)
53. E. Fois, A. Gamba, E. Spano, *J. Phys. Chem. B*, 108, 9557 (2004)
54. G. Kresse, *J. Non-Cryst. Solids* 52, 312 (2002)
55. (a) D. Marx, J. Hutter, *Modern Methods and Algorithms of Quantum Chemistry*, J. Grotendorst (ed.); John von Neumann Institute for Computing: Jüelich, (2000); (b) R. Iftime, P. Minary, M.E. Tuckerman,*Proc. Nate. Acad. Sci. USA*, 102, 19, 6654 (1992)
56. M.C. Payne, M.P. Teter, D.C. Allan, T.A. Arias, J.D. Joannopoulos, *Rev. Mod. Phys.* 64, 1045 (1992)
57. B. Trout, *Adv. Chem. Eng.* 28, 353 (2001)
58. P.T. Cummings, in *Applying Molecular and Materials Modeling*, P.R. Westmoreland, (Ed.), Kluwer, Dordrecht. p. 23 (2002)
59. A.T. Bell, E.J. Maginn, D.N. Theodorou, in *Handbook of Heterogeneous Catalysis*, H.K.G. Ertl, J. Weitkamp (ds.), Wiley-VCH, New York p. 1165
60. S.P. Bates, R.A.van Santen, *Adv. Catal.*, 42, 1 (1998)
61. C.R.A. Catlow, in *Handbook of Heterogeneous Catalysis*, H.K.G. Ertl, J. Weitkamp (eds.), Wiley-VCH, New York, p. 1149 (1997)
62. L.J. Broadbelt, R.Q. Snurr, *Appl. Catal. A*, 200, 23 (2000)
63. M.P. Allen, D.J. Tildesley, *Computer Simulation of Liquids*, Clarendon Press, Oxford (1987)

64. A.R. Leach, *Molecular Modeling: Principles and Applications*, Pearson Education, Harlow (1996)
65. A.K. Rapp, C.J. Casewit, *Molecular Mechanics Across Chemistry*, University Science Books, Sausalito, CA (1997)
66. M. Daw, M. Baskes, *Phys. Rev.* B29, 6443 (1984)
67. M.I. Baskes, *Phys. Rev. B*, 46, 2727 (1992)
68. S.M. Folles, M.I. Baskes, M.S. Daw, *Phys. Rev. B*, 33, 7983 (1986)
69. J.K. Nørskov, *Phys. Rev. B*, 26, 2875 (1982)
70. P. van Beurden, H.G.J. Verhoeven, G.J. Kramer, B.J. Thijsse, *Phys. Rev. B*, 66, 235409(1-11) (2002)
71. P. van Beurden, *Phys. Rev. Lett.* 90, 066106 (2003)
72. (a)J.D. Gale, *J. Chem. Soc. Faraday. Trans.* 93, 629 (1997);
(b)J.D. Gale, *Mol. Simul.* 29, 291 (2003);
(c) http://gulp.curtin.edu.au/index
73. (a) K.F. Czaplewski, R.Q. Snurr, *AIChE J.* 45, 2223 (1999);
(b) B. Smit, T.L. Maesen, *Nature*, 374, 42 (1995)
74. S.P. Bates, et al., *J. Phys. Chem.* 100, 17573 (1996)
75. D. Frenkel, B. Smit, *Understanding Molecular Simulation*, Academic Press, San Diego, CA (1996)
76. A.Z. Panagiotopoulos, *Mol. Phys.* 62, 701 (1987)
77. A.Z. Panagiotopoulos, *Mol. Phys.* 61, 813 (1987)
78. S. Nose, *Mol. Phys.*, 53, 255 (1984)
79. W. G. Hoover, *Phys. Rev. A*, 31, 1695 (1985)
80. J.A. Dumesic, D.F. Rudd, L.M. Aparicio, J.E. Rekoske, A.A. Trevino, *The Microkinetics of Heterogeneous Catalysis*, American Chemical Society, Washington, DC (1993)
81. R.I. Masel, *Principles of Adsorption and Reaction on Solid Surfaces*, Wiley, New York (1996)
82. K. Binder, *Monte Carlo Methods in Statistical Physics, Topics in Current Physics*, Vol. 7, K. Binder (ed.), Springer-Verlag, Berlin (1979)
83. P. Dufour, M. Dumont, V. Chabart, J. Lion, *J. Comput. Chem.* 13, 25 (1989)
84. S.M. Dunn, J.B. Anderson, *J. Chem. Phys.* 102, 2812 (1995)
85. K.A. Fichthorn, E. Gulari, R. M. Ziff, *Surf. Sci.* 243, 273 (1991)
86. A.P.J. Jansen, *Comput. Phys. Commun.* 86, 1 (1994)
87. H.C. Kang, W.H. Weinberg, *Chem. Rev.* 95, 667 (1995)
88. B. Meng, W.H. Weinberg, *J. Chem. Phys.* 100, 5280 (1994)
89. B. Meng, W.H. Weinberg, *J. Chem. Phys.* 102, 9435 (1995)
90. B. Meng, W.H. Weinberg, *Surf. Sci.* 364, 151 (1996)
91. B. Meng, W.H. Weinberg, *Surf. Sci.* 374, 17 (1997)
92. A. Milchev, K. Binder, *Surf. Sci.* 164, 1 (1985)
93. R.M. Nieminen, A.P.J. Jansen, *Appl. Catal. A*, 160, 99 (1997)
94. M. Silverberg, A. Ben-Shaul, *J. Chem. Phys.* 87, 3178 (1987)
95. M. Silverberg, A. Ben-Shaul, *J. Stat. Phys.* 52, 1179 (1988)
96. M. Silverberg,A. Ben-Shaul, *Surf. Sci.*, 214 17 (1989)
97. V.P. Zhdanov, P.R. Norton, *Surf. Sci.*, 350, 271 (1996)
98. V.P. Zhdanov, J.L. Sales, R.O. Unac, *Surf. Sci.* 381, L599 (1997)
99. V.P. Zhdanov, B. Kasemo, *Surf. Sci.* 412–413, 527 (1998)
100. R.M. Ziff, E. Gulari, Y. Barshad, *Phys. Rev. Lett.* 56, 2553 (1986)

101. L.K. Doraiswamy, B.D. Kulkarni, *The Analysis of Chemically Reacting Systems: a Stochastic Approach,* Gordon and Breach Science, New York (1987)
102. V.P. Zhdanov, *Surf. Sci.* 133, 469 (1983)
103. V.P. Zhdanov, *Surf. Sci. Rep.* 12, 183 (1991)
104. V.P. Zhdanov, B. Kasemo, *Physica D,* 70, 383 (1994)
105. V.P. Zhdanov, P.R. Norton, *Surf. Sci.* 312, 441 (1994)
106. V.P. Zhdanov, *Surf. Rev. Lett.* 3, 1555 (1996)
107. V.P. Zhdanov, B. Kasemo, *Surf. Sci.* 405, 27 (1998)
108. V.P. Zhdanov, *Surf. Sci.* 426, 345 (1999)
109. H.J. Kreuzer, *Surf. Sci.* 238, 305 (1990)
110. H.J. Kreuzer, *Appl. Phys. A*, 51, 491 (1990)
111. H.J. Kreuzer, *J. Zhang, Appl. Phys.* A51, 183 (1990)
112. S.H. Payne, H.J. Kreuzer, *Surf. Sci.* 222, 404 (1989)
113. S.H. Payne, H.J. Kreuzer, *Surf. Sci.* 338, 261 (1995)
114. M. Neurock, S. Stark, M.T. Klein, in *Computer-Aided Design of Catalysts,* Marcel Dekker, New York, p. 55 (1993)
115. D.T. Gillespie, *J. Comput. Phys.* 22, 403 (1976)
116. D.T. Gillespie, *J. Phys. Chem.* 81, 2340 (1977)
117. J.B. McDermott, C. Libanati, C. Lamarca, M.T. Klein, *Ind. Eng. Chem. Res.* 29, 22 (1990)
118. C. Libanati, *Thermal Degradation of Poly(Arylether Sulfones),* Thesis University of Delaware (1991)
119. K. Fichthorn, E. Gulari, R. Ziff, *Chem. Eng. Sci.* 44, 1403 (1989)
120. D. Gupta, C.S. Hirtzel, *Chem. Phys. Lett.* 149, 527 (1988)
121. A.S. McLeod, L.F. Gladden, *Catal. Lett.* 43, 189 (1997)
122. J.S. Turner, *J. Phys. Chem.*, 81, 2379 (1977)
123. M.J. Frisch, G.W. Trucks, H.B. Schlegel, G.E. Scuseria, M.A. Robb, J.R. Cheeseman, J.A. Montgomery, Jr., T. Vreven, K.N. Kudin, J.C. Burant, J.M. Millam, S.S. Iyengar, J. Tomasi, V. Barone, B. Mennucci, M. Cossi, G. Scalmani, N. Rega, G.A. Petersson, H. Nakatsuji, M. Hada, M. Ehara, K. Toyota, R. Fukuda, J. Hasegawa, M. Ishida, T. Nakajima, Y. Honda, O. Kitao, H. Nakai, M. Klene, X. Li, J.E. Knox, H.P. Hratchian, J.B. Cross, V. Bakken, C. Adamo, J. Jaramillo, R. Gomperts, R.E. Stratmann, O. Yazyev, A.J. Austin, R. Cammi, C. Pomelli, J.W. Ochterski, P.Y. Ayala, K. Morokuma, G.A. Voth, P. Salvador, J.J. Dannenberg, V.G. Zakrzewski, S. Dapprich, A.D. Daniels, M.C. Strain, O. Farkas, D. K. Malick, A.D. Rabuck, K. Raghavachari, J.B. Foresman, J.V. Ortiz, Q. Cui, A.G. Baboul, S. Clifford, J. Cioslowski, B.B. Stefanov, G. Liu, A. Liashenko, P. Piskorz, I. Komaromi, R.L. Martin, D.J. Fox, T. Keith, M.A. Al-Laham, C.Y. Peng, A. Nanayakkara, M. Challacombe, P.M.W. Gill, B. Johnson, W. Chen, M.W. Wong, C. Gonzalez, J.A. Pople, Gaussian 03, Revision C.02, Gaussian, Wallingford, CT (2004).

Index

Molecular Heterogeneous Catalysis. Rutger Anthony van Santen and Matthew Neurock

ISBN: 3-527-29662-X